GCSE
Combined Science
Foundation Level

GCSE Combined Science is certainly not a Bohr...
and this spectacular all-in-one CGP book puts the cool in molecule!

It's the ultimate revision triple-threat — everything you need to know for Combined Science Biology, Chemistry and Physics. Plus, there are exam-style questions for every topic, a full set of practice papers, *and* a free online edition.

You'll also find links to our fantastic online content, with video solutions for practice questions, as well as Retrieval Quizzes to help you nail down all the facts you need to learn.

Unlock your free online extras!

Just go to **cgpbooks.co.uk/extras** and enter this code or scan the QR codes in the book.

1834 8269 9477 8856

By the way, this code only works for one person. If somebody else has used this book before you, they might have already claimed the Online Edition.

Complete
Revision & Practice

Everything you need to pass the exams!

Contents

Working Scientifically

The Scientific Method 2
Models and Communication 3
Issues Created by Science 4
Risk 5
Designing Investigations 6
Processing Data 9
Presenting Data 10
More on Graphs 11
Rearranging Equations and Units 12
More on Units 13
Drawing Conclusions 14
Uncertainty 15
Evaluations 16

Topic B1 — Cell Biology

Cells 17
Microscopy 19
Warm-Up & Exam Questions 22
Cell Differentiation and Specialisation 23
Cell Specialisation 24
Stem Cells 25
Chromosomes and Mitosis 27
Warm-Up & Exam Questions 29
Diffusion 30
Osmosis 31
Active Transport 33
Exchanging Substances 34
More on Exchanging Substances 36
Warm-Up & Exam Questions 38
Exam Questions 39

Topic B2 — Organisation

Cell Organisation 40
Enzymes 41
Investigating Enzymatic Reactions 43
Enzymes and Digestion 44
Food Tests 45
Warm-Up & Exam Questions 47
Exam Questions 48
The Lungs 49
Circulatory System — The Heart 50
Circulatory System — Blood Vessels 52
Circulatory System — Blood 53
Warm-Up & Exam Questions 54
Exam Questions 55
Cardiovascular Disease 56
More on Cardiovascular Disease 58
Warm-Up & Exam Questions 59
Health and Disease 60
Risk Factors for Non-Communicable Diseases 62
Cancer 64
Warm-Up & Exam Questions 65
Plant Cell Organisation 66
Transpiration and Translocation 67
Transpiration 68
The Rate of Transpiration 69
Warm-Up & Exam Questions 70
Exam Questions 71

Topic B3 — Infection and Response

Communicable Disease 72
Bacterial Diseases 73
Viral Diseases 74
Fungal and Protist Diseases 76
Warm-Up & Exam Questions 77
Fighting Disease 78
Fighting Disease — Vaccination 80
Fighting Disease — Drugs 81
Developing Drugs 82
Warm-Up & Exam Questions 83
Exam Questions 84

Topic B4 — Bioenergetics

Photosynthesis 85
The Rate of Photosynthesis 86
Measuring the Rate of Photosynthesis 88
Warm-Up & Exam Questions 89
Exam Questions 90
Aerobic Respiration 91
Anaerobic Respiration 92
Exercise 93
Metabolism 94
Warm-Up & Exam Questions 95
Revision Summary for Topics B1-4 96

Topic B5 — Homeostasis and Response

Homeostasis 97
The Nervous System 98
Investigating Reaction Time 101
Warm-Up & Exam Questions 102
The Endocrine System 103
Controlling Blood Glucose 105
Warm-Up & Exam Questions 107
Puberty and the Menstrual Cycle 108
The Menstrual Cycle and Controlling Fertility 109
Controlling Fertility 110
More on Controlling Fertility 111
Warm-Up & Exam Questions 112

Topic B6 — Inheritance, Variation and Evolution

DNA 113
Sexual Reproduction 114
Asexual Reproduction 115
Meiosis 116
Fertilisation and Chromosomes 117
X and Y Chromosomes 118
Warm-Up & Exam Questions 119
Genetic Diagrams 120
Inherited Disorders 122
Family Trees 123
Embryo Screening 124
Warm-Up & Exam Questions 125
Variation 126
Mutations 127
Evolution 128
Antibiotic-Resistant Bacteria 130
More on Antibiotic-Resistant Bacteria 131
Warm-Up & Exam Questions 132
Selective Breeding 133
Genetic Engineering 135
Warm-Up & Exam Questions 136
Fossils 137
Classification 138
Warm-Up & Exam Questions 140

Topic B7 — Ecology

Competition 141
Abiotic and Biotic Factors 142
Adaptations 144
Food Chains 145
Warm-Up & Exam Questions 146
Using Quadrats 147
The Water Cycle 149
The Carbon Cycle 150
Warm-Up & Exam Questions 151
Exam Questions 152
Biodiversity and Waste Management 153
Global Warming 154
Deforestation and Land Use 156
Maintaining Ecosystems and Biodiversity 158
Warm-Up & Exam Questions 159
Revision Summary for Topics B5-7 160

Throughout this book you'll see grade stamps like these:

These grade stamps help to show how difficult the questions are.
Remember — to get a top grade you need to be able to answer **all** the questions, not just the hardest ones.

In the real exams, some questions test how well you can write (as well as your scientific knowledge).
In this book, we've marked these questions with an asterisk (*).

Topic C1 — Atomic Structure and the Periodic Table

Atoms 161
Elements 162
Isotopes 163
Compounds 164
Chemical Equations 165
Warm-Up & Exam Questions 167
Exam Questions 168
Mixtures 169
Chromatography 170
Filtration and Crystallisation 171
Simple Distillation 173
Fractional Distillation 174
Warm-Up & Exam Questions 175
The History of the Atom 176
Electronic Structure 178
Development of the Periodic Table 179
The Modern Periodic Table 180
Warm-Up & Exam Questions 181
Metals and Non-Metals 182
Group 1 Elements 183
Group 7 Elements 185
Group 0 Elements 187
Warm-Up & Exam Questions 188
Exam Questions 189

Topic C2 — Bonding, Structure and Properties of Matter

Ions 190
Ionic Bonding 192
Ionic Compounds 193
Warm-Up & Exam Questions 195
Covalent Bonding 196
Warm-Up & Exam Questions 199
Polymers 200
Giant Covalent Structures 201
Metallic Bonding 203
Warm-Up & Exam Questions 204
States of Matter 205
Warm-Up & Exam Questions 208

Topic C3 — Quantitative Chemistry

Relative Formula Mass 209
Conservation of Mass 210
Concentrations of Solutions 212
Warm-Up & Exam Questions 213
Exam Questions 214

Topic C4 — Chemical Changes

Acids, Bases and Their Reactions 215
Warm-Up & Exam Questions 218
Metals and their Reactivity 219
Extracting Metals 221
Warm-Up & Exam Questions 222
Electrolysis 223
Electrolysis of Aqueous Solutions 225
Warm-Up & Exam Questions 227

Topic C5 — Energy Changes

Exothermic and Endothermic Reactions 228
Measuring Energy Changes 229
Reaction Profiles 230
Warm-Up & Exam Questions 231
Revision Summary for Topics C1-5 232

Topic C6 — The Rate and Extent of Chemical Change

Rates of Reaction 233
Factors Affecting Rates of Reaction 234
Warm-Up & Exam Questions 236
Measuring Rates of Reaction 237
Graphs of Reaction Rate Experiments 240
Working Out Reaction Rates 241
Reversible Reactions 242
Warm-Up & Exam Questions 244
Exam Questions 245

Topic C7 — Organic Chemistry

Hydrocarbons 246
Crude Oil 248
Fractional Distillation 249
Cracking 250
Warm-Up & Exam Questions 252
Exam Questions 253

Topic C8 — Chemical Analysis

Purity and Formulations 254
Paper Chromatography 255
Interpreting Chromatograms 256
Tests for Gases 259
Warm-Up & Exam Questions 260
Exam Questions 261

Topic C9 — Chemistry of the Atmosphere

The Evolution of the Atmosphere 262
Climate Change and Greenhouse Gases 264
Carbon Footprints 266
Air Pollution 267
Warm-Up & Exam Questions 268
Exam Questions 269

Topic C10 — Using Resources

Finite and Renewable Resources 270
Resources and Sustainability 271
Reuse and Recycling 272
Life Cycle Assessments 273
Warm-Up & Exam Questions 275
Potable Water and Water Treatment 276
Warm-Up & Exam Questions 280
Exam Questions 281
Revision Summary for Topics C6-10 282

Topic P1 — Energy

Energy Stores 283
Energy Transfer 284
Mechanical Energy Transfer 285
Kinetic and Potential Energy Stores 286
Conservation of Energy 287
Specific Heat Capacity 288
Investigating Specific Heat Capacity 289
Warm-Up & Exam Questions 291
Exam Questions 292
Power 293
Reducing Unwanted Energy Transfers 294
Efficiency 295
Warm-Up & Exam Questions 296
Energy Resources and their Uses 297
Biofuels 298
Wind Power and Solar Power 299
Geothermal and Hydro-electric Power 300
Wave Power and Tidal Barrages 301
Non-Renewables 302
Limitations on the Use of Renewables 303
Warm-Up & Exam Questions 304

Topic P2 — Electricity

Current and Circuit Symbols 305
Charge and Resistance Calculations 306
Ohmic Conductors 307
Investigating Resistance 308
I-V Characteristics 309
Investigating I-V Characteristics 310
Warm-Up & Exam Questions 311
Circuit Devices 312
Sensing Circuits 313
Series Circuits 314
Parallel Circuits 316
Investigating Circuits 318
Warm-Up & Exam Questions 319
Electricity in the Home 320
Power of Electrical Appliances 321
More on Power 322
The National Grid 323
Warm-Up & Exam Questions 325

Topic P3 — Particle Model of Matter

Particle Model 326
Particle Motion in Gases 327
Density of Materials.. 328
Measuring Density.. 329
Internal Energy and Changes of State 330
Specific Latent Heat.. 331
Warm-Up & Exam Questions........................ 332
Exam Questions.. 333

Topic P4 — Atomic Structure

Developing the Model of the Atom............... 334
Isotopes 336
Types of Nuclear Radiation............................ 337
Nuclear Equations 338
Half-Life 340
Irradiation and Contamination....................... 342
Warm-Up & Exam Questions........................ 344
Exam Questions.. 345
Revision Summary for Topics P1-4................ 346

Topic P5 — Forces

Contact and Non-Contact Forces 347
Weight, Mass and Gravity.............................. 348
Resultant Forces and Work Done 349
Warm-Up & Exam Questions........................ 350
Forces and Elasticity 351
Investigating Springs 353
Warm-Up & Exam Questions........................ 355
Distance, Displacement, Speed, Velocity 356
Acceleration 357
Distance-Time Graphs 358
Velocity-Time Graphs 359
Drag and Terminal Velocity 360
Warm-Up & Exam Questions........................ 361
Newton's First and Second Law 362

Newton's Third Law 363
Investigating Motion 364
Warm-Up & Exam Questions........................ 366
Stopping Distance and Thinking Distance..... 367
Braking Distance 368
Reaction Times 369
Warm-Up & Exam Questions........................ 370

Topic P6 — Waves

Wave Basics 371
Transverse and Longitudinal Waves 372
Wave Speed 373
Investigating Waves 374
Refraction 376
Warm-Up & Exam Questions........................ 377
Electromagnetic Waves................................... 378
Uses of EM Waves 379
More Uses of EM Waves 380
Investigating IR Radiation 381
Investigation IR Absorption 382
Dangers of Electromagnetic Waves 383
Warm-Up & Exam Questions........................ 384
Exam Questions.. 385

Topic P7 — Magnetism and Electromagnetism

Magnetism... 386
Electromagnetism 388
Solenoids... 389
Warm-Up & Exam Questions........................ 390
Revision Summary for Topics P5-7................ 391

Practical Skills

Measuring Techniques 392
Safety and Ethics 395
Setting Up Experiments 396
Heating Substances 398
Working with Electronics 399
Sampling 400
Comparing Results 401

Practice Exams

Biology Practice Paper 1 402
Biology Practice Paper 2 416
Chemistry Practice Paper 1 431
Chemistry Practice Paper 2 444
Physics Practice Paper 1 458
Physics Practice Paper 2 473

Answers 487
Glossary 518
Index 534
The Periodic Table and Physics Equations Sheet 538

You'll see **QR codes** throughout the book that you can scan with your smartphone.

A QR code next to a tip box question takes you to a **video** that talks you through solving the question. You can access **all** the videos by scanning this code here.

Video Solutions

A QR code on a 'Revision Summary' page takes you to a **Retrieval Quiz** for that topic. You can access **all** the quizzes by scanning this code here.

Retrieval Quizzes

You can also find the **full set of videos** at cgpbooks.co.uk/GCSEScienceFoundation/Videos and the **full set of quizzes** at cgpbooks.co.uk/GCSEScienceFoundation/Quiz

For useful information about **What to Expect in the Exams** and other exam tips head to cgpbooks.co.uk/GCSEScienceFoundation/Exams

---

Published by CGP

From original material by Richard Parsons.

Editors: Emily Garrett, Rob Hayman, Paul Jordin, Sharon Keeley-Holden, Duncan Lindsay, Sarah Pattison, Rachael Rogers

Contributors: Paddy Gannon

With thanks to Emily Smith for the copyright research.

Stopping distance data used on pages 367 and 484 from the Highway Code.
Contains public sector information licensed under the Open Government Licence v3.0.
http://www.nationalarchives.gov.uk/doc/open-government-licence/version/3/

Printed by Elanders Ltd, Newcastle upon Tyne.
Clipart from Corel®

Illustrations by: Sandy Gardner Artist, email sandy@sandygardner.co.uk

Text, design, layout and original illustrations © Coordination Group Publications Ltd. (CGP) 2021
All rights reserved.

Photocopying more than 5% of this book is not permitted, even if you have a CLA licence.
Extra copies are available from CGP with next day delivery • 0800 1712 712 • www.cgpbooks.co.uk

Working Scientifically

The Scientific Method

*This section **isn't** about how to 'do' science — but it does show you the way **most scientists** work.*

Science is All About **Testing Hypotheses**

Scientists Make an **Observation**

1) Scientists OBSERVE (look at) something they don't understand.
2) They come up with a possible explanation for what they've observed.
3) This explanation is called a HYPOTHESIS.

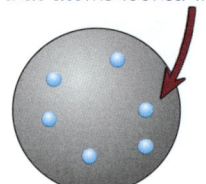

About 100 years ago, scientists thought that atoms looked like this.

They **Test** Their Hypothesis

1) Next, they test whether the hypothesis is right or not.
2) They do this by making a PREDICTION — a statement based on the hypothesis that can be tested.
3) They then TEST this prediction by carrying out experiments.
4) If their prediction is right, this is EVIDENCE that their hypothesis might be right too.

Other Scientists Test the Hypothesis Too

1) Other scientists check the evidence — for example, they check that the experiment was carried out in a sensible way. This is called PEER-REVIEW.
2) Scientists then share their results, e.g. in scientific papers.
3) Other scientists carry out more experiments to test the hypothesis.
4) Sometimes these scientists will find more evidence that the hypothesis is RIGHT.
5) Sometimes they'll find evidence that shows the hypothesis is WRONG.

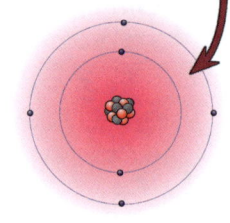

After more evidence was gathered, they changed their hypothesis to this.

The Hypothesis is **Accepted** or **Rejected**

1) If all the evidence that's been found supports the hypothesis, it becomes an ACCEPTED THEORY and goes into textbooks for people to learn.
2) If the evidence shows that the hypothesis is wrong, scientists must:
3) Change the hypothesis, OR
4) Come up with a new hypothesis.

Now we think it's more like this.

Scientific models are constantly being refined...

You can see just how much testing has to be done before something gets accepted as a theory. If scientists aren't busy testing their own hypothesis, then they're busy testing someone else's.

Models and Communication

*Once scientists have made a **new discovery**, they **don't** just keep it to themselves. Oh no. Time to learn about how scientific discoveries are **communicated**, and the **models** that are used to represent theories.*

Theories Can Involve Different Types of Models

1) A model is a simple way of describing or showing what's going on in real life.
2) Models can be used to explain ideas and make predictions. For example:

> - The Bohr model of an atom is a simple picture of what an atom looks like (see p.177).
> - It can be used to explain trends in the periodic table (see p.180 for more).

3) All models have limits — a single model can't explain everything about an idea.

It's Important to Tell People About Scientific Discoveries

1) Scientific discoveries can make a big difference to people's lives.
2) So scientists need to tell the world about their discoveries.
3) They might need to tell people to change their habits, e.g. stop smoking to protect against lung cancer.
4) They might also need to tell people about new technologies. For example:

> The discovery of molecules called fullerenes has led to a new technology that delivers medicine to body cells. Doctors and patients might need to be given information about this technology.

Scientific Evidence can be Presented in a Biased Way

1) Reports about scientific discoveries in the media (e.g. newspapers or television) can be misleading.
2) The data might be presented in a way that's not quite right — or it might be oversimplified.
3) This means that people may not properly understand what the scientists found out.
4) People who want to make a point can also sometimes present data in a biased way (in a way that's unfair or ignores one side of the argument). For example:

> - A scientist may talk a lot about one particular relationship in the data (and not mention others).
> - A newspaper article might describe data supporting an idea without giving any evidence against it.

Companies can present biased data to help sell products...

Sometimes a company may only want you to see half of the story so they present the data in a biased way. For example, a medicines company may want to encourage you to buy their drugs. They might tell you about all the positives, but not report the results of any unfavourable studies.

Working Scientifically

Issues Created by Science

*Science has helped us to **make progress** in loads of areas, from medicine to space travel. But science still has its **issues**. And it **can't answer everything**, as you're about to find out.*

Scientific Developments are Great, but they can Raise Issues

1) Scientific developments include new technologies and new advice.
2) These developments can create issues. For example:

> Economic (money) issues: Society can't always afford to do things scientists recommend, like spend money on green energy sources.

> Social (people) issues: Decisions based on scientific evidence affect people — e.g. should alcohol be banned (to prevent health problems)?

> Personal issues: Some decisions will affect individuals — e.g. people may be upset if a wind farm is built next to their house.

> Environmental issues: Human activity often affects the environment — e.g. some people think that genetically modified crops (see p.135) could cause environmental problems.

Science Can't Answer Every Question — Especially Ethical Ones

1) At the moment scientists don't agree on some things — like what the Universe is made of.
2) This is because there isn't enough data to support the scientists' hypotheses.
3) But eventually, we probably will be able to answer these questions once and for all.
4) Experiments can't tell us whether something is ethically right or wrong. For example, whether it's right for people to use new drugs to help them do better in exams.
5) The best we can do is make a decision that most people are more or less happy to live by.

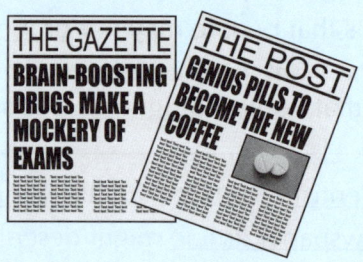

There are often issues with new scientific developments...
The trouble is, there's often no clear right answer where these issues are concerned. Different people have different views, depending on their priorities. These issues are full of grey areas.

Working Scientifically

Risk

Scientific discoveries are often great, but they can prove **risky**. With dangers all around, you've got to be aware of hazards — this includes **how likely** they are to **cause harm** and **how serious** the effects may be.

Nothing is Completely Risk-Free

1) A hazard is something that could cause harm.

2) All hazards have a risk attached to them — this is the chance that the hazard will cause harm.

3) New technology can bring new risks. E.g. scientists are creating technology to capture and store carbon dioxide. But if the carbon dioxide leaked out it could damage soil or water supplies. These risks need to be considered alongside the benefits of the technology, e.g. lower greenhouse gas emissions.

4) To make a decision about activities that involve hazards, we need to think about:
   - the chance of the hazard causing harm,
   - how bad the outcome (consequences) would be if it did.

People Make Their Own Decisions About Risk

1) Not all risks have the same consequences. For example, if you chop veg with a sharp knife you risk cutting your finger, but if you go scuba-diving you risk death.

2) Most people are happier to accept a risk if the consequences don't last long and aren't serious.

3) People tend to think familiar activities are low-risk. They tend to think unfamiliar activities are high-risk. But this isn't always true. For example:

   - Cycling on roads is often high-risk. But it's a familiar activity, so many people are happy to do it.
   - Air travel is actually pretty safe, but a lot of people think it is high-risk.

4) The best way to estimate the size of a risk is to look at data. E.g. you could estimate the risk of a driver crashing by recording how many people in a group of 100 000 drivers crashed their cars over a year.

The pros and cons of new technology must be weighed up...
The world's a dangerous place and it's impossible to rule out the chance of an accident altogether. But if you can recognise hazards and take steps to reduce the risks, you're more likely to stay safe.

Working Scientifically

Designing Investigations

*Dig out your lab coat and dust down your safety goggles... it's **investigation time**. Investigations include **lab experiments** and **studies** done in the **real world**.*

Evidence Can Support or Disprove a Hypothesis

1) Scientists observe things and come up with hypotheses to explain them (see p.2). You need to be able to do the same. For example:

   > Observation: People have big feet and spots. Hypothesis: Having big feet causes spots.

2) To find out if your hypothesis is right, you need to do an investigation to gather evidence.

3) To do this, you need to use your hypothesis to make a prediction — something you think will happen that you can test. E.g. people who have bigger feet will have more spots.

4) Investigations are used to see if there are patterns or relationships between two variables (see below).

Make an Investigation a Fair Test By Controlling the Variables

1) In a lab experiment you usually change one thing (a variable) and measure how it affects another thing (another variable).

   > EXAMPLE: you might change the concentration of a reactant and measure how it affects the temperature change of the reaction.

2) Everything else that could affect the results needs to stay the same. Then you know that the thing you're changing is the only thing that's affecting the results.

   > EXAMPLE continued: you need to keep the volume of the reactants the same. If you don't, you won't know if any change in the temperature is caused by the change in concentration, or the change in volume.

3) The variable that you CHANGE is called the INDEPENDENT variable.
4) The variable you MEASURE is called the DEPENDENT variable.
5) The variables that you KEEP THE SAME are called CONTROL variables.

   > EXAMPLE continued:
   > Independent = concentration
   > Dependent = temperature
   > Control = volume of reactants, pH, etc.

6) Because you can't always control all the variables, you often need to use a CONTROL EXPERIMENT.

7) This is an experiment that's kept under the same conditions as the rest of the investigation, but doesn't have anything done to it. This is so that you can see what happens when you don't change anything.

Evidence Needs to be Repeatable, Reproducible and Valid

1) REPEATABLE means that if the same person does the experiment again, they'll get similar results. To check your results are repeatable, repeat the readings at least three times. Then check the repeat results are all similar.

2) REPRODUCIBLE means that if someone else does the experiment, the results will still be similar. To make sure your results are reproducible, get another person to do the experiment too.

3) VALID results come from experiments that were designed to be a fair test. They're also repeatable and reproducible.

If data is repeatable and reproducible, scientists are more likely to trust it.

Designing Investigations

The Bigger the Sample Size the Better

1) Sample size is how many things you test in an investigation, e.g. 500 people or 20 types of metal.

2) The bigger the sample size the better — to reduce the chance of any weird results.

3) But scientists have to be realistic when choosing how big their sample should be. E.g. if you were studying how lifestyle affects weight it'd be great to study everyone in the UK (a huge sample), but it'd take ages and cost loads.

4) When you choose a sample, you need to make sure you've got a range of different people.

5) For example, both men and women with a range of different ages.

Your Data Should be Accurate and Precise

1) ACCURATE results are results that are really close to the true answer.

2) The accuracy of your results usually depends on your method. You need to make sure you're measuring the right thing.

3) You also need to make sure you don't miss anything that should be included in the measurements. For example:

   If you're measuring the volume of gas released by a reaction, make sure you collect all the gas.

4) PRECISE results are ones where the data is all really close to the mean (average) of your repeated results.

Repeat	Data set 1	Data set 2
1	12	11
2	14	17
3	13	14
Mean	13	14

Data set 1 is more precise than data set 2 — the results are all close to the mean (not spread out).

Your Equipment has to be Right for the Job

1) The measuring equipment you use has to be able to accurately measure the chemicals you're using. E.g. if you need to measure out 11 cm³ of a liquid, use a measuring cylinder that can measure to 1 cm³ — not 5 or 10 cm³.

2) You also need to set up the equipment properly. For example, make sure your mass balance is set to zero before you start weighing things.

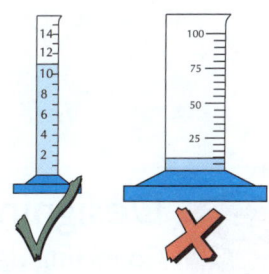

Working Scientifically

Designing Investigations

You Need to Look out for Errors and Anomalous Results

1) The results of your experiment will always vary a bit because of RANDOM ERRORS — for example, mistakes you might make while measuring.
2) You can reduce the effect of random errors by taking repeat readings and finding the mean. This will make your results more precise.
3) If a measurement is wrong by the same amount every time, it's called a SYSTEMATIC ERROR. For example:

> If you measure from the very end of your ruler instead of from the 0 cm mark every time, all your measurements would be a bit small.

4) If you know you've made a systematic error, you might be able to correct it. For example, by adding a bit on to all your measurements.
5) Sometimes you get a result that doesn't fit in with the rest. This is called an ANOMALOUS RESULT.
6) You should try to work out what happened. If you do (e.g. you find out you measured something wrong) you can ignore it when processing your results (see next page).

Investigations Can Have Hazards

1) Hazards from science experiments include things like:

microorganisms (e.g. bacteria)   chemicals   electricity   fire

2) When you plan an investigation you need to make sure that it's safe.
3) You should identify all the hazards that you might come across.
4) Then you should think of ways of reducing the risks. For example:

> - If you're working with sulfuric acid, always wear gloves and safety goggles. This will reduce the risk of the acid burning your skin and eyes.
> - If you're using a Bunsen burner, stand it on a heat proof mat. This will reduce the risk of starting a fire.

There's more on safety in experiments on page 395.

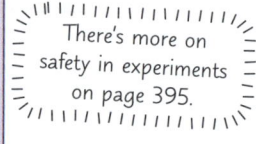

Designing an investigation is an involved process...

Collecting data is what investigations are all about. Designing a good investigation is really important to make sure that any data collected is accurate, precise, repeatable and reproducible.

Processing Data

*Processing your data means doing some **calculations** with it to make it **more useful**.*

Data Needs to be Organised

1) Tables are useful for organising data.
2) When you draw a table use a ruler.
3) Make sure each column has a heading (including the units).

Test tube	Repeat 1 (cm³)	Repeat 2 (cm³)
A	28	37
B	47	51

There are Different Ways of Processing Your Data

1) When you've done repeats of an experiment you should always calculate the mean (a type of average).
2) You might also need to calculate the range (how spread out the data is).

EXAMPLE The results of an experiment to find the volume of gas produced in a reaction are shown in the table below. Calculate the mean volume and the range.

Volume of gas produced (cm³)		
Repeat 1	Repeat 2	Repeat 3
28	37	32

1) To calculate the mean, add together all the data values. Then divide by the total number of values in the sample.

   $(28 + 37 + 32) \div 3$ = 32 cm³

2) To calculate the range, subtract the smallest number from the largest number.

   $37 - 28$ = 9 cm³

3) To find the median, put all your data in order from smallest to largest. The median is the middle value.
4) The number that appears most often is the mode.

   E.g. if you have the data set: 1 2 1 1 3 4 2
   The median is: 1 1 **1** 2 2 3 4. The mode is **1** because 1 appears most often.

   If you have an even number of values, the median is halfway between the middle two values.

5) When calculating any of these values, always ignore any anomalous results.

Round to the Lowest Number of Significant Figures

1st significant figure
0.0307
2nd 3rd

1) The first significant figure of a number is the first digit that's not zero.
2) The second and third significant figures come straight after (even if they're zeros).
3) In any calculation, you should round the answer to the lowest number of significant figures (s.f.) given.
4) If your calculation has more than one step, only round the final answer.

EXAMPLE The mass of a solid is 0.24 g and its volume is 0.715 cm³. Calculate the density of the solid.

Density = 0.24 g ÷ 0.715 cm³ = 0.33566... = 0.34 g/cm³ (2 s.f.)

2 s.f.   3 s.f.

Final answer should be rounded to 2 s.f.

Don't forget your calculator...

In the exam you could be given some data and be expected to process it in some way. Make sure you keep an eye on significant figures in your answers and always write down your working.

Working Scientifically

Presenting Data

Once you've processed your data, e.g. by calculating the mean, you can present your results in a nice **chart** or **graph**. This will help you to **spot any patterns** in your data.

If Your Data Comes in Categories, Present it in a Bar Chart

If the independent variable comes in clear categories (e.g. blood group, types of metal) or can be counted exactly (e.g. number of protons) you should use a bar chart to display the data. Here's what to do:

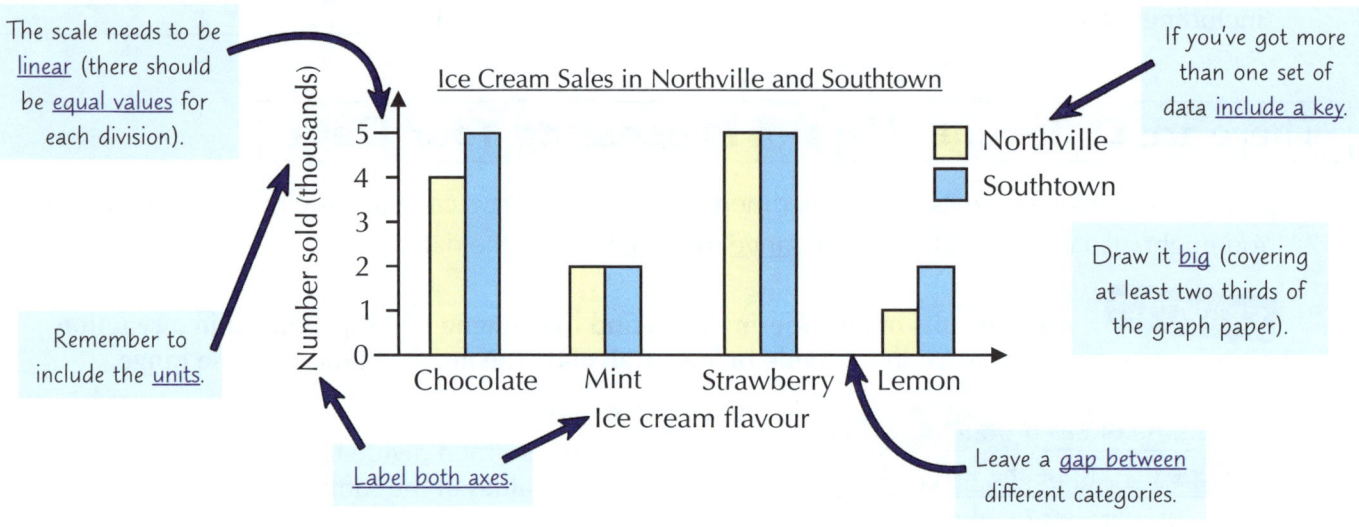

The scale needs to be linear (there should be equal values for each division).

Remember to include the units.

Label both axes.

If you've got more than one set of data include a key.

Draw it big (covering at least two thirds of the graph paper).

Leave a gap between different categories.

If Your Data is Continuous, Plot a Graph

If both variables can have any value within a range (e.g. length, volume) use a graph to display the data. Here are the rules for plotting points on a graph:

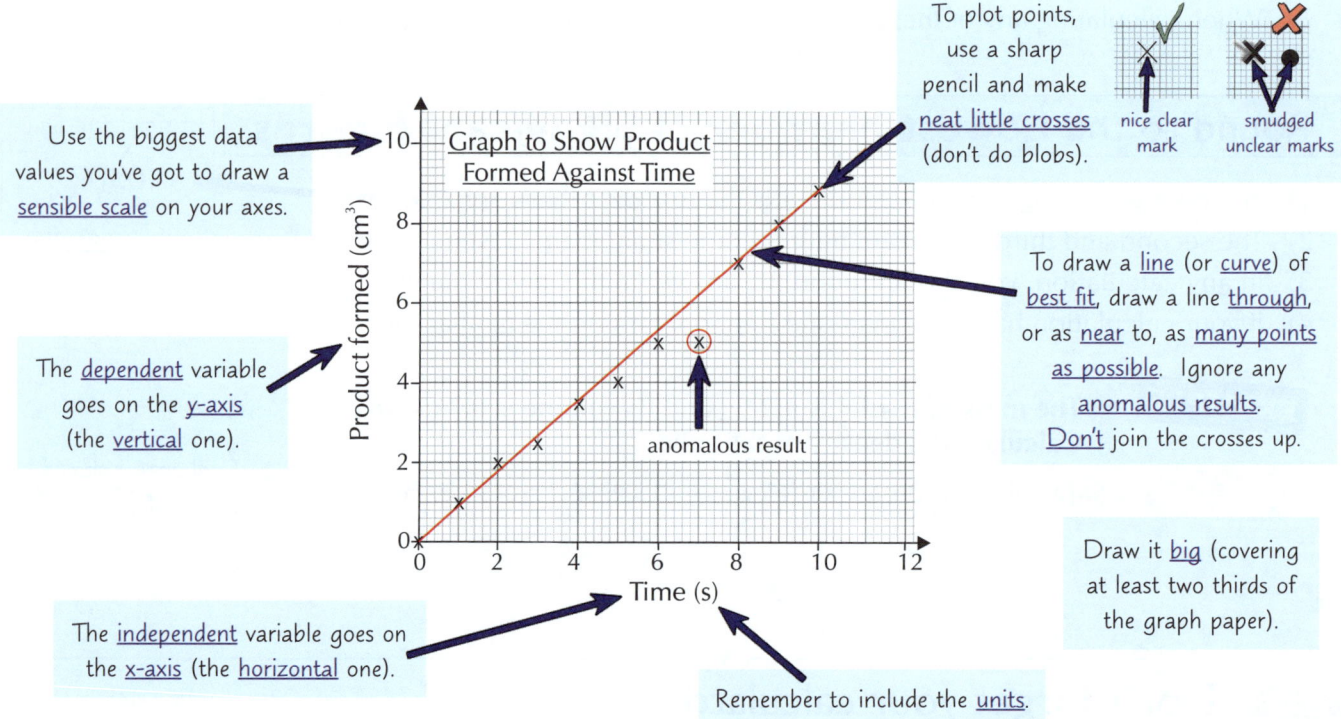

Use the biggest data values you've got to draw a sensible scale on your axes.

The dependent variable goes on the y-axis (the vertical one).

The independent variable goes on the x-axis (the horizontal one).

To plot points, use a sharp pencil and make neat little crosses (don't do blobs).

To draw a line (or curve) of best fit, draw a line through, or as near to, as many points as possible. Ignore any anomalous results. Don't join the crosses up.

Draw it big (covering at least two thirds of the graph paper).

Remember to include the units.

Working Scientifically

More on Graphs

*Graphs aren't just fun to plot, they're also really useful for showing **trends** in your data.*

You Can Calculate the Rate of a Reaction from the Gradient of a Graph

1) This is the formula you need to calculate the gradient (slope) of a graph:

$$\text{gradient} = \frac{\text{change in } y}{\text{change in } x}$$

2) You can use it to work out the rate of a reaction (how quickly the reaction happens).

EXAMPLE The graph shows the volume of gas produced in a reaction against time. Calculate the rate of reaction.

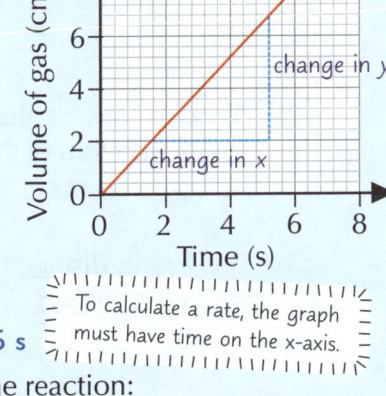

1) To calculate the gradient, pick two points on the line that are easy to read. They should also be a good distance apart.

2) Draw a line down from the higher point. Then draw a line across from the other, to make a triangle.

3) The line drawn down the side of the triangle is the change in y. The line across the bottom is the change in x.

4) Read the x and y values of the points off the graph and work out the change in y and the change in x:

Change in y = 6.8 − 2.0 = 4.8 cm^3 Change in x = 5.2 − 1.6 = 3.6 s

To calculate a rate, the graph must have time on the x-axis.

5) Then put these numbers in the formula above to find the rate of the reaction:

$$\text{Rate} = \text{gradient} = \frac{\text{change in } y}{\text{change in } x} = \frac{4.8 \text{ cm}^3}{3.6 \text{ s}} = 1.3 \text{ cm}^3/\text{s}$$

The units are (units of y)/(units of x). cm^3/s can also be written as cm^3 s^{-1}.

Graphs Show the Relationship Between Two Variables

1) You can get three types of correlation (relationship) between variables:

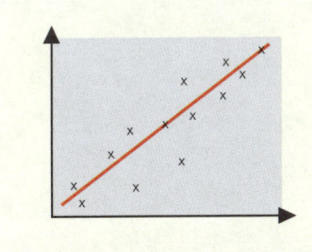

POSITIVE correlation: as one variable increases the other increases.

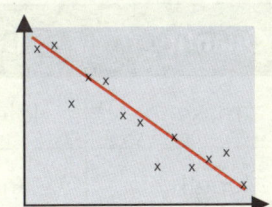

INVERSE (negative) correlation: as one variable increases the other decreases.

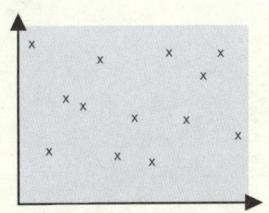

NO correlation: no relationship between the two variables.

2) A correlation doesn't mean the change in one variable is causing the change in the other (see page 14).

Graphs make it much easier to see relationships in data...

If you can't work out what kind of correlation a graph shows, just remember that an inverse correlation gives a line of best fit that slopes downwards, so it looks like the letter N (for Negative).

Working Scientifically

Rearranging Equations and Units

*Graphs and maths skills are all very well, but the numbers don't mean much if you don't get the **units** right.*

You Can **Rearrange** Equations

1) Equations show relationships between variables. For example, speed = $\frac{\text{distance}}{\text{time}}$.
2) The subject of an equation is the variable by itself on one side of the equals sign. So speed is the subject in the equation above.
3) To change the subject of an equation do the same thing to both sides of the equation until you've got the subject you want.

EXAMPLE   **Make distance the subject of the equation above.**

1) Multiply both sides by time.   speed = $\frac{\text{distance}}{\text{time}}$

   ➡ speed × time = $\frac{\text{distance × time}}{\text{time}}$

2) 'Time' is now on the top and the bottom of the fraction, so it cancels out: speed × time = $\frac{\text{distance} \times \cancel{\text{time}}}{\cancel{\text{time}}}$

3) This leaves distance by itself. So it's the subject: **speed × time = distance**

S.I. Units Are Used All Round the World

1) All scientists use the same units to measure their data.
2) These are standard units, called S.I. units.
3) Here are some S.I. units you might see:

Quantity	S.I. Base Unit
mass	kilogram, kg
length	metre, m
time	second, s
temperature	kelvin, K

S.I. units help scientists to compare data...

You can only really compare things if they're in the same units. E.g. if the rate of blood flow was measured in ml/min in one vein and in l/day in another vein, it'd be hard to know which was faster.

Working Scientifically

More on Units

*You can **convert units**, which can save you from having to write a lot of 0's...*

Different Units Help you to Write Large and Small Quantities

1) Quantities come in a huge range of sizes.
2) To make the size of numbers easier to handle, larger or smaller units are used.
3) Larger and smaller units are written as the S.I. base unit with a little word in front (a prefix). Here are some examples of prefixes and what they mean:

Kilogram is an exception. It's an S.I. unit with the prefix already on it.

Prefix	mega (M)	kilo (k)	deci (d)	centi (c)	milli (m)	micro (µ)
How it compares to the base unit	1 000 000 times bigger	1000 times bigger	10 times smaller	100 times smaller	1000 times smaller	1 000 000 times smaller

E.g. 1 kilometre is 1000 metres.   E.g. there are 1000 millimetres in 1 metre.

You Need to be Able to Convert Between Units

You need to know how to convert (change) one unit into another. Here are some useful conversions:

Mass can have units of kg and g.

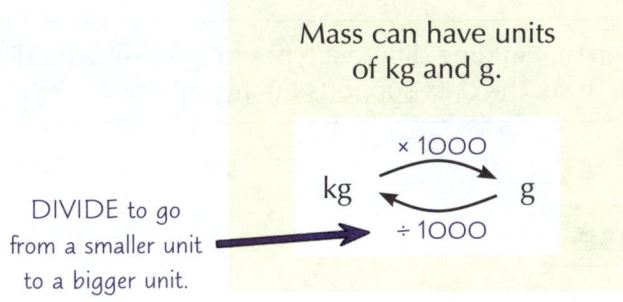

DIVIDE to go from a smaller unit to a bigger unit.

Energy can have units of kJ and J.

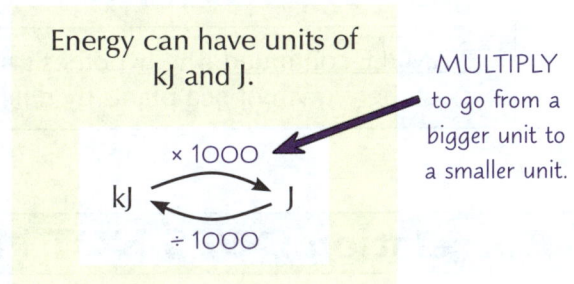

MULTIPLY to go from a bigger unit to a smaller unit.

Length can have lots of units including m, mm and µm.

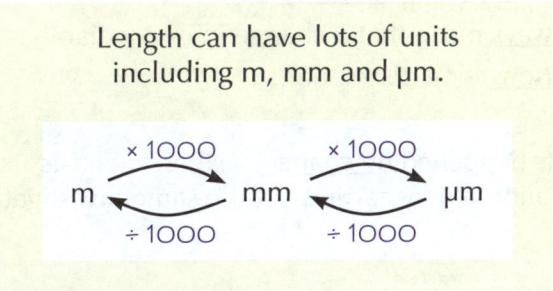

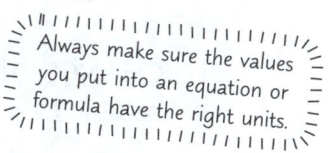

Always make sure the values you put into an equation or formula have the right units.

EXAMPLE
A car has travelled 0.015 kilometres. How many metres has it travelled?

1 km = 1000 m. So to convert from km (a bigger unit) to m (a smaller unit) you need to multiply by 1000.   **0.015 km × 1000 = 15 m**

To convert from bigger units to smaller units...
...you need to multiply, and to convert from smaller units to bigger units, you need to divide. Don't get this the wrong way round or you'll get answers that are far too big or far too small.

Working Scientifically

Drawing Conclusions

*Once you've carried out an experiment and processed your data, it's time to work out **what your data shows**.*

You Can **Only Conclude** What the Data Shows and **NO MORE**

1) To come to a conclusion, look at your data and say what pattern you see.

EXAMPLE: The table on the right shows the heights of pea plant seedlings grown for three weeks with different fertilisers.

Fertiliser	Mean growth / mm
A	13.5
B	19.5
No fertiliser	5.5

CONCLUSION: Pea plant seedlings grow taller over a three week period with fertiliser B than with fertiliser A.

2) It's important that the conclusion matches the data it's based on — it shouldn't go any further.

EXAMPLE continued: You can't conclude that any other type of plant grows taller with fertiliser B than with fertiliser A — the results could be totally different.

3) You also need to be able to use your results to justify your conclusion (i.e. back it up).

EXAMPLE continued: The pea plants grow 6 mm more on average with fertiliser B than with fertiliser A.

4) When writing a conclusion you need to say whether or not the data supports the original hypothesis:

EXAMPLE continued: The hypothesis might have been that adding different types of fertiliser would affect the growth of pea plants by different amounts. If so, the data supports the hypothesis.

Correlation **DOES NOT** Mean **Cause**

1) If two things are correlated, there's a relationship between them — see page 11.
2) But a correlation doesn't always mean that a change in one variable is causing the change in the other.
3) There are three possible reasons for a correlation:

   1) CHANCE: The results happened by chance. Other scientists wouldn't get a correlation if they carried out the same investigation.

   2) LINKED BY A 3RD VARIABLE: There's another factor involved.

   E.g. there's a correlation between water temperature and shark attacks. They're linked by a third variable — the number of people swimming (more people swim when the water's hotter, which means you get more shark attacks).

   3) CAUSE: Sometimes a change in one variable does cause a change in the other. You can only conclude this when you've controlled all the variables that could be affecting the result.

Working Scientifically

Uncertainty

*Uncertainty is how sure you can really be about your data. There's a little bit of **maths** to do, and also a formula to learn. But don't worry too much — it's no more than a simple bit of subtraction and division.*

Uncertainty is the Amount of Error Your Measurements Might Have

1) Measurements you make will have some uncertainty in them (i.e. they won't be completely perfect).

2) This can be due to random errors (see page 8). It can also be due to limits in what your measuring equipment can measure.

3) This means that the mean of your results will have some uncertainty to it.

4) You can calculate the uncertainty of a mean result using this equation:

$$\text{uncertainty} = \frac{\text{range}}{2}$$

The range is the largest value minus the smallest value (p.9).

5) The less precise your results are, the higher the uncertainty will be.

6) Uncertainties are shown using the '±' symbol.

EXAMPLE

The table below shows the results of an experiment to determine the speed of a trolley as it moves along a horizontal surface. Calculate the uncertainty of the mean.

Repeat	1	2	3	mean
Speed (m/s)	2.02	1.98	2.00	2.00

1) First work out the range:

Range = 2.02 − 1.98 = 0.04 m/s

2) Use the range to find the uncertainty:

Uncertainty = range ÷ 2 = 0.04 ÷ 2 = 0.02 m/s

So the uncertainty of the mean = 2.00 ± 0.02 m/s

The smaller the uncertainty, the more precise your results...

Remember that equation for uncertainty. You never know when you might need it — you could be expected to use it in the exams. You need to make sure all the data is in the same units though. For example, if you had some measurements in metres, and some in centimetres, you'd need to convert them all into either metres or centimetres before you set about calculating uncertainty.

Working Scientifically

Evaluations

Hurrah! The end of another investigation. Well, now you have to work out all the things you did **wrong**. That's what **evaluations** are all about I'm afraid. Best get cracking with this page...

Evaluations — Describe **How** it Could be **Improved**

In an evaluation you look back over the whole investigation.

1) You should comment on the method — was it valid?
   Did you control all the other variables to make it a fair test?

2) Comment on the quality of the results — was there enough evidence to reach a valid conclusion?
   Were the results repeatable, reproducible, accurate and precise?

3) Were there any anomalous results? If there were none then say so.
   If there were any, try to explain them — were they caused by errors in measurement?

4) You should comment on the level of uncertainty in your results too.

5) Thinking about these things lets you say how confident you are that your conclusion is right.

6) Then you can suggest any changes to the method that would improve the quality of the results, so you could have more confidence in your conclusion.

7) For example, taking measurements at narrower intervals could give you a more accurate result. E.g.

- Say you do an experiment to find the temperature at which an enzyme works best.
- You take measurements at 30 °C, 40 °C and 50 °C.
  The results show that the enzyme works best at 40 °C.
- To get a more accurate result, you could repeat the experiment and take more measurements around 40 °C. You might then find that the enzyme actually works best at 42 °C.

8) You could also make more predictions based on your conclusion.
   You could then carry out further experiments to test the new predictions.

Always look for ways to improve your investigations

So there you have it — Working Scientifically. Make sure you know this stuff like the back of your hand. It's not just in the lab or the field, when you're carrying out your groundbreaking investigations, that you'll need to know how to work scientifically. You can be asked about it in the exams as well. So swot up...

Topic B1 — Cell Biology

Cells

Cells are the **building blocks** of **every organism on the planet**.

Cells can be **Prokaryotic** or **Eukaryotic**

1) All living things are made of cells.

2) Eukaryotic cells are complex. All animal and plant cells are eukaryotic.

3) Prokaryotic cells are smaller and simpler. Bacteria are prokaryotic cells.

Plant and Animal Cells have Similarities and Differences

The different parts of a cell are called subcellular structures.

Animal Cells

Most animal cells have these subcellular structures:

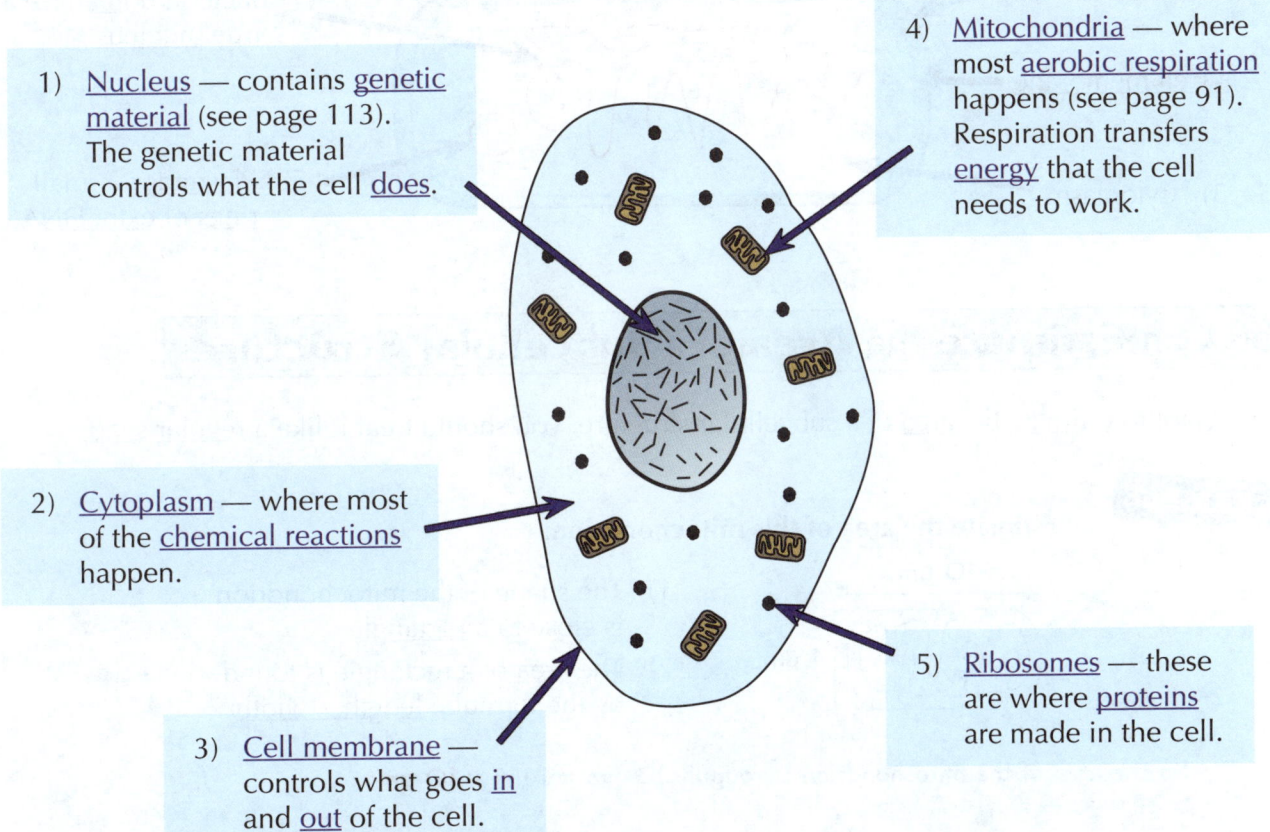

1) Nucleus — contains genetic material (see page 113). The genetic material controls what the cell does.

2) Cytoplasm — where most of the chemical reactions happen.

3) Cell membrane — controls what goes in and out of the cell.

4) Mitochondria — where most aerobic respiration happens (see page 91). Respiration transfers energy that the cell needs to work.

5) Ribosomes — these are where proteins are made in the cell.

Subcellular structures are all the different parts of a cell

Make sure you get to grips with the different subcellular structures that animal cells contain before you move on to the next page. There are more subcellular structures coming up that you need to know...

Topic B1 — Cell Biology

Cells

Plant Cells

Plant cells usually have all the bits that animal cells have. They also have:

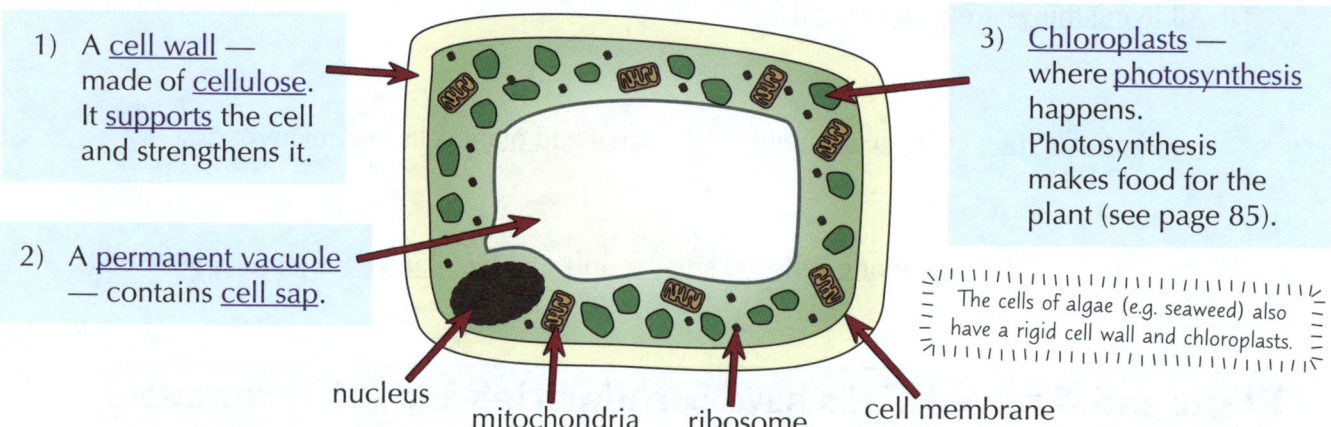

1) A cell wall — made of cellulose. It supports the cell and strengthens it.

2) A permanent vacuole — contains cell sap.

3) Chloroplasts — where photosynthesis happens. Photosynthesis makes food for the plant (see page 85).

The cells of algae (e.g. seaweed) also have a rigid cell wall and chloroplasts.

Labels: nucleus, mitochondria, ribosome, cell membrane

Bacterial Cells Have These Subcellular Structures:

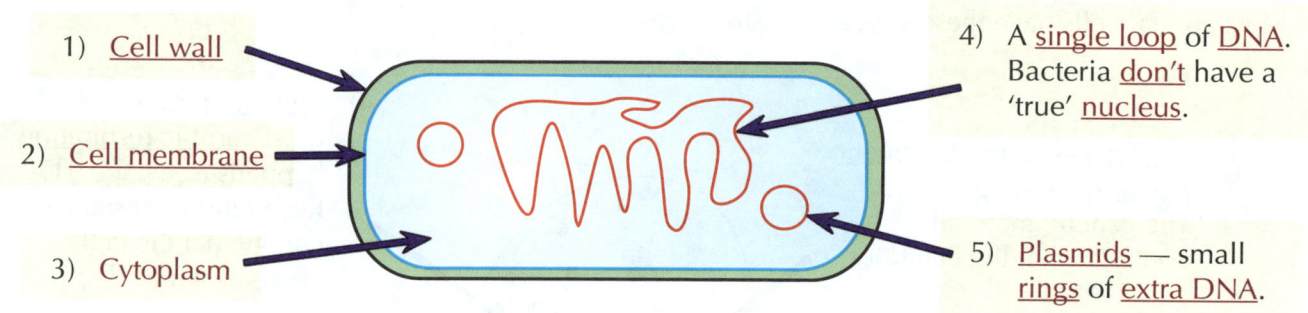

1) Cell wall
2) Cell membrane
3) Cytoplasm
4) A single loop of DNA. Bacteria don't have a 'true' nucleus.
5) Plasmids — small rings of extra DNA.

You Can Estimate the Area of a Subcellular Structure

If you want to estimate the area of a subcellular structure, you should treat it like a regular shape:

EXAMPLE

Estimate the area of this mitochondrion:

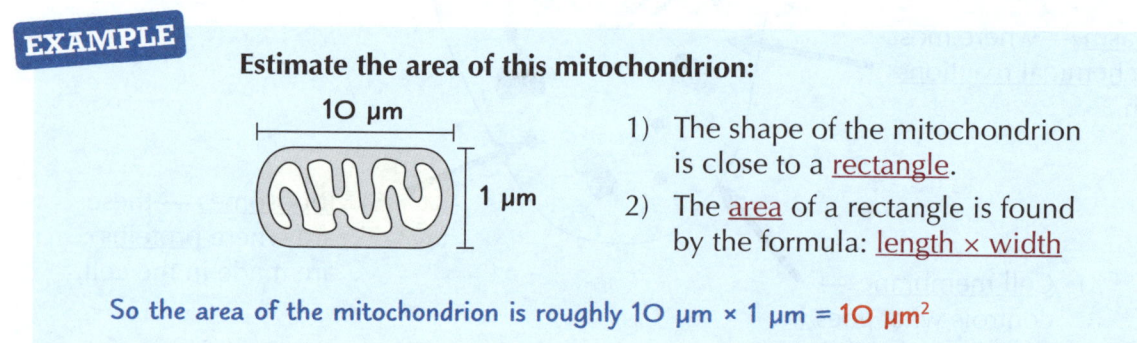

1) The shape of the mitochondrion is close to a rectangle.
2) The area of a rectangle is found by the formula: length × width

So the area of the mitochondrion is roughly 10 µm × 1 µm = **10 µm²**

There's quite a bit to learn in biology — but that's life, I guess...
On these pages are a typical animal cell, plant cell and bacterial cell. Make sure you're familiar with all their structures. A good way to check that you know what all the bits and pieces are is to copy out the diagrams and see if you can remember all the labels. No cheating.

Topic B1 — Cell Biology

Microscopy

Microscopes are pretty important for biology. So here are a couple of pages all about them...

Microscopes **Magnify** Things (Make Them Look **Bigger**)

1) The ways we can use microscopes have developed over the years. This is because technology and knowledge have improved.
2) Light microscopes can be used to look at cells. They let us see large subcellular structures (like the nucleus).
3) Electron microscopes have a higher resolution than light microscopes — they show things in more detail.
4) Electron microscopes also have a higher magnification than light microscopes. They can let us see really small things like ribosomes and plasmids.
5) Electron microscopes were invented after light microscopes. They helped scientists understand more about subcellular structures.

See the next page for how to use a light microscope.

You Need to be Able to Use the **Formula** for **Magnification**

Magnification is how many times bigger the image is than the real thing. You can work out the magnification of an image using this formula:

$$\text{magnification} = \frac{\text{image size}}{\text{real size}}$$

Image size and real size should have the same units.

EXAMPLE

The width of a cell is 0.02 mm. The width of its image under a microscope is 8 mm. What magnification was used to view the cell?

magnification = 8 mm ÷ 0.02 mm = × 400

You Can Write Numbers in **Standard Form**

Standard form is useful for writing very big or very small numbers in a simpler way.

EXAMPLE Write 0.0025 mm in standard form.

1) The first number needs to be between 1 and 10 so the decimal point needs to move after the '2'.
2) Count how many places the decimal point has moved — this is the power of 10.
3) The power of 10 is positive if the decimal point is moved to the left. It's negative if the decimal point has moved to the right. Here, the decimal point has moved right, so it needs a minus sign.

0.0025 → 2.5
1 2 3

The decimal point has moved 3 places = 10^3

2.5×10^{-3}

Standard form saves you from having to write a lot of zeros

Remember, when you're writing a number in standard form, you've got to move the decimal point to just after the first digit that's not a zero. E.g. 342 000 is 3.42×10^5, and 0.0009513 is 9.513×10^4.

Q1 A cheek cell is viewed under a microscope with × 40 magnification. The image of the cell is 2.4 mm wide. Calculate the real width of the cheek cell. Give your answer in μm. [2 marks]

Topic B1 — Cell Biology

# Microscopy

So you know what microscopes do... now you need to know how to use one.

You Need to Prepare Your Slide

1) Add a drop of water to the middle of a clean slide.

2) Cut up an onion and take off one layer.

3) Use tweezers to peel off some epidermal tissue (the clear 'skin') from the bottom of the layer.

4) Using the tweezers, place the skin into the water on the slide.

5) Add a drop of iodine solution. Iodine solution is a stain. Stains can make different parts of a cell easier to see.

6) Place a cover slip on top. Try not to get any air bubbles under it.

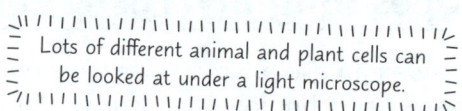
Lots of different animal and plant cells can be looked at under a light microscope.

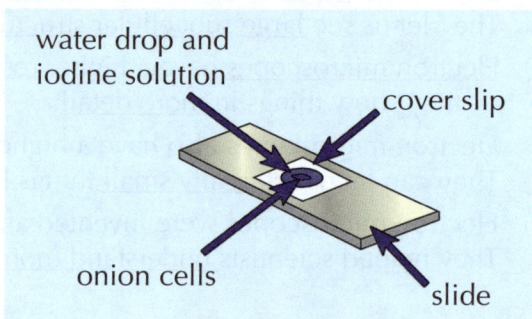

Know the Parts of a Light Microscope

To look at your prepared slides, you need to know how to use a light microscope. Here are the main parts you'll use:

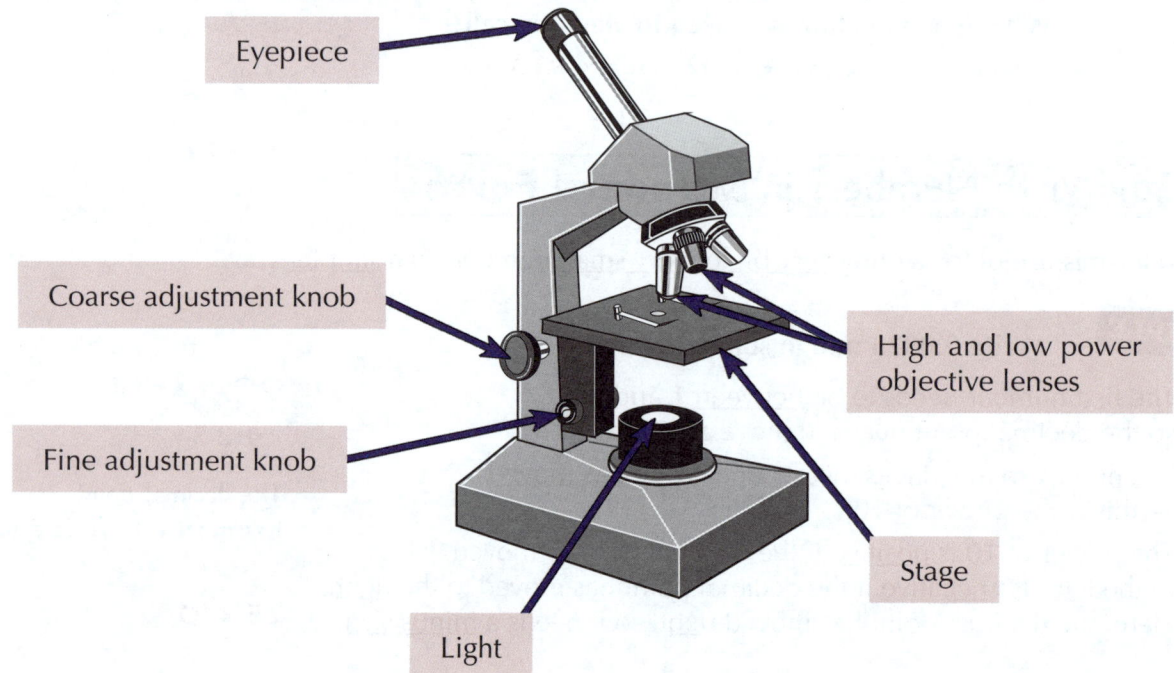

Stains can make subcellular structures easier to see

Carry on to the next page for how to use the microscope above to view your specimen.

Topic B1 — Cell Biology

Microscopy

Use a **Light Microscope** to Look at Your **Slide**

1) Clip the slide onto the stage.

2) Select the objective lens with the lowest magnification.

3) Use the coarse adjustment knob to move the stage up to just below the objective lens.

4) Look down the eyepiece. Move the stage downwards until the image is roughly in focus.

5) Move the fine adjustment knob, until you get a clear image of what's on the slide.

6) If you want a bigger image, use an objective lens with a higher magnification and refocus.

Draw Your Observations **Neatly** with a **Pencil**

1) You should use a pencil with a sharp point to draw what you see under the microscope.
2) Use smooth lines to draw the outlines of the main features (e.g. nucleus, chloroplasts).
3) Don't do any shading or colouring in.
4) Label the features with straight lines. Make sure the lines don't cross over each other.
5) The drawing should take up at least half the space available.
6) Include a title and a scale.
7) Write down the magnification that it was observed under.

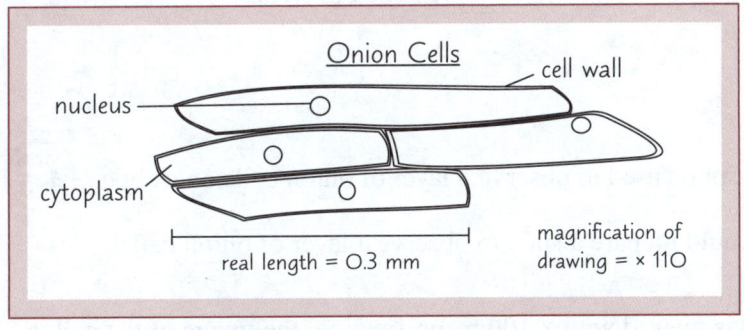

You can measure the real size of the cells using a ruler that fits onto your microscope (see p.394).

 **Your microscope might look a bit different**
The appearance of light microscopes can vary (e.g. they might have two eyepieces rather than one) but they should have the same basic features shown on the previous page.

Topic B1 — Cell Biology

Warm-Up & Exam Questions

So, hopefully you've read the last five pages. But could you cope if a question on cells or microscopes came up in the exam? See what happens when you try out these questions...

Warm-Up Questions

1) Name the subcellular structures where aerobic respiration takes place.
2) Give two ways in which animal cells are different from plant cells.
3) True or false? Bacterial cells have a nucleus.
4) What type of microscope should be used to look at ribosomes and plasmids?
5) Write the number 0.00045 µm in standard form.

Exam Questions

1 Which of the following cells is an example of a prokaryotic cell? Tick **one** box. *(Grade 3-4)*

☐ bacterial cell   ☐ animal cell   ☐ plant cell   ☐ sperm cell

[1 mark]

2 **Figure 1** shows a typical plant cell. *(Grade 3-4)*

Figure 1

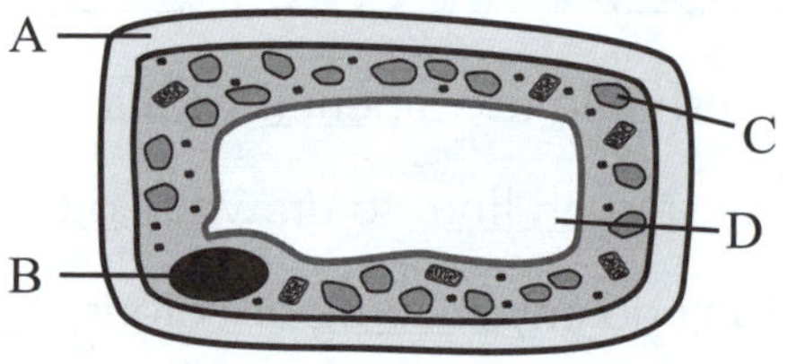

2.1 Which label points to a chloroplast? Tick **one** box.

☐ A   ☐ B   ☐ C   ☐ D

[1 mark]

2.2 What is the function of a chloroplast? Tick **one** box.

☐ allows photosynthesis to take place   ☐ contains genetic material   ☐ contains cell sap

[1 mark]

2.3 **Figure 1** also shows ribosomes.
What is the function of a ribosome? Tick **one** box.

☐ aerobic respiration   ☐ making proteins   ☐ storing extra DNA

[1 mark]

PRACTICAL

3 A light microscope can be used to observe a layer of onion cells on a slide. *(Grade 4-5)*

3.1* Describe how you would prepare a slide to observe a layer of onion cells.

[4 marks]

3.2 When the onion cell is viewed with × 100 magnification, the image of the cell is 7.5 mm wide.

Calculate the real width of the onion cell in micrometres (µm). Complete the following steps.

Calculate the real width in mm using the formula: $\text{real size} = \dfrac{\text{image size}}{\text{magnification}}$

Convert mm to µm.

[2 marks]

Topic B1 — Cell Biology

Cell Differentiation and Specialisation

*Cells **don't** all look the **same**. They have **different structures** to suit their **different functions**.*

Specialised Cells are Cells that Carry Out a Specific Function

1) The process by which cells change to become specialised is called differentiation.

2) As cells change, they develop different subcellular structures.
They turn into different types of cells. This allows them to carry out specific functions.

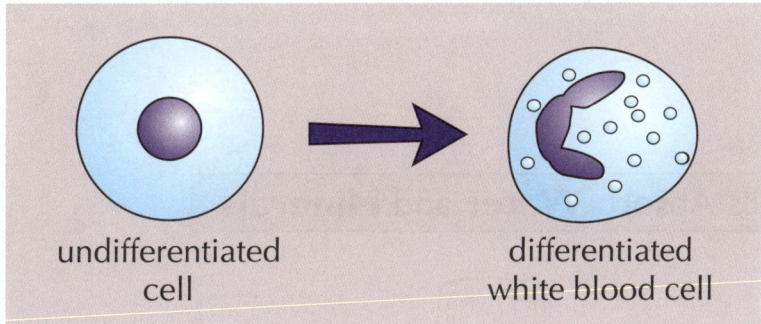

3) Most differentiation occurs as an organism develops.

4) Most animal cells can only differentiate at an early stage of the animal's life.

5) But lots of plant cells can differentiate for the whole of the plant's life.

6) The cells that differentiate in adult animals are mainly used for repairing and replacing cells.

7) Some cells are undifferentiated — they are called stem cells.
There's more about them on pages 25-26.

There Are Many Examples of Specialised Cells...

Sperm Cells Take the Male DNA to the Egg

1) A sperm cell has a tail to help it swim to the egg.
2) It has a lot of mitochondria (see p.17). These provide energy for swimming.

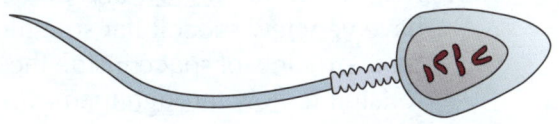

Nerve Cells Carry Electrical Signals Around the Body

1) Nerve cells are long to cover more distance in the body.
2) They have branches at the end to connect to other nerve cells.

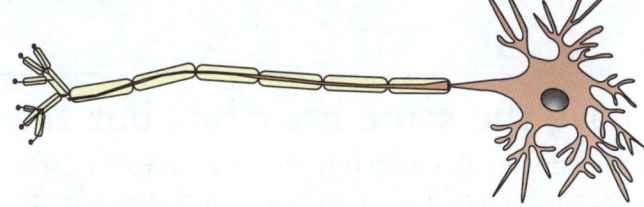

Topic B1 — Cell Biology

Cell Specialisation

Muscle Cells Contract (Shorten)

1) Muscle cells are long so they have space to contract.
2) They have lots of mitochondria. These provide energy for contracting.

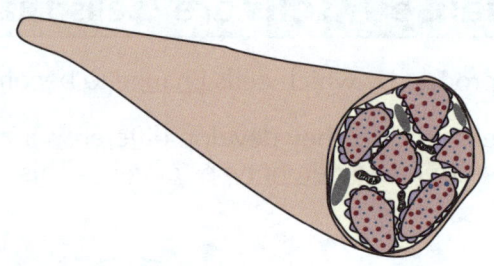

Root Hair Cells Absorb Water and Minerals

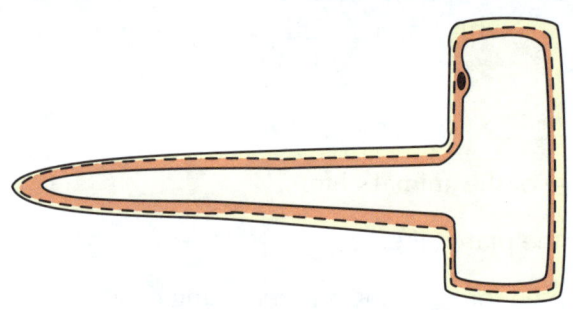

1) Root hair cells grow into long "hairs" that stick out into the soil.
2) This gives the plant a big surface area for absorbing water and mineral ions from the soil.

Phloem Cells Transport Food and Xylem Cells Transport Water

1) Phloem and xylem cells form phloem and xylem tubes.
2) To form the tubes, the cells are long and joined end to end.
3) Xylem cells are hollow and phloem cells have very few subcellular structures. So there's lots of space inside the cells for stuff to flow through them.

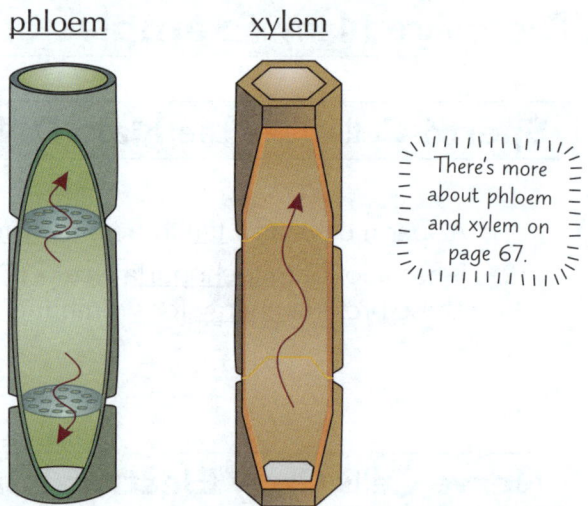

There's more about phloem and xylem on page 67.

Cells have the same basic bits but are specialised for their function

Not all cells contain all of the bits shown on pages 17-18. This is because some specialised cells don't have a use for certain subcellular structures — it depends on their function. For example, root hair cells grow underground in the soil. They don't need chloroplasts because they don't photosynthesise.

Topic B1 — Cell Biology

Stem Cells

*Stem cell research has exciting **possibilities**, but **not everyone agrees** with some of the uses of stem cells.*

Stem Cells can Differentiate into Different Types of Cells

1) Cells differentiate (change) to become specialised for their job (see page 23).

2) Undifferentiated cells are called stem cells.

3) Stem cells can produce lots more undifferentiated cells and differentiate into different types of cell.

4) Stem cells found in early human embryos are called embryonic stem cells.

   An embryo is an unborn baby at an early stage of growth.

5) Embryonic stem cells can turn into any kind of cell at all.

6) Adults also have stem cells. They're only in certain places in the body, like bone marrow (a tissue inside bones).

7) Adult stem cells can only produce certain types of specialised cell, e.g. blood cells.

8) Stem cells from embryos and bone marrow can be cloned (copied) in a lab. The cloned cells can be used in medicine or research.

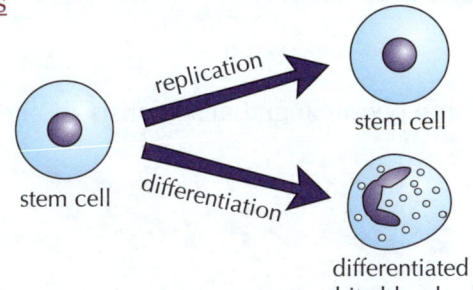

Stem Cells May Be Able to Cure Many Diseases

1) Embryonic stem cells could be used to replace faulty cells in sick people.

> E.g. you could make nerve cells for people with paralysis (where they can't move part of the body due to an injury to their spine) or insulin-producing cells for people with diabetes (see page 106).

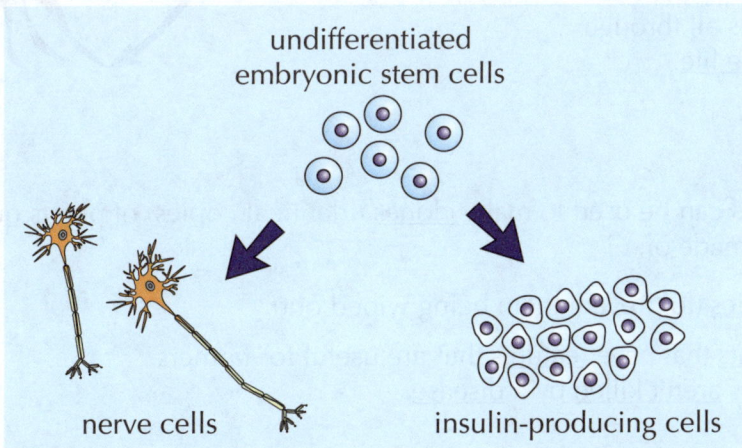

2) It's possible to make an embryo that has the same genes as a patient. This is called therapeutic cloning.

3) This means that the stem cells from the embryo wouldn't be rejected by the patient's body.

4) However, there are risks involved in using stem cells in medicine. For example, the stem cells could be infected with a virus. The virus could be passed on to a patient and make them sicker.

Topic B1 — Cell Biology

Stem Cells

Some People Are Against Stem Cell Research

1) Some people feel embryos shouldn't be used for research because each one could be a human life.

2) Others think that curing patients who are suffering is more important than the rights of embryos.

3) They argue that the embryos used in the research are usually unwanted ones from fertility clinics. If they weren't used for research, they would probably just be destroyed.

4) Some people feel that scientists should be finding other sources of stem cells.

Stem Cells Can Produce Identical Plants

1) Plants have tissues called meristems. Meristems are where growth occurs — in the tips of roots and shoots.

2) The meristems contain stem cells that can differentiate into any type of plant cell. They can do this all through the plant's entire life.

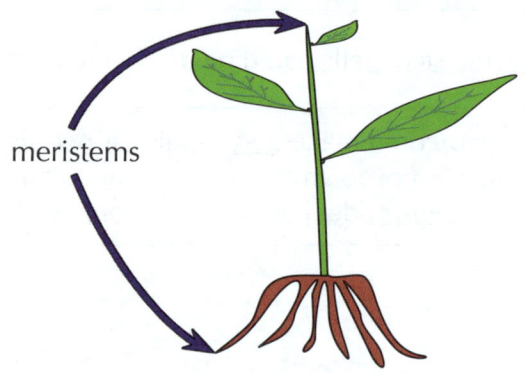

meristems

3) These stem cells can be used to make clones (identical copies) of plants quickly and cheaply. Clones can be made of:

- rare species (to prevent them being wiped out).
- crop plants that have features that are useful for farmers, e.g. plants aren't killed by a disease.

Getting stem cells from other places would avoid the disagreements

Scientists are looking into ways of getting human stem cells without using embryos. Whatever your opinion of stem cell research is, it's good to know what their uses are and the arguments for and against using them.

Topic B1 — Cell Biology

Chromosomes and Mitosis

*In order to survive and grow, our cells have got to be able to **divide**. And that means our DNA does as well...*

Chromosomes Contain Genetic Information

1) The nucleus of a cell contains chromosomes.

2) Chromosomes are coiled up lengths of DNA molecules.

3) Each chromosome carries a large number of genes.

4) Different genes control the development of different characteristics, e.g. hair colour.

5) Body cells normally have two copies of each chromosome.

6) There are 23 pairs of chromosomes in a human cell.

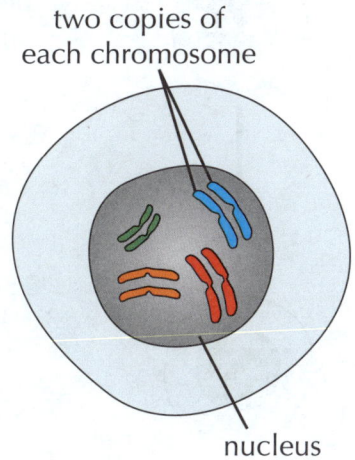

The Cell Cycle Makes New Cells

1) Body cells in multicellular organisms (e.g. like you, me or a plant) divide to make new cells. This is part of a series of stages called the cell cycle.

2) The stage of the cell cycle when the cell divides is called mitosis.

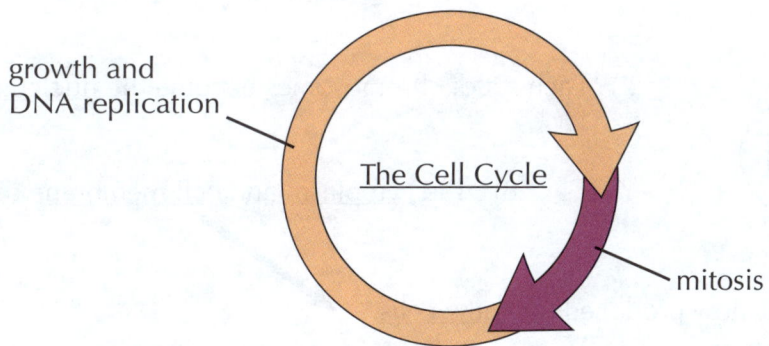

3) Multicellular organisms use mitosis to grow and develop.

4) You need to know about the main stages of the cell cycle shown on the next page.

The cell cycle is important for growth and repair

When a cell goes through the cell cycle, you end up with two cells where you originally had just one. The body controls which cells divide and when — if this control fails, it can result in cancer (see page 64).

Topic B1 — Cell Biology

Chromosomes and Mitosis

There are two main stages of the **cell cycle**...

Growth and DNA Replication

Before it divides:

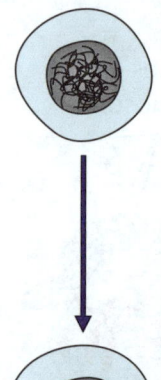

1) The cell grows and increases the amount of subcellular structures such as ribosomes and mitochondria (see page 17).

2) The DNA is replicated (copied) — so there's one copy for each new cell.

3) The DNA forms X-shaped chromosomes. Each 'arm' of the chromosome is an exact copy of the other.

The left arm has the same DNA as the right arm of the chromosome.

Mitosis

The cell is now ready for dividing...

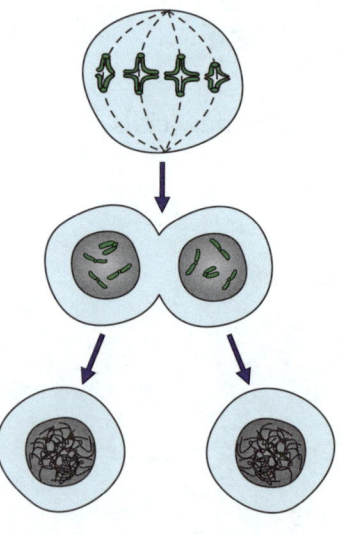

4) The chromosomes line up at the centre of the cell.

5) The two arms of each chromosome are pulled apart to opposite ends of the cell.

6) This divides the nucleus.

7) Each set of chromosomes become the nucleus of a new cell.

8) The cytoplasm and cell membrane divide.

The cell has now produced two new cells.
- They both contain the same DNA — they're identical.
- They're also identical to the original cell.

Mitosis produces two identical cells

Mitosis can seem tricky at first. But don't worry — just go through it slowly, one step at a time.

Q1 A student looks at cells in the tip of a plant root under a microscope. She counts 11 cells that are undergoing mitosis and 62 cells that are not.
  a) Calculate the percentage of cells that are undergoing mitosis. [1 mark]
  b) Suggest how the student can tell whether a cell is undergoing mitosis or not. [1 mark]

Q1 Video Solution

Topic B1 — Cell Biology

Warm-Up & Exam Questions

There's only one way to do well in the exam — learn the facts and then practise lots of exam questions to see what it'll be like on the big day. We couldn't have made it easier for you — so do it.

Warm-Up Questions

1) What is the name of the process where cells become specialised for their job?
2) Describe how a root hair cell is specialised for its function.
3) How many copies of each chromosome does a normal body cell have?
4) Where in the cell are chromosomes found?
5) True or false? The cells produced in mitosis are genetically identical.

Exam Questions

1 Draw **one** line from each cell type to its adaptation.

Cell type   **Adaptation**

Sperm cell   Many mitochondria to provide energy for contracting

                   Very few subcellular structures, so there is more space inside the cell

Nerve cell

                   A tail for swimming

Muscle cell   Branches to connect to other cells

[3 marks]

2 Stem cells are cells produced by both plants and animals that can develop into any type of cell.

2.1 Draw **two** arrows on **Figure 1** to show two places where stem cells are produced in plants.

Figure 1

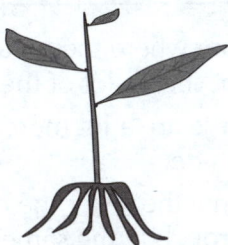

[2 marks]

2.2 Describe **one** way that plant stem cells can be used by farmers to increase crop yields.

[2 marks]

2.3 Stem cells might be useful for the treatment of human diseases.
Give **one** health condition that could possibly be treated with stem cells.

[1 mark]

Topic B1 — Cell Biology

Diffusion

*Diffusion is **really important** in living organisms — it's how a lot of **substances** get **in** and **out** of cells.*

Don't Be Put Off by the Fancy Word

1) "Diffusion" is the movement of particles from where there are lots of them to where there are fewer of them.

2) You have to learn this fancy way of saying it:

> DIFFUSION is the SPREADING OUT of particles from an area of HIGHER CONCENTRATION to an area of LOWER CONCENTRATION.

3) Diffusion happens in solutions and gases. For example, the smell of perfume diffuses through the air in a room:

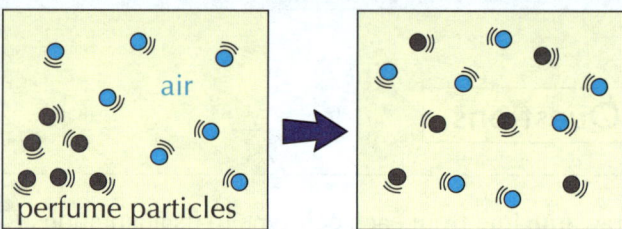

4) The difference in concentration is called the concentration gradient. The bigger the difference in concentration, the faster the diffusion rate.

5) A higher temperature will also give a faster diffusion rate. This is because the particles have more energy, so move around faster.

Cell Membranes Are Kind of Clever...

Oxygen is needed for aerobic respiration — see page 91.

1) Cell membranes let stuff diffuse in and out of the cell.

2) Only very small molecules can fit through cell membranes, e.g. oxygen, glucose, amino acids and water.

3) Big molecules like starch and proteins can't fit through the membrane:

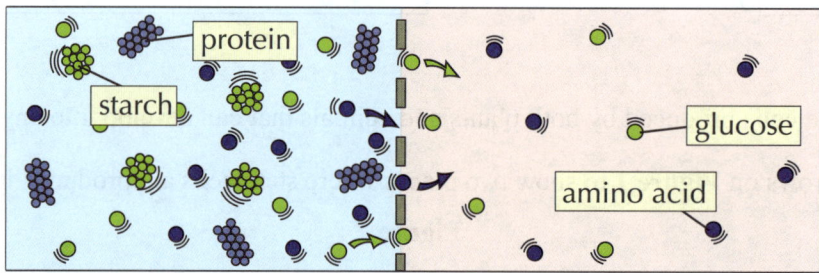

4) Molecules flow through the membrane from where there's a higher concentration (a lot of them) to where there's a lower concentration (not such a lot of them).

5) They actually move both ways — but if there are a lot more particles on one side of the membrane, there's a net (overall) movement from that side.

6) The larger the surface area of the membrane, the faster the diffusion rate. This is because more particles can pass through at the same time.

Diffusion is just particles spreading out

Really tiny particles can go through cell membranes to even up the concentration on either side.

Q1 Video Solution

Q1 A student adds a drop of ink to a glass of cold water.
    a) What will the student observe to happen to the drop of ink. Explain your answer. [2 marks]
    b) How might the observation differ if the ink was added to a glass of warm water? [1 mark]

Q2 Explain how the surface area of a membrane affects the rate of diffusion. [2 marks]

Osmosis

*If you've got your head round **diffusion**, osmosis will be a **breeze**.
If not, have another look at the previous page...*

Osmosis is the Movement of Water Molecules

> OSMOSIS is the movement of water molecules across a partially permeable membrane from a less concentrated solution to a more concentrated solution.

1) A partially permeable membrane is just one with very small holes in it.

2) Tiny molecules (like water) can pass through it, but bigger molecules (e.g. sucrose) can't.

3) Water molecules actually pass both ways through the membrane during osmosis.

4) But overall, the water molecules move from the less concentrated solution (where there are lots of water molecules) to the more concentrated solution (where there are fewer water molecules).

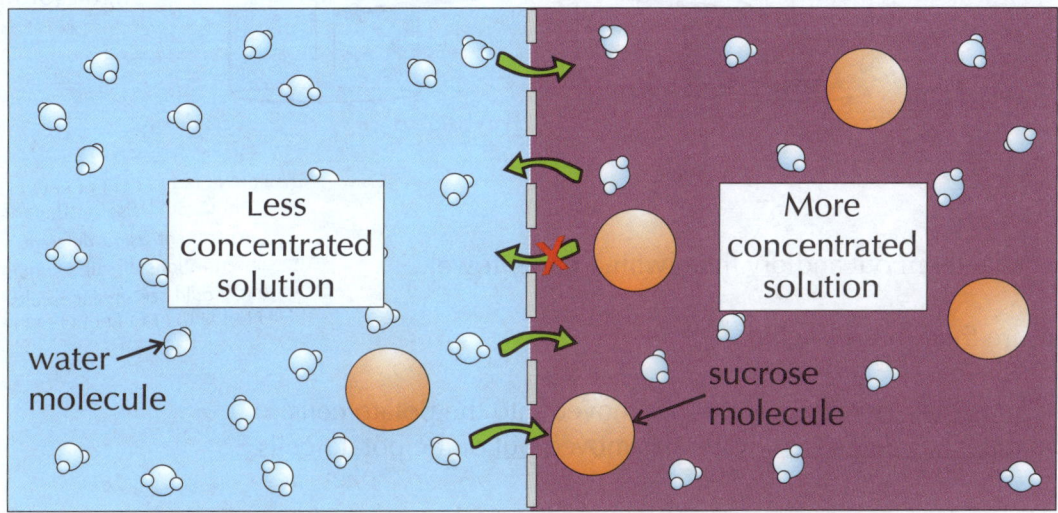

5) This means the more concentrated solution gets more dilute.

6) The water acts like it's trying to "even up" the concentration either side of the membrane.

Diffusion is movement from where there's lots to where there's few...
...so osmosis is really just a fancy word for the diffusion of water molecules. It's simple really.

Topic B1 — Cell Biology

Osmosis

*There's an **experiment** you can do to show osmosis at work.*

You can Observe the Effect of Sugar Solutions on Plant Tissue

1) First, cut up a potato into cylinders with the same length and width.

2) Then get two beakers — one with pure water and another with a very concentrated sugar solution (e.g. 1 mol/dm³).

3) You can also have a few other beakers with less concentrated sugar solutions (e.g. 0.2 mol/dm³, 0.4 mol/dm³, etc).

4) Measure the mass of each potato cylinder, then put one in each beaker.

5) Leave the potato cylinders for twenty four hours.

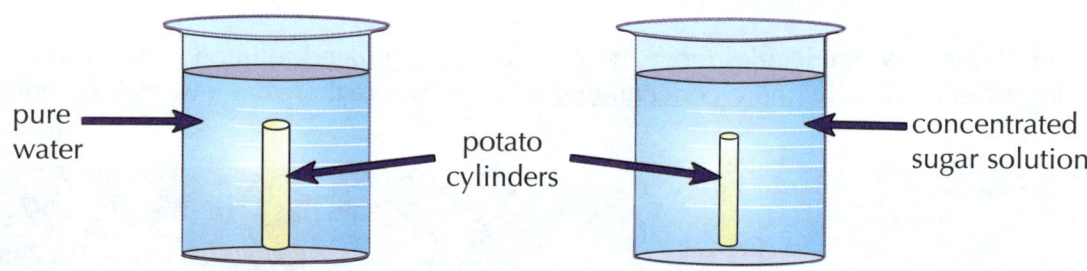

6) Then take them out and dry them with a paper towel.

7) Measure their masses again.

8) If the mass has increased, water has moved into the potato cells.
   If the mass has decreased, water has moved out of the potato cells.

9) You can calculate the percentage change in mass for each potato cylinder — see p.401. This means you can compare the effects of each sugar solution.

10) The only thing you should change in this experiment is the concentration of the sugar solution. Everything else (e.g. volume of solution, temperature, time, type of sugar used) should stay the same.

You could also do this experiment using different concentrations of salt solution. You should see similar results.

Osmosis is the reason why it's bad to drink sea-water...

The high salt content means you end up with a much lower water concentration in your blood and tissue fluid than in your cells. All the water is sucked out of your cells and they shrivel and die.

Q1 Explain what will happen to the mass of a piece of potato added to a concentrated salt solution.

[2 marks]

Topic B1 — Cell Biology

Active Transport

*Sometimes substances need to be absorbed from an area where they are in **low concentration** into an area where they are in **high concentration**. This happens by a process called **active transport**.*

Root Hairs Take In Minerals and Water

1) Plant roots are covered in millions of root hair cells.
2) These cells stick out into the soil.
3) The "hairs" give the roots a large surface area.
4) This is useful for absorbing water and mineral ions from the soil.
5) Plants need mineral ions for healthy growth.

root hair cell

Root Hairs Take in Minerals Using Active Transport

1) The concentration of minerals is usually higher in the root hair cells than in the soil around them.
2) So the root hair cells can't use diffusion to take up minerals from the soil.
3) They use active transport instead.
4) Active transport allows the plant to absorb minerals from a very dilute solution in the soil — it moves the minerals against the concentration gradient.
5) But active transport needs ENERGY from respiration to make it work.

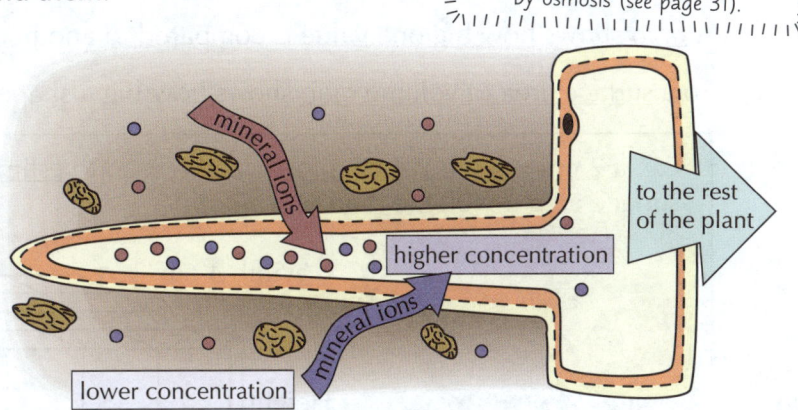

Water is taken into root hair cells by osmosis (see page 31).

We Need Active Transport to Stop Us Starving

1) The body needs to absorb nutrients (e.g. glucose and amino acids) from food to survive.
2) The nutrients have to move from the gut into the blood.
3) When there's a higher concentration of nutrients in the gut, they diffuse into the blood.
4) Sometimes there's a lower concentration of nutrients in the gut than there is in the blood.
5) The body uses active transport to move the nutrients (like glucose) from a lower concentration in the gut to a higher concentration in the blood.
6) This means glucose can be taken into the blood against the concentration gradient. The glucose is then transported to cells, where it's used for respiration (see page 91).

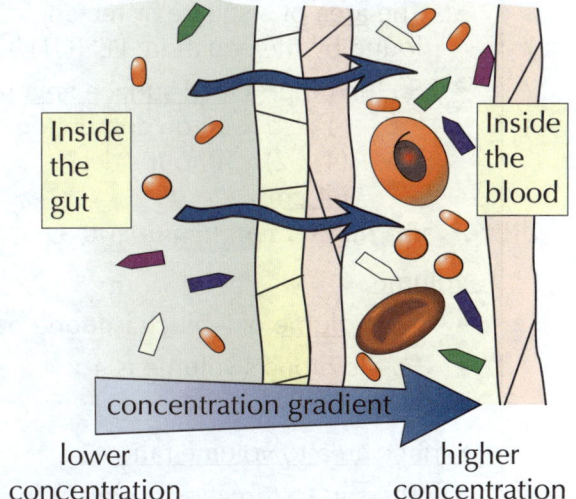

Active transport uses energy

An important difference between active transport and diffusion is that active transport uses energy.

Topic B1 — Cell Biology

Exchanging Substances

*How easily stuff **moves** between an **organism** and its **environment** depends on its **surface area to volume ratio**.*

Organisms **Exchange Substances** with their **Environment**

1) Cells can use diffusion to take in substances from the environment, such as oxygen.
2) They also use diffusion to get rid of waste products, such as:

   - Carbon dioxide (from respiration).
   - Urea (from the breakdown of proteins) — urea diffuses from cells into the blood plasma (see p.53). It is then removed from the body by the kidneys.

3) How easy it is for an organism to exchange (swap) substances with its environment depends on the organism's surface area to volume ratio.

You Can **Calculate** an Organism's **Surface Area to Volume Ratio**

1) A ratio shows how big one value is compared to another.
2) So a surface area to volume ratio shows how big a shape's surface is compared to its volume.

> E.g. a 2 cm × 4 cm × 4 cm block can be used to estimate the surface area to volume ratio of a hippo:
>
>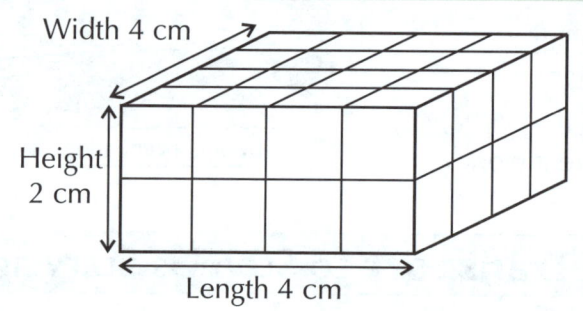
>
> Width 4 cm
> Height 2 cm
> Length 4 cm
>
> Surface area
> - The area of a square or rectangle is found by the equation: LENGTH × WIDTH.
> - So the hippo's total surface area is:
>   (4 × 4) × 2 (top and bottom surfaces)
>   + (4 × 2) × 4 (four sides)
>   = 64 cm².
>
> Volume
> - The volume of a block is found by the equation: LENGTH × WIDTH × HEIGHT.
> - So the hippo's volume is 4 × 4 × 2 = 32 cm³.
>
> Surface area to volume ratio
> - The surface area to volume ratio (SA : V) of the hippo can be written as 64 : 32.
> - To get the ratio so that volume is equal to one, divide both sides of the ratio by the volume.
>
>   64 ÷ 32 = 2   32 ÷ 32 = 1   So the SA : V of the hippo is 2 : 1.

Exchanging Substances

*Now for some more on **surface area to volume ratios**, and why they're important in biology...*

Larger Objects Usually Have Smaller Surface Area to Volume Ratios

1) A 1 cm × 1 cm × 1 cm block can be used to estimate the surface area to volume ratio of a mouse.

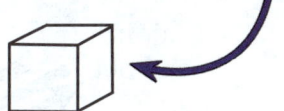

It's found that the SA : V of the mouse is 6 : 1.

2) The larger the organism, the smaller its surface area is compared to its volume.

> Example: The surface area of the mouse is six times its volume.
> The surface area of the hippo (see previous page) is only two times its volume.
>
> So the hippo has a smaller surface area compared to its volume.

3) The smaller its surface area compared to its volume, the harder it is for an organism to exchange substances with its environment.

Multicellular Organisms Need Exchange Surfaces

1) Single-celled organisms have a large surface area compared to their volume.
2) So, they can exchange all the substances they need across their surface (the cell membrane).
3) Multicellular organisms (such as animals) have a smaller surface area compared to their volume.
4) They can't normally exchange enough substances across their outside surface alone.
5) Instead, multicellular organisms have specialised exchange surfaces — see the next page and page 37 for some examples.
6) They also have transport systems that carry substances to and from their exchange surfaces.
7) The exchange surfaces are ADAPTED to allow enough of different substances to pass through:

> - They have a thin membrane (so substances only have a short distance to diffuse).
> - They have a large surface area (so lots of a substance can diffuse at once).
> - Exchange surfaces in animals have lots of blood vessels (so stuff can get into and out of the blood quickly).
> - Gas exchange surfaces in animals (e.g. alveoli) are ventilated too — air moves in and out.

Surface area to volume ratios crop up a lot in biology...

...so it's a good idea to try to understand them now. Remember that, generally speaking, a smaller object has a larger surface area to volume ratio than a bigger object.

Q1 A bacterial cell can be represented by a 2 µm × 2 µm × 1 µm block. Calculate the cell's surface area to volume ratio. [3 marks]

Topic B1 — Cell Biology

More on Exchanging Substances

Here are a couple of examples of *exchange surfaces* inside your body...

Gas Exchange Happens in the Lungs

1) Oxygen (O_2) and carbon dioxide (CO_2) are exchanged in the lungs.

2) The lungs contain millions of little air sacs called alveoli. This is where gas exchange happens.

3) The alveoli are specialised for the diffusion of oxygen and carbon dioxide. They have:
   - A large surface area.
   - Very thin walls (so gases don't have far to diffuse).
   - A good blood supply.

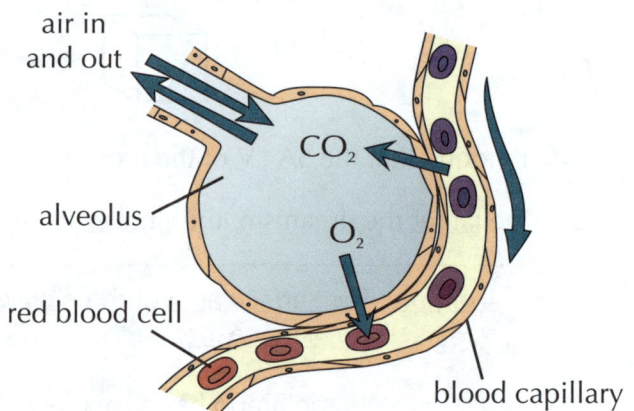

The Villi Provide a Really Big Surface Area

1) The inside of the small intestine is covered in millions of villi.

2) They increase the surface area so that digested food is absorbed more quickly into the blood.

3) They have:
   - a single layer of surface cells,
   - a very good blood supply.

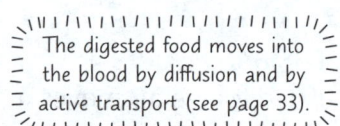

The digested food moves into the blood by diffusion and by active transport (see page 33).

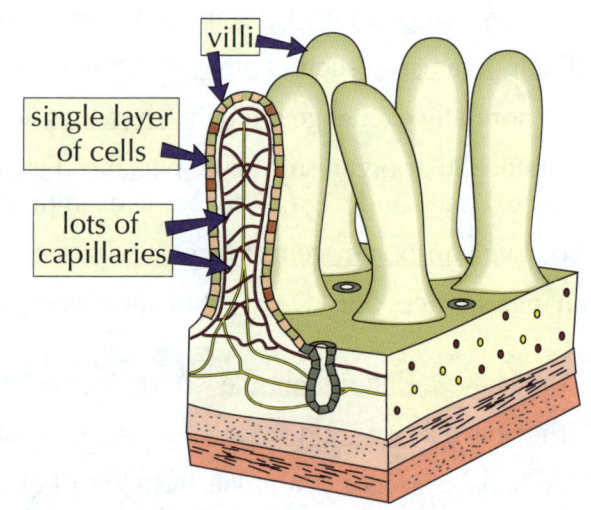

Humans need alveoli for gas exchange

Q1 Give one way in which alveoli are adapted for gas exchange. [1 mark]

Q2 Coeliac disease causes inflammation of the small intestine, which can damage the villi. Suggest why a person with coeliac disease might have low levels of iron in their blood. [2 marks]

Topic B1 — Cell Biology

More on Exchanging Substances

*More stuff on exchange surfaces for diffusion now — only this time they're in **plants** and **fish**.*

The Structure of Leaves Lets Gases Diffuse In and Out of Cells

1) Plant leaves need to take in carbon dioxide for photosynthesis, and get rid of oxygen and water vapour.
2) The underneath of the leaf is an exchange surface. It's covered in small holes called stomata.
3) Carbon dioxide diffuses through the stomata into the leaf.
4) Oxygen and water vapour diffuse out through the stomata.
5) The size of the stomata are controlled by guard cells — see page 68.
6) The flattened shape of the leaf increases the area of its exchange surface.
7) The walls of the cells inside the leaf are another exchange surface. Gases diffuse into and out of the cells through these walls.
8) There are air spaces inside the leaf to increase the area of these surfaces.

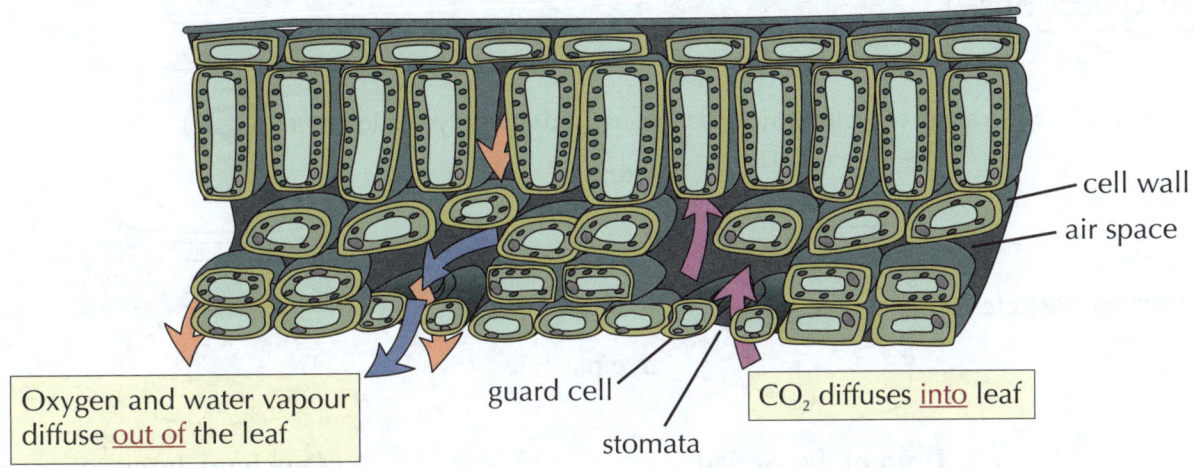

Oxygen and water vapour diffuse out of the leaf

guard cell

stomata

CO_2 diffuses into leaf

cell wall

air space

Gills Have a Large Surface Area for Gas Exchange

1) The gills are the gas exchange surface in fish.
2) Water (containing oxygen) flows into the fish's mouth and passes out through the gills.
3) In the gills, oxygen diffuses from the water into the blood. Carbon dioxide diffuses from the blood into the water.
4) The gills are made up of lots of thin plates. This gives them a large surface area for gases to be exchanged.
5) The plates have lots of blood capillaries. So they have a good blood supply to speed up diffusion.
6) They also have a thin layer of surface cells. So the gases only have to diffuse a short distance.

Exchange surfaces make diffusion quick and easy

Multicellular organisms are really well adapted for getting the substances they need to their cells. It makes sense — if they couldn't do this well, they'd die out. A large surface area is a key way that organisms' exchange surfaces are made more effective — molecules can only diffuse through a membrane when they're right next to it, and a large surface area means that a lot more molecules are close to the membrane.

Topic B1 — Cell Biology

Warm-Up & Exam Questions

Question time again — Warm-Up first, then Exam (or the other way round if you want to be different).

Warm-Up Questions

1) What is the word for the spreading out of particles from an area of higher concentration to an area of lower concentration?
2) True or false? Larger objects usually have smaller surface area to volume ratios than smaller objects.
3) How do villi in the small intestine help food to be absorbed more quickly?
4) Name one other exchange surface in humans.
5) What is the name of the small holes for exchanging gases on the underside a leaf?
6) Give one way in which the structure of a gill is adapted for effective gas exchange.

Exam Questions

1   **Figure 1** shows a cup of water which has just had a drop of dye added to it. *Grade 3-4*

Figure 1

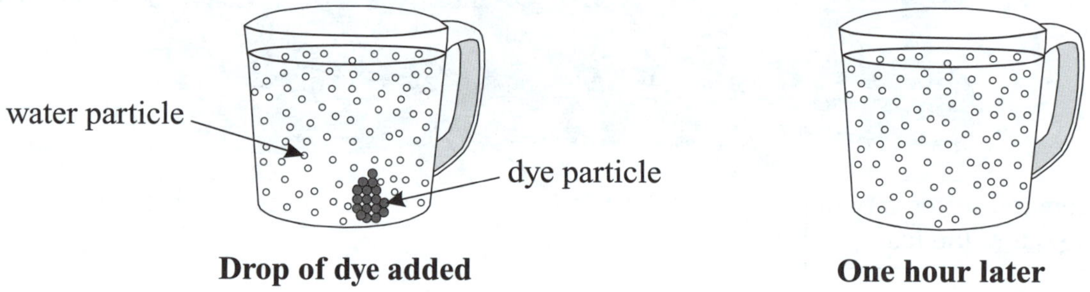

water particle

dye particle

Drop of dye added   **One hour later**

1.1   Complete **Figure 1** by drawing the molecules of dye in the cup after one hour.

[1 mark]

1.2   Describe the effect that a lower temperature would have on diffusion.

[1 mark]

2   Active transport is an important form of transport in organisms. *Grade 3-4*

2.1   Which of the following is a correct description of active transport?
Tick **one** box.

☐ The movement of substances from an area of higher concentration to an area of lower concentration, without requiring energy.

☐ The movement of substances from an area of lower concentration to an area of higher concentration, without requiring energy.

☐ The movement of substances from an area of higher concentration to an area of lower concentration, requiring energy.

☐ The movement of substances from an area of lower concentration to an area of higher concentration, requiring energy.

[1 mark]

Topic B1 — Cell Biology

Exam Questions

2.2 Which of the following processes involve active transport?
Tick **two** boxes.

☐ Movement of mineral ions from the soil into root hair cells

☐ Movement of carbon dioxide from the air into the stomata of leaves

☐ Movement of oxygen from the water into the blood of fish

☐ Movement of nutrients, such as glucose, from the gut into the blood

[2 marks]

3 **Figure 2** shows a tank divided in two by the structure labelled **X**. *Grade 3-4*
Osmosis will occur between the two sides of the tank.

Figure 2

(diagram showing tank with side A containing sucrose molecules and water molecules, side B containing water molecules, separated by structure X)

○ Water molecule
● Sucrose molecule

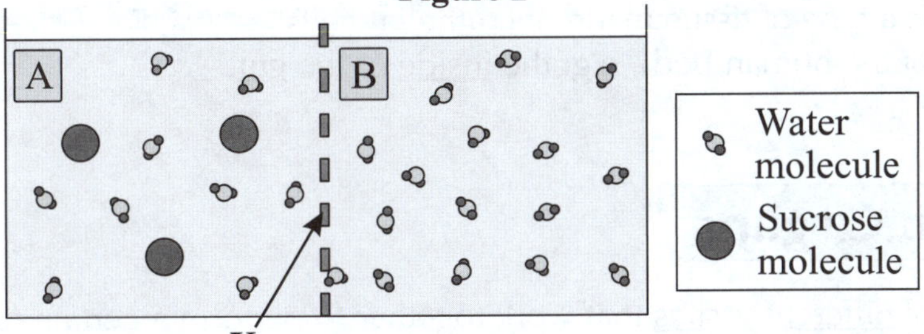

3.1 Name the structure labelled **X** on **Figure 2**.

[1 mark]

3.2 Which of the following will happen to the level of liquid on side **B**? Tick **one** box.

☐ The liquid level on side B will remain the same.

☐ The liquid level on side B will fall.

☐ The liquid level on side B will rise.

[1 mark]

PRACTICAL

4 In an experiment, four 50 mm long cylinders were cut from a fresh potato. *Grade 4-5*
The cylinders were then placed in different sugar solutions.
After 24 hours the potato cylinders were removed and measured. The results are shown in **Table 1**.

Table 1

Cylinder	1	2	3	4
Length after 24 hours (mm)	40	43	51	55
Change in length (mm)				+5

4.1 Complete **Table 1** by calculating the change in length for cylinders 1-3.

[1 mark]

4.2 Explain the change that occurred to the length of cylinder 4.

[2 marks]

Topic B1 — Cell Biology

Topic B2 — Organisation

Cell Organisation

Some organisms are made of lots of cells. To get a working organism, these cells need to be organised.

Large Multicellular Organisms are Made Up of Organ Systems

1) Cells are the basic building blocks that make up all living organisms.
2) Specialised cells carry out a particular function (see p.23-24).
3) These specialised cells form tissues, which form organs, which form organ systems (see below).
4) Large multicellular organisms (e.g. humans) have different systems inside them for exchanging and transporting materials.

Epithelial cell

Similar Cells Make Up Tissues

1) A tissue is a group of similar cells that work together to carry out a function.
2) E.g. epithelial tissue is a type of tissue made of epithelial cells. It covers some parts of the human body, e.g. the inside of the gut.

Epithelial tissue

Tissues Make Up Organs

1) An organ is a group of different tissues that work together to perform a certain function.
2) For example, the stomach is an organ. Epithelial tissue lines the inside and outside of the stomach.

Stomach

Organs Make Up Organ Systems

1) An organ system is a group of organs working together to perform a function.
2) The digestive system is an organ system found in humans and other mammals.
3) It breaks down and absorbs food.
4) It's made up of these organs:

You need to know where these organs are on a diagram.

Digestive system

Salivary glands — Produce digestive juices.

Liver — Produces bile.

Large intestine — Absorbs water from undigested food, leaving faeces (poo).

Stomach — Digests food.

Pancreas — Produces digestive juices.

Small intestine — Digests food and absorbs soluble food molecules, e.g. glucose.

5) Organ systems work together to make entire organisms.

Remember — cells, tissues, organs, organ systems

OK, so from this page you should know that cells are organised into tissues, the tissues into organs, the organs into organ systems and the organ systems into a whole organism.

Enzymes

Chemical reactions are what make you work. And *enzymes* are what make them work.

Enzymes Are Catalysts

1) Living things have tons of reactions going on inside their cells.

2) These reactions are controlled by enzymes.

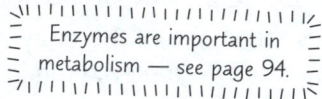
Enzymes are important in metabolism — see page 94.

3) Enzymes are large proteins.

4) They speed up reactions inside living things by acting as catalysts:

> A CATALYST is a substance which INCREASES the speed of a reaction, without being CHANGED or USED UP in the reaction.

Enzymes Have Special Shapes

1) Chemical reactions usually involve things either being split apart or joined together.
2) Every enzyme has an active site with a unique shape.
3) The substance involved in the reaction has to fit into the active site for the enzyme to work.
4) So enzymes are really picky — they usually only catalyse one specific reaction.
5) This diagram shows the 'lock and key' model of enzyme action:

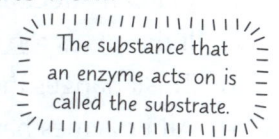
The substance that an enzyme acts on is called the substrate.

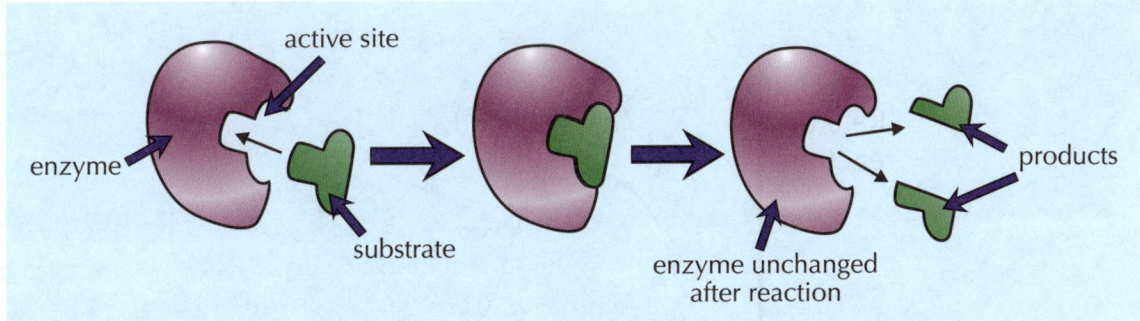

This is a useful model but it's a bit simpler than how enzymes actually work.

Enzymes speed up chemical reactions

Just like you've got to have the correct key for a lock, you've got to have the right substrate for an enzyme. As you can see in the diagram above, if the substrate doesn't fit, the enzyme won't catalyse the reaction...

Topic B2 — Organisation

Enzymes

*Enzymes are clearly very clever, but they need just the right **conditions** if they're going to work properly.*

Enzymes Need the Right Temperature...

1) Temperature affects the rate of a reaction involving an enzyme.
2) A higher temperature increases the rate at first.
3) But if it gets too hot, some of the bonds holding the enzyme together break.
4) This changes the shape of the enzyme's active site, so the substrate won't fit any more — the enzyme is denatured.
5) All enzymes have a temperature that they work best at — their optimum temperature.

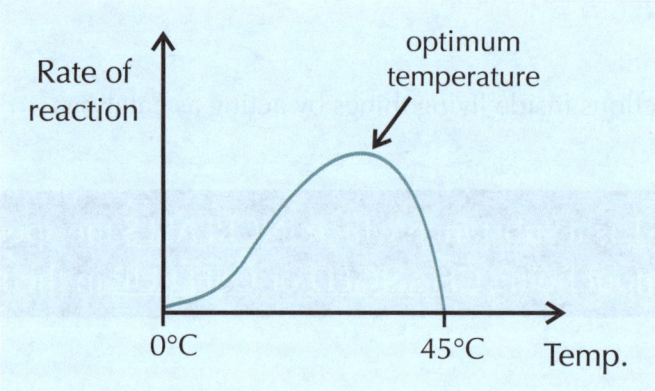

... and the Right pH

1) pH can affect the rate of a reaction involving an enzyme.
2) If the pH is too high or too low, it affects the bonds holding the enzyme together.
3) This changes the shape of the active site, and denatures the enzyme.
4) All enzymes have a pH that they work best at — their optimum pH.

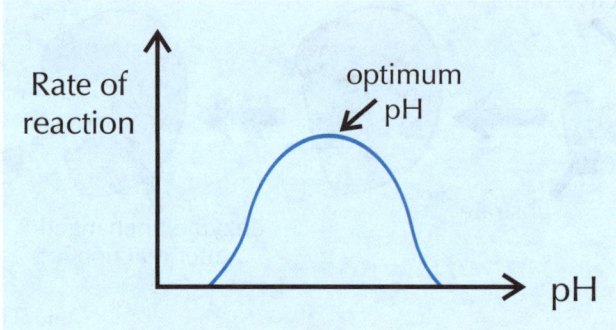

Most enzymes catalyse just one reaction

The optimum temperature for most human enzymes is around normal body temperature. And stomach enzymes work best at low pH, but the enzymes in your small intestine like a higher pH.

Topic B2 — Organisation

Investigating Enzymatic Reactions

*Give this page a read and you'll soon know how to investigate the effect of **pH** on the rate of **enzyme activity**.*

You Can Investigate the Effect of pH on Enzyme Activity **PRACTICAL**

1) The enzyme amylase catalyses the breakdown of starch to sugar.
2) You can detect starch using iodine solution — if starch is present, the iodine solution will change from browny-orange to blue-black.

1) Put a drop of iodine solution into every well of a spotting tile.
2) Set up a water bath at 35 °C.
   (You could use a Bunsen burner and a beaker of water, or an electric water bath.)
3) Add some amylase solution and a buffer solution with a pH of 5 to a boiling tube.
4) Put the boiling tube in the water bath and wait for five minutes.
5) Add some starch solution to the boiling tube, mix, and start a stop clock.
6) Every 30 seconds, take a sample from the boiling tube using a dropping pipette.
7) Put a drop of the sample into a well on the spotting tile.
8) When the iodine solution stays browny-orange, all the starch in the sample has been broken down. Record how long this takes.

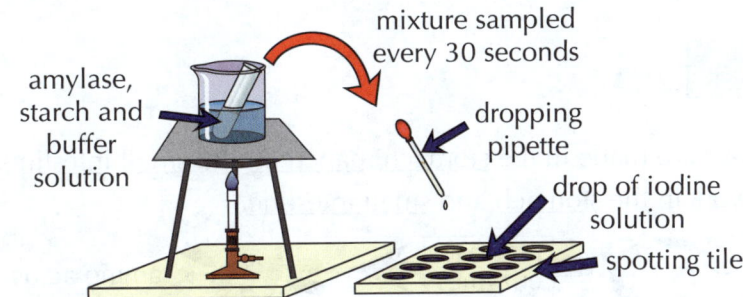

9) Repeat the experiment with buffer solutions of different pH values.
10) As the pH changes, the time it takes for the starch to be broken down should also change.
11) Remember to control any variables each time you repeat the experiment. This will make it a fair test. For example, the concentration and volume of the amylase solution should always be the same.

Here's How to Calculate the Rate of Reaction

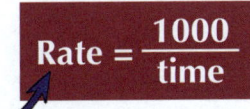

1) Rate is a measure of how much something changes over time.
2) For the experiment above, you can calculate the rate of reaction using this formula:

EXAMPLE The time taken for amylase to break down all of the starch in a solution was 90 seconds. Calculate the rate of the reaction. Write your answer in s^{-1}.

Rate of reaction = 1000 ÷ time = 1000 ÷ 90 s
= 11 s^{-1}

 just means 'per second'.

You can investigate other factors too...

You could easily adapt this experiment to investigate how factors other than pH affect the rate of amylase activity. For example, you could use a water bath to investigate the effect of temperature.

Q1 Calculate the rate of a reaction that finished in 2.5 minutes.
    Give your answer in s^{-1}. [2 marks]

Topic B2 — Organisation

Enzymes and Digestion

The **enzymes** used in **digestion** are produced by **cells**. They're released into the **gut** to mix with food.

Digestive Enzymes Break Down Big Molecules

1) <u>Starch</u>, <u>proteins</u> and <u>fats</u> are BIG molecules.
2) They're <u>too big</u> to pass through the walls of the digestive system.
3) So <u>digestive enzymes</u> break these BIG molecules down into smaller ones.
4) These smaller, <u>soluble</u> molecules can <u>easily</u> be <u>absorbed</u> into the <u>bloodstream</u>.

Carbohydrases

- <u>Amylase</u> is an example of a <u>carbohydrase</u>.
- Amylase is made in the <u>salivary glands</u>, <u>pancreas</u> and <u>small intestine</u>.
- It works in the <u>mouth</u> and <u>small intestine</u>.

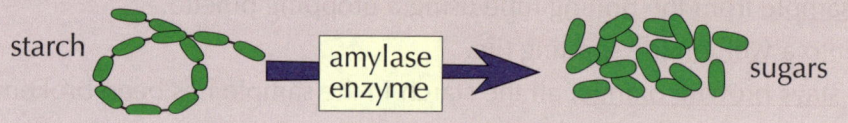

Starch is a carbohydrate.

Proteases

- Proteases are made in the <u>stomach</u>, <u>pancreas</u> and <u>small intestine</u>.
- They work in the <u>stomach</u> and <u>small intestine</u>.

Lipases

- Lipases are made in the <u>pancreas</u> and <u>small intestine</u>.
- They work in the <u>small intestine</u>.

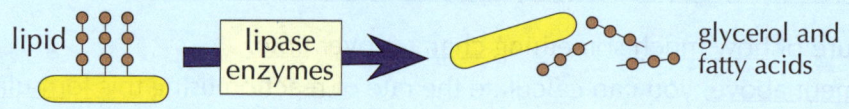 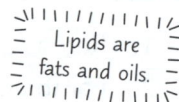

Lipids are fats and oils.

5) The <u>products</u> of digestion can be used to make <u>new carbohydrates</u>, <u>proteins</u> and <u>lipids</u>.
6) <u>Glucose</u> is a <u>sugar</u> produced by digestion. Some of it is used in <u>respiration</u> (see p.91-92).

Bile Neutralises the Stomach Acid and Emulsifies Fats

1) Bile is <u>produced</u> in the <u>liver</u>. It's <u>stored</u> in the <u>gall bladder</u> before it's released into the <u>small intestine</u>.
2) Bile is <u>alkaline</u>. It <u>neutralises</u> hydrochloric acid (from the stomach) and makes conditions <u>alkaline</u>.
3) The enzymes in the small intestine <u>work best</u> in these alkaline conditions.
4) Bile also <u>emulsifies</u> fats. Emulsify means that it breaks the fats down into <u>tiny droplets</u>. This gives a <u>bigger surface area</u> of fat for lipase to work on. This makes its digestion <u>faster</u>.

Food Tests

*There are some clever ways to **identify** what type of **food molecule** a sample contains.*

Prepare Your Food Sample First

For each test, you need to prepare a food sample. It's the same each time though — here's what you'd do:

1) Get a piece of food and break it up using a pestle and mortar.
2) Transfer the ground up food to a beaker and add some distilled water.
3) Give the mixture a good stir with a glass rod to dissolve some of the food.
4) Filter the solution using a funnel lined with filter paper. This will get rid of the solid bits of food.

Use the Benedict's Test to Test for Sugars

Glucose is a reducing sugar.

The Benedict's test is used to test for a type of sugar called a reducing sugar. Here's how you do it:

1) Prepare a food sample and transfer 5 cm^3 to a test tube.
2) Prepare a water bath so that it's set to 75 °C.
3) Add some Benedict's solution to the test tube (about 10 drops) using a pipette.
4) Place the test tube in the water bath using a test tube holder. Leave it in there for 5 minutes.
5) If the food sample contains a reducing sugar, the solution in the test tube will change from the normal blue colour to green, yellow or brick-red. The colour change depends on how much sugar is in the food.

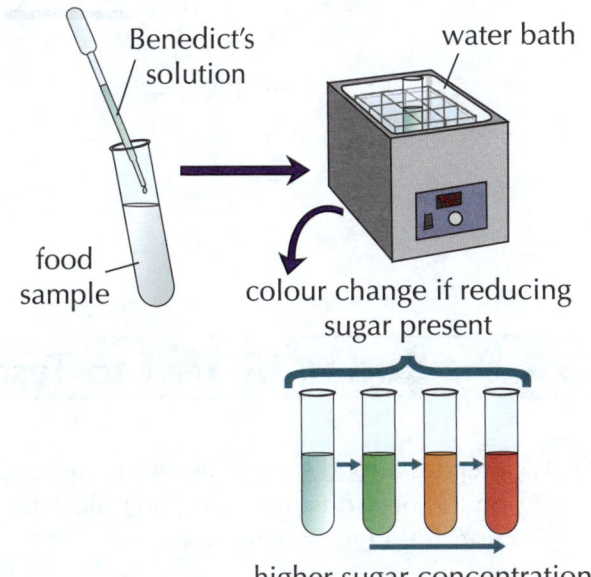

higher sugar concentration

Use Iodine Solution to Test for Starch

1) Make a food sample and transfer 5 cm^3 to a test tube.
2) Add a few drops of iodine solution. Gently shake the tube to mix the contents.
3) If the sample contains starch, the colour of the solution will change from browny-orange to black or blue-black.

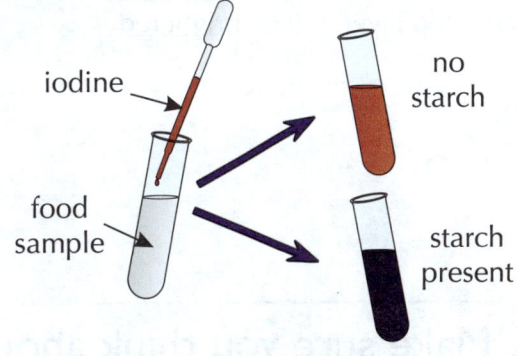

There are four food molecules you should know how to test for

You're halfway through them now. Turn the page to read about the other two — proteins and lipids...

Topic B2 — Organisation

Food Tests

*There are a couple more **food tests** coming up on this page — for proteins and for lipids. As with the other tests, you need to use the method on the previous page to prepare a **sample** of your food first.*

Use the Biuret Test to Test for Proteins

1) Prepare a sample of your food and transfer 2 cm³ to a test tube.
2) Add 2 cm³ of biuret solution to the sample. Mix the contents of the tube by gently shaking it.
3) If the food sample contains protein, the solution will change from blue to purple.

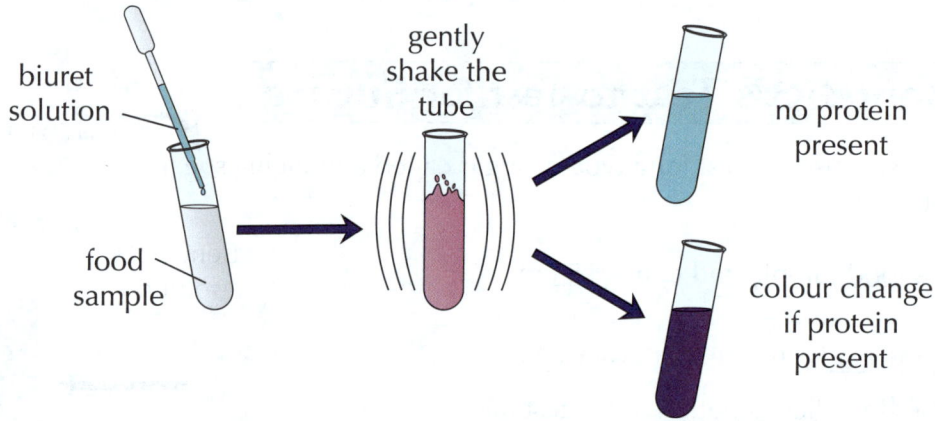

Use the Sudan III Test to Test for Lipids

1) Prepare a food sample using the method on the previous page but don't filter it. Transfer 5 cm³ to a test tube.
2) Add 3 drops of Sudan III stain solution to the test tube. Gently shake the tube.
3) If the sample contains lipids, the mixture will separate out into two layers. The top layer will be bright red.

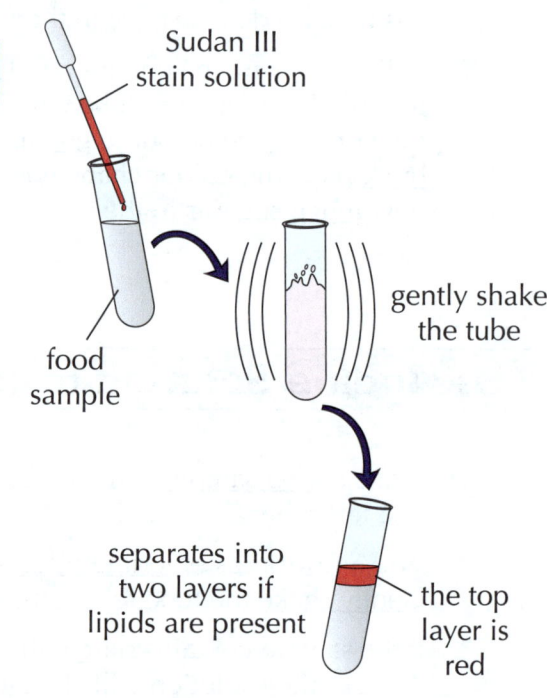

Make sure you think about all of the hazards...

Iodine is an irritant to the eyes, and the chemicals in the biuret solution are dangerous, so wear safety goggles for food tests. If you spill any of the chemicals on your skin, wash it off straight away. Be careful around the water bath in the Benedict's test, too. And if that's not enough to be cautious about, Sudan III stain solution is flammable, so keep it away from any lit Bunsen burners.

Topic B2 — Organisation

Warm-Up & Exam Questions

Doing well in exams isn't just about remembering all the facts, although that's important. You have to get used to the way the exam questions are phrased and make sure you always read them carefully.

Warm-Up Questions

1) Is the stomach a tissue or an organ?
2) What is the name for the part of an enzyme that a substrate fits into?
3) Which enzyme digests starch?
4) What are the products when lipids are broken down?
5) True or false? Protease enzymes are made in the liver.
6) Describe how you would prepare a food sample before testing it for the presence of different food molecules.

Exam Questions

1   **Figure 1** shows the human digestive system. *Grade 1-3*

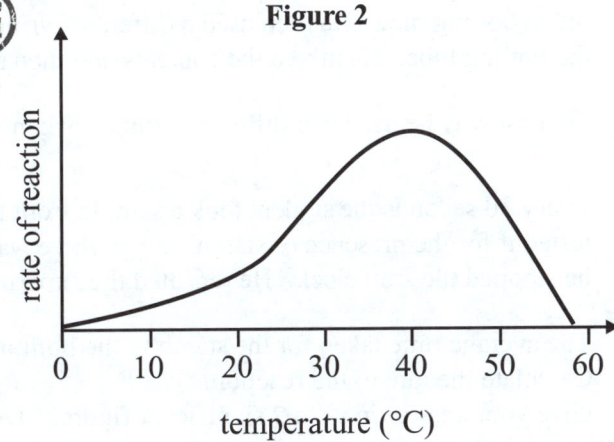
Figure 1

1.1   Which label, **A-D**, shows the place where bile is produced?
[1 mark]

1.2   Use words from the box to complete the sentences below.

| acidic | alkaline | neutral |
| lipids | starch | proteins |

Bile makes conditions in the small intestine ....

Bile also emulsifies ....

[2 marks]

2   **Figure 2** shows the effect of temperature on the action of an enzyme.

2.1   At what temperature does the enzyme work best?
[1 mark]

2.2   What name is given to the temperature at which an enzyme works best?
[1 mark]

Topic B2 — Organisation

Exam Questions

3 Figure 3 represents the action of an enzyme in catalysing a biological reaction.

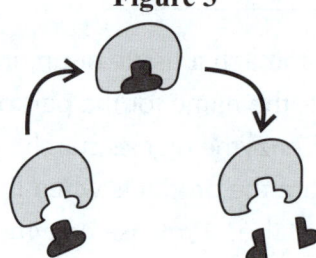

Figure 3

3.1 In terms of the enzyme's shape, explain why an enzyme only catalyses one reaction.

[1 mark]

3.2 The optimum pH of the enzyme is pH 7. Explain what effect a very low pH would have on the activity of the enzyme.

[2 marks]

PRACTICAL

4 A student wanted to know which substances were present in a food sample. She prepared a solution containing the food, and added some of the solution to each of three test tubes. She then added different chemicals to each test tube, to test for different food molecules. Her results are shown in **Table 1**.

Table 1

Test tube	Chemical added to test tube	Description of solution in test tube
A	Iodine	Blue-black
B		Blue
C	Sudan III stain solution	In two layers. Top layer red.

4.1 Complete the table to show what solution would have been added to test tube **B** to test for proteins.

[1 mark]

4.2 Does the food sample contain starch? Explain your answer.

[1 mark]

4.3 Does the food sample contain lipids? Explain your answer.

[1 mark]

PRACTICAL

5 A student was investigating the effect of pH on the rate of amylase activity. He used a syringe to put amylase solution and a buffer solution with a pH of 6 into a boiling tube. He then used a different syringe to add a starch solution to the boiling tube. He mixed the contents and then started a stop clock.

5.1 Suggest why he used two different syringes when adding substances to the boiling tube.

[1 mark]

Every 30 seconds the student took a sample from the boiling tube and tested it for the presence of starch. When there was no starch present he stopped the stop clock. He repeated the experiment three times.

5.2 The average time taken for the starch in the boiling tube to be broken down was 60 seconds. Calculate the rate of the reaction.
Give your answer in s^{-1} to 2 significant figures. Use the formula: rate = $\frac{1000}{time}$

[2 marks]

Topic B2 — Organisation

The Lungs

You need **oxygen** to supply your **cells** for **respiration** (see p.91-92). You also need to get rid of **carbon dioxide**. This all happens in your **lungs** when you breathe air in and out.

Air Moves In and Out of the Lungs

1) This diagram shows the structure of the lungs:
2) The air that you breathe in goes through the trachea.
3) Then it passes through the bronchi, then the bronchioles and ends up in the alveoli (small air sacs).

One bronchus, two bronchi. One alveolus, many alveoli.

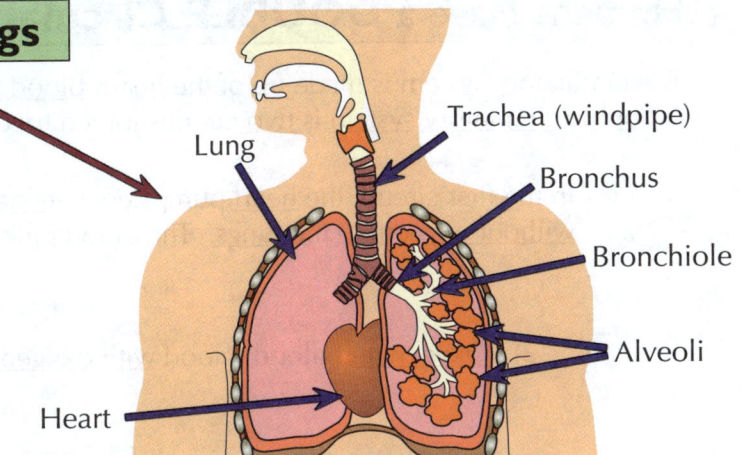

Alveoli Carry Out Gas Exchange

1) Alveoli in the lungs are surrounded by blood capillaries.
2) Blood comes into the lungs through the capillaries. It contains lots of carbon dioxide and very little oxygen.
3) Oxygen diffuses (see p.30) out of the air in the alveolus (where there's a higher concentration) into the blood (where there's a lower concentration).
4) Carbon dioxide diffuses out of the blood (higher concentration) into the air in the alveolus (lower concentration).
5) The blood then leaves the lungs and travels around the body.

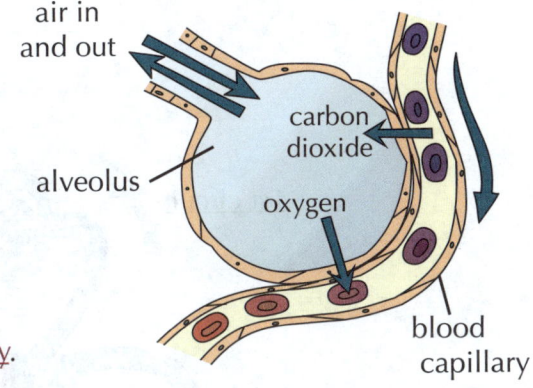

You Can Calculate the Breathing Rate

Breathing rate is how fast a person moves air in and out of their lungs. Here's how to calculate it:

EXAMPLE

Bob takes 91 breaths in 7 minutes. Calculate his average breathing rate in breaths per minute.

breaths per minute = number of breaths ÷ number of minutes
= 91 ÷ 7
= **13 breaths per minute**

Take a deep breath...

...and make sure you get to grips with the layout of the innards of your chest.

Q1 Aaqib jogged around the park for 8 minutes. During his jog he took 304 breaths. Calculate his average breathing rate in breaths per minute. [1 mark]

Topic B2 — Organisation

Circulatory System — The Heart

The circulatory system carries **food** and **oxygen** to every cell in the body.
It also carries **waste products** to where they can be removed from the body.

Humans Have a DOUBLE Circulatory System

The circulatory system is made up of the heart, blood vessels and blood.
A double circulatory system is two circuits joined together:

1) In the first circuit, the heart pumps deoxygenated blood (blood without oxygen) to the lungs. The blood picks up oxygen in the lungs.

2) Oxygenated blood (blood with oxygen) then returns to the heart.

3) In the second circuit, the heart pumps oxygenated blood around all the other organs of the body. This delivers oxygen to the body cells.

4) Deoxygenated blood returns to the heart to be pumped out to the lungs again.

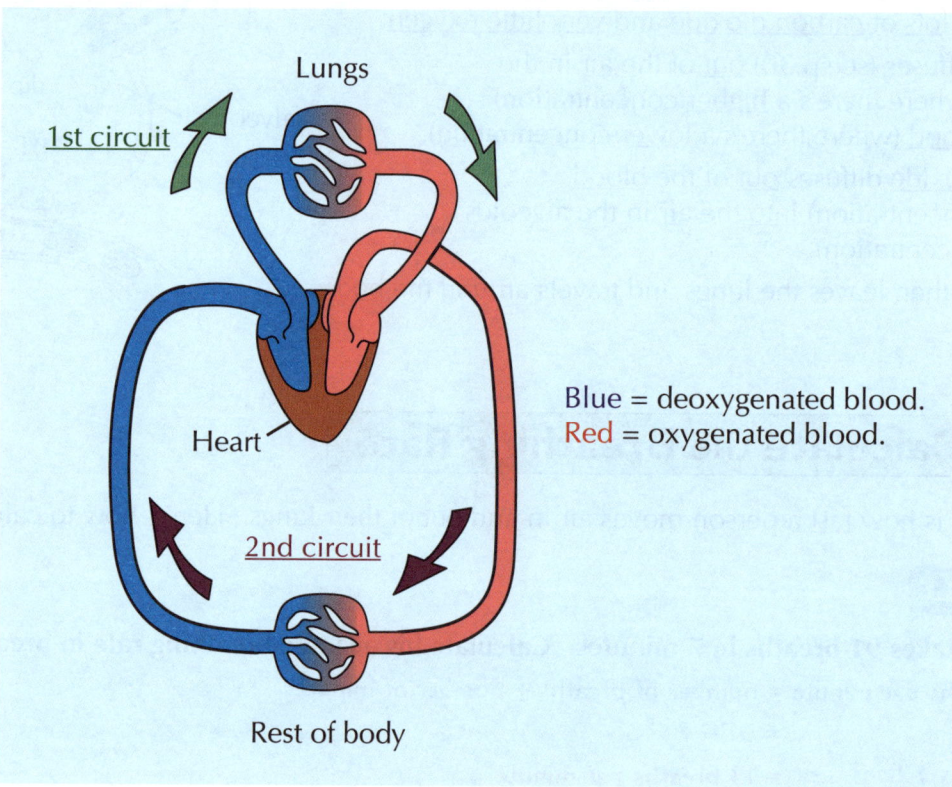

Blue = deoxygenated blood.
Red = oxygenated blood.

First, blood goes to the lungs to pick up oxygen...

...then it takes the oxygen around the rest of the body to all of the cells. The next few pages are also about the circulatory system, so make sure you understand what a double circulatory system is before moving on.

Topic B2 — Organisation

Circulatory System — The Heart

*Now you know how the circulatory system works **overall**, it's time to look at the individual **parts** of it. First up, the **heart**...*

The Heart Pumps Blood Around The Body

1) The heart is an organ with four chambers. The walls of the chambers are mostly made of muscle tissue.
2) This muscle tissue is used to pump blood around the body. Here's how:

> 1) Blood flows into the two atria from the vena cava and the pulmonary vein.
> 2) The atria pump the blood into the ventricles.
> 3) The ventricles pump the blood out of the heart:
>   - Blood from the right ventricle goes through the pulmonary artery to the lungs.
>   - Blood from the left ventricle goes through the aorta to the rest of the body.
> 4) The blood then flows to the organs through arteries, and returns through veins (see next page).
> 5) The atria fill again — the whole cycle starts over.

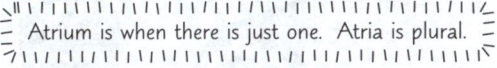
Atrium is when there is just one. Atria is plural.

3) The valves in the heart stop the blood flowing backwards.
4) The heart also needs its own supply of oxygenated blood.
5) It gets oxygenated blood from arteries called coronary arteries. These branch off the aorta and surround the heart.

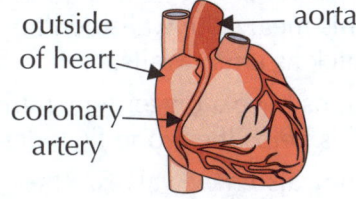

The Heart Has a Pacemaker

1) Your resting heart rate is controlled by a group of cells in the right atrium wall.
2) These cells act as a pacemaker — they tell the heart when to pump blood.
3) A pacemaker that doesn't work properly causes an irregular heartbeat. An artificial pacemaker (a small electrical device) can be used to keep the heart beating regularly.

 **Make sure you know the names of the parts of the heart**
The heart diagram on this page is really important. Try copying it out a few times to get yourself really familiar with it. If you get your atria and ventricles mixed up, think about how the shape of the heart comes to a point at the bottom, just like a letter V, which is where the ventricles are.

Topic B2 — Organisation

Circulatory System — Blood Vessels

Blood needs a good set of 'tubes' to carry it round the body. Here's a page on the different types:

Arteries Carry Blood Under Pressure

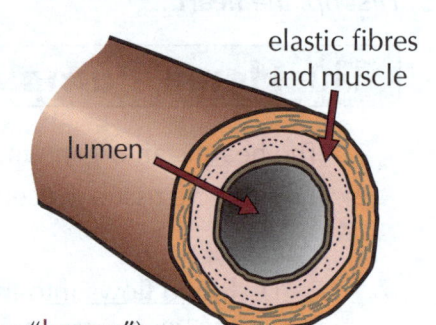

1) Arteries carry blood away from the heart.
2) The heart pumps the blood out at high pressure.
3) So artery walls are strong and elastic.
4) They have thick layers of muscle to make them strong.
5) They also have elastic fibres to allow them to stretch and spring back.
6) The walls are thick compared to the size of the hole down the middle (the "lumen").

Capillaries are Really Small

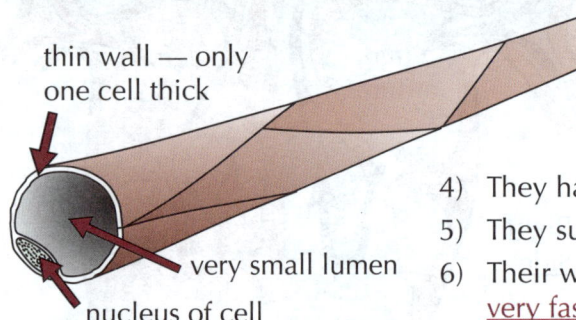

1) Arteries branch into capillaries.
2) Capillaries are really tiny — too small to see.
3) They carry the blood really close to every cell in the body to exchange substances with them.
4) They have gaps in their walls, so substances can diffuse in and out.
5) They supply food and oxygen, and take away waste like CO_2.
6) Their walls are usually only one cell thick. This means that diffusion is very fast because there is only a short distance for molecules to travel.

Veins Take Blood Back to the Heart

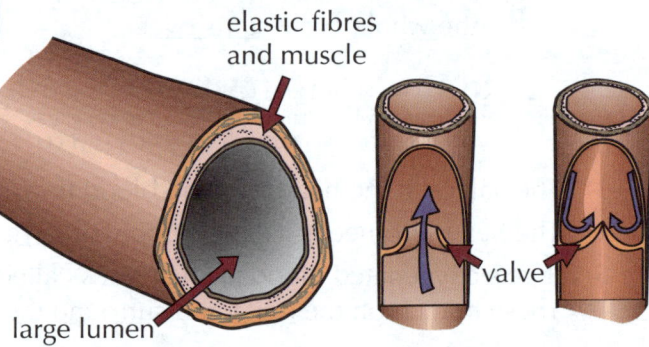

1) Capillaries join up to form veins.
2) The blood is at lower pressure in the veins. This means the walls don't need to be as thick as artery walls.
3) Veins have a bigger lumen than arteries. This helps the blood flow despite the lower pressure.
4) They also have valves. These help keep the blood flowing in the right direction.

You Can Calculate the Rate of Blood Flow

The rate of blood flow is the amount of blood that passes through a blood vessel in a given time. Here's how to calculate it:

EXAMPLE 1300 ml of blood passed through an artery in 4 minutes.
Calculate the rate of blood flow through the artery in ml per minute.

rate of blood flow = volume of blood ÷ number of minutes
= 1300 ÷ 4 = 325 ml per minute

Veins have valves — arteries carry blood away from the heart

Q1   2.175 litres of blood passed through a vein in 8.7 minutes.
Calculate the rate of blood flow through the vein in ml/min.   [2 marks]

Q2   Describe how the features of veins help them to carry blood back to the heart.   [2 marks]

Topic B2 — Organisation

Circulatory System — Blood

Blood is a **tissue** (see p.40). One of its jobs is to act as a huge **transport** system. It has four main parts...

Red Blood Cells Carry Oxygen

1) The job of red blood cells is to carry oxygen from the lungs to all the cells in the body.
2) Their shape gives them a large surface area for absorbing oxygen.
3) They contain a red substance called haemoglobin.
4) Haemoglobin is the stuff that allows red blood cells to carry oxygen.
5) Red blood cells don't have a nucleus — this leaves more space for carrying oxygen.

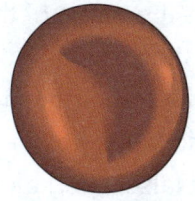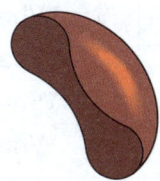

White Blood Cells Defend Against Infection

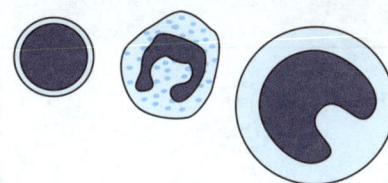

1) White blood cells are part of your immune system — see page 78.
2) Some can change shape to gobble up unwelcome microorganisms.
3) Others produce molecules called antibodies and antitoxins to defend against microorganisms.
4) Unlike red blood cells, they do have a nucleus.

Platelets Help Blood Clot

1) These are small fragments of cells. They have no nucleus.
2) They help the blood to clot (clump together) at a wound.
3) This stops all your blood pouring out.
4) It also stops any microorganisms getting in.

Plasma is the Liquid That Carries Everything in Blood

This is a pale straw-coloured liquid. It carries:
1) Red and white blood cells and platelets.
2) Food molecules (like glucose and amino acids).
3) Waste products (like carbon dioxide and urea).
4) Hormones.
5) Proteins.

Blood — red blood cells, white blood cells, platelets and plasma

Blood tests can be used to diagnose loads of things — not just disorders of the blood. This is because the blood transports so many chemicals produced by so many organs.

Q1 Describe the purpose of platelets in blood. [1 mark]

Q2 State the function of the cell labelled X in the image on the right. [1 mark]

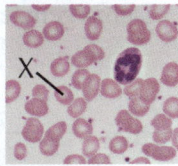

Topic B2 — Organisation

Warm-Up & Exam Questions

There are some nice diagrams to learn on the previous few pages. If you don't bother, you'll feel pretty silly if you turn over the exam paper and the first question asks you to label a diagram of the heart. Just saying... Anyway, let's see if these questions get your blood pumping...

Warm-Up Questions

1) What is the name of the tubes that the trachea splits into?
2) What is the function of the coronary arteries?
3) What does an artificial pacemaker do?
4) What do veins do?
5) True or false? Red blood cells don't have a nucleus.

Exam Questions

1   **Figure 1** shows the human heart and four blood vessels, as seen from the front. The left ventricle and the right atrium have been labelled.

Figure 1

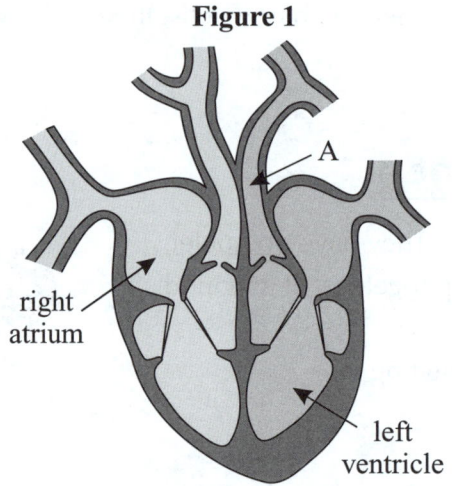

1.1 Name the part labelled **A**.

[1 mark]

1.2 What is the function of the left ventricle? Tick **one** box.

☐ It pumps blood to the lungs.

☐ It pumps blood around the body.

☐ It prevents the backflow of blood.

[1 mark]

1.3 Complete the sentence.

Deoxygenated blood enters the right atrium through the ....

[1 mark]

Topic B2 — Organisation

Exam Questions

2 Draw **one** line from each part of the blood to its function.

part of blood | function

red blood cell — carrying everything in the blood

plasma — carrying oxygen

platelets — helping the blood to clot

[2 marks]

3 The cell shown in **Figure 2** transports oxygen around the body.

Figure 2

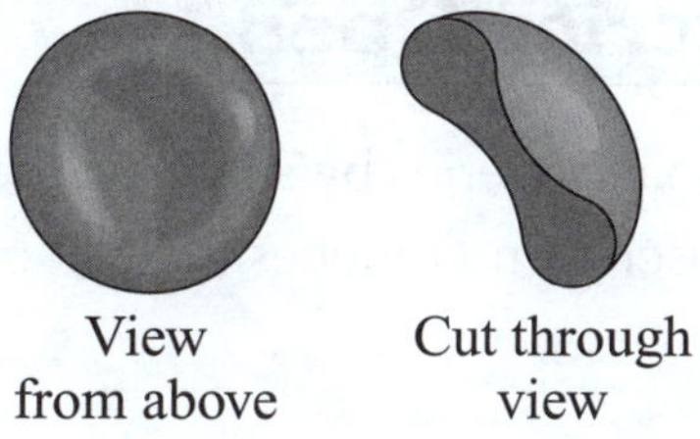

View from above | Cut through view

Explain how the shape of the cell in **Figure 2** is adapted for transporting oxygen.

[1 mark]

4 A student ran for 12 minutes.

4.1 During this 12 minute run, the student took 492 breaths.
Calculate his average breathing rate in breaths per minute.

[1 mark]

4.2 The student also measured his heart rate before and during his run.
Before his run, the student's heart rate was at its natural resting rate.
Outline how natural resting heart rate is controlled.

[1 mark]

5 Blood cells are carried in the bloodstream inside blood vessels.

5.1* Capillaries are one type of blood vessel.
Explain how the structure of a capillary allows it to carry out its function.

[4 marks]

5.2 Blood flows through different types of blood vessels at different rates.
The volume of blood that passed through an artery in 150 seconds was 1155 ml.
Calculate the rate of blood flow through the artery in ml/min.

[2 marks]

Topic B2 — Organisation

Cardiovascular Disease

Cardiovascular diseases are diseases of the **heart or blood vessels**. One example is **coronary heart disease**.

Coronary Heart Disease is a Disease of the Coronary Arteries

1) The coronary arteries supply the heart muscle with blood.
2) Coronary heart disease is when layers of fatty material (called fatty deposits) build up in the coronary arteries. This causes the arteries to become narrow.
3) This reduces the blood flow to the heart muscle.
4) This means less oxygen can get to the heart muscle. This can result in a heart attack.

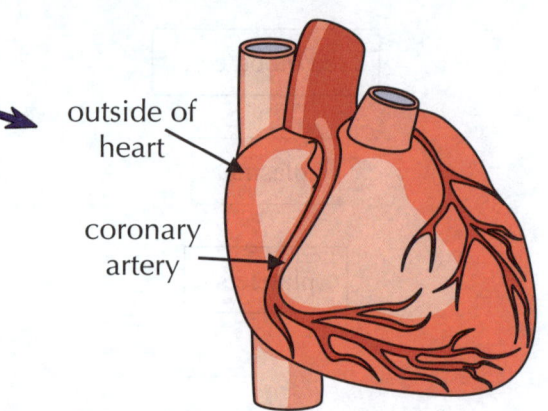

Stents Keep Coronary Arteries Open

1) Stents are tubes that are put inside coronary arteries by surgery. They keep the arteries open.
2) This allows blood to reach the heart muscles and reduces the risk of a heart attack.

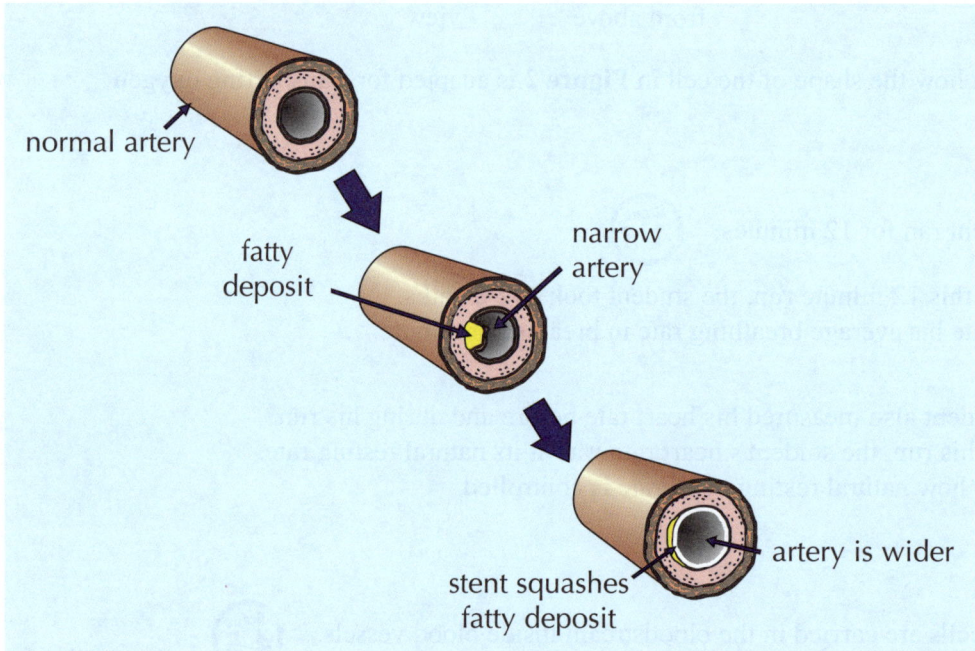

3) Stents are effective for a long time. Recovery time from the surgery is also quite quick.
4) But there are risks. These include having a heart attack during the operation, or getting an infection after surgery. Patients may also develop a blood clot near the stent.

Coronary heart disease is a type of cardiovascular disease

Coronary heart disease is caused by the arteries being blocked by fatty material — if the heart muscle can't get enough oxygen, then it can't work properly. And if the heart can't work properly, well, you're in trouble.

Cardiovascular Disease

You've read about stents, so now it's time for a second treatment for coronary heart disease — **statins**.

Statins Reduce Cholesterol in the Blood

1) Cholesterol is a lipid that your body needs.
2) However, too much cholesterol can cause fatty deposits to form inside arteries.
3) Statins are drugs that can reduce the amount of cholesterol in the blood.
4) This slows down the rate of fatty deposits forming.

Advantages
1) Statins reduce the risk of strokes, coronary heart disease and heart attacks.
2) Some studies suggest that statins may also help prevent some other diseases.

Disadvantages
1) Statins must be taken regularly over a long time. A person could forget to take them.
2) Statins can cause unwanted side effects, for example, headaches.
3) The effect of statins isn't instant. It takes time for their effect to work.

Faulty Heart Valves Can Be Replaced

1) The valves (see p.51) in the heart can be damaged by heart attacks, infection or old age.
2) This may cause the valve to stiffen, so it won't open properly.
3) A valve may also become leaky — blood flows in both directions instead of just forward.
4) This means that blood doesn't flow around the body as well as normal.
5) Damaged valves may be replaced by biological valves — valves from humans or other mammals (e.g. cows or pigs). Or they can be replaced by mechanical valves — these are man-made valves.
6) Replacing a valve is less risky than a heart transplant (see next page). But as it requires surgery, there is a risk of bleeding and infection. There can also be problems with blood clots.

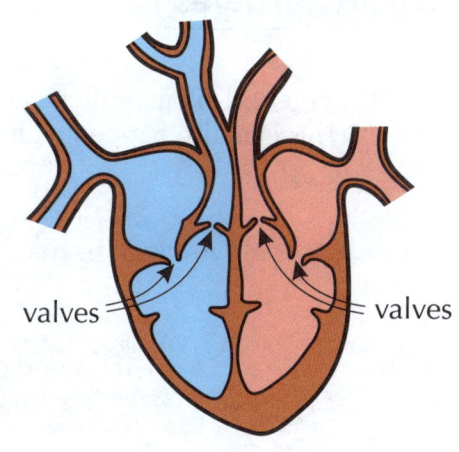

valves — valves

There are risks with any treatment for cardiovascular disease

Anything involving surgery on the heart is going to be risky, but even taking statins comes with a risk of side effects. Doctors and patients have to weigh up the benefits against the risks.

Topic B2 — Organisation

More on Cardiovascular Disease

*One final page on cardiovascular disease. Having the valves replaced isn't the only surgery for cardiovascular disease — when someone has heart failure, surgeons can **replace** the **whole heart**.*

An **Artificial Heart** Can **Pump Blood** Round the Body

1) A heart transplant is when a person's heart is replaced by a donor heart (a heart from someone who has recently died).

2) This can happen if someone has heart failure. Heart failure is when the heart can't pump enough blood.

3) The lungs may also be replaced if they are diseased.

4) If a donor heart isn't available, doctors may fit an artificial heart (a machine that pumps blood around the body).

5) Artificial hearts can be used to keep a person alive until a donor heart is available. Or they can help a person recover by allowing the heart to rest and heal.

6) Sometimes artificial hearts are permanent, so a donor heart isn't needed anymore.

7) Here are some of the advantages and disadvantages of artificial hearts:

Advantage

Artificial hearts are made from metals or plastics. This makes them less likely to be attacked by the body's immune system (see page 78) than a donor heart.

Disadvantages

1) Surgery to fit an artificial heart can lead to bleeding and infection. (This can also happen with transplant surgery.)

2) Artificial hearts don't work as well as healthy natural ones.

3) Blood doesn't flow through artificial hearts as smoothly as through a natural heart. This can cause blood clots and lead to strokes.

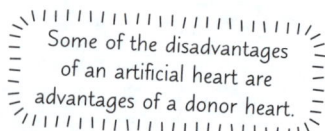
Some of the disadvantages of an artificial heart are advantages of a donor heart.

4) The patient has to take drugs to thin their blood. This means they can bleed a lot more than is usual if they have an accident.

Don't lose heart...

You could be asked to evaluate treatments for cardiovascular disease. Don't panic — just use any information you're given and your own knowledge to weigh up the advantages and disadvantages. Make sure your answer doesn't just focus on one side — e.g. don't just talk about the advantages and ignore the disadvantages. You should also include a conclusion that you can back up.

Topic B2 — Organisation

Warm-Up & Exam Questions

Hopefully I've persuaded you by now that it's a good idea to try these questions. Believe me, when you're sitting with your real exam paper in front of you, you'll feel so much better knowing that you've already been through loads of practice questions. So off you go...

Warm-Up Questions

1) Which vessels are affected in coronary heart disease?
2) If someone has heart failure, and there is no donor heart available, how might they be treated?
3) What is an artificial heart?

Exam Questions

1   There are different ways to treat coronary heart disease.

1.1 Complete **Table 1** to give the names of the treatments described.

Table 1

Description of treatment	Name of treatment
Tubes that are put inside arteries to keep them open	
Drugs that reduce cholesterol in the blood	

[2 marks]

1.2 Suggest **one** disadvantage to a patient of taking drugs to reduce blood cholesterol.

[1 mark]

2   A patient is taken into hospital. They are diagnosed as having a leaky heart valve.
The doctor decides that a surgeon should replace the valve.

2.1 Give the **two** types of valve that the surgeon might use.

[2 marks]

2.2 Suggest **one** risk of replacing the faulty valve.

[1 mark]

3   **Figure 1** shows a cross-section of a blood vessel in someone
with coronary heart disease.

3.1 Name the substance labelled **X** on **Figure 1**.

[1 mark]

3.2 Explain how the presence of this substance can
affect oxygen delivery to the heart muscle.

[2 marks]

Figure 1

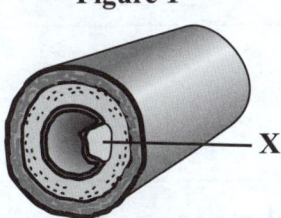

Topic B2 — Organisation

Health and Disease

Try as we might, it's unlikely that we'll be in tip-top condition for all of our lives — **disease** tends to get us all at some point. There are lots of **different types** of diseases we could get...

Diseases are a Major Cause of Ill Health

1) Health is the state of physical and mental wellbeing.

2) This means that both the body and mind are well.

3) Diseases are often responsible for causing ill health.

4) Diseases can be communicable or non-communicable:

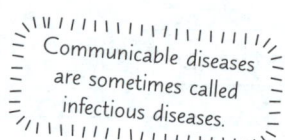
Communicable diseases are sometimes called infectious diseases.

Communicable Diseases

1) These are diseases that can spread from person to person or between animals and people.
2) Communicable diseases can be caused by bacteria, viruses, parasites or fungi.
3) Measles and malaria are examples of communicable diseases. See pages 73-76 for more.

Non-Communicable Diseases

1) These are diseases that cannot spread between people or between animals and people.
2) Coronary heart disease (see page 56) is an example of a non-communicable disease.

Different Types of Disease Can Interact

Sometimes a disease can cause other physical and mental health issues. Here are a few examples:

1) The immune system helps to fight off pathogens (see page 78). Some people have problems with their immune system. This makes them more likely to suffer from communicable diseases.
2) An immune system reaction (caused by a pathogen) may lead to an allergic reaction, such as a skin rash. Or it may worsen the symptoms of asthma for asthma sufferers.
3) Viruses infect cells in the body. This can lead to some types of cancer.
4) Physical health problems may also lead to mental health problems. For example, a person may become depressed if they can't carry out everyday activities because of ill health.

Pathogen is just the fancy term for a microorganism that can cause disease.

Communicable diseases can spread...

...but non-communicable diseases can't. Remember that — it's really important.

Topic B2 — Organisation

Health and Disease

Ill health isn't just about having a disease — there are plenty of **other causes**.
And then there's the **cost** of ill health to consider too — there might be more to it than you'd first thought...

Other Factors Can Also Affect Your Health

There are plenty of factors other than diseases that can also affect your health. For example:

1) A poor diet can affect your physical and mental health. A good diet is balanced and provides your body with everything it needs, in the right amounts.

2) Being constantly under lots of stress can lead to poor health.

3) Your life situation can affect your health. This is because it affects how easily you can access medicine or things that prevent you from getting ill. E.g. being able to buy condoms to prevent the spread of some sexually transmitted diseases.

Non-Communicable Diseases Can Be Costly

The Human Cost

1) Tens of millions of people around the world die from non-communicable diseases every year.
2) People with these diseases may have a lower quality of life or a shorter lifespan — this is the human cost.

The Financial Cost

1) The financial cost of researching and treating these diseases is huge.
2) It can also be expensive for individuals if they have to move or adapt their home because of a disease. If a person has to give up work or if they die, then their family's income will be reduced.
3) A reduction in the number of people able to work can also affect a country's economy.

Lots of things affect health, and ill health can be costly

A human cost is the effect something has on humans. A financial cost is to do with how much spending something results in. When you're studying biology, you'll come across lots of things that have a human cost or a financial cost (or both). Some things can have quite far-reaching knock-on effects.

Topic B2 — Organisation

Risk Factors for Non-Communicable Diseases

You've probably heard the term '*risk factor*' before. These next couple of pages have lots of info on them. There's nothing too tricky, but there's quite a bit to read — take it slowly and make sure it goes in.

Risk Factors Increase Your Chance of Getting a Disease

1) Risk factors are things that are linked to an increased chance of getting a certain disease.

2) However, risk factors don't mean that someone will definitely get the disease.

3) They can be:
   - part of a person's lifestyle (for example, how much exercise they do),
   - substances in a person's environment (e.g. air pollution),
   - substances in a person's body (e.g. asbestos fibres in the lungs can cause cancer).

4) Many non-communicable diseases are caused by several risk factors that interact with each other.

5) Lifestyle factors can have different effects locally, nationally and globally.
   - Globally, non-communicable diseases are more common in developed countries. This is because people in developed countries generally earn more and can buy high-fat food.
   - Nationally, cardiovascular disease, obesity and Type 2 diabetes are more common in poorer areas. This is because people in poorer areas are more likely to smoke, have a poor diet and not exercise.
   - Your individual choices affect how common a disease is locally.

Some Risk Factors Can Cause a Disease Directly

Some risk factors are able to directly cause a disease. For example:

Smoking can cause cardiovascular disease, lung disease and lung cancer. It damages the walls of arteries and the lining of the lungs.

Lifestyle can have a big impact on a person's health

Smoking makes a person more likely to get certain diseases. The next page has some more risk factors.

Risk Factors for Non-Communicable Diseases

Here are a few more risk factors that can directly cause disease:

Obesity may cause Type 2 diabetes by making the body less sensitive or resistant to (not affected by) insulin.

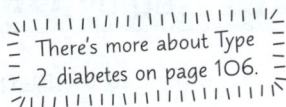
There's more about Type 2 diabetes on page 106.

Drinking too much alcohol can damage the brain and the liver.

Smoking and drinking alcohol when pregnant can cause health problems for the unborn baby.

Cancer can be caused by exposure to certain substances or radiation. Things that cause cancer are known as carcinogens. Ionising radiation (e.g. from X-rays) is an example of a carcinogen.

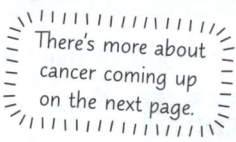
There's more about cancer coming up on the next page.

Risk Factors Can be Identified Using Correlation

Some risk factors don't directly cause a disease.
BUT there is a correlation between the risk factor and the disease.

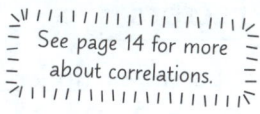
See page 14 for more about correlations.

For example, a lack of exercise and a high fat diet are risk factors for cardiovascular disease, but they can't cause the disease. It's the resulting high cholesterol levels (see p.57) that can cause it.

It's hard to avoid all risk factors of disease...

...but remember that risk factors that cause disease don't mean you'll definitely get the disease. They just increase the chance of it happening. Also, remember that not all risk factors cause disease. Many are just correlated with the disease, meaning there is a relationship between them.

Topic B2 — Organisation

Cancer

*The more we understand **cancer**, the better our chances of **avoiding** and **beating** it.*

Cancer is Caused by Uncontrolled Cell Growth and Division

1) Changes in cells can lead to uncontrolled growth and division. This results in a tumour (a mass of cells).
2) Tumours can be benign or malignant:

1) Benign tumours are masses of abnormal cells. 2) They stay in one place (usually within a membrane). 3) They don't invade other parts of the body. 4) This type isn't normally dangerous, and the tumour isn't cancerous.	1) Malignant tumours spread to other parts of the body. 2) The cells can break off and travel in the bloodstream. 3) The cells get into healthy tissues and form secondary tumours. 4) Malignant tumours are dangerous and can be fatal — they are cancers.

Risk Factors Can Increase the Chance of Some Cancers

Scientists have identified lots of risk factors for cancers. For example:

Lifestyle Factors

1) Smoking — Smoking is linked to many types of cancer.
2) Obesity — Obesity has also been linked to many different cancers.
3) Viral infection — Infection with some viruses can increase the chances of developing certain types of cancer.
4) UV exposure — The Sun produces UV radiation. This radiation has been linked to an increased chance of developing skin cancer.

Genetic Factors

1) Genes are passed on (inherited) from parent to offspring — see page 114.
2) Sometimes you can inherit faulty genes that make you more likely to get cancer.

People are Now More Likely to Survive Cancer

People have become more likely to survive cancer because:

1) Treatments have improved.
2) Doctors can diagnose cancer earlier.
3) More people are being screened (tested) for cancer.
4) People know more about the risk factors for cancer.

Knowing about the risk factors can help to protect people
E.g. people are now more aware that wearing sun block reduces the risk of skin cancer from UV radiation.

Topic B2 — Organisation

Warm-Up & Exam Questions

It's time for some more questions — don't just assume that you've remembered everything you just read on the past few pages. Give these a go, and then go back over anything that you struggled with.

Warm-Up Questions

1) True or false? Health is the state of physical wellbeing only.
2) What does it mean if a disease is 'communicable'?
3) True or false? Physical health problems may lead to mental health problems.
4) What disease can uncontrolled cell division lead to?

Exam Questions

1   Diseases can be communicable or non-communicable.
    What is meant by a non-communicable disease?
    Tick **one** box.

    ☐ A disease that can be spread between people or between animals and people.

    ☐ A disease that cannot be spread between people or between animals and people.

    ☐ A disease cause by uncontrolled cell growth and division.

    ☐ A disease caused by the build-up of fatty material in the coronary arteries.

    [1 mark]

2   Tumours can be either benign or malignant.

    Complete **Table 1** to show whether each statement is true for benign tumours, malignant tumours, or both.
    Put **one or two** ticks in each row.

    Table 1

	Benign	Malignant
The tumour is made up of a mass of cells formed by uncontrollable division and growth.		
The tumour cells can break off and travel into the bloodstream.		
The tumour is cancerous.		

    [3 marks]

3   Many diseases have risk factors.

3.1 What is meant by the term 'risk factor'?

    [1 mark]

3.2 Give **one** risk factor for lung disease.

    [1 mark]

3.3 Name **one** carcinogen that is a risk factor for cancer.

    [1 mark]

Plant Cell Organisation

*Just like in animals, plant cells are also organised. Here are a few examples of plant **tissues** and **organs**.*

Plant Cells Are Organised Into Tissues And Organs

1) Plants are made of organs. These organs work together to make organ systems.
2) For example, stems, roots and leaves are all plant organs. They work together to transport (carry) substances around the plant.
3) Plant organs are made of tissues. Examples of plant tissues are:

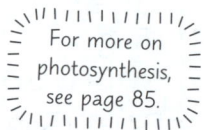
For more on photosynthesis, see page 85.

- Epidermal tissue — this covers the whole plant.
- Palisade mesophyll tissue — this is the part of the leaf where most photosynthesis happens.
- Spongy mesophyll tissue — this is the part of the leaf that has big air spaces. This allows gases to diffuse in and out of cells.
- Xylem and phloem — these transport things like water, mineral ions and food around the roots, stems and leaves (see next page).
- Meristem tissue — this is found at the growing tips of shoots and roots.

Leaves Contain Epidermal, Mesophyll, Xylem and Phloem Tissue

1) The leaf is where photosynthesis and gas exchange happens in a plant.
2) The structures of the tissues in a leaf are related to their function:

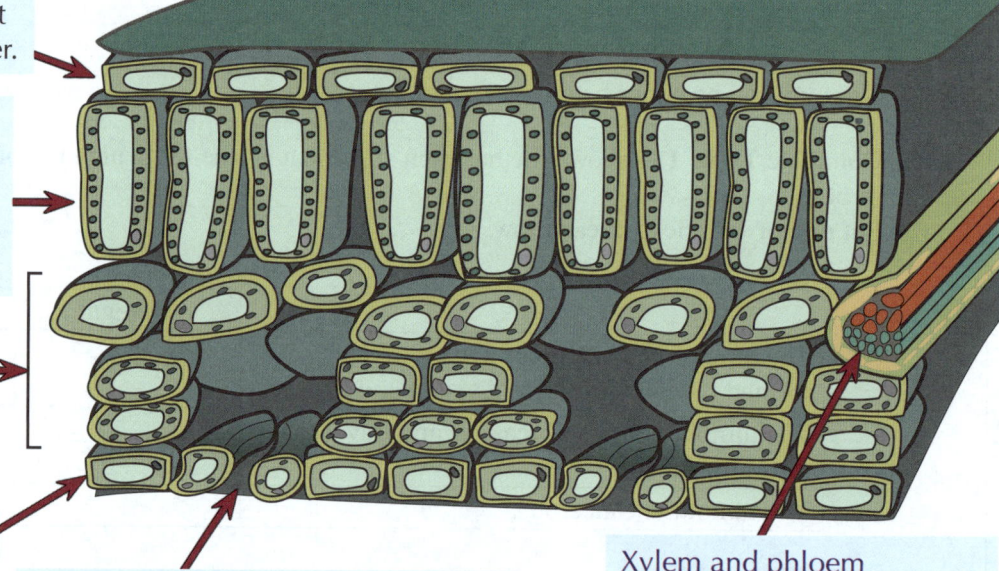

Upper epidermis
This layer is transparent (see-through). This lets light through to the palisade layer.

Palisade layer
This has lots of chloroplasts (see page 18). They are near the top of the leaf to get more light.

Spongy mesophyll
This contains air spaces which increase the rate of diffusion of gases.

Lower epidermis

Stomata
These let gases diffuse into and out of the leaf. They are opened and closed in response to the environment. This is controlled by guard cells.

Xylem and phloem
These bring water and nutrients to the leaf and take away glucose produced by photosynthesis. They also support the leaf.

Each tissue in a leaf is adapted for its function
There are a lot of weird names here, so make sure you spend plenty of time on this page. Try drawing your own leaf diagram. Label it with the different tissues and describe each type.

Topic B2 — Organisation

Transpiration and Translocation

*Flowering plants have **two** separate types of vessel — **xylem** and **phloem** — for transporting stuff around.*
***Both** types of vessel go to **every part** of the plant, but they are totally **separate**.*

Phloem Tubes Transport Food

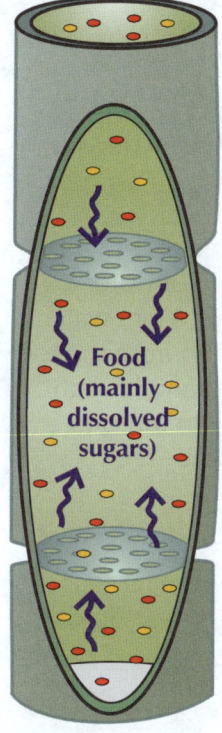

1) Phloem tubes are made of elongated (stretched out) living cells.

2) There are end walls between the cells. These have pores (small holes) to allow cell sap to flow through.

3) Plants make food substances (e.g. dissolved sugars) in their leaves.

4) Phloem tubes transport these food substances around the plant for immediate use or for storage.

5) The transport goes in both directions.

6) This process is called translocation.

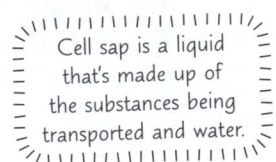

Cell sap is a liquid that's made up of the substances being transported and water.

Xylem Tubes Take Water Up

1) Xylem tubes are made of dead cells.

2) The cells are joined together with a hole down the middle.

3) There are no end walls between the cells.

4) The cells are strengthened with a material called lignin.

5) Xylem tubes carry water and mineral ions from the roots to the stem and leaves.

6) The movement of water from the roots, through the xylem and out of the leaves is called the transpiration stream (see next page).

Xylem vessels carry water, phloem vessels carry sugars

Make sure you don't get your phloem mixed up with your xylem. To help you to learn which is which, you could remember that phloem transports substances in both directions, but xylem only transports things upwards — xy to the sky. It might just bag you a mark or two on exam day...

Topic B2 — Organisation

Transpiration

*If you don't water a house plant for a few days it starts to go all droopy. Plants need **water**.*

Transpiration is the Loss of Water from the Plant

1) Transpiration is caused by evaporation and diffusion of water from a plant's surface (mainly the leaves).
2) Here's how it happens:

Evaporation is when water turns from a liquid into a gas. See page 30 for more on diffusion.

1) Water evaporates from the leaves and diffuses into the air.

2) This creates a slight shortage of water in the leaf. More water is drawn up from the rest of the plant through the xylem tubes to replace it.

3) This means more water is drawn up from the roots.

Head back to page 33 to see how root hair cells are adapted for taking up water.

3) There's a constant stream of water through the plant. This is called the transpiration stream.

Guard Cells Control Gas Exchange and Water Loss

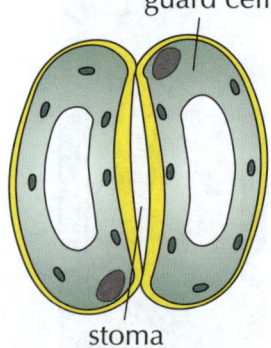

guard cell

stoma (plural — stomata)

1) Water is lost from a plant's leaves through the stomata.
2) Stomata are surrounded by guard cells.
3) These change shape to control the size of the stomata.
4) When the plant has lots of water the guard cells fill with it and get fat. This makes the stomata open so gases can be exchanged for photosynthesis.
5) When the plant is short of water, the guard cells lose water and become floppy. This makes the stomata close. This helps stop too much water vapour escaping.
6) There are usually more stomata on the bottoms of leaves than on the tops. This is because the lower surface is cooler — so less water gets lost.

Transpiration involves evaporation and diffusion

A big tree loses about a thousand litres of water from its leaves every single day — it's a fact. That's as much water as the average person drinks in a whole year, so the roots have to be very effective at drawing in water from the soil. Which is why they have all those root hairs, you see.

Topic B2 — Organisation

The Rate of Transpiration

*The **rate of transpiration** varies according to the **environmental conditions**...*

Transpiration Rate is Affected by Four Main Things:

Air Flow

1) The more windy it is, the faster transpiration happens.
2) Fast moving air means that water vapour around the leaf is swept away.
3) This means there's a higher concentration of water vapour inside the leaf compared to outside. So water will diffuse out of the leaf more quickly.

Temperature

1) The warmer it is, the faster transpiration happens.
2) This is because the water particles have more energy. So they evaporate and diffuse out of the stomata faster.

Many factors affect transpiration rate by affecting the rate of diffusion of water. There's more about diffusion on page 30.

Humidity

1) If the air is humid there's a lot of water in it already.
2) This means there isn't much of a difference between the inside and the outside of the leaf.
3) This means that diffusion will not happen very fast.
4) The drier the air around a leaf, the faster transpiration happens.

Light Intensity

1) The brighter the light, the greater the transpiration rate.
2) Photosynthesis can't happen in the dark, so stomata begin to close as it gets darker.
3) When the stomata are closed, very little water can escape.

Transpiration is fastest when it's windy, warm, dry and bright

Make sure you know these four factors, and that you understand why they affect transpiration rates.

Q1   Aloe vera plants grow in hot, dry areas. Primroses grow in cool, wet areas. Predict which plant will have fewer stomata per cm² on the underside of its leaves. Explain your answer.   [2 marks]

Q2   Explain how low light intensity affects the rate of transpiration.   [3 marks]

Q1 Video Solution

Topic B2 — Organisation

Warm-Up & Exam Questions

Just a few simple Warm-Up Questions and a few slightly harder Exam Questions stand between you and mastering cell organisation and transport in plants...

Warm-Up Questions

1) Where is meristem tissue found in a plant?
2) Which layer of plant tissue contains lots of chloroplasts?
3) True or false? Substances pass in both directions through xylem vessels.
4) State one of the main factors that affects the rate of transpiration in plants.

Exam Questions

1 Leaves contain many types of tissue. *(Grade 3-4)*

1.1 Which type of plant tissue contains air spaces for the diffusion of gases?
Tick **one** box.

☐ epidermal tissue

☐ xylem tissue

☐ spongy mesophyll tissue

[1 mark]

1.2 Which type of plant tissue forms the transparent layer covering the outside of plant?
Tick **one** box.

☐ epidermal tissue

☐ palisade mesophyll tissue

☐ spongy mesophyll tissue

[1 mark]

2 **Figure 1** shows a vessel that transports cell sap. *(Grade 3-4)*

Figure 1

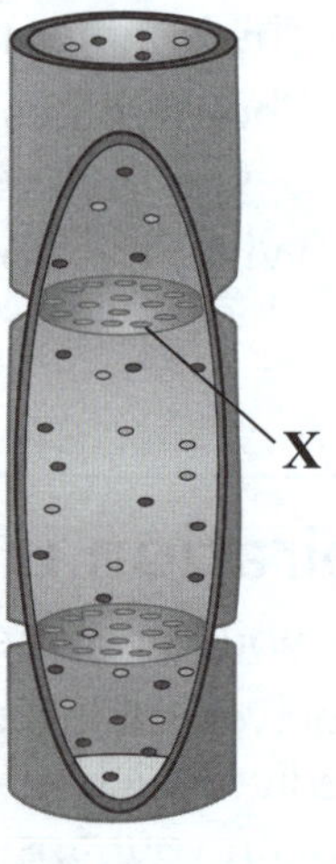

2.1 Name the type of vessel shown in **Figure 1**.

[1 mark]

2.2 Explain why the vessel has the pores labelled **X** on **Figure 1**.

[1 mark]

2.3 Name the movement of cell sap through the plant's transport vessels.

[1 mark]

Topic B2 — Organisation

Exam Questions

3 Stomata are found on the surface of leaves.

3.1 Complete the sentence below about stomata.

Stomata are opened and closed by cells called ....

[1 mark]

3.2 **Figure 2** shows a stoma in different conditions.

Figure 2

Condition **A**   Condition **B**

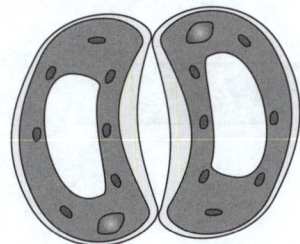 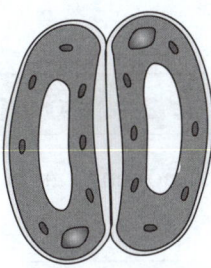

Which condition, **A** or **B**, shows a stoma when the plant is short of water? Explain your answer.

[1 mark]

4 Plants absorb water and mineral ions through their root hair cells.
Xylem vessels transport water and mineral ions from the roots of a plant to the leaves.

4.1 Describe the structure of xylem vessels.

[3 marks]

4.2 Name the process of water transport through a plant.

[1 mark]

5 The water loss from a plant during two different days is shown on the graph in **Figure 3**.
One of the days was hot and dry, and the other day was cold and wet.

5.1 Which line, **A** or **B**, shows water loss on the hot, dry day? Explain your answer.

[1 mark]

5.2* Explain how the rate of water loss from a plant would be affected by windy weather.

[4 marks]

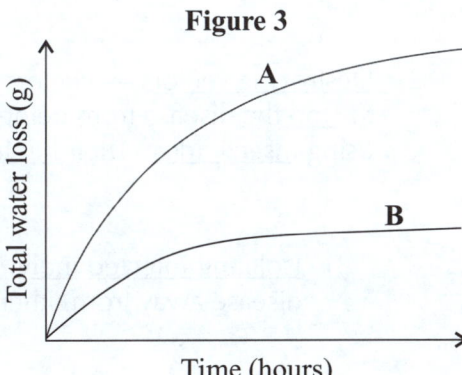

Figure 3

Topic B2 — Organisation

Topic B3 — Infection and Response

Communicable Disease

*If you're hoping I'll ease you gently into this new topic... no such luck. Straight on to the **baddies** of biology.*

There Are Several Types of Pathogen

1) Pathogens are microorganisms that enter the body and cause disease.
2) They cause communicable (infectious) diseases.
3) Communicable diseases are diseases that can spread (see p.60).
4) Both plants and animals can be infected by pathogens.
5) There are four main types of pathogens:

   Bacteria   Viruses   Protists   Fungi

Pathogens Can Be Spread in Different Ways

Here are a few ways that pathogens can be spread:

Air

Pathogens can be carried in the air and can then be breathed in. Some pathogens are carried in the air in droplets made when you cough or sneeze.

Water

Some pathogens can be picked up by drinking or bathing in dirty water.

Direct Contact

Some pathogens can be picked up by touching surfaces they're on (e.g. the skin).

The Spread of Disease Can Be Reduced or Prevented

There are things that we can do to reduce or prevent the spread of disease, such as...

1) Being hygienic (clean) — For example, washing your hands before making food can stop you spreading pathogens onto the food and infecting a person who eats it.

2) Destroying vectors — Vectors are organisms that spread disease. Killing them helps to stop the disease from being passed on. Vectors that are insects can be killed using insecticides. Their habitats can also be destroyed so that they can't breed.

3) Isolating infected individuals — If you keep someone who has a communicable disease away from other people, it prevents them from passing it on to anyone else.

4) Vaccination (see page 80) — Vaccinations make it less likely that people and animals will get a communicable disease and pass it on to others.

Bacterial Diseases

*First up from the **pathogen** hall of fame are... **bacteria**.*

Bacteria are Very Small Living Cells

1) Bacteria reproduce rapidly inside your body.
2) They can make you feel ill by producing toxins (poisons).
3) Toxins damage your cells and tissues.

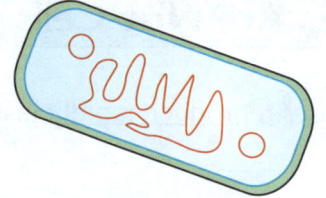

Salmonella and Gonorrhea Are Two Bacterial Diseases

Salmonella

1) *Salmonella* is a type of bacteria. It causes food poisoning.

2) Infected people can suffer from fever, stomach cramps, vomiting and diarrhoea. These symptoms are caused by toxins from the bacteria.

3) You can get *Salmonella* food poisoning by:

- Eating food that's got *Salmonella* bacteria in it already, e.g. eating chicken that caught the disease whilst it was alive.
- Eating food that has been made where the bacteria is present, e.g. in an unclean kitchen or on the hands of the person making the food.

4) In the UK, most poultry (e.g. chickens and turkeys) are given a vaccination against *Salmonella*. This is to control the spread of the disease.

Gonorrhoea

1) Gonorrhoea is caused by bacteria.
2) Gonorrhoea is a sexually transmitted disease (STD).
3) STDs are passed on by sexual contact, e.g. having unprotected sex.
4) A person with gonorrhoea will get pain when they urinate (wee). Another symptom is a thick yellow or green discharge (fluid) from the vagina or the penis.
5) Gonorrhoea used to be treated with an antibiotic called penicillin. There are now new strains (types) of gonorrhoea that are resistant to (not killed by) penicillin. So this antibiotic doesn't work anymore.
6) To prevent the spread of gonorrhoea:

- People can be treated with other antibiotics,
- People should use barrier methods of contraception (see page 110), such as condoms.

Topic B3 — Infection and Response

Viral Diseases

Viruses may be tiny but there are lots of *diseases* caused by them.

Viruses Are Not Cells — They're Much Smaller

1) Viruses reproduce rapidly inside your body.

2) They live inside your cells.

3) Inside your cells, they make lots of copies of themselves.

4) The cells will usually then burst, releasing all the new viruses.

5) This cell damage is what makes you feel ill.

A virus

A body cell

Different Viruses Cause Different Diseases, Such as...

Measles

1) Measles is a viral disease. It is spread by droplets from an infected person's sneeze or cough.

2) People with measles develop a red skin rash.

3) They'll also show signs of a fever (a high temperature).

4) Measles can be very serious. People can die from measles if there are complications (problems).

5) Because of this, most people are vaccinated against measles when they're young.

Topic B3 — Infection and Response

Viral Diseases

HIV

1) HIV is a virus spread by sexual contact or by exchanging bodily fluids (e.g. blood). This can happen when people share needles when taking drugs.

2) To start with, HIV causes flu-like symptoms for a few weeks.

3) After that the person doesn't usually have any symptoms for several years.

4) HIV can be controlled with antiretroviral drugs. These stop the virus copying itself in the body.

5) If it's not controlled, the virus attacks the immune cells (see page 78).

6) If the body's immune system is badly damaged, it can't cope with other infections or cancers. At this stage, the virus is known as late stage HIV infection or AIDS.

Tobacco Mosaic Virus

1) Tobacco mosaic virus (TMV) is a virus that affects many species of plants, e.g. tomatoes.

2) It causes parts of the leaves to become discoloured. This gives them a mosaic pattern.

3) The discoloured leaves have less chlorophyll to absorb light (see p.85).

4) This means less photosynthesis happens in the leaves, so the plant can't make enough food to grow.

Don't be put off by something you haven't heard of...
If you're given some information about a disease you've never heard of before, don't panic. You just need to use what you know and apply it to the disease in the question. For example, if the disease is viral, just use what you know about viruses to answer the question.

Fungal and Protist Diseases

*Sorry — I'm afraid there are some more **diseases** to learn about here...*

Rose Black Spot is a Fungal Disease

1) Rose black spot is a disease caused by a fungus.
2) The fungus causes purple or black spots on the leaves of rose plants. The leaves can then turn yellow and drop off.
3) This means that less photosynthesis can happen, so the plant doesn't grow very well.
4) It is spread in water or by the wind.
5) Gardeners can treat the disease using fungicides (chemicals that kill fungi).
6) They can also strip the affected leaves off the plant. These leaves then need to be destroyed so that the fungus can't spread to other rose plants.

Malaria is a Disease Caused by a Protist

1) Malaria is caused by a protist.
2) Part of the protist's life cycle takes place inside the mosquito.
3) The mosquitoes are vectors. They help spread malaria like this...

- The mosquitoes pick up the protist when they feed on an infected animal.
- The mosquitoes don't get malaria.
- They pass on the protist to other animals (like us) when they bite them.
- These animals get malaria.

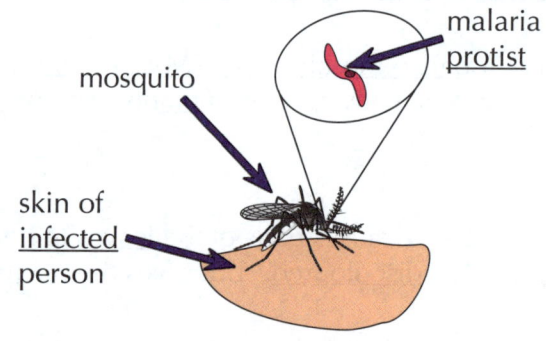

4) Malaria causes repeating episodes of fever. People can die from malaria.
5) The spread of malaria can be reduced by stopping the mosquitoes from breeding.
6) People can be protected from mosquito bites using mosquito nets.

Hang in there, this stuff is pretty gross, but it's nearly over...
Try drawing out a table with columns for 'disease', 'type of pathogen it's caused by', 'symptoms' and 'how it's spread', then fill it in for all the diseases on this page and the previous three. See how much you can write down without looking back at the page.

Topic B3 — Infection and Response

Warm-Up & Exam Questions

Have a go at these questions to test whether you know about each of the diseases covered on the previous pages, including their symptoms and how they are spread.

Warm-Up Questions

1) True or false? A communicable disease is a disease that can spread.
2) How is gonorrhoea passed between individuals?
3) Where do viruses live inside the human body?
4) What symptom of measles is shown on the skin?
5) Which disease causes the leaves of tomato plants to become discoloured?
6) What effect does rose black spot disease have on plants?

Exam Questions

1   Which of the following statements about malaria is **not** correct? Tick **one** box.

☐ People with malaria can have repeating episodes of fever.
☐ Mosquitos are the vectors of malaria.
☐ Malaria is caused by a virus.
☐ People can be protected from mosquito bites by using mosquito nets.

[1 mark]

2   The spread of disease can be reduced or prevented in many ways.

2.1 Which method is used to prevent the spread of *Salmonella* between people? Tick **one** box.

☐ hand-washing   ☐ stopping vectors from breeding   ☐ vaccinating people

[1 mark]

2.2 Which method is used to help prevent the spread of measles between people? Tick **one** box.

☐ hand-washing   ☐ stopping vectors from breeding   ☐ vaccinating people

[1 mark]

3   Diseases are often recognised by their symptoms.

3.1 Describe the first symptoms of HIV infection.

[1 mark]

3.2 Give **one** symptom of gonorrhoea.

[1 mark]

3.3 A person has food poisoning caused by *Salmonella*. Give **one** symptom that they may have.

[1 mark]

Topic B3 — Infection and Response

Fighting Disease

The human body has some pretty neat features when it comes to **fighting disease**.

Your Body Has a Pretty Good Defence System

The human body has got features that stop a lot of nasties getting inside. For example:

1) The skin — It stops pathogens getting inside you. It also releases substances that kill pathogens.

2) Nose hairs — They trap particles that could contain pathogens.

3) Mucus (snot) — The trachea and bronchi (airways — see page 49) release mucus to trap pathogens.

4) Cilia (hair-like structures) — The trachea and bronchi are lined with cilia. They move the mucus up to the back of the throat where it can be swallowed.

5) Stomach acid — The stomach makes hydrochloric acid. This kills pathogens in the stomach.

Your Immune System Can Attack Pathogens

1) If pathogens do make it into your body, your immune system kicks in to destroy them.
2) The most important part of your immune system is the white blood cells.
3) When they come across an invading pathogen they have three lines of attack:

Phagocytosis

White blood cells can engulf (surround) pathogens and digest them. This is called phagocytosis.

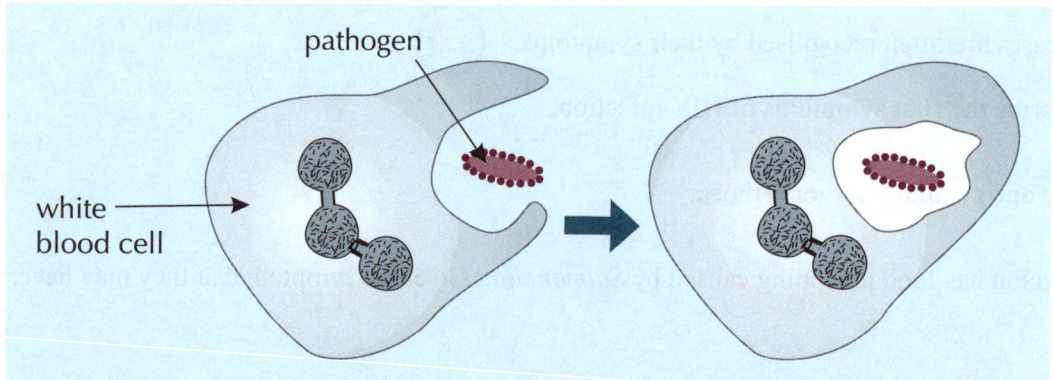

Topic B3 — Infection and Response

Fighting Disease

Producing Antibodies

1) Every invading pathogen has unique molecules on its surface. These molecules are called antigens.

2) When some types of white blood cell come across a foreign antigen (i.e. one they don't know), they will start to make antibodies.

3) Antibodies lock onto the invading pathogens. The antibodies made are specific to that type of antigen — they won't lock on to any others.

4) The antibodies make sure the pathogens can be found and destroyed by other white blood cells.

5) If the person is infected with the same pathogen again, the white blood cells will rapidly make the antibodies to kill it. This means the person is naturally immune to that pathogen and won't get ill.

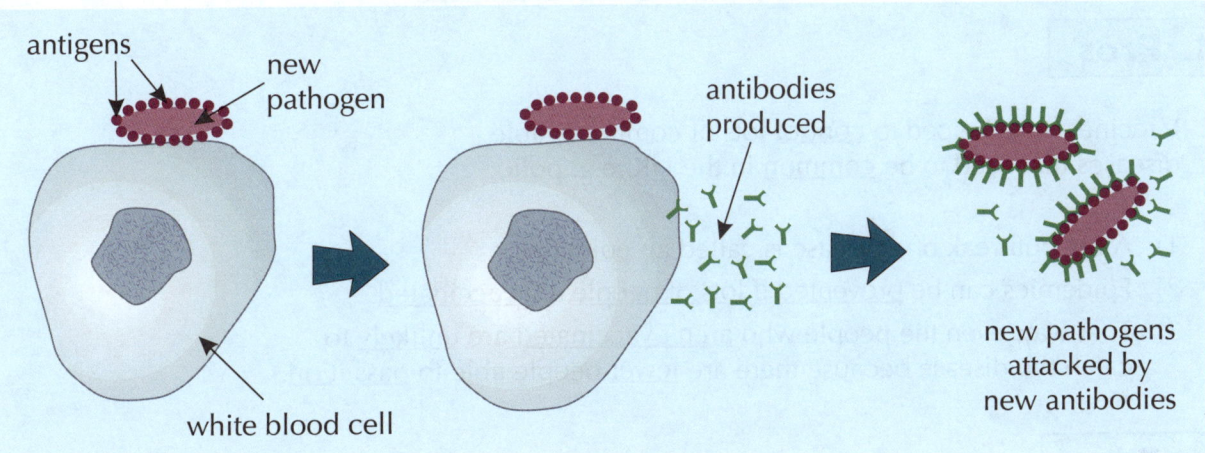

Producing Antitoxins

These stop toxins produced by the invading bacteria from working.

Fighting disease is one thing the body is really good at...

The immune system attacks pathogens that get inside the body. There are three ways that white blood cells kill pathogens — phagocytosis, making antibodies and making antitoxins. Make sure you know them all.

Topic B3 — Infection and Response

Fighting Disease — Vaccination

Vaccinations mean we don't always have to treat a disease — we can **stop** the disease in the first place.

Vaccination — Can Protect from Future Infections

1) Vaccinations involve injecting small amounts of dead or inactive pathogens into the body.
2) These pathogens have antigens on their surface.
3) The antigens cause your white blood cells to produce antibodies to attack the pathogens.
4) If you're infected with the same pathogen later, your white blood cells quickly produce lots of antibodies.
5) These antibodies can kill the pathogen — so vaccination makes you less likely to get ill.

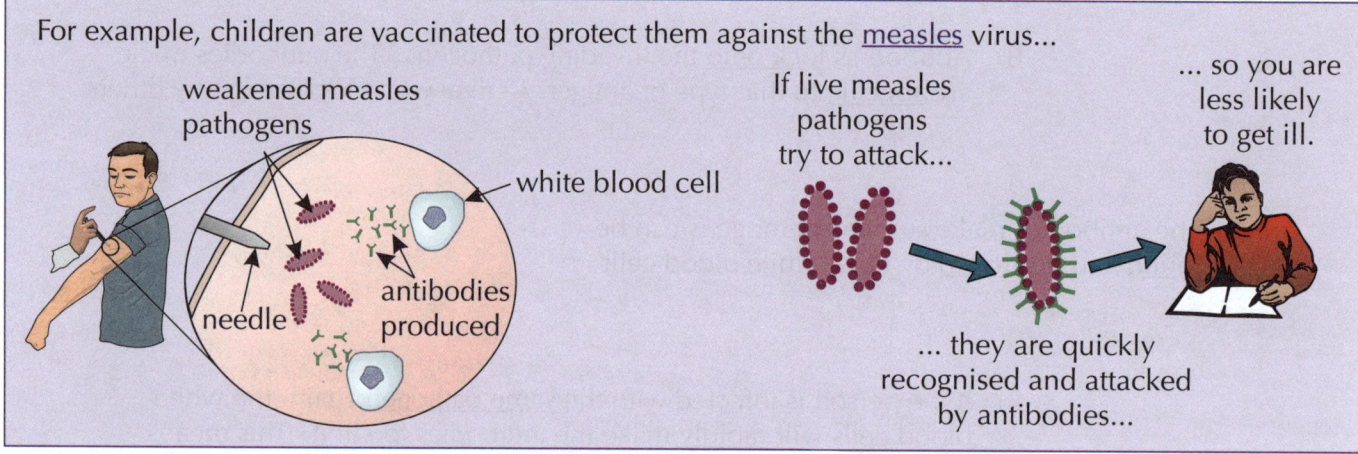

There are Pros and Cons of Vaccination

1. Pros

Vaccines have helped to control lots of communicable diseases that used to be common in the UK, e.g. polio.

1) A big outbreak of a disease is called an epidemic.
2) Epidemics can be prevented if lots of people are vaccinated.
3) That way, even the people who aren't vaccinated are unlikely to catch the disease because there are fewer people able to pass it on.

2. Cons

1) Vaccines don't always work — sometimes they don't give you immunity.
2) You can sometimes have a bad reaction to a vaccine (e.g. swelling or a fever).

Prevention is better than cure...

Although vaccinations aren't perfect, it's better to have a vaccine than risk catching a nasty disease.

Q1 Basia is vaccinated against flu and Cassian isn't. They are both exposed to a flu virus. Cassian falls ill whereas Basia doesn't. Explain why. [2 marks]

Topic B3 — Infection and Response

Fighting Disease — Drugs

You've probably had to take some sort of **medicine** if you've been ill, e.g. cough remedies, painkillers.

Some Drugs Get Rid of Symptoms — Others Cure the Problem

1) Some drugs help to get rid of the symptoms of a disease, e.g. painkillers reduce pain.
2) But these drugs don't kill the pathogens that cause the disease.

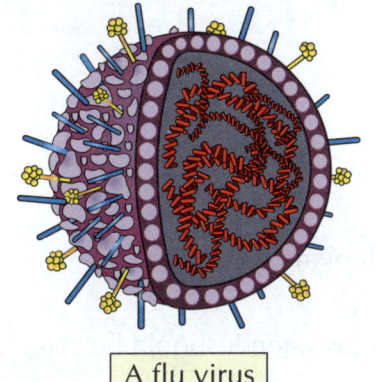

A flu virus

3) Antibiotics (e.g. penicillin) kill bacteria.
4) Different antibiotics kill different types of bacteria, so it's important to be treated with the right one.
5) The use of antibiotics has greatly reduced the number of deaths from communicable diseases caused by bacteria.
6) Antibiotics don't destroy viruses (e.g. flu viruses).
7) Viruses reproduce using your own body cells. This makes it very difficult to develop drugs that destroy the virus without killing the body's cells.

Bacteria Can Become Resistant to Antibiotics

1) Bacteria can mutate (change).
2) Some of these mutations cause the bacteria to become resistant to (not be killed by) an antibiotic.
3) Resistant strains (types) of bacteria, e.g. MRSA, have increased as a result of natural selection (see page 130).

Many Drugs First Came From Plants

1) Plants produce chemicals to defend themselves against pests and pathogens.
2) Some of these chemicals can be used as drugs to treat human diseases or relieve symptoms.
3) A lot of our medicines were found by studying plants used in old-fashioned cures. For example:

   1) Aspirin is used as a painkiller. It was made from a chemical found in willow.
   2) Digitalis is used to treat heart conditions. It was made from a chemical found in foxgloves.

4) Some drugs have come from microorganisms. For example:

   1) Alexander Fleming found that a type of mould (called *Penicillium*) makes a substance that kills bacteria.
   2) This substance is called penicillin.
   3) Penicillin is used as an antibiotic.

5) These days, new drugs are made by the pharmaceutical industry (companies that make and sell drugs).
6) The drugs are made by chemists in labs.
7) The process of making the drugs still might start with a chemical taken from a plant.

Topic B3 — Infection and Response

Developing Drugs

*New drugs are always being developed. But before they can be given to people like you and me, they have to go through **a lot** of **tests**. This is what usually happens...*

There Are Different Stages in the Development of New Drugs

1) Once a possible drug has been discovered, it needs to be developed.
2) This involves preclinical and clinical testing.

Preclinical Testing

1) Drugs are first tested on human cells and tissues in the lab.
2) Next the drug is tested on live animals. This is to find out:

- Its efficacy (whether the drug works and has the effect you're looking for).
- Its toxicity (how harmful it is and whether it has any side effects).
- Its dosage (the concentration of the drug that works best and how often it should be taken).

Clinical Testing

If the drug passes the tests on animals then it's tested on human volunteers in a clinical trial.

1) First, the drug is tested on healthy volunteers. This is to make sure it doesn't have any harmful side effects when the body is working normally.
2) At the start of the trial, a very low dose of the drug is given. This dose is increased little by little.
3) If these results are good, the drugs can be tested on patients (people with the illness).
4) The optimum dose is found — this is the dose of drug that is the most effective and has few side effects.
5) To test how well the drug works, patients are put into two groups...

Group 1 is given the new drug.

Group 2 is given a placebo (a substance that's like the drug being tested but doesn't do anything).

6) The doctor compares the two groups of patients to see if the drug makes a real difference.
7) Clinical trials are blind — the patient doesn't know whether they're getting the drug or the placebo.
8) In fact, they're often double-blind — neither the patient nor the doctor knows who's taken the drug and who's taken the placebo until all the results have been gathered.
9) The results of these tests aren't published until they've been through peer review. This helps to prevent false claims.

Peer review is when other scientists check the work — see page 2.

The placebo effect doesn't work with revision...

... you can't just expect to get a good mark and then magically get it. I know, I know, there's a lot of information to take in on this page, but just read it through slowly. There's nothing too tricky here — it's just a case of going over it again and again until you've got it all firmly lodged in your memory.

Warm-Up & Exam Questions

It's easy to think you've learnt everything in the section until you try the Warm-Up Questions.
Don't panic if there's a bit you've forgotten, just go back over that bit until it's firmly fixed in your brain.

Warm-Up Questions

1) How does the skin defend the body against pathogens?
2) What is the role of the immune system?
3) What is meant by the term 'antigen'?
4) How do bacteria become resistant to antibiotics?
5) What is meant by the efficacy of a drug?

Exam Questions

1 Many of the drugs used today first came from plants or microorganisms.

1.1 Draw **one** line from each drug to where it was originally extracted from.

Aspirin		Mould
Digitalis		Willow
Penicillin		Foxgloves

[2 marks]

1.2 State what aspirin is used for.

[1 mark]

1.3 State what penicillin is used for.

[1 mark]

2 The human immune system fights pathogens using a number of different methods.
One process for destroying pathogens is shown in **Figure 1**.

Figure 1

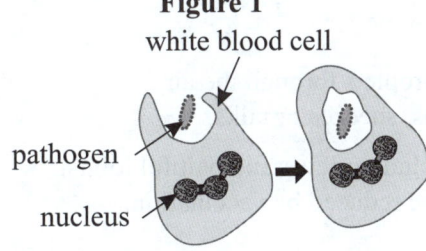

2.1 Name the process shown in **Figure 1**.

[1 mark]

2.2 Antibodies play a role in the immune response. Complete the sentences about antibodies.

Antibodies are produced by ....

They attach to specific antigens on the surface of the ....

[2 marks]

Topic B3 — Infection and Response

Exam Questions

3 A scientist is carrying out a clinical trial. *(Grade 3-4)*

3.1 What is a drug tested on in a clinical trial? Tick **one** box.

☐ human cells
☐ human volunteers
☐ live animals
☐ human tissue

[1 mark]

3.2 The clinical trial is double-blind. What is meant by 'double-blind'? Tick **one** box.

☐ The patient does not know whether they are receiving the drug or placebo, but the doctor does.

☐ All the patients are given a placebo first, followed by the drug.

☐ Neither the patient or the doctor know who is receiving the drug and who is receiving the placebo.

☐ All the patients are given the drug first, followed by a placebo.

[1 mark]

4 There are many different lines of defence in the human body that help to prevent pathogens from entering the blood. *(Grade 4-5)*

4.1 What is the role of the hairs and mucus in the nose?

[1 mark]

4.2 How do the cilia in the trachea and bronchi help to defend the body?

[1 mark]

4.3 What does the stomach produce to kill pathogens?

[1 mark]

5 Rubella is a communicable viral disease. *(Grade 4-5)*

The rubella virus is spread in droplets through the air when an infected person coughs, sneezes or talks.

The virus causes symptoms including fever and painful joints. The spread of the disease can be reduced by vaccination.

5.1* Explain how being vaccinated against rubella can prevent a person from catching the disease. In your answer, suggest why vaccinating a large number of people reduces the risk of someone who hasn't been vaccinated from catching rubella.

[6 marks]

5.2 Suggest **one** reason why some individuals may choose not to receive a vaccination against a disease.

[1 mark]

Topic B4 — Bioenergetics

Photosynthesis

*First, the **photosynthesis equation**. Then onto how plants use **glucose**...*

Photosynthesis Produces Glucose Using Light

1) Photosynthesis uses energy to change carbon dioxide and water into glucose and oxygen.
2) It takes place in chloroplasts in plant cells.
3) Chloroplasts contain chlorophyll that absorbs light.
4) Energy is transferred to the chloroplasts from the environment by light.
5) Photosynthesis is an endothermic reaction.
   This means that energy is transferred from the environment during the reaction.
6) You need to learn the word equation for photosynthesis:

carbon dioxide + water $\xrightarrow{\text{light}}$ glucose + oxygen

7) You also need to know the chemical symbols for the substances involved in photosynthesis:

carbon dioxide: CO_2 water: H_2O glucose: $C_6H_{12}O_6$ oxygen: O_2

Plants Use Glucose in Five Main Ways...

1) For respiration — This transfers energy from glucose (see p.91). This allows the plants to change the rest of the glucose into other useful substances.

2) For making cell walls — Glucose is changed into cellulose for making strong plant cell walls (see p.18).

3) For making amino acids — Glucose is combined with nitrate ions to make amino acids. Nitrate ions are absorbed from the soil. Amino acids are used to make proteins.

4) Stored as oils or fats — Glucose is turned into lipids (fats and oils) for storing in seeds.

5) Stored as starch — Glucose is turned into starch and stored in roots, stems and leaves.
   • Plants can use this starch when photosynthesis isn't happening.
   • Starch is insoluble (it can't be dissolved).
   • Being insoluble makes starch much better for storing than glucose. This is because a cell with lots of glucose in would draw in loads of water and swell up.

The Rate of Photosynthesis

The **rate** of photosynthesis can be **affected** by **a few** different things. These are called **limiting factors**...

Light, Temperature and CO₂ Affect the Rate of Photosynthesis

1) The rate of photosynthesis is affected by intensity of light (how bright the light is), concentration of CO_2 and temperature.

2) Any of these things can become the limiting factor of photosynthesis.

3) A limiting factor is something that stops photosynthesis from happening any faster.

4) Chlorophyll can also be a limiting factor of photosynthesis.
   - The amount of chlorophyll in a plant can be affected by disease.
   - It can also be affected by changes in the environment, such as a lack of nutrients.
   - These factors can cause chloroplasts to become damaged or to not make enough chlorophyll.
   - This means they can't absorb as much light. The rate of photosynthesis is reduced.

Three Important Graphs for Rate of Photosynthesis

1. Not Enough Light Slows Down the Rate of Photosynthesis

1) At first, the more light there is, the faster photosynthesis happens.

2) This means the rate of photosynthesis depends on the amount of light. Light is the limiting factor.

3) After a certain point the graph flattens out. Here photosynthesis won't go any faster — even if you increase the light intensity.

4) This is because light is no longer the limiting factor. Now it's either the temperature or the amount of carbon dioxide that's the limiting factor.

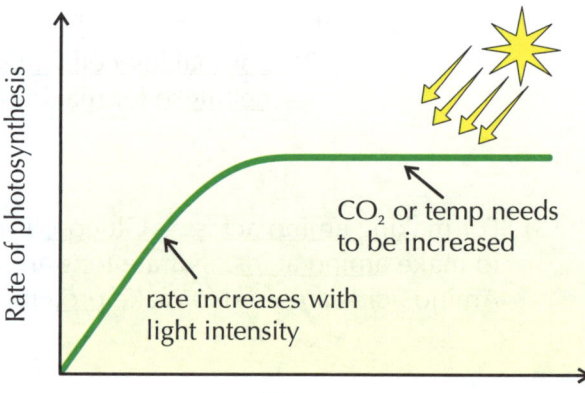

The amount of light is called light intensity.

'Photo' means light and 'synthesis' means putting together...

...so photosynthesis means 'putting together using light'. And the thing being put together is glucose. Well, I guess that's one way of remembering it... (Maybe just learn the word equation instead.)

Topic B4 — Bioenergetics

The Rate of Photosynthesis

2. Too Little Carbon Dioxide Also Slows it Down

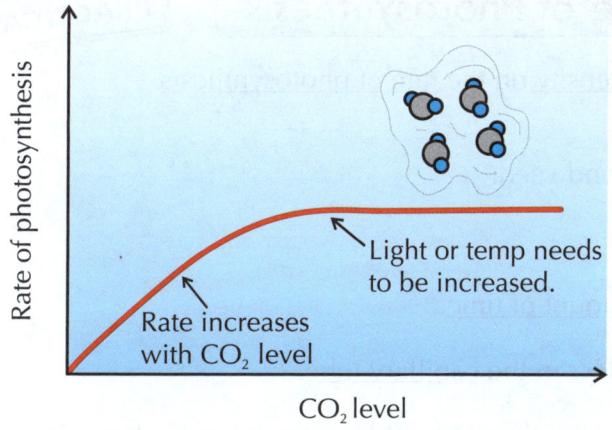

1) The more carbon dioxide (CO_2) there is, the faster photosynthesis happens.
2) This means the amount of CO_2 is the limiting factor.
3) After a certain point, photosynthesis won't go any faster because CO_2 is no longer the limiting factor.
4) If there's plenty of light and carbon dioxide then it must be the temperature that's the limiting factor.

3. The Temperature has to be Just Right

1) Usually, if the temperature is the limiting factor it's because it's too low.

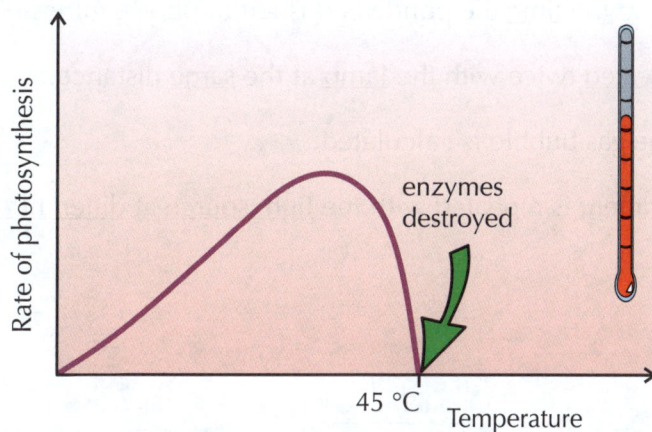

2) This is because the enzymes (see page 41) needed for photosynthesis work more slowly at low temperatures.

3) But if the plant gets too hot, photosynthesis won't happen at all.

4) This is because the enzymes are damaged if the temperature's too high (over about 45 °C).

Graphs, graphs and more graphs

Q1 a) The graph on the right shows how light intensity affects the rate of photosynthesis. State at which point (A-D) the light intensity stops being the limiting factor. [1 mark]
b) Suggest what the limiting factor could be at point D. [1 mark]

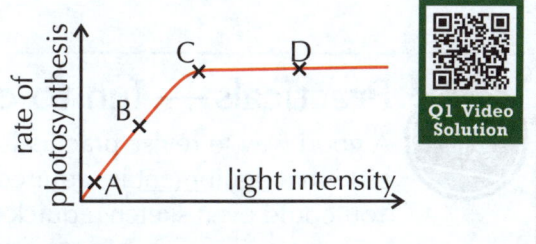

Topic B4 — Bioenergetics

Measuring the Rate of Photosynthesis

It's practical time again. This one lets you see how changing **light intensity** affects the **rate of photosynthesis**.

Oxygen Production Shows the Rate of Photosynthesis

Pondweed can be used to measure the effect of light intensity on the rate of photosynthesis. Here's how the experiment works:

1) A ruler is used to measure a set distance from the pondweed.

2) A light is placed at that distance.

3) The pondweed is left to photosynthesise for a set amount of time.

4) As it photosynthesises, the oxygen released will collect in the capillary tube.

5) At the end of the experiment, the syringe is used to draw the gas bubble in the tube up alongside a ruler.

6) The length of the gas bubble is measured.

7) The length of the gas bubble tells you how much oxygen has been produced during that amount of time. This means that the longer the gas bubble, the faster the rate of photosynthesis.

8) For this experiment, any variables that could affect the results should be controlled. E.g. the temperature and the time the pondweed is left to photosynthesise.

9) The experiment is repeated twice with the lamp at the same distance.

10) The mean length of the gas bubble is calculated.

11) Then the whole experiment is repeated with the light source at different distances from the pondweed.

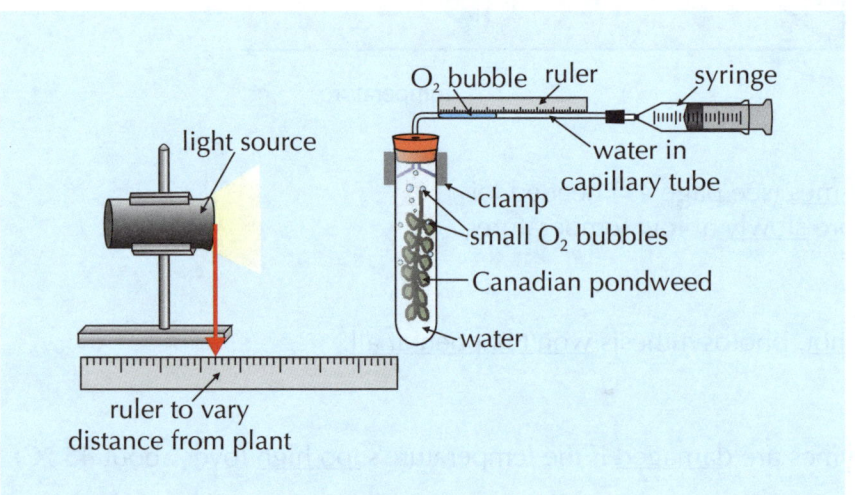

You can compare the results at different light intensities by giving the rate as the length of the bubble per unit time, e.g. cm/min.

 **Practicals — fun to carry out, not so fun to answer questions on**
A good way to revise practicals is to scribble down as much as you can remember. List what equipment is required and write out the method step-by-step. You could even sketch a quick diagram of the apparatus all set up and practise labelling it.

Topic B4 — Bioenergetics

Warm-Up & Exam Questions

Time for a break in the topic and some questions. Do them now, whilst all that learning is fresh in your mind. Using that knowledge will help you to remember it all, and that's what this game is all about.

Warm-Up Questions

1) What substance do chloroplasts contain that absorbs light?
2) Where does the energy for photosynthesis come from?
3) True or false? Photosynthesis is an exothermic reaction.
4) True or false? Glucose is used for respiration.

Exam Questions

1 Photosynthesis involves a number of substances. *Grade 1-3*

1.1 Draw **one** line from each substance to its chemical symbol.

glucose O_2

oxygen H_2O

water $C_6H_{12}O_6$

[2 marks]

1.2 Carbon dioxide is another substance involved in photosynthesis.
What is the chemical symbol for carbon dioxide?

[1 mark]

2 Photosynthesis produces glucose using light. *Grade 3-4*

2.1 Complete the word equation for photosynthesis.

                                  light
carbon dioxide + ... → glucose + ...

[2 marks]

2.2 Plants use glucose to make a substance which strengthens their cell walls.
Which of the following substances strengthens cells walls? Tick **one** box.

☐ cellulose   ☐ oils   ☐ starch   ☐ fats

[1 mark]

2.3 Which of the following is another way that plants use glucose? Tick **one** box.

☐ making nitrates   ☐ storage as oils   ☐ making chlorophyll

[1 mark]

Topic B4 — Bioenergetics

Exam Questions

3 Plants store glucose as starch. *Grade 3-4*

3.1 Which of the following is a characteristic of starch that makes it suitable for storage? Tick **one** box.

☐ it contains carbon ☐ it is soluble ☐ it is insoluble ☐ starch molecules are large

[1 mark]

3.2 Explain why plants use starch as a source of glucose when photosynthesis is **not** happening.

[1 mark]

4 A student investigated the effect of increasing carbon dioxide concentration on the rate of photosynthesis of a plant. The results are shown in **Figure 1**. *Grade 4-5*

Figure 1

(Graph showing rate of photosynthesis vs carbon dioxide concentration, with point A on the rising portion and point B on the plateau)

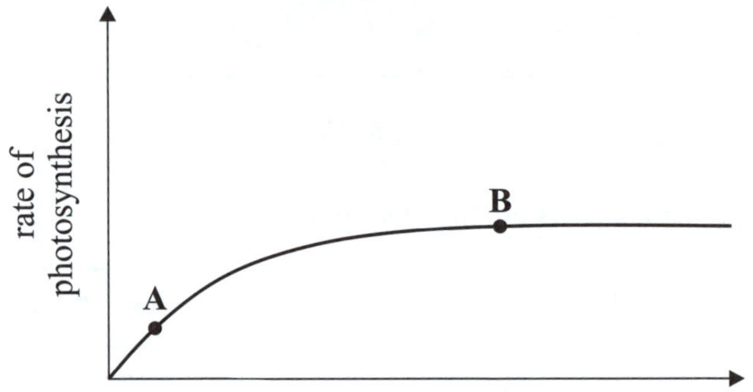

4.1 Describe the trend shown on the graph at point **A**.

[1 mark]

4.2 Give the limiting factor at point **A** on the curve.

[1 mark]

4.3 Explain why the curve is flattening out at point **B**.

[2 marks]

4.4 Explain why low temperatures limit the rate of photosynthesis.

[1 mark]

PRACTICAL

5 A student did an experiment to see how the rate of photosynthesis depends on light intensity. **Figure 2** shows some of her apparatus. *Grade 4-5*

5.1 How can the student measure the rate of photosynthesis?

[1 mark]

Figure 2 (diagram showing test tube with gas bubbles, light source arrow pointing to pond plant)

5.2 State the dependent variable in this experiment.

[1 mark]

5.3 State the independent variable in this experiment.

[1 mark]

5.4 State **one** factor that should be kept constant during this experiment.

[1 mark]

Topic B4 — Bioenergetics

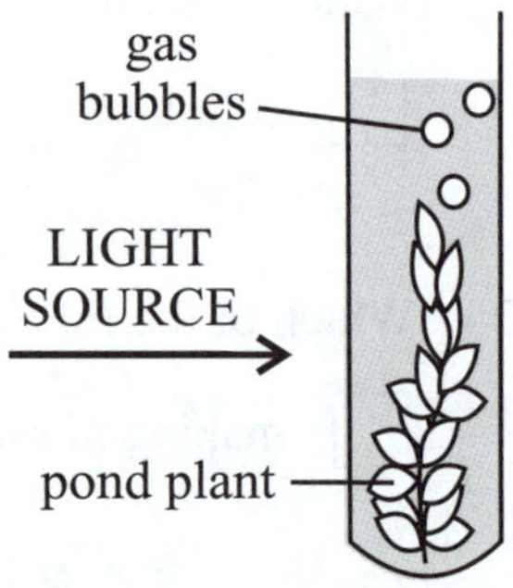

Aerobic Respiration

*You need **energy** to keep your body going. Energy comes from **food**, and it's **transferred** by **respiration**.*

Respiration is NOT "Breathing In and Out"

1) All living things respire.
2) Respiration is the process of transferring energy from the breakdown of glucose (a sugar).
3) Respiration goes on in every cell in your body all the time.
4) The energy transferred from respiration is used for all living processes (everything a cell needs to do).

> RESPIRATION is the process of TRANSFERRING ENERGY FROM GLUCOSE, which goes on IN EVERY CELL.

5) Respiration is exothermic. This means it transfers energy to the environment.

Respiration Transfers Energy for All Kinds of Things

Here are three examples of how organisms use the energy transferred by respiration:

1) To build up larger molecules from smaller ones.
2) In animals, to move about.
3) In mammals and birds, to keep warm.

Aerobic Respiration Needs Plenty of Oxygen

1) Aerobic respiration is respiration using oxygen.
2) Aerobic respiration goes on all the time in plants and animals.
3) Most of the reactions in aerobic respiration happen inside mitochondria (see page 17).
4) You need to learn the overall word equation for respiration:

> glucose + oxygen ⟶ carbon dioxide + water

5) You also need to know the chemical symbols for the substances involved:
glucose: $C_6H_{12}O_6$ oxygen: O_2 carbon dioxide: CO_2 water: H_2O

Topic B4 — Bioenergetics

Anaerobic Respiration

Anaerobic respiration is just as important as aerobic respiration — especially when there's **not enough oxygen**...

Anaerobic Respiration is Used if There's Not Enough Oxygen

1) When you do hard exercise, your body sometimes can't supply enough oxygen to your muscles.
2) When this happens, they start doing anaerobic respiration as well as aerobic respiration.
3) Anaerobic respiration is the incomplete breakdown of glucose (the glucose isn't broken down properly).
4) Here's the word equation for anaerobic respiration in muscle cells:

glucose ⟶ lactic acid

5) Anaerobic respiration does not transfer anywhere near as much energy as aerobic respiration.
6) This is because the glucose has not combined with oxygen like it does in aerobic respiration.
7) The posh way of saying this is that the oxidation of glucose is not complete.

Anaerobic Respiration in Plants and Yeast is Slightly Different

1) Plants and yeast cells can respire without oxygen too.
2) Here is the word equation for anaerobic respiration in plants and yeast cells:

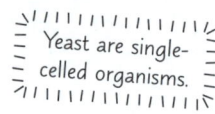

Yeast are single-celled organisms.

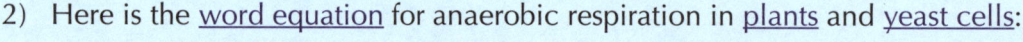

glucose ⟶ ethanol + carbon dioxide

3) Anaerobic respiration in yeast cells is called fermentation.
4) In the food and drinks industry, fermentation by yeast is of great value.
5) It's used to make bread. It's the carbon dioxide from fermentation that makes bread rise.
6) It's used to make alcoholic drinks (beer and wine). It's the fermentation process that produces alcohol.

Respiration releases energy from glucose

So... respiration is a pretty important thing — the energy transferred from glucose is used to make molecules that our cells need. When it comes to this topic, make sure you know the word equations from this page and the previous page and can compare the processes of aerobic and anaerobic respiration.

Exercise

*When you **exercise**, your body responds in different ways to get enough **energy** to your **cells**.*

When You Exercise You Respire More

1) Muscles need energy from respiration to contract (shorten).
2) When you exercise, some of your muscles contract more often. This means you need more energy.
3) This energy comes from increased respiration.
4) The increase in respiration in your cells means you need to get more oxygen into them. To do this:

   1) Your breathing rate (how fast you breathe) increases.
   2) Your breath volume (how deep the breaths you take are) increases.
   3) Your heart rate (how fast your heart beats) increases.

5) Increasing your breathing rate and breath volume gets oxygen into your blood quicker. Blood containing oxygen is called oxygenated blood.
6) Your heart rate increases to get this oxygenated blood around the body faster.

An unfit person's heart rate goes up a lot more during exercise than a fit person, and they take longer to recover.

Hard Exercise Can Lead to Anaerobic Respiration

1) When you do really hard exercise, your body can't supply oxygen to your muscles quickly enough.
2) This means your muscles start doing anaerobic respiration (see the previous page).
3) This is NOT the best way to transfer energy from glucose. This is because lactic acid builds up in the muscles, which gets painful.
4) Long periods of exercise also cause muscle fatigue. This is when the muscles get tired and stop contracting efficiently.

Anaerobic Respiration Leads to an Oxygen Debt

1) After anaerobic respiration stops, you'll have an "oxygen debt".
2) An oxygen debt is the amount of extra oxygen your body needs after exercise.
3) Your lungs, heart and blood couldn't keep up with the demand for oxygen earlier on. So you have to "repay" the oxygen that you didn't get to your muscles in time.
4) This means you have to keep breathing hard for a while after you stop.
5) This gets more oxygen into your blood, which is transported to the muscle cells.

Very hard exercise = anaerobic respiration = oxygen debt

Q1 A scientist measured the concentration of lactic acid in her blood after walking for 5 minutes. She also measured the concentration of lactic acid in her blood after running for 5 minutes. Suggest why the concentration of lactic acid in her blood was higher after running than after walking.

[3 marks]

Topic B4 — Bioenergetics

Metabolism

*Metabolism is going on **all of the time**. Right now. And now. Even now. Okay, you get the picture. Time to read all about it.*

Metabolism is ALL the Chemical Reactions in an Organism

1) In a cell there are lots of chemical reactions happening all the time.

2) These reactions are controlled by enzymes.

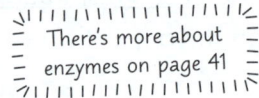
There's more about enzymes on page 41

3) In some of these reactions, larger molecules are made from smaller ones. For example:

> 1) Lots of small glucose (sugar) molecules are joined together in reactions to form:
> - starch (a storage molecule in plant cells),
> - glycogen (a storage molecule in animal cells),
> - cellulose (a component of plant cell walls).
>
> 2) Lipid molecules are each made from one molecule of glycerol and three fatty acids.
>
> 3) Glucose is combined with nitrate ions to make amino acids. These are then made into proteins.

4) In other reactions, larger molecules are broken down into smaller ones. For example:

> 1) Glucose is broken down in respiration.
> - Respiration transfers energy to power all the reactions in the body that make molecules.
>
> 2) Excess protein is broken down in a reaction to produce urea. Urea is then excreted in urine.

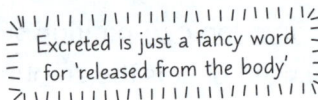
Excreted is just a fancy word for 'released from the body'

5) The sum (total) of all of the reactions that happen in a cell or the body is called its metabolism.

It's still going on now

Remember, the energy for metabolism comes from respiration. This energy allows cells to make larger molecules from smaller ones, and to break larger molecules down into smaller ones. Enzymes are key to metabolism, so if you need a reminder about them, now is a good time to head back to page 41.

Topic B4 — Bioenergetics

Warm-Up & Exam Questions

You know the drill by now — work your way through the Warm-Up questions, then the Exam Questions.

Warm-Up Questions

1) True or false? All living things respire.
2) What is anaerobic respiration in yeast cells called?
3) Give one way that the body gets more oxygen into cells during exercise.
4) What is excess protein broken down into?

Exam Questions

1 In the human body, respiration may be aerobic or anaerobic at different times. *Grade 3-4*

1.1 Which of the following is the word equation for anaerobic respiration in humans. Tick **one** box.

☐ glucose → lactic acid + carbon dioxide ☐ glucose → lactic acid

☐ glucose → ethanol + carbon dioxide ☐ glucose → ethanol

[1 mark]

1.2 The body uses anaerobic respiration during hard exercise.
Complete the sentences about anaerobic respiration during exercise.

Muscles start using anaerobic respiration when they don't get enough ....

This causes a build up of ....

After anaerobic respiration stops, the body is left with an oxygen ....

[3 marks]

2 Metabolism is a process in which larger molecules are made or broken down. *Grade 3-4*

Which of the following molecules is **not** formed by joining lots of glucose molecules together? Tick **one** box.

☐ starch ☐ glycogen ☐ protein ☐ cellulose

[1 mark]

3 Respiration is a process carried out by all living cells.
It can take place aerobically or anaerobically. *Grade 4-5*

3.1 Give **two** differences between aerobic and anaerobic respiration.

[2 marks]

3.2 Complete the word equation for aerobic respiration.

glucose + ... $\xrightarrow{\text{light}}$ carbon dioxide + ...

[2 marks]

Topic B4 — Bioenergetics

Revision Summary for Topics B1-4

Well, it's all over for Topics B1-4 folks — I know you'll miss them, so here are some questions on them...
- Try these questions and tick off each one when you get it right.
- When you're completely happy with a topic, tick it off.

For even more practice, try the Retrieval Quizzes for Topics B1-4 — just scan the QR codes!

Topic B1 — Cell Biology (p.17-37) ☐
1) Name five subcellular structures that both plant and animal cells have.
2) Where is the genetic material found in: a) animals cells b) bacterial cells?
3) Give two ways that a sperm cell is adapted for swimming to an egg.
4) Give one way that embryonic stem cells could be used to cure diseases.
5) What are chromosomes?
6) What is diffusion?
7) What type of molecules move by osmosis?
8) Give three ways that exchange surfaces can be adapted for diffusion.
9) Give two ways that the villi in the small intestine are adapted for absorbing digested food.

Topic B2 — Organisation (p.40-69) ☐
10) What is a tissue?
11) What does it mean when an enzyme has been 'denatured'?
12) Describe how you could investigate the effect of pH on the rate of amylase activity.
13) Name the solution that you would use to test for the presence of sugars in a food sample.
14) Where does gas exchange happen in the lungs?
15) How are arteries adapted to carry blood away from the heart?
16) Why do red blood cells not have a nucleus?
17) What is the difference between biological and mechanical replacement heart valves?
18) Give an example of different types of disease interacting in the body.
19) Give two lifestyle factors that increase the chance of cancer.
20) Explain how the structure of the palisade layer in a leaf is related to its function.
21) What is the function of phloem tubes?

Topic B3 — Infection and Response (p.72-82) ☐
22) Give one way that pathogens can be spread.
23) Why does tobacco mosaic virus affect photosynthesis?
24) What are the vectors for malaria?
25) Give three ways that white blood cells can defend against pathogens.
26) Why is it difficult to develop drugs that kill viruses?

Topic B4 — Bioenergetics (p.85-94) ☐
27) Where in a plant cell does photosynthesis take place?
28) Why would a low concentration of chlorophyll limit photosynthesis?
29) Describe how you could measure the effect of light intensity on the rate of photosynthesis.
30) What is the word equation for anaerobic respiration in yeast cells?
31) Name two products of the food and drink industry that fermentation is needed for.

Topic B5 — Homeostasis and Response

Homeostasis

*Homeostasis — a word that strikes fear into the heart of many a GCSE student. But it's really not that bad at all. This page is a brief **introduction** to the topic, so you need to **nail all of this** before you can move on.*

Homeostasis — Keeping Conditions Inside Your Body Steady

1) Homeostasis is the fancy word for keeping the conditions in your body and cells at the right level. This happens in response to changes inside and outside of the body.
2) This is really important because your cells need the right conditions to work properly.
3) This includes having the right conditions for enzymes to work (see p.41).

Your Body Uses Control Systems for Homeostasis

1) You have loads of control systems that keep the conditions in your body steady. For example, they keep your body temperature, blood glucose level and water level steady.
2) These control systems are automatic — you don't have to think about them.
3) They can control conditions in the body using the nervous system or hormones.
4) Control systems are made up of three main parts:

   - receptors,
   - coordination centres (including the brain, spinal cord and pancreas),
   - effectors.

5) When the level of something (e.g. blood glucose) gets too high or too low, its control system brings it back to normal.

> If the level is too HIGH, the control system DECREASES the level.
> If the level is too LOW, the control system INCREASES the level.

6) Here's how a control system works:

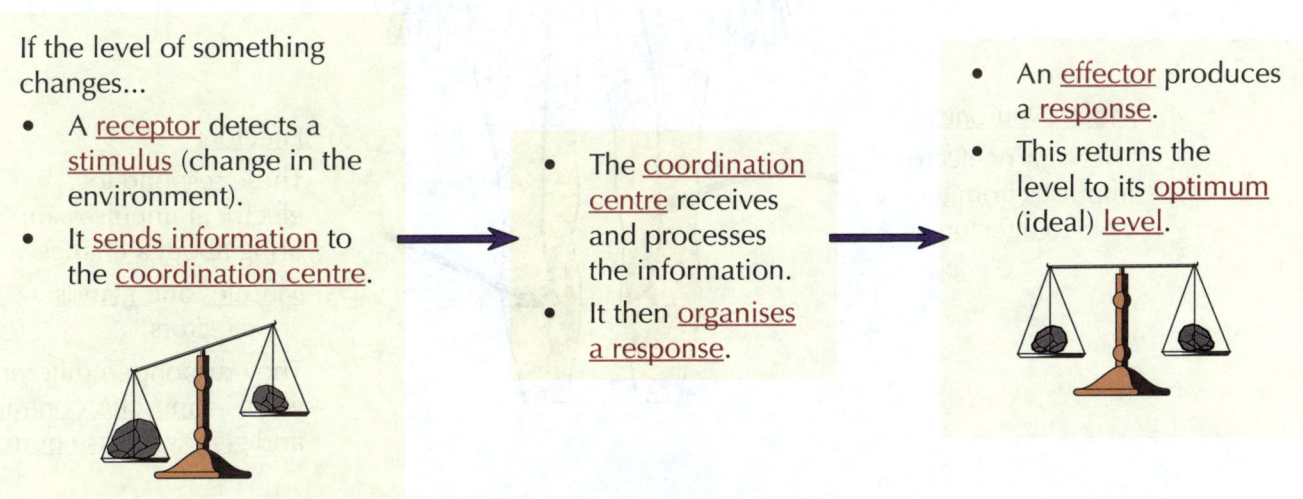

If the level of something changes...
- A receptor detects a stimulus (change in the environment).
- It sends information to the coordination centre.

- The coordination centre receives and processes the information.
- It then organises a response.

- An effector produces a response.
- This returns the level to its optimum (ideal) level.

Homeostasis is always happening without us thinking about it

Homeostasis is really important for keeping processes in your body working. It does this by keeping everything at the right level. Make sure you know what receptors, coordination centres and effectors do.

The Nervous System

*Organisms need to **respond to stimuli** (changes in the environment). That's where the **nervous system** comes in — it picks up information from the environment and brings about a **response**.*

The Nervous System Detects and Reacts to Stimuli

1) The nervous system means that humans can react to their surroundings and coordinate their behaviour.
2) The nervous system is made up of different parts:

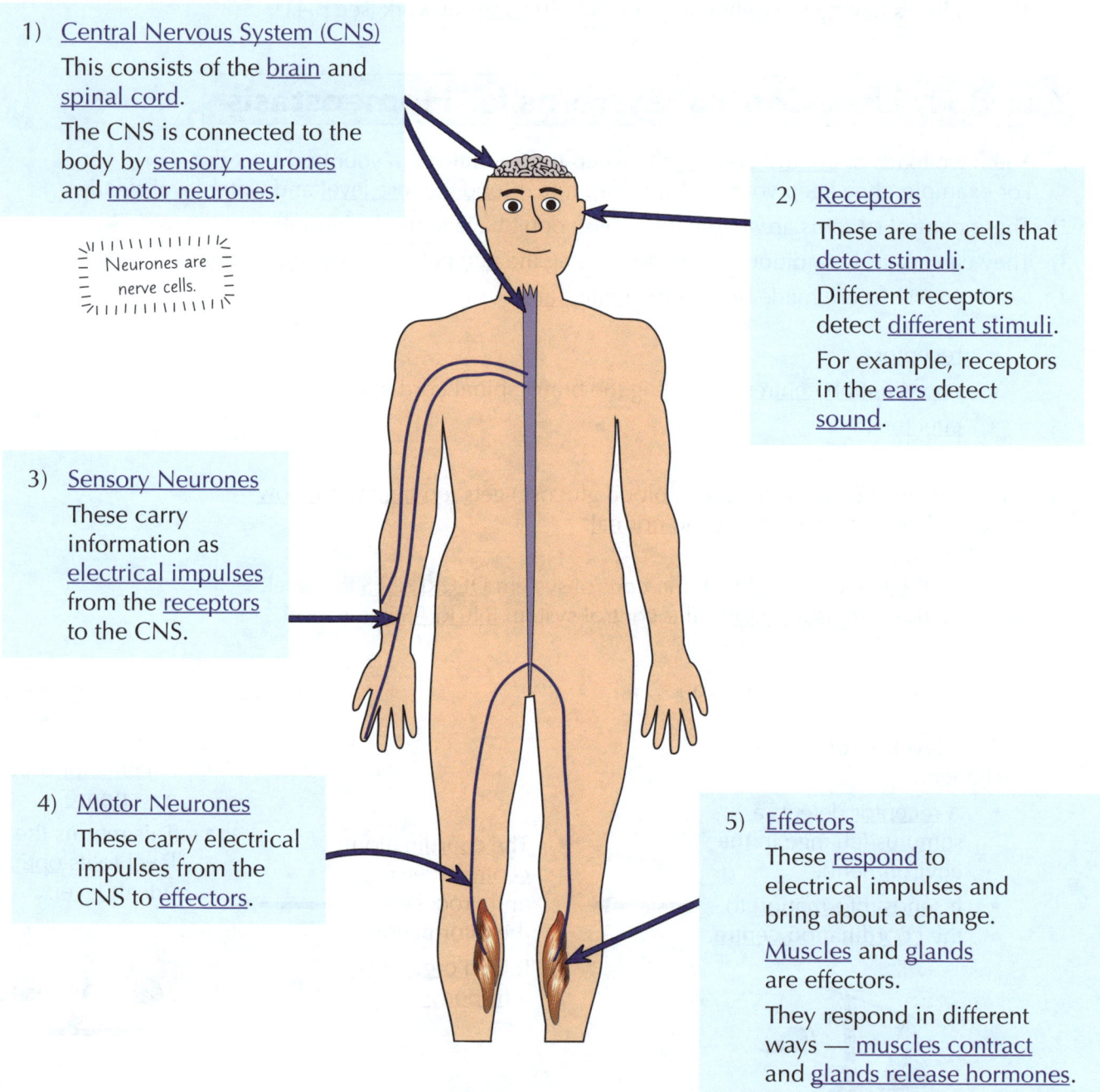

1) Central Nervous System (CNS)
   This consists of the brain and spinal cord.
   The CNS is connected to the body by sensory neurones and motor neurones.

   Neurones are nerve cells.

2) Receptors
   These are the cells that detect stimuli.
   Different receptors detect different stimuli.
   For example, receptors in the ears detect sound.

3) Sensory Neurones
   These carry information as electrical impulses from the receptors to the CNS.

4) Motor Neurones
   These carry electrical impulses from the CNS to effectors.

5) Effectors
   These respond to electrical impulses and bring about a change.
   Muscles and glands are effectors.
   They respond in different ways — muscles contract and glands release hormones.

Learn the different parts of the nervous system

Don't be confused by the terms 'nervous system' and 'central nervous system'. The 'nervous system' includes all of the parts above, but the 'central nervous system' means just the brain and spinal cord.

Topic B5 — Homeostasis and Response

The Nervous System

Now for some detail about how the parts of the nervous system on the previous page work together.

The Central Nervous System (CNS) Coordinates the Response

1) The CNS is a coordination centre.
2) It receives information from the receptors and then coordinates a response (decides what to do about it).
3) The response is carried out by effectors.

> For example, a small bird is eating some seed...
> 1) ...when it spots a cat coming towards it (this is the stimulus).
> 2) The receptors in the bird's eye are stimulated (activated).
> 3) Sensory neurones carry the information from the receptors to the CNS.
> 4) The CNS decides what to do about it.
> 5) The CNS sends information to the muscles in the bird's wings (the effectors) along motor neurones.
> 6) The muscles contract and the bird flies away to safety.
>
> Stimulus → Receptor → Sensory neurone → CNS → Motor neurone → Effector → Response
>
> receptors in the eye

Synapses Connect Neurones

1) A synapse is where two neurones join together.
2) The electrical impulse is passed from one neurone to the next by chemicals.
3) These chemicals move across the gap.
4) The chemicals set off a new electrical impulse in the next neurone.

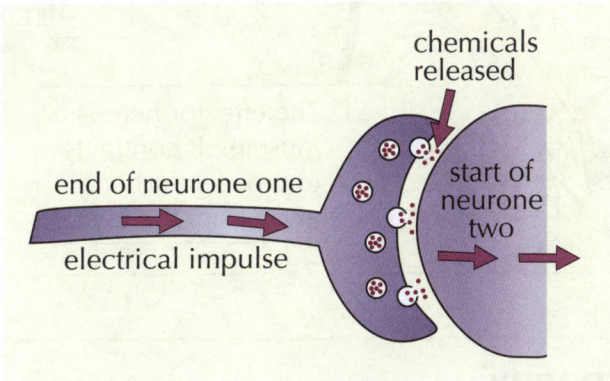

end of neurone one — electrical impulse — chemicals released — start of neurone two

Don't let the thought of exams play on your nerves...

Cover up the page and practise writing out the order of events from stimulus to response. Pay attention to names of the neurones — you don't want to be getting them mixed up in the exam.

Topic B5 — Homeostasis and Response

The Nervous System

Neurones transmit information **very quickly** to and from the brain, and your brain **quickly decides** how to respond to a stimulus. But **reflexes** are even quicker...

Reflexes Help Prevent Injury

1) Reflexes are automatic responses — you don't have to think about them.

   Reflexes are also called reflex reactions.

2) This makes them really quick.

3) They can help stop you getting injured.

4) The passage of information in a reflex (from receptor to effector) is called a reflex arc.

5) The neurones in reflex arcs go through the spinal cord or through an unconscious part of the brain (part of the brain not involved in thinking).

Here's an example of how a reflex arc would work if you were stung by a bee:

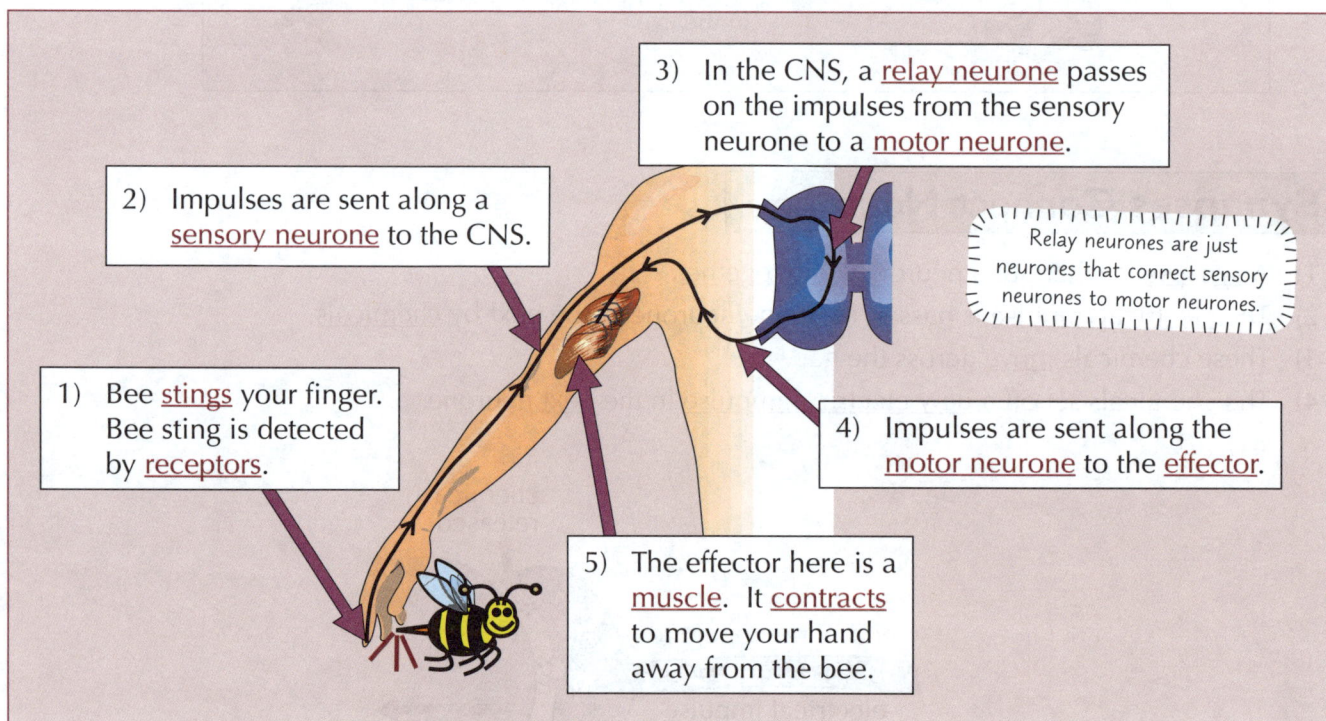

1) Bee stings your finger. Bee sting is detected by receptors.
2) Impulses are sent along a sensory neurone to the CNS.
3) In the CNS, a relay neurone passes on the impulses from the sensory neurone to a motor neurone.
4) Impulses are sent along the motor neurone to the effector.
5) The effector here is a muscle. It contracts to move your hand away from the bee.

Relay neurones are just neurones that connect sensory neurones to motor neurones.

Reflexes are super speedy

Reflexes don't involve your conscious brain at all when a quick response is essential.

Q1 What is a reflex action? [1 mark]

Q2 A chef touches a hot pan. A reflex reaction causes him to immediately move his hand away.
    a) State the effector in this reflex reaction. [1 mark]
    b) Describe the pathway of the reflex from stimulus to effector. [4 marks]

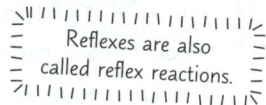

Q2 Video Solution

Investigating Reaction Time — PRACTICAL

*Reaction time is the time it takes to **respond to a stimulus** — it's often **less** than a **second**. It can be **affected** by factors such as **age**, **gender** or **drugs**.*

Reaction Time is How Quickly You Respond

1) Reaction time is the time it takes to respond to a stimulus.
2) It's often less than a second. This means it may be measured in milliseconds (ms).
3) It can be affected by factors such as age, gender or drugs.

You Can Measure Reaction Time

Caffeine is a drug. It can speed up a person's reaction time.
The effect of caffeine on reaction time can be measured like this...

1) The person being tested should sit with their arm resting on the edge of a table.
2) Hold a ruler upright between their thumb and forefinger. Make sure that the zero end of the ruler is level with their thumb and finger. Don't let them grip the ruler.
3) Then let go without giving any warning.
4) The person being tested should try to catch the ruler as quickly as they can.
5) Reaction time is measured by the number on the ruler where it's caught.

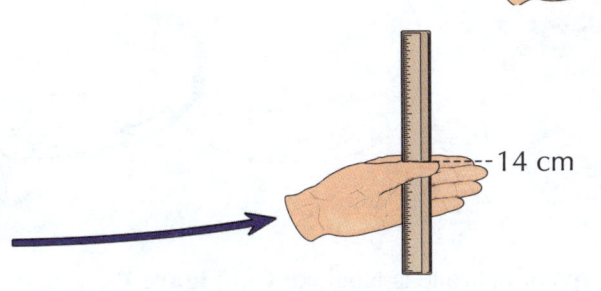

- The number should be read from the top of the person's thumb.
- The higher the number, the slower their reaction time.

6) Repeat the test several times then calculate the mean distance that the ruler fell.
7) Now give the person being tested a caffeinated drink (e.g. cola).
8) After 10 minutes, repeat steps 1 to 6.
9) You need to control any variables to make sure that this is a fair test. For example:

- Use the same person to catch the ruler each time.
- That person should always use the same hand to catch the ruler.
- The ruler should always be dropped from the same height.

As with any practical, you need to control the variables

Q1 Some students investigated the effect of an energy drink on reaction time. They measured their reaction times using a computer test. They had to click the mouse when the screen changed from red to green. Each student repeated the test five times before having an energy drink, and five times afterwards.
  a) The results for one of the students before having the energy drink were as follows:
     242 ms, 256 ms, 253 ms, 249 ms, 235 ms. Calculate the mean reaction time. [2 marks]
  b) Suggest two variables that the students needed to control during their investigation. [2 marks]

Q1 Video Solution

Topic B5 — Homeostasis and Response

Warm-Up & Exam Questions

Welcome to some questions. There are quite a few of them, but that's because they're pretty important...

Warm-Up Questions

1) State one thing that is controlled by homeostasis in the human body.
2) What name is given to the connection between two neurones?
3) True or false? You don't have to think about reflex responses.
4) Students are investigating reaction time by measuring how quickly a ruler can be caught before and after drinking caffeine. Give one variable that would have to be controlled.

Exam Questions

1   A man picked up a plate in the kitchen without realising it was hot, then immediately dropped it. **Figure 1** shows the reflex arc for this incident.

Figure 1

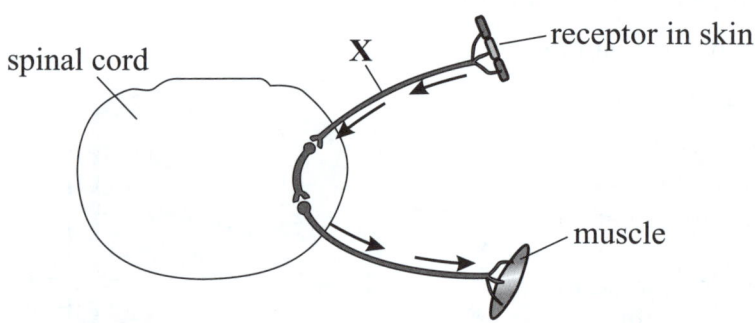

What type of neurone is labelled **X** in **Figure 1**?
Tick **one** box.

☐ motor neurone   ☐ sensory neurone   ☐ relay neurone

[1 mark]

PRACTICAL

2   A student is taking part in an experiment to test reaction times. Every time a red triangle appears on the computer screen in front of her, she has to click the mouse.

2.1   Suggest what the stimulus is in this experiment.

[1 mark]

2.2   Suggest what the receptors are in this experiment.

[1 mark]

2.3   Suggest what the effectors are in this experiment.

[1 mark]

2.4   The student took the test three times. Her reaction time in test 1 was 328 ms. Her reaction time in test 2 was 346 ms. Her mean reaction time was 343 ms. Calculate her reaction time for test 3.

[2 marks]

Topic B5 — Homeostasis and Response

The Endocrine System

*The other way to send information around the body (apart from along nerves) is by using **hormones**.*

Hormones Are Chemical Messengers Sent in the Blood

1) Hormones are chemicals released by glands. They're released directly into the blood.

2) These glands are called endocrine glands. They make up your endocrine system.

3) Hormones are carried in the blood to other parts of the body.

4) They only affect particular cells in particular organs (called target organs).

Hormones and Nerves Have Differences

Hormones and nerves do similar jobs — they both carry information and instructions around the body. But there are some important differences between them:

Nerves

1) Very FAST action.

2) Act for a very SHORT TIME.

3) Act on a very PRECISE AREA.

Hormones

1) SLOWER action.

2) Act for a LONG TIME.

3) Act in a more GENERAL way.

If you're not sure whether a response is nervous or hormonal, have a think about the speed of the reaction and how long it lasts.

Nerves, hormones — no wonder revision makes me tense...

Hormones control various organs and cells in the body, though they tend to control things that aren't immediately life-threatening (so things like sexual development, blood sugar level, water content, etc.).

Topic B5 — Homeostasis and Response

The Endocrine System

Hormones are released by **endocrine glands**. There are a few examples you need to learn on this page.

Endocrine Glands Are Found in Different Places in The Body

PITUITARY GLAND
1) Sometimes called the 'master gland'.
2) This is because it produces many hormones that regulate body conditions.
3) These hormones act on other glands. They make the glands release hormones that bring about change.

THYROID
1) Produces thyroxine.
2) This is involved in regulating things like the rate of metabolism, heart rate and temperature.

OVARIES (females only)
1) Produce oestrogen.
2) This is involved in the menstrual cycle (see page 108).

ADRENAL GLAND
1) Produces adrenaline.
2) This is used to prepare the body for a 'fight or flight' response.

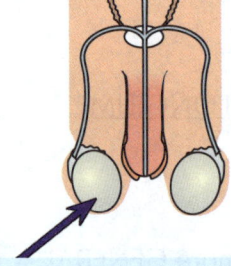

THE PANCREAS
1) Produces insulin.
2) This is used to regulate the blood glucose level (see next page).

TESTES (males only)
1) Produce testosterone.
2) This controls puberty and sperm production in males (see page 108).

 **You need to know where these glands are in the body**
Draw a rough outline of the human body and see if you can draw each of these endocrine glands onto it in the right place. Don't forget you need to remember both the testes and the ovaries.

Topic B5 — Homeostasis and Response

Controlling Blood Glucose

*You should remember from page 97 that **homeostasis** is all about keeping conditions inside the body **stable**. Blood glucose is controlled as part of homeostasis — **insulin** is an important **hormone** in this.*

Insulin Reduces the Blood Glucose Level

1) Eating carbohydrates puts glucose (a type of sugar) into the blood.

2) Glucose is removed from the blood by cells (which use it for energy).

3) When you exercise, a lot more glucose is removed from the blood.

4) Changes in the blood glucose concentration are monitored and controlled by the pancreas.

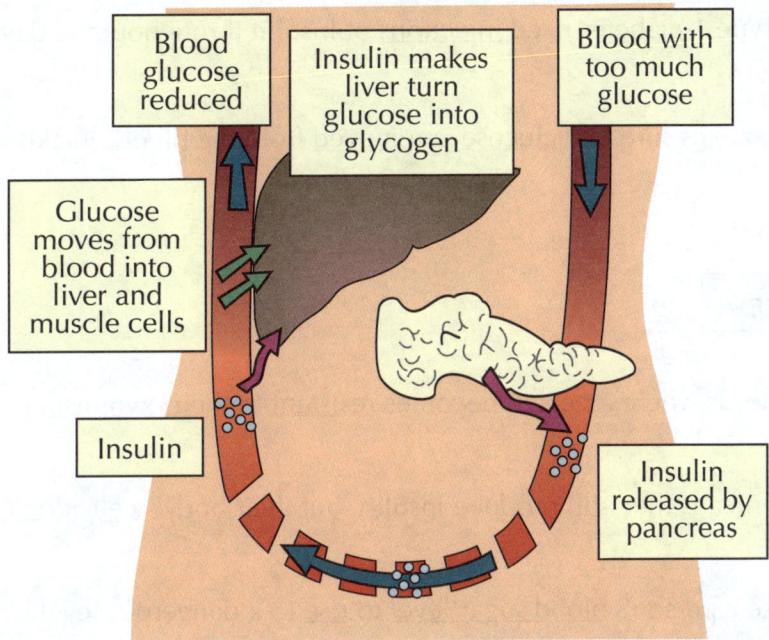

5) If blood glucose concentration gets too high, the pancreas releases the hormone insulin.

6) Insulin causes glucose to move into cells (so it removes glucose from the blood).

7) Glucose can be stored as glycogen.

8) Glucose is converted to glycogen in liver and muscle cells.

Glucose in the blood needs to be kept at a safe level

This stuff can seem a bit confusing at first, but if you learn that diagram, it should get a bit easier.

Q1 The graph shows the changes in concentration of glucose and insulin in a person's blood over time, after they ate a meal. Which curve represents insulin? Explain your answer. [2 marks]

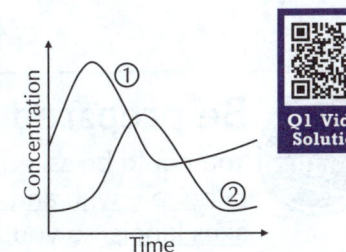

Topic B5 — Homeostasis and Response

Controlling Blood Glucose

Sometimes, homeostasis goes wrong. **Diabetes** is an example of this.

With Diabetes, You Can't Control Your Blood Sugar Level

There are two types of diabetes:

Type 1 Diabetes

1) Type 1 diabetes is where the pancreas does not produce enough insulin. It may not produce any insulin at all.

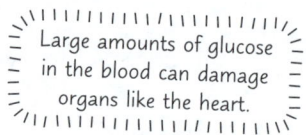
Large amounts of glucose in the blood can damage organs like the heart.

2) This means a person's blood glucose level can rise to a level that can kill them.

3) People with Type 1 diabetes need injections of insulin throughout the day.

4) This makes sure that glucose is removed from the blood quickly after the food is digested.

Type 2 Diabetes

1) Type 2 diabetes is where a person becomes resistant to their own insulin.

2) This means they still produce insulin, but their body's cells don't respond properly to it.

3) This can cause a person's blood sugar level to rise to a dangerous level.

4) Being obese (very overweight) can increase your chance of developing Type 2 diabetes.

5) Type 2 diabetes can be controlled by eating a carbohydrate-controlled diet.

6) This is a diet where the amount of carbohydrates eaten is carefully measured.

7) Type 2 diabetes can also be controlled by taking regular exercise.

EXAM TIP — **Be prepared to interpret graphs in the exam**
You could be asked to interpret a graph showing the effects of insulin on the blood sugar levels of people with and without diabetes. Don't panic — just study the graph carefully (including the axes labels) so you know exactly what it's showing you. Then apply your blood sugar knowledge.

Warm-Up & Exam Questions

If these questions don't get your adrenaline pumping, I don't know what will. Better get started...

Warm-Up Questions

1) How do hormones travel to their target organs?
2) Which gland produces thyroxine?
3) True or false? Type 2 diabetes is where the pancreas does not produce enough insulin.
4) Give one way that Type 2 diabetes can be controlled.

Exam Questions

1 Hormones are produced in endocrine glands.

Complete **Table 1** to show which endocrine glands the hormones are released from.

Table 1

Hormone	Gland the hormone is released from
Testosterone	
	Adrenal gland
Oestrogen	

[3 marks]

2 **Figure 1** shows how the body responds when the glucose concentration of the blood gets too high.

Figure 1

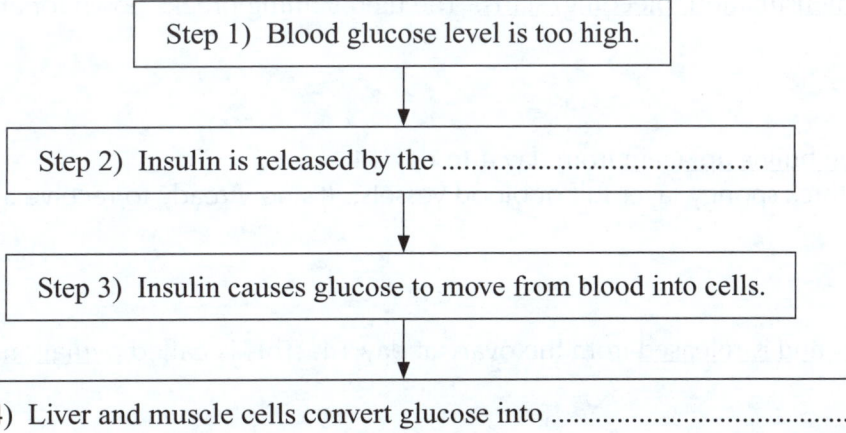

Complete **Step 2** and **Step 4** in **Figure 1**.

[2 marks]

Topic B5 — Homeostasis and Response

Puberty and the Menstrual Cycle

*The monthly **release of an egg** from a woman's ovaries is part of the **menstrual cycle**.*

Hormones Cause Sexual Characteristics To Develop at Puberty

1) At puberty, your body starts releasing sex hormones.
2) These sex hormones trigger secondary sexual characteristics. For example, the development of facial hair in men and breasts in women.
3) Female sex hormones also cause eggs to mature (develop) in women.
4) In men, the main reproductive hormone is testosterone. It's produced by the testes. It stimulates sperm production.
5) In women, the main reproductive hormone is oestrogen. It's produced by the ovaries. Oestrogen is involved in the menstrual cycle.

The Menstrual Cycle Has Four Stages

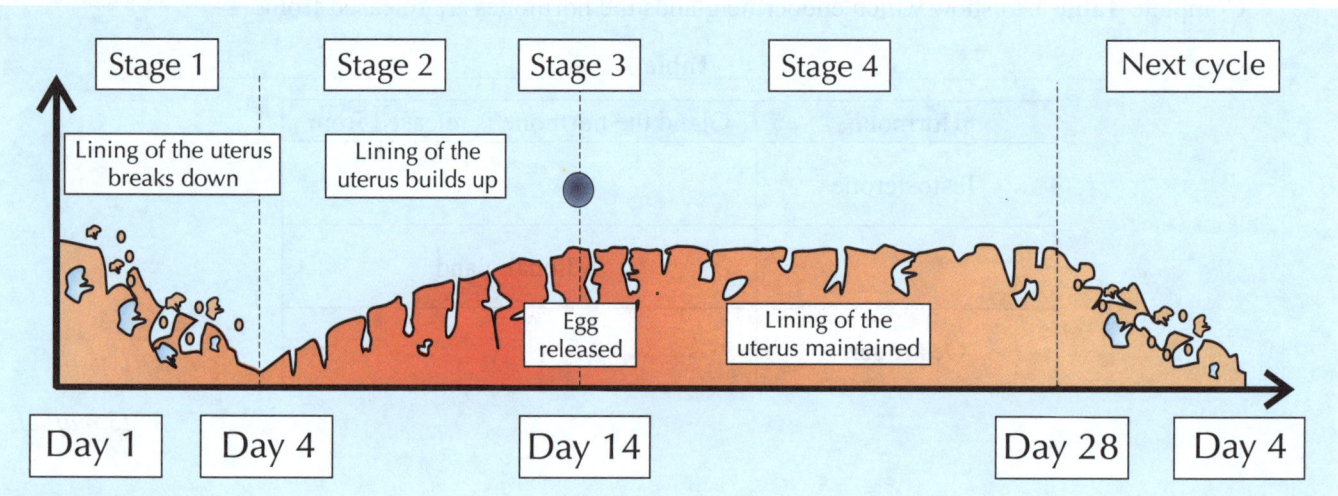

Stage 1

Day 1 is when menstruation (bleeding) starts. The uterus lining breaks down for about four days.

Stage 2

The uterus lining builds up again from day 4 to day 14.
It builds into a thick spongy layer full of blood vessels. It's now ready to receive a fertilised egg.

Stage 3

An egg develops and is released from the ovary at day 14. This is called ovulation.

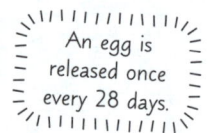

An egg is released once every 28 days.

Stage 4

The wall is then maintained (kept the same) for about 14 days until day 28.
If no fertilised egg has landed on the uterus wall by day 28,
the spongy lining starts to break down. The whole cycle starts again.

The Menstrual Cycle and Controlling Fertility

A set of **hormones** controls the **menstrual cycle**. Some of these hormones can be used to **prevent pregnancy**.

The Menstrual Cycle is Controlled by Four Hormones...

HORMONE	WHAT THE HORMONE DOES
FSH (Follicle-Stimulating Hormone)	Causes an egg to mature in one of the ovaries.
LH (Luteinising Hormone)	Causes the release of an egg (ovulation).
Oestrogen	These hormones are involved in the growth and maintenance of the uterus lining.
Progesterone	

Hormones Can Be Used to Reduce Fertility

1) Fertility is how easy it is for a woman to get pregnant.
2) Contraceptives are things that prevent pregnancy.
3) Hormones can be used in contraceptives — these are called hormonal contraceptives.

Oral Contraceptives Contain Hormones

1) Oral contraceptives are taken through the mouth as pills.

2) They stop the hormone FSH from being released.

3) This stops eggs maturing.

4) Oral contraceptives are over 99% effective at preventing pregnancy.

5) But they can have bad side effects.
   For example, they can cause headaches and make you feel sick.

The hormones of the menstrual cycle can be controlled
Learn this stuff until you know what hormone does what, and how oral contraceptives work.

Topic B5 — Homeostasis and Response

Controlling Fertility

*Oral contraceptives aren't the only way that hormones can be used to **control fertility**.*

Some **Hormonal** Contraceptives **Release Progesterone**

1) Some hormonal contraceptives work by slowly releasing progesterone.
2) This stops eggs from maturing or being released from the ovaries.
3) Examples of contraceptives that work this way are:

Contraceptive patch
1) This is a small patch that is stuck to the skin.
2) It lasts one week.

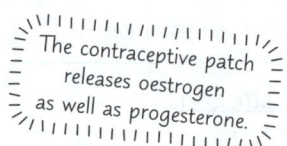
The contraceptive patch releases oestrogen as well as progesterone.

Contraceptive implant
1) This is inserted under the skin of the arm.
2) An implant can last for three years.

Contraceptive injection
Each dose lasts two to three months.

Some **Intrauterine Devices** Contain **Hormones**

1) An intrauterine device (IUD) is a T-shaped device that's inserted into the uterus (womb).
2) It can stop fertilised eggs from implanting in the uterus wall.
3) Some types of IUD release a hormone.

Barriers Stop Egg and Sperm **Meeting**

1) Non-hormonal contraceptives (types that don't use hormones) stop the sperm from getting to the egg.
2) Barrier methods are one type of non-hormonal contraceptive. For example:

Condoms
1) Condoms are worn over the penis during sexual intercourse.
2) Female condoms are worn inside the vagina.
3) Condoms are the only form of contraception that will protect against sexually transmitted diseases.

There are many different types of contraceptives
There are a lot of options when it comes to contraception. There are even more on the next page...

Topic B5 — Homeostasis and Response

More on Controlling Fertility

Using condoms isn't the only **barrier** method of contraception — **diaphragms** can also stop the sperm from reaching the egg.

Diaphragms

1) A diaphragm is a shallow plastic cup that fits over the entrance to the uterus.
2) It has to be used with spermicide (a chemical that disables or kills the sperm).
3) Spermicide can be used alone as a form of contraception. But when used alone, it is not as effective (it's only about 70-80% effective at preventing pregnancy).

There are Other Ways to Avoid Pregnancy

Sterilisation

1) In females, sterilisation involves cutting or tying the fallopian tubes (tubes that connect the ovaries to the uterus).
2) In males, it involves cutting or tying the sperm ducts (tubes between the testes and the penis).
3) Sterilisation is permanent (lasts for life).

Natural Methods

1) Pregnancy may be avoided by not having sexual intercourse when a woman is at the stage of the menstrual cycle when she is most likely to get pregnant.
2) It's popular with people who think that hormonal and barrier methods are unnatural.
3) But it's not very effective.

Abstinence

1) The only way to be sure that sperm and egg don't meet is to not have intercourse.
2) This is called abstinence.

Some methods of contraception are more effective than others

You might be asked to evaluate the different hormonal and non-hormonal methods of contraception in your exam. If you do, make sure you weigh up and write about the pros AND the cons of each method.

Topic B5 — Homeostasis and Response

Warm-Up & Exam Questions

Right then, another lot of pages down. Now there's just the small matter of answering some questions...

Warm-Up Questions

1) Name the hormone that stimulates sperm production in males.
2) Name the hormone that causes an egg to mature in the ovary.
3) What hormones does the contraceptive patch contain?
4) What is spermicide?
5) True or false? Only males can be sterilised to avoid pregnancy?

Exam Questions

1   During puberty, secondary sex characteristics develop. *(Grade 1-3)*

1.1   Give **one** example of a secondary sex characteristic.

[1 mark]

1.2   In females, ovulation begins to occur at puberty.
How often does ovulation usually occur?

[1 mark]

2   There are several methods that can be used to avoid pregnancy. *(Grade 1-3)*

Which of the following is a barrier method of contraception?
Tick **one** box.

☐ sterilisation

☐ diaphragm

☐ contraceptive implant

☐ contraceptive injection

[1 mark]

3   The menstrual cycle is controlled by several different hormones. *(Grade 4-5)*

3.1   What does the hormone oestrogen do?

[1 mark]

3.2   Which hormone causes the release of an egg?

[1 mark]

3.3   Hormones can also be used in contraception.
Explain how oral contraceptives prevent pregnancy.

[2 marks]

Topic B5 — Homeostasis and Response

Topic B6 — Inheritance, Variation and Evolution

DNA

The first step in understanding **genetics** is getting to grips with **DNA**.

Chromosomes Are Really Long Molecules of DNA

1) DNA is the chemical that all of the genetic material in a cell is made up from.
2) It contains all the instructions to put an organism together and make it work.
3) A DNA molecule is made up of two strands of DNA coiled together. They make a double helix (a double-stranded spiral).
4) A DNA strand is a polymer. A polymer is something made up of lots of smaller pieces joined together.
5) DNA is found in the nucleus of animal and plant cells.
6) It's found in really long structures called chromosomes.

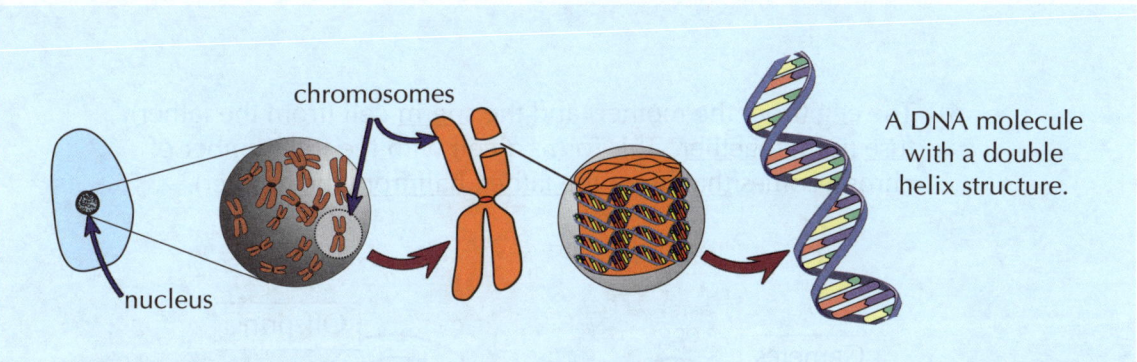

A DNA molecule with a double helix structure.

A Gene Codes for a Specific Protein

1) A gene is a small section of DNA found on a chromosome.
2) Each gene codes for a particular sequence of amino acids.
3) These amino acids are joined together to make a protein.

Every Organism Has a Genome

1) Genome is just the fancy term for all of the genetic material in an organism.
2) Scientists have worked out the whole human genome.
3) Understanding the human genome is really important for medicine. This is because:

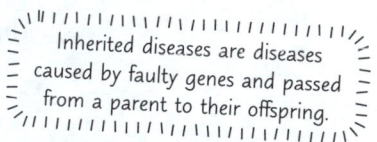

Inherited diseases are diseases caused by faulty genes and passed from a parent to their offspring.

1) Scientists can find genes in the genome that are linked to different types of disease.
2) If scientists know which genes are linked to inherited diseases, they can understand them better. This could help us to develop treatments.
3) Scientists can look at tiny differences in the genomes of different people. This can help them find out about the migration (movement) of certain populations of people around the world over history.

Sexual Reproduction

*Reproduction is **very important** for all species — it's how they **pass** on their **genes** to the next generation.*

Sexual Reproduction Produces Genetically Different Cells

1) Sexual reproduction is where genes from two organisms (a father and a mother) are mixed.

2) The mother and father produce gametes (sex cells). E.g. egg and sperm cells in animals.

3) The gametes are produced by meiosis (see page 116). Each gamete contains half the number of chromosomes of a normal cell.

4) The egg (from the mother) and the sperm cell (from the father) fuse (join) together. This forms a cell with the full number of chromosomes (half from the father, half from the mother).

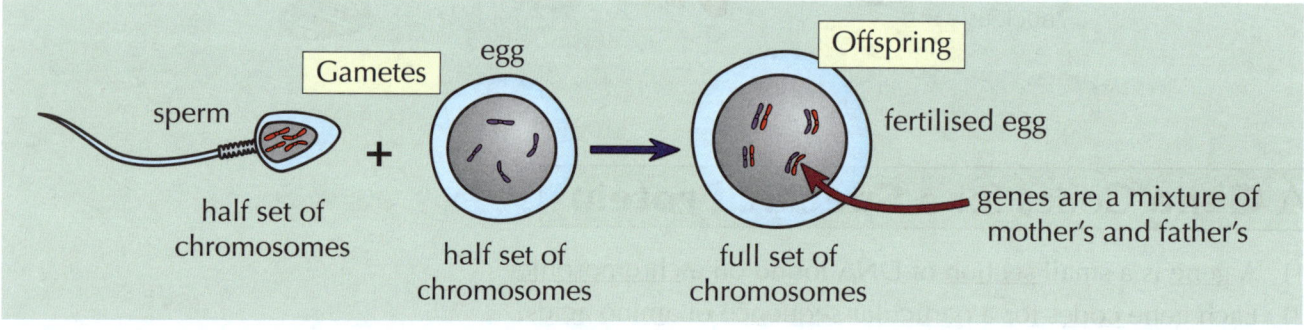

5) The offspring receives a mixture of genes, so inherits features from both parents.

6) This mixture of genes produces variation in the offspring.

7) Flowering plants can reproduce in this way too. Their gametes are egg cells and pollen.

Sexual reproduction mixes genes from two organisms...
When you're revising, make sure that you've got your head around sexual reproduction, before moving on to asexual reproduction — that way you don't confuse the two.

Asexual Reproduction

*Some organisms use **asexual** reproduction to **pass** on their **genes**...*

Asexual Reproduction Produces Genetically Identical Cells

1) Asexual reproduction happens by mitosis.

2) One parent cell makes a new cell by dividing in two (see page 28).

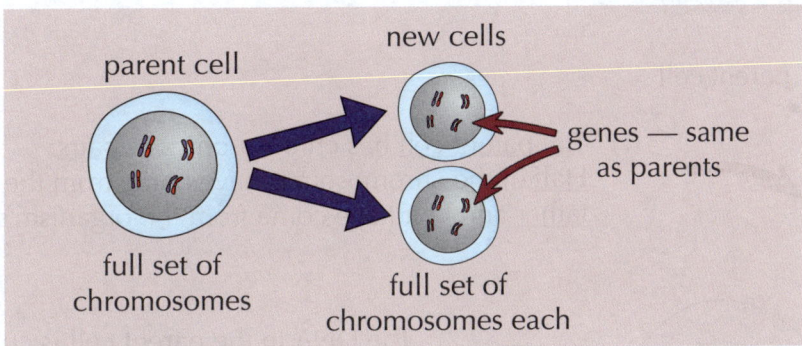

3) There's no fusion of gametes.

4) So there's no mixing of genes.

5) This means there's no genetic variation in the new cells.

6) Each new cell is genetically identical to the parent cell — it has exactly the same genes. The new cell is a clone.

You might need to reproduce these facts in the exam...
The main messages about reproduction are that: 1) sexual reproduction needs two parents and forms cells that are genetically different to the parents, so there's lots of genetic variation. And 2) asexual reproduction needs just one parent to make genetically identical cells, so there's no genetic variation in the offspring.

Topic B6 — Inheritance, Variation and Evolution

Meiosis

*Time now to learn about how **sperm** and **egg** cells are made...*

Gametes Are Produced by Meiosis

1) Gametes only have half the number of chromosomes of normal cells.

2) To make gametes, cells divide by meiosis.

3) In humans, meiosis only happens in the reproductive organs (the ovaries in females and testes in males).

4) Here's how meiosis happens:

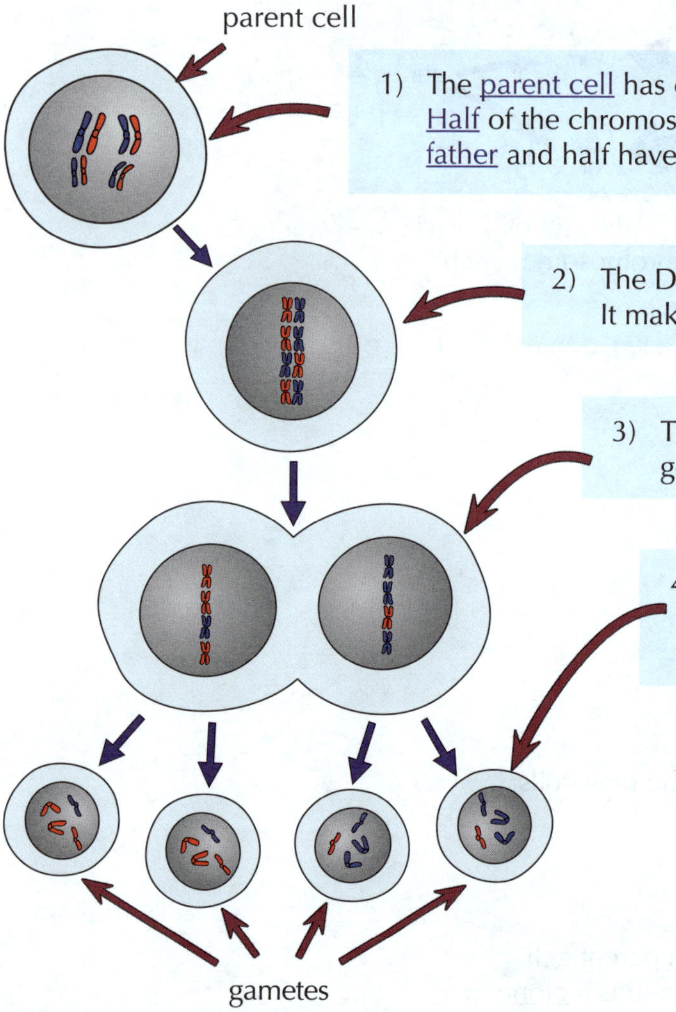

parent cell

1) The parent cell has chromosomes in pairs. Half of the chromosomes have come from the organism's father and half have come from the organism's mother.

2) The DNA in the parent cell is copied. It makes X-shaped chromosomes.

3) The cell divides. Each new cell gets half of the chromosomes.

4) Each cell divides again. The X-shaped chromosomes are pulled apart.

5) You end up with four new daughter cells. These are the gametes. Each gamete:
- only has a single set of chromosomes,
- is genetically different (each has a different mix of the mother's and father's chromosomes).

gametes

Gametes only have one set of chromosomes...

In humans, meiosis only occurs in reproductive organs, for making gametes.

Q1 Human body cells contain 46 chromosomes each. The graph on the right shows how the mass of DNA per cell changed as some cells divided by meiosis in a human ovary. How many chromosomes were present in each cell when they reached stage 6? [1 mark]

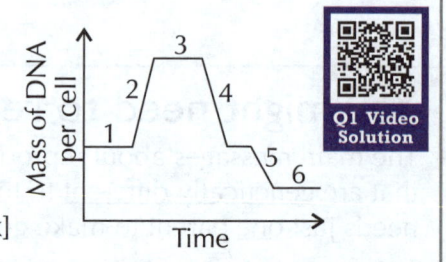

Topic B6 — Inheritance, Variation and Evolution

Fertilisation and Chromosomes

Now for a bit more about *gametes* and two *very* important little *chromosomes*...

Gametes Fuse to Make a New Cell

1) During fertilisation, two gametes fuse together (see p.114). This makes a new cell.

2) This new cell has the normal number of chromosomes.

3) The new cell divides by mitosis many times to produce lots of new cells. This forms an embryo.

4) As the embryo develops, these cells differentiate (see page 23). The cells become different types of specialised cell that make up a whole organism.

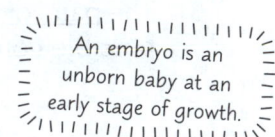
An embryo is an unborn baby at an early stage of growth.

Your Chromosomes Control Whether You're Male or Female

1) There are 23 pairs of chromosomes in every human body cell.
2) 22 are matched pairs of chromosomes that just control your characteristics.
3) The 23rd pair are labelled XY or XX.
4) They're the two chromosomes that decide your sex (whether you turn out male or female).

All males have an X and a Y chromosome: XY
The Y chromosome causes male characteristics.

All females have two X chromosomes: XX
The XX combination allows female characteristics to develop.

5) Each sperm has either an X or a Y chromosome.
6) All egg cells have an X chromosome.

Now that I have your undivided attention...
Make sure you know what the X and Y chromosomes are before you head on to the next page...

Topic B6 — Inheritance, Variation and Evolution

X and Y Chromosomes

Genetic Diagrams Show the Possible Gamete Combinations

To find the probability (chance) of getting a boy or a girl, you can draw a genetic diagram. This type of genetic diagram is called a Punnett square.

1) Put the possible gametes (eggs or sperm) from one parent down the side. Put those from the other parent along the top.

2) Then in each middle square you fill in the letters from the top and side that line up with that square.

3) The pairs of letters in the middle show the possible combinations of the gametes.

4) There are two XX results and two XY results.

5) This means that there's the same probability of getting a boy or a girl — each one has a 1 in 2 chance (which is the same as 50%).

female gametes (eggs) X X
male gametes (sperm) X Y

	X	X
X	XX	XX
Y	XY	XY

possible combinations of gametes... ...two males (XY) and two females (XX).

There's More Than One Type of Genetic Diagram

The other type of genetic diagram looks a bit more complicated, but it shows exactly the same thing.

1) At the top are the parents.

2) The middle circles show the possible gametes that are formed. One gamete from the female combines with one gamete from the male (during fertilisation).

3) The criss-cross lines show all the possible ways the X and Y chromosomes could combine.

4) The possible offspring you could get are shown in the bottom circles.

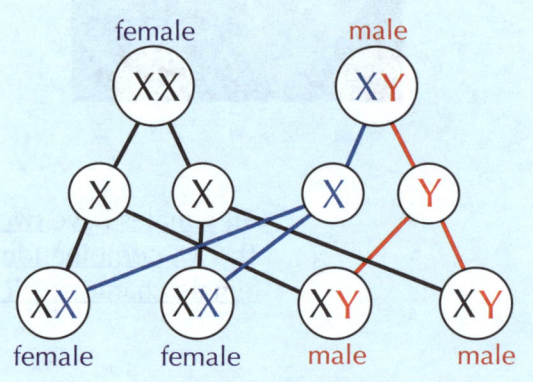

female XX — male XY
X X X Y
XX XX XY XY
female female male male

These diagrams aren't as scary as they look...

Most genetic diagrams you'll see in exams concentrate on a gene, instead of a chromosome. But it's pretty much the same. Don't worry — there are loads of other examples on pages 121-122.

Topic B6 — Inheritance, Variation and Evolution

Warm-Up & Exam Questions

It's time to see how much you picked up about meiosis, reproduction and sex chromosomes...

Warm-Up Questions

1) True or false? There is variation in the offspring of sexual reproduction.
2) How many cell divisions take place in meiosis?
3) What combination of sex chromosomes do human females have?

Exam Questions

1 Sexual reproduction involves gametes fusing together to form offspring. This is shown in **Figure 1**. The number of chromosomes in each cell is incomplete.

Figure 1

sperm + egg → fertilised egg

Chromosomes: ......23...... ....

Complete **Figure 1** to show the number of chromosomes in an egg cell and in a fertilised egg.

[2 marks]

2 An organism's genetic material is made up of a chemical called DNA.

2.1 Which of the following describes the structure of DNA? Tick **one** box.

☐ A protein made up of two strands.   ☐ A polymer made up of two strands.
☐ A protein made up of four strands. ☐ A polymer made up of four strands.

[1 mark]

2.2 Which of the following contains the largest amount of an organism's DNA? Tick **one** box.

☐ A gene   ☐ Its genome   ☐ A chromosome

[1 mark]

2.3 Explain the relationship between DNA and the proteins produced by an organism.

[3 marks]

3 Some species of worm can produce offspring through a process called fragmentation. In this process, bits of the parent's body break off and go through cell division by mitosis, to develop into complete organisms.

3.1 What term is used to describe this form of reproduction?

[1 mark]

3.2 Suggest how the chromosomes in the offspring will compare to those of the parent worm.

[1 mark]

Topic B6 — Inheritance, Variation and Evolution

Genetic Diagrams

For those of you expecting to see a **diagram** or two on a page called 'Genetic Diagrams', prepare to be disappointed. You need to understand a bit more about what genetic diagrams **show** to start with...

Different Genes Control Different Characteristics

1) Some characteristics are controlled by a single gene. For example:

- mouse fur colour
- red-green colour blindness in humans.

2) However, most characteristics are controlled by several genes.

All Genes Exist in Different Versions Called Alleles

1) You have two alleles of every gene in your body — one on each chromosome in a pair.

2) If the two alleles are the same, then the organism is homozygous for that characteristic.

3) If the two alleles are different, then the organism is heterozygous for that characteristic.

4) Some alleles are dominant (these are shown with a capital letter on genetic diagrams, e.g. 'C'). Some alleles are recessive (these are shown by a small letter on genetic diagrams, e.g. 'c').

5) For an organism to show a recessive characteristic, both its alleles must be recessive (e.g. cc). But to show a dominant characteristic, only one allele needs to be dominant (e.g. either CC or Cc).

6) The mix of alleles you have is called your genotype.

7) Your alleles determine your characteristics. The characteristics you have is called your phenotype.

There are lots of fancy words to learn on this page...
Make sure you fully understand what all the different terms on this page mean (i.e. genes, alleles, homozygous, heterozygous, dominant, recessive, genotype and phenotype). You'll feel much more comfortable going into the exam knowing that these words aren't going to trip you up.

Topic B6 — Inheritance, Variation and Evolution

Genetic Diagrams

*This page is all about how **characteristics** are **inherited** — it involves drawing **genetic diagrams**.*

Genetic Diagrams Can Show How Characteristics are Inherited

You can use genetic diagrams to show how single genes for characteristics are inherited (passed from parents to offspring). For example:

> 1) An allele that causes hamsters to have superpowers is recessive ("b").
> 2) Normal hamsters don't have superpowers due to a dominant allele ("B").
> 3) Two homozygous hamsters (BB and bb) are crossed (bred together). A genetic diagram shows what could happen:

A hamster with the genotype BB or Bb will be normal. A hamster with the genotype bb will have superpowers.

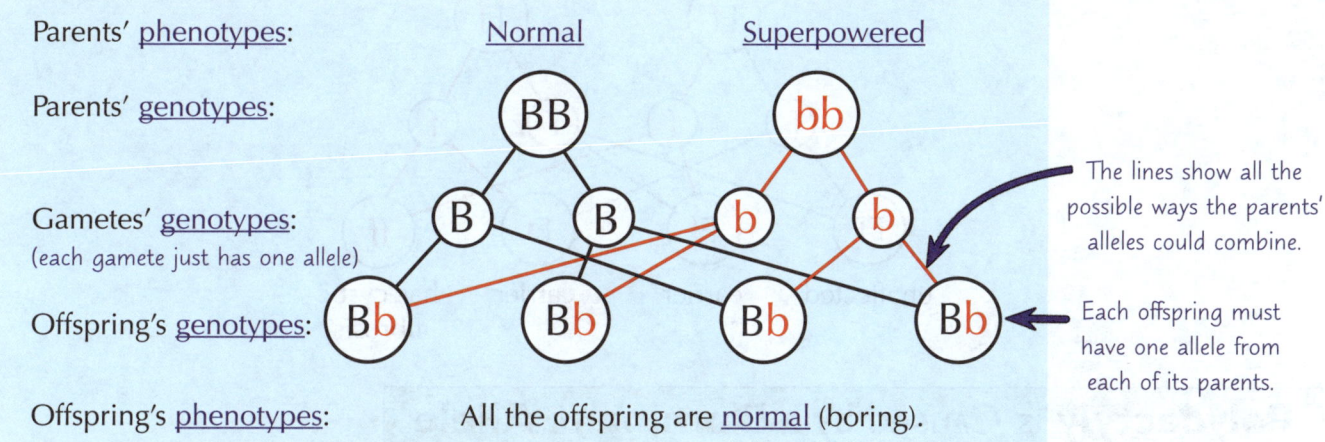

Parents' phenotypes: Normal   Superpowered
Parents' genotypes: BB   bb
Gametes' genotypes: B B b b
(each gamete just has one allele)
Offspring's genotypes: Bb Bb Bb Bb
Offspring's phenotypes: All the offspring are normal (boring).

The lines show all the possible ways the parents' alleles could combine.

Each offspring must have one allele from each of its parents.

Punnett Squares are Another Type of Genetic Diagram

You can also show genetic crosses in a Punnett square.

This Punnett square shows a cross between two heterozygous hamsters (Bb and Bb):

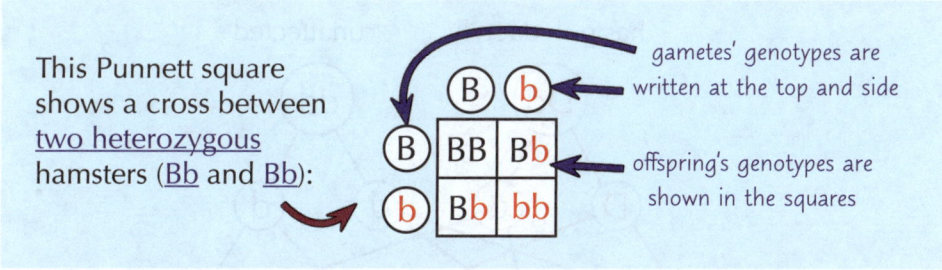

gametes' genotypes are written at the top and side

offspring's genotypes are shown in the squares

> 1) There's a 3 in 4 (75%) chance that offspring will be normal.
> 2) There's a 1 in 4 (25%) chance that offspring will have superpowers.
> 3) This gives a 3 normal : 1 superpowers ratio (3:1).

You can master genetic diagrams — you just need to practise them...

You should know how to produce and understand both of these types of genetic diagram before exam day.

Topic B6 — Inheritance, Variation and Evolution

Inherited Disorders

*Inherited disorders are **health conditions**. They are caused by **inheriting** faulty **alleles**.*

Cystic Fibrosis is Caused by a Recessive Allele

Cystic fibrosis is an inherited disorder of cell membranes.
1) The allele which causes cystic fibrosis is a recessive allele, 'f'.
2) Because it's recessive, people with only one copy of the allele won't have the disorder — they're known as carriers.
3) For a child to have the disorder, both parents must be either carriers or have the disorder themselves.
4) As the diagram shows, there's a 1 in 4 chance of a child having the disorder if both parents are carriers.

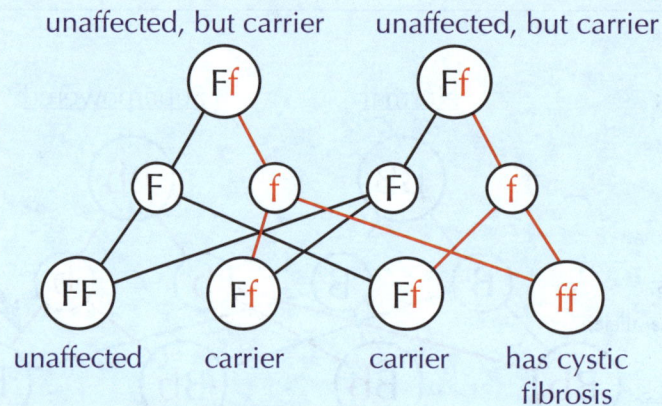

Polydactyly is Caused by a Dominant Allele

Polydactyly is an inherited disorder where a baby's born with extra fingers or toes.
1) The disorder is caused by a dominant allele, 'D'.
2) This means that it can be inherited if just one parent carries the faulty allele.
3) The parent that has the faulty allele will have the disorder too since the allele is dominant.
4) As the diagram shows, there's a 50% chance of a child having the disorder if one parent has one D allele.

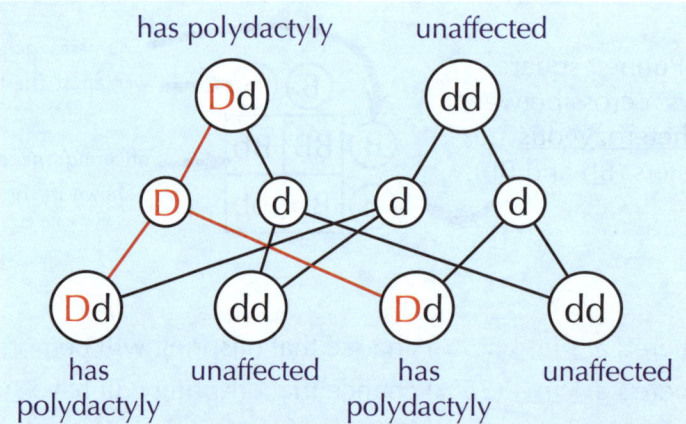

There are no carriers if a dominant allele is the cause...

The important bits to take away from this page are that the allele for cystic fibrosis is recessive and the allele for polydactyly is dominant.

Q1 The Punnett square on the right shows the possible inheritance of polydactyly from two parents. Give the probability that their offspring will have polydactyly. [1 mark]

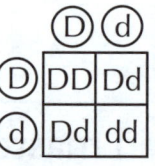

Topic B6 — Inheritance, Variation and Evolution

Family Trees

*Just when you thought you'd finished with **genetic diagrams**, **family trees** show up...*

Family Trees Show the Inheritance of Alleles

1) The diagram below is a family tree for cystic fibrosis (see previous page).

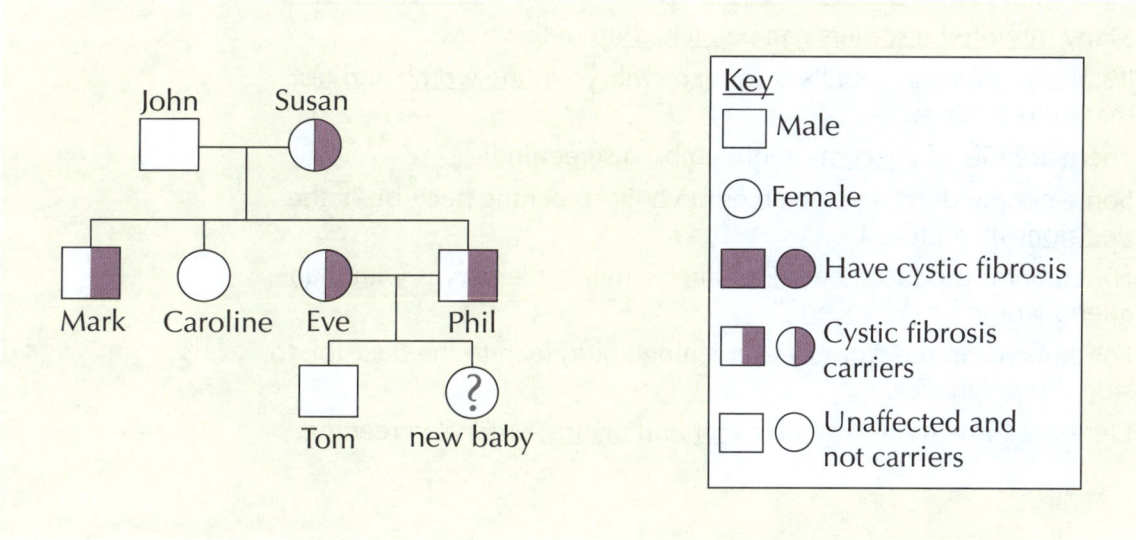

2) From the family tree, you can tell that the allele for cystic fibrosis <u>isn't</u> dominant. This is because plenty of the family <u>carry</u> the allele but <u>don't</u> have the disorder.

3) There is a <u>25% chance</u> that the <u>new baby</u> will have the <u>disorder</u> and a <u>50% chance</u> that it will be a <u>carrier</u>. This is because both of the baby's parents are <u>carriers</u> (Eve and Phil are both Ff).

4) The case of the new baby is just the same as in the genetic diagram on page 122. The baby could be <u>unaffected</u> (FF), a <u>carrier</u> (Ff) or <u>have</u> cystic fibrosis (ff).

It's enough to make you go cross-eyed...

If you're struggling to work out a family tree in the exam, try writing each person's genotype onto it — it might help you to understand it more easily.

Q1 Cystic fibrosis is caused by a recessive allele, f. The dominant allele is F. The family tree on the right shows the inheritance of cystic fibrosis. What is Tamsin's genotype? Use the key above to help you.

[1 mark]

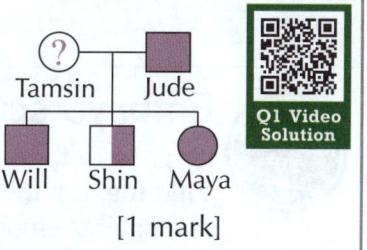

Topic B6 — Inheritance, Variation and Evolution

Embryo Screening

Embryos can be screened for disorders, but not everyone agrees with it...

Embryos Can Be Screened for Inherited Disorders

1) During *in vitro* fertilisation (IVF), embryos are fertilised in a lab and then put in the mother's womb.
2) Before they are put into the mother, scientists can remove a cell from each embryo and look at its genes. This is called embryo screening.
3) Many inherited disorders can be picked up in this way.
4) It's also possible to get DNA from an embryo in the womb and test that for disorders.
5) There are lots of concerns about embryo screening.
6) Some people don't agree with embryonic screening because of the decisions it can lead to.
7) For embryos produced by IVF — after screening, embryos with 'bad' alleles would be destroyed.
8) For embryos in the womb — screening could lead to the decision to stop the pregnancy.
9) Here are some more arguments for and against embryo screening:

For Embryonic Screening

1) It will help to stop people suffering.
2) Treating disorders costs a lot of money.
3) There are laws to stop it going too far. At the moment parents cannot even select the sex of their baby (unless it's for health reasons).

Against Embryonic Screening

1) It suggests that people with genetic problems are not wanted. This could lead to them being treated unfairly.
2) There may come a point when people want to screen their embryos so they can pick the features they prefer. E.g. they want a certain eye colour or hair colour.
3) Screening is expensive.

Embryo screening — it's a tricky one...

It's great to think that we might be able to stop people from having inherited disorders that cause suffering, but there are many concerns to think about too. Try writing a balanced argument for and against embryo screening — it's good practice.

Topic B6 — Inheritance, Variation and Evolution

Warm-Up & Exam Questions

There's no better preparation for exam questions than doing... err... practice exam questions. Hang on, what's this I see...

Warm-Up Questions

1) What is a different version of a gene called?
2) What does phenotype mean?
3) What is polydactyly?
4) Give one argument for embryo screening.

Exam Questions

1 Draw **one** line from each genetic term to its definition. *Grade 1-3*

heterozygous	having two of the same allele
homozygous	the mix of alleles in an organism
genotype	having two different alleles

[2 marks]

2 Cystic fibrosis is a genetic disorder caused by recessive alleles. *Grade 4-5*

F = the normal allele   **f** = the faulty allele that causes cystic fibrosis

Figure 1 is an incomplete Punnett square showing the possible inheritance of cystic fibrosis from one couple.

Figure 1

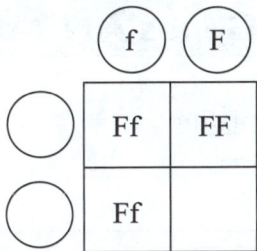

2.1 Complete the Punnett square to show the missing offspring's genotype and the missing genotypes of the gametes.

[2 marks]

2.2 What proportion of the possible offspring are homozygous?

[1 mark]

2.3 State the phenotypes of the parents.

[2 marks]

Variation

You'll probably have noticed that not all people are identical. There are reasons for this.

Organisms of the Same Species Have Differences

1) Different species look... well... different — my dog definitely doesn't look like a daisy.
2) But even organisms of the same species will usually look at least slightly different.
3) These differences are called the variation within a species.
4) Variation can be huge within a population.

Variation Can be Genetic

Variation can be genetic — this means it's caused by differences in genes that are inherited.

Variation Can Also be Environmental

Variation can also be environmental — this means it's caused by the conditions in which an organism lives. For example:

1) A plant grown on a nice sunny windowsill could grow healthy and green.
2) The same plant grown in darkness would grow tall and spindly and its leaves would turn yellow.

Most Characteristics are Due to Genes AND the Environment

Most variation in phenotype is caused by a mixture of genes and the environment. For example:

1) The maximum height that an animal or plant could grow to is determined by its genes.
2) But whether it actually grows that tall depends on its environment (e.g. how much food it gets).

You can't blame all of your faults on your parents...

The genes that you inherit from your parents have a really important role in controlling what characteristics you have. However, the conditions in which you live usually affect your characteristics too.

Mutations

*Sometimes the **sequence** of **DNA** can be changed. These changes are called **mutations**. Read on...*

Mutations are Changes to the Genome

1) Sometimes, a gene can mutate.

2) A mutation is a random change in an organism's DNA that can be inherited.

3) Mutations occur continuously.

4) Mutations mean that the gene is changed. This produces a genetic variant (a different form of the gene).

5) Most genetic variants have very little or no effect on an organism's phenotype (its characteristics).

6) Some variants have a small effect on the organism's phenotype. They alter the individual's characteristics but only slightly. For example:

   - Some characteristics (e.g. eye colour) are controlled by more than one gene.
   - A mutation in one of the genes may change the eye colour a bit, but the difference might not be huge.

7) Very rarely, variants can have such a big effect that they lead to a new phenotype, e.g. cystic fibrosis.

8) A new phenotype may be useful if the environment that an organism lives in changes.

9) This is because sometimes a new phenotype makes an individual more suited to a new environment.

10) If this happens, the mutation can become common throughout the species relatively quickly. This happens by natural selection — see the next page.

Topic B6 — Inheritance, Variation and Evolution

Evolution

*Evolution is very important. Without it we wouldn't have the great **variety of life** we have on Earth today.*

> THEORY OF EVOLUTION: All of today's species have evolved from simple life forms that first started to develop over three billion years ago.

Only the **Fittest Survive**

1) Charles Darwin came up with a really important theory about evolution — it's called evolution by natural selection. It works like this:

   1) Organisms in a species show wide variation in their characteristics.

   2) Organisms have to compete for resources in an ecosystem.

   3) This means organisms with characteristics that make them better adapted to their environment will be better at competing with other organisms.

   4) These organisms are more likely to survive and reproduce.

   5) So the genes for the useful characteristics are more likely to be passed on to their offspring.

   6) Over time, useful characteristics become more common in the population and the species changes. This is evolution.

2) Darwin's theory wasn't perfect. At the time he couldn't explain how new characteristics appeared or were passed on. Nowadays we have evidence to back up Darwin's theory, such as:

   1) The discovery of genetics — it showed that characteristics are passed on in an organism's genes. It also showed that genetic variants (see page 127) produce the characteristics (phenotypes) that are better adapted to the environment.
   2) Fossils — by looking at fossils of different ages (the fossil record), scientists could see how changes in organisms developed slowly over time.
   3) Antibiotic resistance — how bacteria are able to evolve to become resistant to antibiotics also further supports evolution by natural selection (see page 130).

Natural selection — the fittest pass on their genes...

Natural selection's all about the organisms with the best characteristics surviving to pass on their genes so that the whole species ends up adapted to its environment.

Q1 The sugary nectar in some orchid flowers is found at the end of a long tube behind the flower. There are moth species with long tongues that can reach the nectar. Explain how natural selection could have led to the moths developing long tongues. [4 marks]

Topic B6 — Inheritance, Variation and Evolution

Evolution

*Species need to **continue evolving** in order to **survive**. Sometimes this evolution creates a whole **new species**, but if a species can't evolve fast enough it might **die out** completely.*

Evolution Can Lead to New Species Developing

1) Over a long period of time, the phenotype of organisms can change a lot because of natural selection.

2) Sometimes, the phenotype can change so much that a completely new species is formed.

3) New species develop when populations of the same species change so much that they can't breed with each other to produce fertile offspring.

Extinction is When No Individuals of a Species Are Left

Species become extinct for these reasons:

1) The environment changes too quickly (e.g. their habitat is destroyed).

2) A new predator kills them all (e.g. humans hunting them).

3) A new disease kills them all.

4) They can't compete with another (new) species for food.

5) A catastrophic event happens that kills them all (e.g. a volcanic eruption).

Evolution's happening all the time...
Many species evolve so slowly that there are no big changes in them within our lifetime. However, some species (e.g. bacteria) reproduce really quickly, so we're able to watch evolution in action.

Topic B6 — Inheritance, Variation and Evolution

Antibiotic-Resistant Bacteria

The discovery of **antibiotics** was a huge benefit to medicine — but they might not be a **permanent solution**.

Bacteria can Evolve and Become Antibiotic-Resistant

1) Antibiotics are drugs that kill bacteria.
2) Bacteria can become resistant to antibiotics by natural selection.
3) Here's what happens:

There's more about natural selection on page 128.

> 1) Bacteria can develop random mutations (changes) in their DNA.
> 2) These can lead to the bacteria being resistant to (not killed by) a particular antibiotic.
> 3) These new strains (types) of bacteria are called antibiotic-resistant bacteria.
> 4) The ability to resist antibiotics is a big advantage for the bacteria.
> 5) It means that the bacteria are able to survive in a host who's being treated to get rid of an infection.
> 6) So the antibiotic-resistant bacteria can reproduce many more times.
> 7) They pass on their gene for antibiotic resistance to their offspring.
> 8) The gene for antibiotic resistance becomes more common in the population over time — the bacteria have evolved.

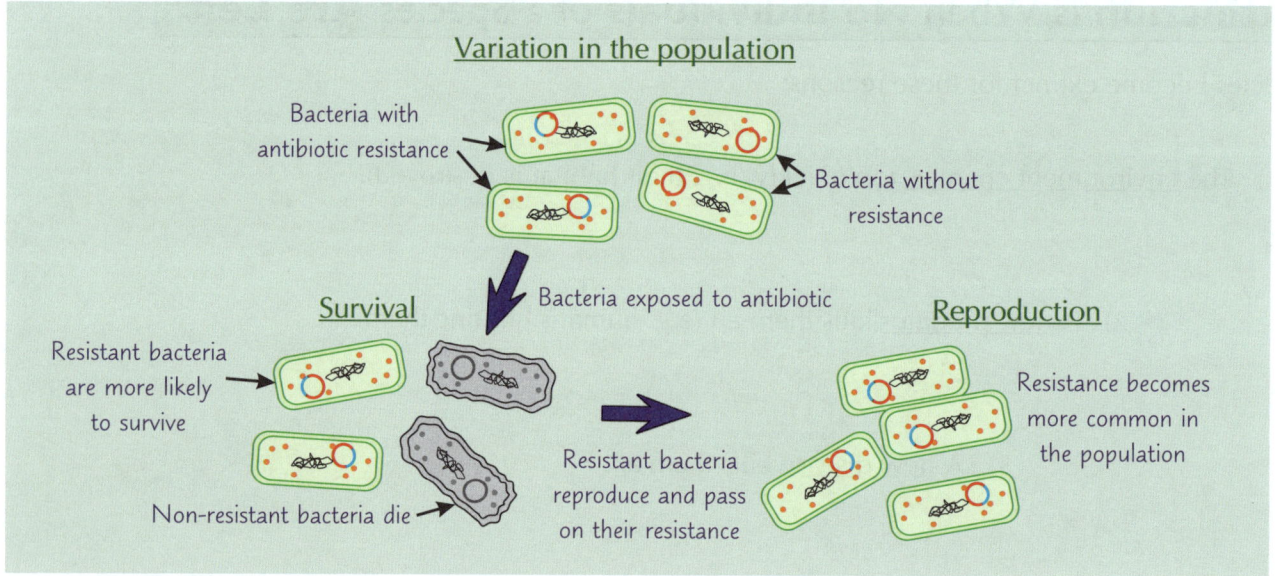

4) Because bacteria are so rapid at reproducing, they can evolve quite quickly.
5) The antibiotic-resistant bacteria keep reproducing. This increases the population size of the antibiotic-resistant strain.

Antibiotic Resistant Bacteria Spread Easily

1) Antibiotic-resistant bacteria are a problem because:

> - There is no effective treatment for the infection.
> - People are not immune to the new strain.

2) This means that the antibiotic-resistant strain is able to easily spread between people.

Topic B6 — Inheritance, Variation and Evolution

More on Antibiotic-Resistant Bacteria

*Antibiotic-resistance is getting **worse**. But there are ways to **fight it**...*

Antibiotic Resistance is Becoming More Common

1) The problem of antibiotic resistance is getting worse because:

   - Antibiotics are being overused.
   - People aren't using antibiotics correctly.

2) 'Superbugs' (bacteria that are resistant to most known antibiotics) are becoming more common.

   E.g. MRSA is a relatively common 'superbug' that's really hard to get rid of.

Antibiotics Need to Be Used Sensibly

There are a few things that can be done to avoid antibiotic-resistant bacteria forming:

1) Doctors should only prescribe antibiotics when they really need to.
   - They shouldn't be prescribed for non-serious conditions or infections caused by viruses.

Antibiotics don't kill viruses.

2) You should take all the antibiotics a doctor prescribes for you.
   - Taking the full course makes sure that all the bacteria are destroyed.
   - This means that there are none left to mutate and develop into antibiotic-resistant strains.

3) The use of antibiotics by farmers should be restricted because:
   - In farming, antibiotics can be given to animals to prevent them becoming ill and make them grow faster.
   - This can lead to the development of antibiotic-resistant strains of bacteria in the animals.
   - The antibiotic-resistant bacteria can then spread to humans.

We Can't Make New Antibiotics Fast Enough

1) Drug companies are working on developing new antibiotics that kill the resistant strains.
2) But there are problems:

   - The rate of development is slow.
   - The process is really expensive.

3) This means that we're unlikely to be able to keep up with the demand for new drugs to fight new antibiotic-resistant strains.

Topic B6 — Inheritance, Variation and Evolution

Warm-Up & Exam Questions

You need to test your knowledge with a few Warm-Up Questions, followed by some Exam Questions...

Warm-Up Questions

1) True or false? There is usually only a small amount of variation within a population.
2) True or false? Evolution never leads to new species developing.
3) What is meant by the term 'extinction'?
4) What is an antibiotic?

Exam Questions

1   **Figure 1** shows a type of stingray. The stingray's appearance looks like a flat rock. It spends most of its time on a rocky sea bed.

1.1   Use words from the box to complete the sentences about the evolution of the stingray's appearance.

Figure 1

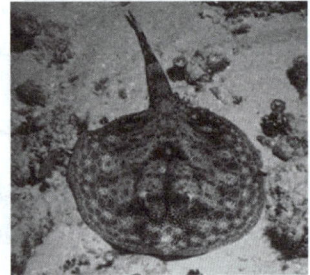

survive	generation	population
die	variation	similarity

The appearance of this stingray's ancestors showed ....

The ancestors that looked like flat rocks were hidden, so were more likely to ....

They were more likely to reproduce and pass their genes on to the next ....

[3 marks]

1.2   Suggest what caused some of the ancestors to look more like flat rocks than others.

[1 mark]

2   Helen and Stephanie are identical twins. This means they have identical DNA.

2.1   Helen weighs 7 kg more than Stephanie. Explain whether this is due to genes, environmental factors or both.

[2 marks]

2.2   Stephanie has a birthmark on her shoulder. Helen doesn't. Explain how this shows that birthmarks are **not** caused by genes.

[1 mark]

Topic B6 — Inheritance, Variation and Evolution

Selective Breeding

'Selective breeding' sounds like it could be a tricky topic, but it's actually quite simple. You take the **best** plants or animals and breed them together to get the best possible **offspring**. That's it.

Selective Breeding is Very Simple

1) Selective breeding is when humans choose which plants or animals are going to breed.

2) Organisms are selectively bred to develop features that are useful or attractive. For example:

   Selective breeding is also known as 'artificial selection'.

   - Animals that produce more meat or milk.
   - Crops with disease resistance (that are not killed by disease).
   - Dogs with a good, gentle personality.
   - Decorative plants with big or unusual flowers.

3) This is the basic process involved in selective breeding:

   1) From your existing plants or animals select the ones which have the feature you're after.
   2) Breed them with each other.
   3) Select the best of the offspring, and breed them together.
   4) Continue this process over several generations. Eventually, all offspring will have the feature you want.

4) Selective breeding is nothing new — people have been doing it for thousands of years.

5) This is how we ended up with edible crops from wild plants and domesticated animals like cows and dogs.

Selective breeding is just breeding the best to get the best...

Different breeds of dog came from selective breeding. E.g. somebody thought 'I like this small, yappy wolf — I'll breed it with this other one'. After thousands of generations, we got poodles.

Q1 Explain how you could selectively breed for floppy ears in rabbits. [4 marks]

Topic B6 — Inheritance, Variation and Evolution

Selective Breeding

Here's an Example of Selective Breeding:

A farmer might want his cattle to produce more meat.

1) Genetic variation means some cattle will have better characteristics for producing meat than others, e.g. a larger size.
2) The farmer could select the largest cows and bulls and breed them together.
3) He could then select the largest offspring and breed them together.
4) After several generations, he would get cows with a very high meat yield.

Selective Breeding Has Disadvantages

1) The main problem with selective breeding is that it reduces the number of different alleles in a population.

There's more on alleles on page 120.

2) This is because the "best" animals or plants are always used for breeding, and they are all closely related — this is known as inbreeding.

3) This means there's more chance of selectively bred organisms having health problems caused by their genes, e.g. they may inherit harmful genetic defects.

4) There can also be serious problems if a new disease appears.

5) This is because it's less likely that individuals in the population will have alleles that make them resistant to the disease.

6) So, if one individual is affected by the disease, the rest are also likely to be affected.

Selective breeding has its pros and cons...

Selective breeding has already been producing good results for thousands of years, but it's still important that farmers and other people working in agriculture are aware of the disadvantages. They need to look at the evidence and weigh up the pros and cons before coming to a decision.

Topic B6 — Inheritance, Variation and Evolution

Genetic Engineering

As well as selective breeding, humans can also use **genetic engineering** to **control** an organism's **features**.

Genetic Engineering Involves Changing an Organism's DNA

1) Genetic engineering is used to give organisms new and useful characteristics.
2) It involves cutting a gene out of one organism and putting it into another organism's cells.
3) Organisms that have had a new gene inserted are called genetically modified (GM) organisms.

Genetic Engineering is Useful in Agriculture and Medicine

In Agriculture:

1) Crops can be genetically engineered — this makes genetically modified (GM) crops.
2) They may be genetically engineered to be resistant to herbicides (chemicals that kill plants). This means that farmers can spray their crops to kill weeds, without affecting the crop itself.
3) Crops can also be genetically engineered to be resistant to insects or disease. Or they can be made to grow bigger and better fruit.
4) These things can increase crop yield (the amount of food produced).

In Medicine:

1) Bacteria can be genetically engineered to produce human insulin. This can be used to treat diabetes (see p.106).
2) Treatments using genetic modification for inherited diseases are being researched.

But There are Some Concerns About Genetic Engineering

There are concerns about using genetic engineering in animals:

1) It can be hard to predict how changing an animal's DNA will affect the animal.
2) Many genetically modified embryos don't survive.
3) Some genetically modified animals also suffer from health problems later in life.

There are also concerns about growing GM crops:

1) Some people say that growing GM crops will affect the number of wild flowers. This could also affect the population of insects.
2) Some people are worried that we might not understand the effects of GM crops on human health.

Topic B6 — Inheritance, Variation and Evolution

Warm-Up & Exam Questions

By doing these Warm-Up and Exam Questions, you'll soon find out if you've got the basic facts straight.

Warm-Up Questions

1) True or false? Genetic engineering involves changing an organism's DNA.
2) What is the name given to crops that have been genetically engineered?
3) Name one useful product that humans have genetically modified bacteria to produce.

Exam Questions

1   A farmer wants to use selective breeding to improve disease resistance in her crops.

Use the words from the box to complete the sentences about the disadvantages of selective breeding.

| chromosomes | health problems | better characteristics |
| alleles | more resistant | less resistant |

In a population of selectively bred plants, there will be fewer different ....

This means if a new disease appears, the plants may be ....

Due to inbreeding, there's also more chance of selectively bred plants having ....

[3 marks]

2   Organisms can be genetically modified.
    This means an organism's genes can be altered to alter its characteristics.

2.1 Plants can be genetically modified to become more resistant to disease.
    Suggest **one** other useful way that plants can be genetically modified.

[1 mark]

2.2 Some people think that it is wrong to genetically modify crop plants.
    Give **one** objection that a person might have.

[1 mark]

3   Cows can be selectively bred to produce offspring that produce a high milk yield.
    Table 1 shows the average milk yield over three generations for a population of cows.

Table 1

Generation	Average milk yield per cow in litres per year
1	5000
2	5375
3	5750

3.1 Calculate the percentage change in average milk yield from generation **2** to generation **3**.

[2 marks]

3.2* Describe the method the farmer used to selectively breed cows to produce a higher milk yield.

[4 marks]

Topic B6 — Inheritance, Variation and Evolution

Fossils

*Fossils are great. If they're **well-preserved**, you can see what really old creatures **looked** like.*

Fossils are the Remains of Plants and Animals

1) Fossils are the remains of organisms from many thousands of years ago. They're found in rocks.
2) They provide the evidence that organisms lived ages ago.
3) Fossils can tell us a lot about how much or how little organisms have changed (evolved) over time.
4) Fossils form in rocks in one of three ways:

1. From gradual replacement by minerals

1) Things like teeth, shells and bones don't easily decay.
2) This means they can last a long time when buried.
3) When they do decay, they get replaced by minerals.
4) The minerals form a rock-like substance shaped like the original hard part.

(Most fossils happen this way.)

2. From casts and impressions

1) Fossils can be formed when an organism is buried in a soft material like clay. The clay hardens around it and the organism decays. The organism leaves a cast of itself. An animal's burrow or a plant's roots can also be preserved as casts.
2) Things like footprints are pressed into soft materials. This leaves an impression when they harden.

3. From preservation in places where no decay happens

1) Decay microbes only work if there's oxygen, moisture, warmth and the right pH.
2) In some substances these conditions aren't all present, so decay doesn't happen. For example, there's no oxygen or moisture in amber so decay organisms can't survive.

A preserved organism in amber.

But No One Knows How Life Began

1) Fossils show how much or how little different organisms have changed (evolved) as life has developed on Earth over millions of years.
2) There are lots of hypotheses (see p.2) suggesting how life first came into being. For example:

- Maybe the first life forms appeared in a swamp (or under the sea) here on Earth.
- Or maybe simple carbon molecules were brought here on comets and developed into simple life forms.

But no one really knows.

3) These hypotheses can't be supported or disproved because there's a lack of valid evidence.
4) There's a lack of evidence because many early organisms were soft-bodied. Soft tissue tends to decay away completely. So the fossil record is incomplete (unfinished).
5) Plus, fossils that did form millions of years ago may have been destroyed by geological activity. E.g. the movement of tectonic plates may have crushed fossils already formed in the rock.

Topic B6 — Inheritance, Variation and Evolution

Classification

*People really seem to like **putting things** into **groups** — biologists certainly do anyway...*

Classification is Organising Living Organisms into Groups

1) In the past, organisms were classified according to characteristics you can see (like number of legs). They were also classified by the structures that make them up (like mitochondria in cells).
2) The more similar two organisms appeared, the more closely related they were thought to be.
3) These characteristics were used to classify organisms in the five kingdom classification system.
4) In this system, living things are divided into five groups called kingdoms. These are:

- Animals — fish, mammals, reptiles, etc.
- Plants — grasses, trees, etc.
- Fungi — mushrooms and toadstools, yeasts, all that mouldy stuff on your loaf of bread.
- Prokaryotes — all single-celled organisms without a nucleus.
- Protists — eukaryotic single-celled organisms.

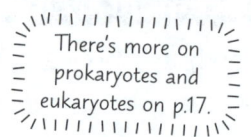
There's more on prokaryotes and eukaryotes on p.17.

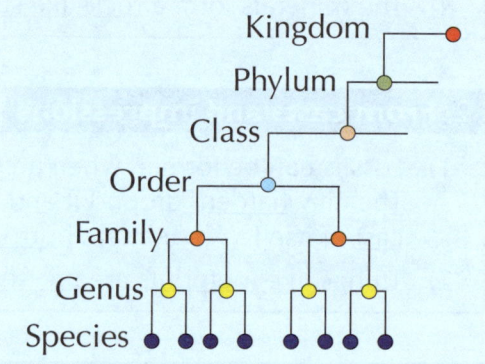

5) The kingdoms are then split into smaller and smaller groups.
6) These groups are phylum, class, order, family, genus and species.
7) The five kingdom classification system was made up by Carl Linnaeus.

Classification Systems Change Over Time

1) Over time, our knowledge of the processes taking place inside organisms has developed.
2) Microscopes have also improved over time. This has allowed us to find out more about the internal structures of organisms.
3) Using this new knowledge, scientists made new models of classification.
4) One of these new models was the three-domain system. It was made up by Carl Woese.
5) He used evidence from analysing chemicals to come up with the system.
6) It showed him that some species were less closely related than first thought.
7) Here's how the three-domain system works:

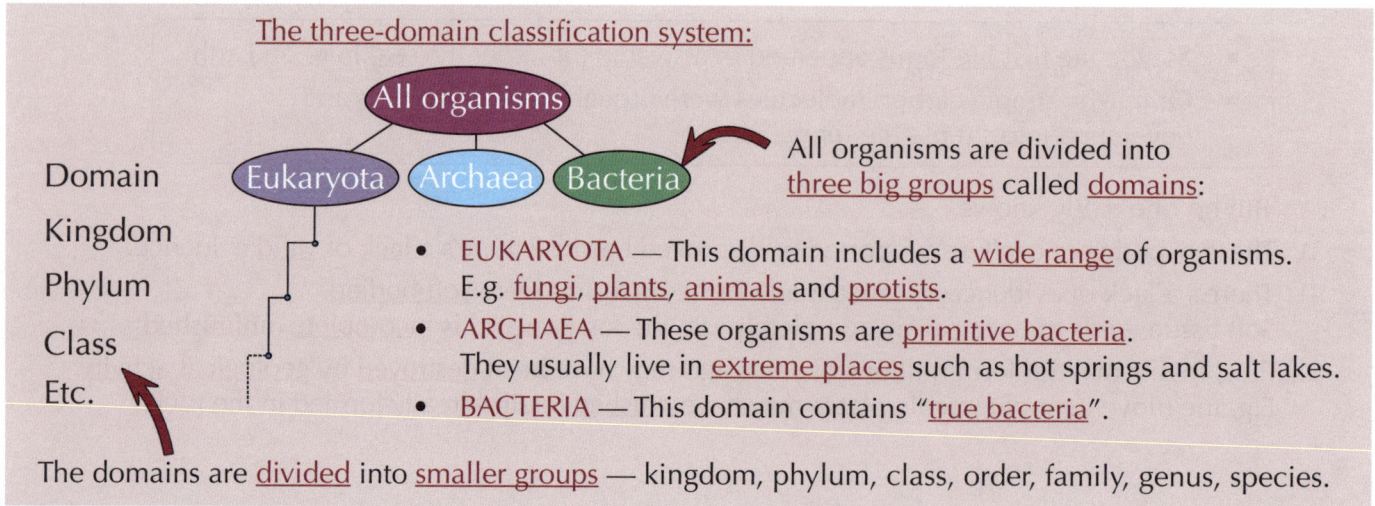

The three-domain classification system:

All organisms are divided into three big groups called domains:

- EUKARYOTA — This domain includes a wide range of organisms. E.g. fungi, plants, animals and protists.
- ARCHAEA — These organisms are primitive bacteria. They usually live in extreme places such as hot springs and salt lakes.
- BACTERIA — This domain contains "true bacteria".

The domains are divided into smaller groups — kingdom, phylum, class, order, family, genus, species.

Topic B6 — Inheritance, Variation and Evolution

Classification

*A bit of a **Latin** lesson for you now. And a diagram of a **funny-looking tree**.*

Organisms Are **Named** According to the **Binomial System**

1) In the binomial system, every organism is given its own two-part Latin name.
2) The first part refers to the genus that the organism belongs to. This gives you information on the organism's ancestry (the organisms it's related to).
3) The second part refers to the species.

E.g. humans are known as *Homo sapiens*.

'*Homo*' is the genus...   ...and '*sapiens*' is the species.

Evolutionary Trees Show **Relationships**

1) Evolutionary trees show how scientists think different species are related.
2) They show common ancestors and relationships between species.
3) The more recent the common ancestor, the more closely related the two species. Also, the more characteristics they are likely to share.
4) Scientists look at lots of different types of data to work out these relationships. For example:

- For living organisms, they use current classification data.
- For extinct species, they use information from the fossil record (see page 128).

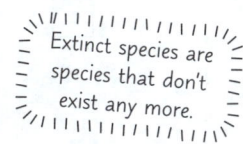
Extinct species are species that don't exist any more.

5) Here's an example of an evolutionary tree:

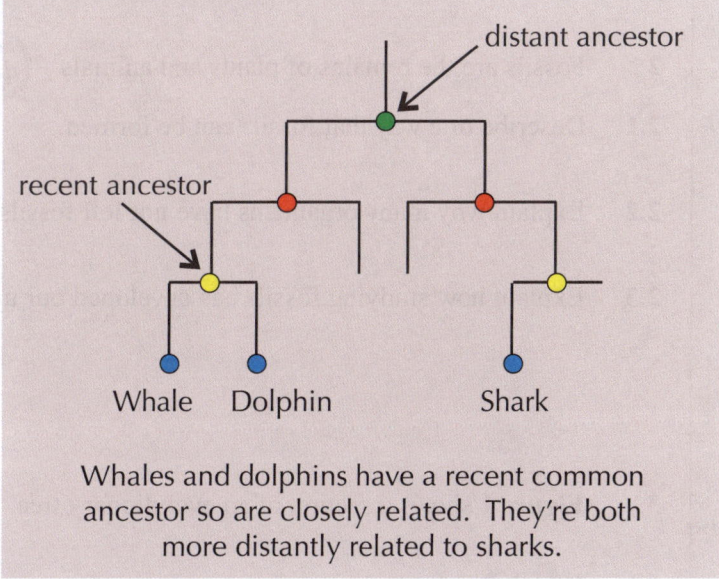

Whales and dolphins have a recent common ancestor so are closely related. They're both more distantly related to sharks.

Binomial system — uh oh, sounds like maths...

Sometimes, the genus in a binomial name is abbreviated to a capital letter with a full stop after it.

Q1 The evolutionary tree on the right shows the relationship between four species, A-D. Which two species shown in the tree are the most closely related? [1 mark]

Q2 What genus does the Eurasian beaver, *Castor fiber*, belong to? [1 mark]

Warm-Up & Exam Questions

The end of the topic is in sight now — just a few more questions to check you've been paying attention.

Warm-Up Questions

1) True or false? The fossil record is incomplete.
2) True or false? Carl Linnaeus classified organisms based on their characteristics.
3) In the binomial naming system, what does the first part of a name refer to?
4) In the binomial naming system, what does the second part of a name refer to?

Exam Questions

1 Scientists organise living organisms into groups. *(Grade 1-3)*

1.1 Which of the following is **not** a domain of the three-domain classification system? Tick **one** box.

☐ Plants   ☐ Eukaryota   ☐ Prokaryotes   ☐ Archaea

[1 mark]

1.2 Which of the following scientists developed the three-domain classification system? Tick **one** box.

☐ Charles Darwin   ☐ Carl Linnaeus   ☐ Carl Woese

[1 mark]

2 Fossils are the remains of plants and animals. *(Grade 4-5)*

2.1 Describe **one** way that fossils can be formed.

[1 mark]

2.2 Explain why many organisms have **not** left fossils behind.

[2 marks]

2.3 Explain how studying fossils has developed our understanding of evolution.

[1 mark]

3 **Figure 1** shows a section of an evolutionary tree. *(Grade 4-5)*

Figure 1

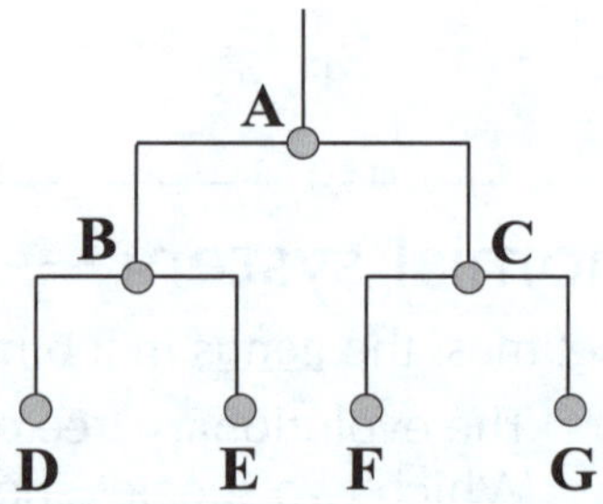

3.1 Which species is the most recent common ancestor of Species **F** and Species **G**?

[1 mark]

3.2 Would you expect Species **D** to look similar to Species **E**? Give a reason for your answer.

[1 mark]

Topic B6 — Inheritance, Variation and Evolution

Topic B7 — Ecology

Competition

*Organisms **interact** with **each other** and their **environment**. This is what **ecology** is all about.*

Learn These Words Before You Start

1) Habitat — the place where an organism lives.
2) Population — all the organisms of one species in a habitat.
3) Community — all the populations of different species in a habitat.
4) Ecosystem — the interaction of a community of organisms with the non-living parts of their environment.

Organisms Compete for Resources to Survive

1) Resources are things that organisms need from their environment and other organisms to survive and reproduce:

   - Animals need food, territory (space) and mates.
   - Plants need light, water, space and mineral ions.

2) Organisms compete with other species (and members of their own species) for the same resources.

Organisms in a Community are Interdependent

1) In a community, different species depend on each other for things like food, shelter, pollination and seed dispersal. This is called interdependence.

2) This means that a big change in one part of an ecosystem (e.g. a species being removed) can affect the whole community.

3) The diagram on the right shows part of a food web (a diagram of what eats what) from a stream.

4) If all the stonefly larvae die, then for example:

   - There would be less food for water boatmen, so their population might decrease.
   - The blackfly larvae would not have to compete with the stonefly larvae for food (algae) so their population might increase.

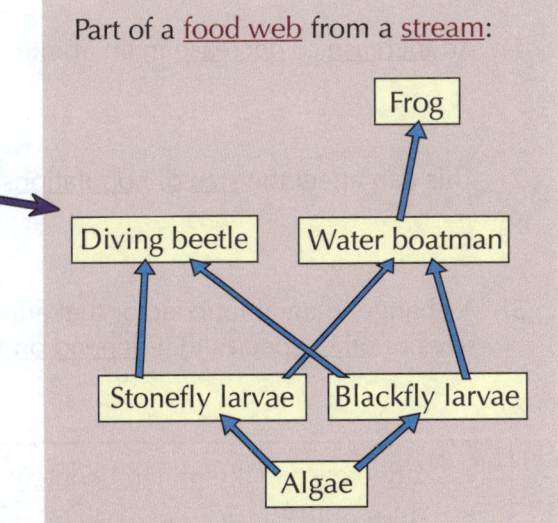

Part of a food web from a stream:

5) In stable communities, all the species and environmental factors are in balance. This means that the population sizes stay about the same.

Survival — the prize for being a winner

Q1   Give three resources that plants compete for. [3 marks]

Q2   Look at the food web above. Suggest why the number of water boatmen might increase if all of the diving beetles died. [2 marks]

Q2 Video Solution

Abiotic and Biotic Factors

The **environment** in which organisms live **changes** all the time. The **things that change** are either *abiotic* (non-living) or *biotic* (living) factors. These changes can have a big **effect** on a community...

Abiotic Factors Can Change in an Ecosystem

Abiotic factors are the non-living factors in an environment. For example:

1) Moisture level

2) Light intensity

3) Temperature

4) Carbon dioxide level (for plants)

5) Wind intensity and direction

6) Oxygen level (for animals that live in water)

7) Soil pH and mineral content

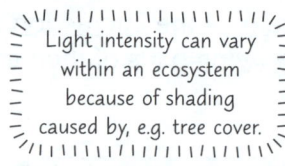

Light intensity can vary within an ecosystem because of shading caused by, e.g. tree cover.

Changes in Abiotic Factors Can Affect Populations

1) An increase or decrease in an abiotic factor is a change in the environment.

2) This can affect the size of populations in a community.

3) A change in an abiotic factor that affects one species could also affect the population sizes of other species that depend on them (see previous page). For example:

- A decrease in the mineral content of the soil could affect the growth of a plant species.
- This could cause a decrease in the population size of the plant species.
- A decrease in the plant population could affect any animal species that depend on it for food.

A = not, biotic = living, so abiotic means non-living

Some human activities can affect the abiotic factors of ecosystems — see pages 153, 156 and 157 for more. As you can see from this page, this can affect some organisms directly and other organisms indirectly.

Topic B7 — Ecology

Abiotic and Biotic Factors

*The previous page shows how abiotic factors can affect the **populations** in an ecosystem, but changes in **biotic factors** can also have big consequences. This page has a few examples to show you how.*

Biotic Factors Can Also Change in an Ecosystem

1) Biotic factors are the living factors in an environment.
2) A change in a biotic factor could affect the population size of some species. This could then affect species that depend on them (see page 141).
3) Biotic factors include:

1. Competition

One species may outcompete another so that numbers are too low to breed.

- Red and grey squirrels live in the same habitat and eat the same food.
- Grey squirrels outcompete the red squirrels for food and shelter.
- So the population of red squirrels is decreasing.

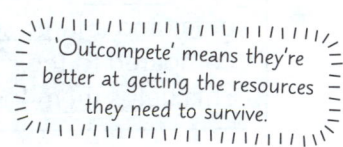
'Outcompete' means they're better at getting the resources they need to survive.

2. Availability of Food

If there is less food available, the population size will decrease.

3. New Predators

Predators are animals that kill other animals

A new predator could cause a decrease in the prey population. There's more about predator-prey populations on page 145.

4. New Pathogens

Pathogens are microorganisms that cause disease.

A new pathogen could quickly decrease the population of an affected species.

Changing biotic factors — it's like dominoes...

Learn the list of factors here, as well as on the previous page. I reckon this is a prime time for shutting the book, scribbling them all down and then checking how you did.

Topic B7 — Ecology

Adaptations

*Life exists in so many **different environments**. It's all because of the **adaptations** that organisms have...*

Adaptations Allow Organisms to Survive

1) Organisms, including microorganisms, are adapted to survive in the conditions of their environment.
2) This means they have special features that suit their environment.
3) These features are called adaptations. Adaptations can be:

1. Structural

These are features of an organism's body structure — such as shape or colour. For example:

Animals that live in hot places (like camels) have a thin layer of fat and a large surface area compared to their volume. This helps them lose heat.

Arctic animals (like the Arctic fox) have white fur so they can't be seen against the snow. This helps them avoid predators and sneak up on prey.

2. Behavioural

These are ways that organisms behave.

E.g. many species (e.g. swallows) migrate (move away) to warmer climates during the winter. So they avoid the problems of living in cold conditions.

3. Functional

These are things that go on inside an organism's body.
They can be related to processes like metabolism (all the chemical reactions happening in the body).

E.g. desert animals make sure they don't lose too much water. They produce very little sweat and small amounts of concentrated urine (wee without much water in it).

Extremophiles Live in Extreme Places

Some microorganisms (e.g. bacteria) are extremophiles — they're adapted to live in extreme conditions. For example:

- at high temperatures (e.g. in super hot volcanic vents)
- in places with a high salt concentration (e.g. very salty lakes)
- at high pressure (e.g. in deep sea vents).

Organisms can adapt to life in really extreme environments

Q1 The diagram on the right shows a penguin. Some of its adaptations are labelled. Penguins live in the cold, icy environment of the Antarctic. They swim in the sea to hunt for fish to eat. Some penguins huddle together in groups to keep warm.
   a) What type of adaptation is being described when penguins 'huddle together'? [1 mark]
   b) Use the labels on the diagram to explain one way that the penguin is adapted to its environment. [2 marks]

Topic B7 — Ecology

Food Chains

*Remember **food webs** from page 141? Well, **food chains** are a similar idea — read on to find out more...*

Food Chains Show What's Eaten by What

1) Food chains always start with a producer.
2) Producers make (produce) their own food using energy from the Sun.
3) Producers are usually green plants or algae — they make glucose by photosynthesis (see page 85).
4) Some of this glucose is used to make the plant's biomass — its mass of living material.
5) Biomass is passed along a food chain when an organism eats another organism.
6) Consumers are organisms that eat other organisms:

- Primary consumers eat producers.
- Secondary consumers eat primary consumers.
- Tertiary consumers eat secondary consumers.

7) Here's an example of a food chain:

Populations of Prey and Predators Go in Cycles

For more about a stable community see page 141.

1) Consumers that hunt and kill other animals are called predators.
2) The animals they eat are called prey.
3) In a stable community, the population size of a species is limited by the amount of food it has. So the population size of predators is affected by the number of their prey.

1) Foxes are predators. Rabbits are their prey.
2) If the number of rabbits increases, then the number of foxes will increase.
3) This is because there is more food for the foxes.
4) But as the number of foxes increases, then the number of rabbits will decrease.
5) This is because more rabbits will be eaten by the foxes.

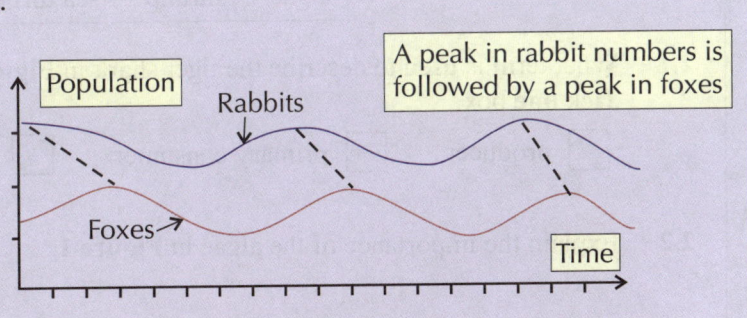

A peak in rabbit numbers is followed by a peak in foxes

4) It takes a while for one population to respond to changes in the other one.

E.g. the number of foxes goes up after the number of rabbits goes up. This is because it takes time for the foxes to reproduce.

A food chain shows part of a food web

Q1 Look at the following food chain for a particular area: grass → grasshopper → rat → snake
   a) Name the producer in the food chain. [1 mark]
   b) How many consumers are there in the food chain? [1 mark]
   c) Name the primary consumer in the food chain. [1 mark]
   d) All the rats in the area are killed.
      Explain two effects that this could have on the food chain. [4 marks]

Q1 Video Solution

Topic B7 — Ecology

Warm-Up & Exam Questions

This ecology topic's a long one — so make sure you've really got these first few pages stuck in your head before moving on and learning the rest. These questions should help you out.

Warm-Up Questions

1) Animals compete with one another for food. Give one other factor that animals compete for.
2) Name two abiotic factors.
3) True or false? Food chains always start with a primary consumer.
4) What would happen to the size of a predator population if there was less prey available?

Exam Questions

1   The Amazon rainforest is the biggest tropical rainforest in the world. *Grade 3-4*

1.1   The Amazon rainforest is a habitat for many different species.
Describe what is meant by a habitat.
[1 mark]

1.2   Which word can be used to describe all the different species living in a habitat?
Tick **one** box.

☐ population   ☐ community   ☐ ecosystem   ☐ distribution
[1 mark]

2   **Figure 1** shows a food chain for a particular area: *Grade 4-5*

Figure 1

algae → shrimp → sea turtle → tiger shark

2.1   What term is used to describe the tiger shark in **Figure 1**?
Tick **one** box.

☐ producer   ☐ primary consumer   ☐ secondary consumer   ☐ tertiary consumer
[1 mark]

2.2   Explain the importance of the algae in **Figure 1**.
[2 marks]

3   The Harris's antelope squirrel lives in hot deserts in parts of the USA and Mexico. *Grade 4-5*
It has grey fur, small ears and does not sweat.
It has sharp claws which enable it to dig burrows, in which it lives.
Above ground, during the hottest parts of the day it often holds its large tail over its head, or lies in the shade with its limbs spread out wide.

3.1   Give **one** functional adaptation the Harris's antelope squirrel has to its environment.
[1 mark]

3.2   Give **two** behavioural adaptations the Harris's antelope squirrel has to its environment.
[2 marks]

Topic B7 — Ecology

Using Quadrats

*This is where the **fun** starts. Studying **ecology** gives you the chance to **rummage around** in bushes, get your hands **dirty** and look at some **real organisms**, living in the wild.*

Differences in the Environment Affect Where Organisms Live

1) The distribution of an organism is where an organism is found.
2) Where an organism is found is affected by biotic and abiotic factors (see pages 142-143).
3) An organism might be more common in one area than another due to differences in factors between the two areas. For example, in a field, you might find daisies are more common in the open than under trees, because there's more light.
4) To study the distribution of an organism you can use quadrats or transects (see next page).

Use Quadrats to Study The Distribution of Small Organisms

Here's how to compare how common an organism is in two different areas — these are called sample areas.

A quadrat (square frame)

1) Place a quadrat on the ground in the first sample area. It needs to be placed at random (see p.400).
2) Count all the organisms you're interested in within the quadrat.
3) Repeat steps 1 and 2 as many times as you can.
4) Work out the mean number of organisms per quadrat within the first sample area.

> **EXAMPLE** Fatima counted the number of daisies in 7 quadrats within her first sample area. She recorded the following results: 18, 20, 22, 23, 23, 23, 25
>
> Here the MEAN is: $\dfrac{\text{TOTAL number of organisms}}{\text{NUMBER of quadrats}} = \dfrac{154}{7}$ = **22 daisies per quadrat**

5) Repeat steps 1 to 4 in the second sample area.
6) Finally compare the two means. E.g. you might find 2 daisies per quadrat in a shady area, and 22 daisies per quadrat (lots more) in a sunny area.

The population size of an organism is sometimes called its abundance.

You Can Work Out the Population Size of an Organism

> **EXAMPLE** Students used quadrats, each with an area of 0.25 m², to randomly sample daisies in a field. They found a mean of 10 daisies per quadrat. The field's area was 800 m². Estimate the population of daisies in the field.
>
> 1) Divide the area of the habitat by the quadrat size.
>   800 ÷ 0.25 = 3200
>
> 2) Multiply this by the mean number of organisms per quadrat.
>   3200 × 10 = **32 000 daisies in the field**

Don't try to study elephants using a quadrat

Q1 A field was randomly sampled for tulips using a quadrat with an area of 0.25 m². The field had an area of 1200 m². A mean of 0.75 tulips was found per quadrat. Estimate the total population of tulips.   [2 marks]

Topic B7 — Ecology

Using Quadrats

*So, now you think you've learnt **all about** distribution. Well **hold on** — there's more **ecology fun** to be had.*

Use **Transects** to **Study** The **Distribution** of Organisms

You can use lines called transects to help find out how organisms are distributed across an area.
E.g. if an organism becomes more or less common as you move from a hedge towards the middle of a field.
Here's what to do:

1) Mark out a line in the area you want to study using a tape measure.
2) Collect data along the line by either:

- Counting all the organisms you're interested in that touch the line.
- Or by using quadrats (see previous page) placed along the line.

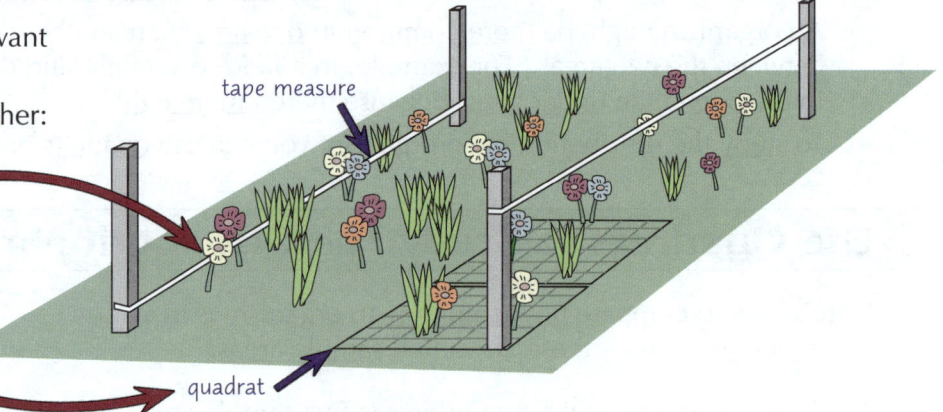

You Can **Estimate** the **Percentage Cover** of a **Quadrat**

1) Sometimes it can be difficult to count all of the organisms in a quadrat (e.g. if they're grass).
2) In this case, you can find the percentage cover instead.
3) This means estimating the percentage area of the quadrat that the organisms cover.
4) You can do this by counting the number of little squares they cover.

EXAMPLE

Some students were measuring the distribution of an organism across a school playing field. They placed quadrats at regular intervals along a transect.
Below is a picture of one of the quadrats.
Calculate the percentage cover of the organism in this quadrat.

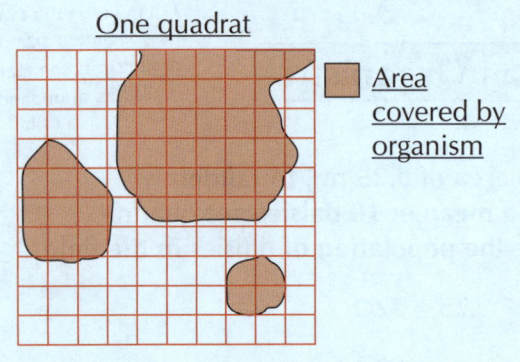

One quadrat — Area covered by organism

1) Count the number of squares covered by the organism. You count a square if it's more than half covered.

47 squares are covered by the organism.

2) Make this into a percentage:
- Divide the number of squares covered by the organism by the total number of squares in the quadrat (100).
- Multiply the result by 100.

(47/100) × 100
= 0.47 × 100 = 47%

You don't need fancy kit to study the distribution of organisms

So if you want to measure the distribution of a organism across an area, you could use a transect. You can use them alone or along with quadrats. Using percentage cover instead of number of organisms is a good way of studying the distribution of plants, as there may be too many to count.

Topic B7 — Ecology

The Water Cycle

The **amount** of water on Earth is pretty much **constant** — but **where** it is changes. Water moves between **rivers**, **lakes**, **oceans** and the **atmosphere** in what's known as the **water cycle**.

The Water Cycle Means Water is Constantly Recycled

1) Energy from the Sun makes water evaporate from the land and sea. This turns the water into water vapour.

2) Water also evaporates from plants — this is called transpiration (see p.68).

3) The warm water vapour is carried upwards. When it gets higher up, the water vapour cools. It condenses to form clouds.

4) Water falls from the clouds as precipitation (usually rain, but sometimes snow or hail). Precipitation provides fresh water for plants and animals:

As warm water vapour rises it cools down and forms clouds.

Plants

1) Some water is absorbed by the soil. Plants take up the water through their roots.
2) Plants need water for things like photosynthesis (p.85).
3) Some water becomes part of the plants' tissues. It's passed to animals when plants are eaten.

Animals

1) Animals need water for the chemical reactions in their bodies.
2) They return water to the soil and atmosphere in their waste (e.g. sweat and urine).

5) Water that doesn't get absorbed by the soil will run off into streams and rivers.

6) The water drains back into the sea. Then it evaporates all over again.

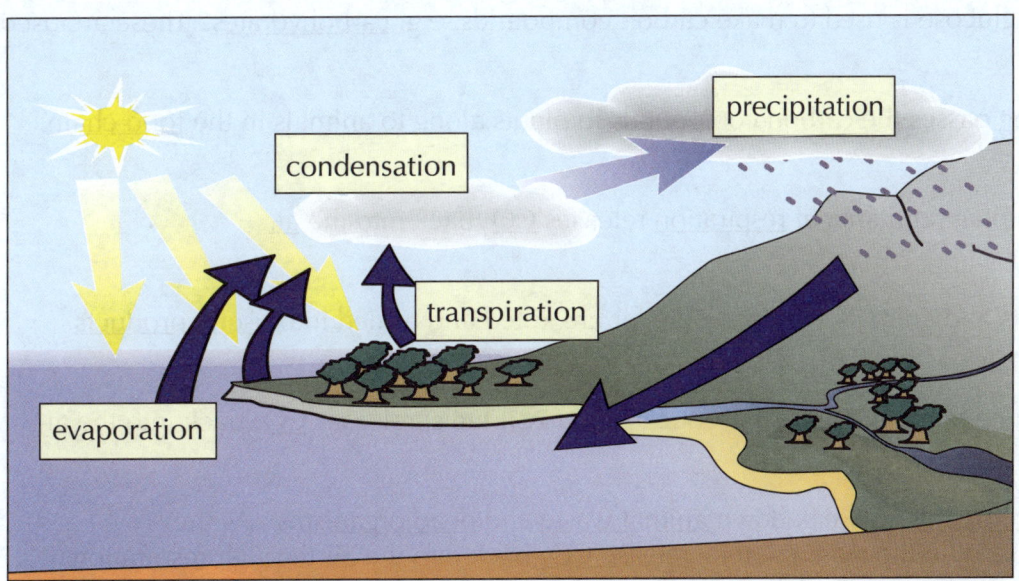

Topic B7 — Ecology

The Carbon Cycle

*All the **nutrients** in our environment get **recycled** — there's a balance between what **goes in** and what **goes out**.*

Materials are Recycled by Decay

1) Living things are made of materials they take from the world around them.

> E.g. plants take up mineral ions from the soil.
> - These are used to make molecules that make up the plant.
> - The molecules are passed up the food chain when the plant is eaten.

2) These materials are returned to the environment in waste products, or when dead organisms decay.
3) Materials decay because they're broken down by microorganisms.
4) Decay puts stuff that plants need to grow (e.g. mineral ions) back into the soil — they are recycled.

The Constant Cycling of Carbon is called the Carbon Cycle

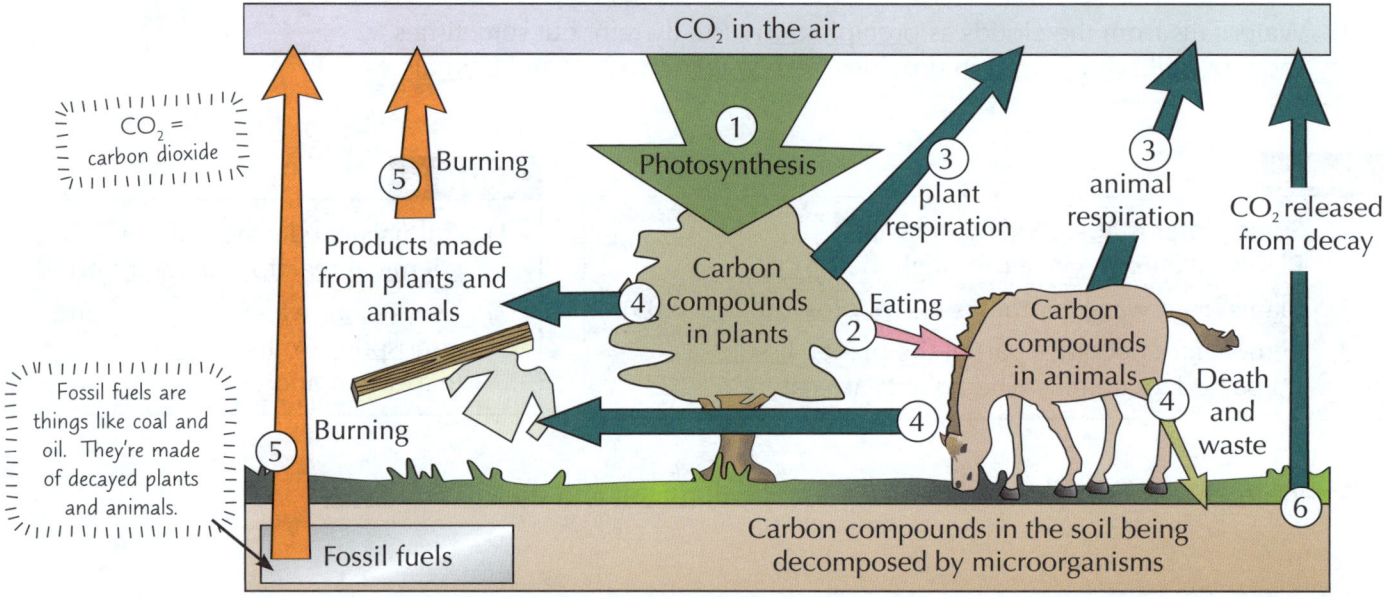

① Plants take in CO_2 from the air during photosynthesis. They use the carbon in CO_2 to make glucose. This glucose is used to make carbon compounds, e.g. carbohydrates. These are used for growth.

② Eating passes the carbon compounds in plants along to animals in the food chain.

③ Both plant and animal respiration releases CO_2 back into the air.

See p.91-92 for more on respiration.

④ Plants and animals eventually die, or are killed and turned into useful products.

⑤ Burning plant and animal products (and fossil fuels) releases CO_2 back into the air.

⑥ Microorganisms break down animal waste and dead organisms. As they break down the material, they release CO_2 back into the air through respiration.

Topic B7 — Ecology

Warm-Up & Exam Questions

You can't just stare at these pages and expect all of the information to go in. Especially the practical pages with the maths examples. Do these questions to see how well you really know the stuff.

Warm-Up Questions

1) Give one piece of equipment that you could use to study the distribution of an organism.
2) Name one type of precipitation.
3) What is the role of microorganisms in the carbon cycle?
4) How is the carbon in fossil fuels returned to the atmosphere?

Exam Questions

1   The water cycle describes the constant movement of water molecules on the Earth.

Use words from the box below to complete the sentences about the water cycle.

| condenses   precipitation   transpiration   evaporate   condensation   condense   evaporates |

Energy from the sun makes water ....

When warm water vapour gets higher up, the water vapour cools and ....

Water falls from the clouds as ....

[3 marks]

PRACTICAL

2   Some students studied the distribution of poppies across a field next to a wood. A sketch of the area is shown in **Figure 1**.

Figure 1

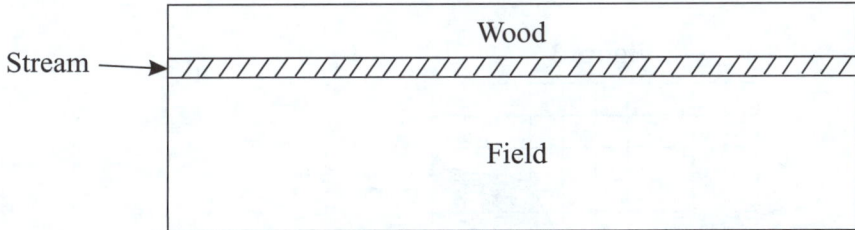

The students' results are shown in **Table 1**.

Table 1

Number of poppies per m²	5	9	14	19	26
Distance from wood (m)	2	4	6	8	10

2.1   Describe the trend shown in **Table 1**.

[1 mark]

2.2   The students suggest that light intensity may affect the distribution of the poppies.
Give **one** other factor that could affect the distribution of the poppies.

[1 mark]

Exam Questions

3 **Figure 2** shows a simplified version of the carbon cycle.

Figure 2

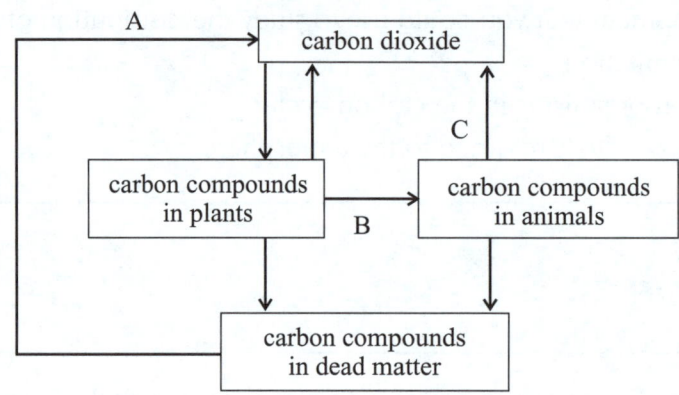

3.1 Name the process that is occurring at stage **B**.

[1 mark]

3.2 Name the process that is occurring at stage **C**.

[1 mark]

3.3 Explain how carbon is released from dead matter in the soil (stage **A**).

[2 marks]

3.4 Name the only process in the carbon cycle that removes carbon dioxide from the air.

[1 mark]

PRACTICAL

4 A group of students decided to study the distribution of a grass species in a field. **Figure 3** shows a sketch of the grass species in one of their quadrats.

Figure 3

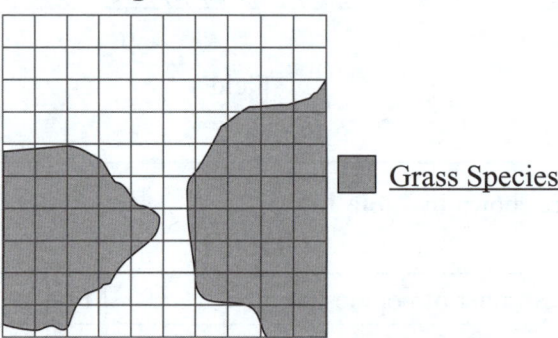

Grass Species

4.1 Use **Figure 3** to estimate the percentage cover of the grass species in the quadrat.

[1 mark]

4.2 Give **one** reason why the students may have decided to estimate the percentage cover of the grass.

[1 mark]

Biodiversity and Waste Management

*Unfortunately, human activity can **negatively affect** the **planet** and its **variety of life**.*

Earth's **Biodiversity** is Important

> Biodiversity is the variety of different species of organisms on Earth, or within an ecosystem.

1) Different species depend on each other for different things in an ecosystem (see page 141).
2) Different species can also help keep the conditions in their environment right for each other, e.g. they can help keep the soil at the right pH.
3) So having a high biodiversity can mean that an ecosystem is more stable.
4) For the human species to survive, it's important that a good level of biodiversity is maintained.
5) Lots of human actions are reducing biodiversity (see below, and pages 155-157).
6) It's only recently that we've started taking measures to stop biodiversity decreasing.

More People Means Greater Demands on the **Environment**

1) The population of the world is increasing very quickly.
2) More people need more resources to survive.
3) People are also demanding a higher standard of living. This means that more people want luxuries that make life more comfortable, e.g. cars, computers, etc.
4) This means that we use more raw materials and more energy to make things.
5) So resources are being used more quickly than they are being replaced.

We're Also Producing **More Waste**

1) As we make more things, we produce more waste. This includes waste chemicals.
2) This waste can cause harmful pollution if it's not handled properly.
3) Pollution kills plants and animals. This reduces biodiversity.
4) Pollution can affect:

Water
- Sewage and toxic chemicals from industry can pollute lakes, rivers and oceans.
- Fertilisers (and other chemicals) used on land can be washed into water.
- This will affect the plants and animals that rely on these sources of water for survival.

Land
- We use toxic chemicals for farming (e.g. pesticides and herbicides).
- We dump a lot of household waste in landfill sites.

Air
- Smoke and acidic gases can pollute the air if they are released into the atmosphere.

Topic B7 — Ecology

Global Warming

*You might remember the **carbon cycle** from p.150. Well, carbon dioxide has an important role in keeping the Earth **warm enough** for life. It's not so good when there's **too much** of it in the atmosphere though...*

Carbon Dioxide and Methane Trap Energy from the Sun

1) Gases in the Earth's atmosphere trap energy from the Sun.

2) These gases mean that not all of the energy is lost into space. This helps to keep the Earth warm.

3) These gases are called greenhouse gases. Without them the Earth would be very cold.

4) But the levels of two greenhouse gases are increasing — carbon dioxide (CO_2) and methane.

5) The increasing levels of greenhouse gases are causing the Earth to heat up — this is global warming.

Topic B7 — Ecology

Global Warming

The Earth is getting **warmer**. Climate scientists study how global warming is **changing** our planet, and the **effects** on people and nature. Sadly, it's not as simple as everyone having nicer summers.

The **Results** of **Global Warming** Could be Pretty **Serious**

There are several reasons to be worried about global warming. Here are a few:

Flooding

1) Higher temperatures cause seawater to expand and ice to melt. This causes the sea level to rise.
2) Sea level rises lead to flooding of low-lying land.
3) This results in the loss of habitats (where organisms live).

Changes in the distribution of species

1) Global warming is causing changes in rainfall and temperature in many areas.
2) This is causing the distribution (spread) of many animal and plant species to change.
3) E.g. in areas that are becoming warmer:
   - Species that do well in warm conditions are spreading further.
   - Species that need cooler temperatures have a smaller area to live in.

Less biodiversity

1) Some species may not be able to survive a change in the climate.
2) These species might become extinct.
3) This could reduce biodiversity (see p.153).

Changes in migration patterns

1) There have been changes in migration patterns (where animals move to during different seasons).
2) E.g. some birds migrate further north, as more northern areas are getting warmer.

We need to act fast, before the worst effects of climate change

Global warming is rarely out of the news. Most scientists accept that it's happening and that human activity has caused most of the recent warming, based on the evidence that has so far been collected. However, they don't know exactly what the effects will be and scientists will have to collect more data before these questions can really be answered.

Topic B7 — Ecology

Deforestation and Land Use

Trees and *peat bogs* trap carbon dioxide and **lock it up**. The problems start when it **escapes**...

Humans Use **Lots of Land** for **Lots of Purposes**

1) We use land for things like building, quarrying, farming and dumping waste.
2) This means that there's less land available for other organisms.
3) Sometimes, the way we use land has a bad effect on the environment.

Destroying Peat Bogs Adds More CO_2 to the Atmosphere

1) Bogs are areas of land that are acidic and waterlogged.

2) Plants that live in bogs don't fully decay when they die. The partly-rotted plants build up to form peat.

3) So the carbon in the plants is stored in the peat.

4) Peat bogs can be drained, so the peat can be sold to gardeners as compost.

5) When the peat is drained, microorganisms can break it down. They release carbon dioxide (CO_2) when they respire.

6) Peat can also be sold as a fuel. CO_2 is released when the peat is burned.

7) Destroying the bogs reduces the area of the habitat.

8) This reduces the number of animals, plants and microorganisms that live there. So it reduces biodiversity.

We need to use land, but we also need a healthy environment

We can't really avoid using land — we need it to grow enough food or build enough houses for people. The human population is increasing so it's likely that we'll use even more land in the future. We'll have to find a way to manage land use to reduce the negative effects on the environment.

Topic B7 — Ecology

Deforestation and Land Use

Many parts of the world have **already** been changed lots by deforestation. For example, much of the **UK** used to be covered in forests. Deforestation can be **bad news** for several reasons.

Deforestation Means Chopping Down Trees

Deforestation is the cutting down of forests. It is done for many reasons, like:

- To clear land for farming (e.g. cattle or rice crops) to provide more food.
- To grow crops to make biofuels.

Deforestation Can Cause Many Problems

Deforestation causes big problems when it's done on a large-scale, e.g. cutting down rainforests in tropical areas. These are:

Less carbon dioxide taken in

- Trees take in carbon dioxide from the atmosphere during photosynthesis (page 85).
- So cutting down trees means that less carbon dioxide is removed from the atmosphere.
- Trees 'lock up' some of the carbon in their wood. Removing trees means that less is locked up.

More carbon dioxide released

- Carbon dioxide is released when trees are burnt to clear land.
- Microorganisms feeding on dead wood release carbon dioxide through respiration (p.91).

More CO_2 in the atmosphere causes global warming (see p.154).

Less biodiversity

- Habitats like forests can contain many species of plants and animals — they have high biodiversity.
- When forests are destroyed, many species may become extinct (p.129). This reduces biodiversity.

Not a very cheerful page, I know...

Make sure you can link together all the information on pages 154-157 — for example, how deforestation and peat burning can contribute to global warming, and how this might affect biodiversity. In the exam, you might get an extended response question that requires you to draw on several different areas of knowledge like this.

Maintaining Ecosystems and Biodiversity

It's really important that biodiversity is **maintained** as damage to ecosystems or populations of species can be **hard to undo**. This page is about some of the different **methods** that can be used to maintain biodiversity.

There are Programmes to Protect Ecosystems and Biodiversity

1) Human activities can reduce biodiversity and damage ecosystems.
2) In some areas, programmes to minimise the damage have been set up by concerned citizens and scientists. Here are a few examples:

1. Breeding Programmes

1) Animal species that are at risk of dying out are called endangered species.
2) They can be bred in captivity.
3) This makes sure some individuals will survive if the species dies out in the wild.
4) Individuals can sometimes be released into the wild. This can be to boost a population or replace one that's been wiped out.

Pandas are an endangered species. Many efforts have been made to breed pandas in captivity.

2. Habitat Protection

Protecting and regenerating (rebuilding) rare habitats helps to protect the species that live there.

3. Reintroducing Hedgerows and Field Margins

1) Field margins are areas of land around the edges of fields where wild flowers and grasses are left to grow.
2) Hedges can be planted around fields to form hedgerows.
3) Hedgerows and field margins provide a habitat for lots of types of organisms.
4) This is very useful for fields that only have one type of crop. This is because these fields have very low biodiversity.

4. Recycling

1) This reduces the amount of waste that gets dumped in landfill sites.
2) This could reduce the amount of land taken over for landfill. So ecosystems can be left alone.

5. Government Programmes

1) Deforestation increases the amount of carbon dioxide in the atmosphere (see previous page).
2) Some governments have made rules to reduce deforestation.
3) They have also made rules to reduce the amount of carbon dioxide released by businesses.
4) This could help to stop global warming increasing (see page 154).

Warm-Up & Exam Questions

I hope you've got all of that important information in your head. There's a lot to remember here, so have a flick back when you're doing these questions in case you've forgotten any little details.

Warm-Up Questions

1) What is meant by 'biodiversity'?
2) Give one way in which land can become polluted.
3) Give two greenhouse gases.
4) True or false? Destroying peat bogs adds more CO_2 to the atmosphere.
5) What is meant by the term 'deforestation'?

Exam Questions

1   Ecosystems and biodiversity can be protected by programmes set up by scientists.

1.1 Hedgerows can be reintroduced around fields in order to protect biodiversity.

Give **one** reason why reintroducing hedgerows around fields is important for biodiversity.

[1 mark]

1.2 Breeding programmes can be used to protect endangered species.

Explain how breeding programmes can be used to prevent an endangered species from dying out.

[3 marks]

2   Humans are producing increasing amounts of waste.
This has negative impacts for biodiversity and the environment.

2.1 Give **two** types of waste that pollute the air.

[2 marks]

2.2 Describe **one** way in which waste produced by humans pollutes water.

[1 mark]

2.3 Give **two** reasons why humans are producing increasingly more waste.

[2 marks]

2.4 Recycling reduces the amount of waste that gets dumped at landfill sites.
Explain how this helps to protect biodiversity.

[1 mark]

3* Most scientists accept that global warming is happening.
They believe that it could cause major changes to our environment.

Explain why deforestation contributes to global warming.

[6 marks]

Topic B7 — Ecology

Revision Summary for Topics B5-7

That's Topics B5-7 done. I bet you're in the mood for a list of revision questions now. You're in luck.
- Try these questions and tick off each one when you get it right.
- When you're completely happy with a topic, tick it off.

For even more practice, try the Retrieval Quizzes for Topics B5-7 — just scan the QR codes!

Topic B5 — Homeostasis and Response (p.97-111) ☑
1) What is homeostasis?
2) What makes up the central nervous system?
3) What is the purpose of a reflex reaction?
4) Name one factor that can affect reaction time.
5) Give two differences between nervous and hormonal responses.
6) What does luteinising hormone (LH) do?
7) What hormone does the contraceptive injection contain?
8) Which of the following is a hormonal contraceptive — condom, implant or diaphragm?

Topic B6 — Inheritance, Variation and Evolution (p.113-139) ☑
9) What do genes code for?
10) What is the name for all of the genetic material in an organism?
11) Name the male and female gametes of animals.
12) What is the probability that offspring will have the XX combination of sex chromosomes?
13) Which of these genotypes is heterozygous — FF, Ff or ff?
14) Name one inherited disorder.
15) Give one argument against screening embryos for inherited disorders.
16) What is variation?
17) Name the process by which evolution happens.
18) What leads to the formation of antibiotic-resistant strains of bacteria?
19) What is selective breeding?
20) What is genetic engineering?
21) What are fossils?
22) What is the smallest group in the Linnaean system of classification?

Topic B7 — Ecology (p.141-158) ☑
23) Define 'ecosystem'.
24) What things do plants compete for in an ecosystem?
25) What are biotic factors?
26) What is meant by the term 'biomass'?
27) What do primary consumers eat?
28) Explain how a quadrat can be used to investigate the distribution of clover plants in two areas.
29) What role do plants have in the carbon cycle?
30) Suggest why it's important to have high biodiversity in an ecosystem.
31) Give one way global warming could reduce biodiversity.
32) Explain why the destruction of peat bogs adds more carbon dioxide to the atmosphere.

Topic C1 — Atomic Structure and the Periodic Table

Atoms

*All substances are made of **atoms**. They're really **tiny** — too small to see, even with your microscope. Atoms are so tiny that a **50p piece** contains about 77 400 000 000 000 000 000 000 of them.*

Atoms Contain **Protons**, **Neutrons** and **Electrons**

Atoms have a radius of about 0.1 nanometres (that's 1×10^{-10} m). There are a few different modern models of the atom — but chemists tend to like the model on the right best.

A nanometre (nm) is 0.000000001 m. Shown in standard form, that's 1×10^{-9} m. Standard form is used for showing really large or really small numbers.

The **Nucleus**

1) It's in the middle of the atom.
2) It contains protons and neutrons.
3) The nucleus has a radius of around 1×10^{-14} m (that's around 1/10 000 of the atomic radius).
4) It has a positive charge because of the protons.
5) Almost the whole mass of the atom is in the nucleus.

The **Electrons**

1) Move around the nucleus in electron shells (levels).
2) They're negatively charged.
3) Electrons have almost no mass.

Particle	Relative Mass	Relative Charge
Proton	1	+1
Neutron	1	0
Electron	Very small	−1

You need to know the charges of protons, neutrons and electrons.
You also need to know how heavy they are compared to each other (their relative masses).
- Protons are heavy (compared to electrons) and positively charged.
- Neutrons are heavy (compared to electrons) and neutral.
- Electrons are tiny and negatively charged.

Atomic Number and **Mass Number** Describe an Atom

1) The nuclear symbol of an atom tells you its atomic (proton) number and mass number.

atomic number = number of protons
mass number = number of protons + number of neutrons
number of neutrons = mass number − atomic number

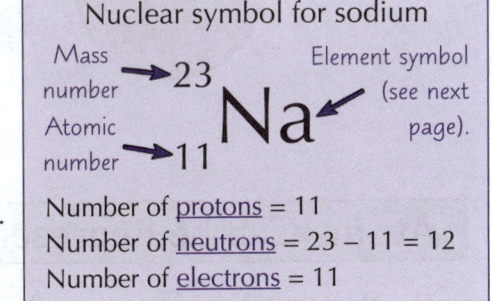

Nuclear symbol for sodium
Mass number → 23
Atomic number → 11
Na Element symbol (see next page).

Number of protons = 11
Number of neutrons = 23 − 11 = 12
Number of electrons = 11

2) Atoms have no charge overall. They're neutral.
3) This is because they have the same number of protons as electrons. So, in an atom...

number of electrons = atomic number

4) The charge on the electrons is the same size as the charge on the protons, but opposite. So they cancel out.
5) In an ion, the number of protons doesn't equal the number of electrons. This means it has an overall charge. For example, an ion with a 2− charge, has two more electrons than protons.

In a positive ion: number of electrons = atomic number − charge
In a negative ion: number of electrons = atomic number + charge

An ion is an atom or group of atoms that has lost or gained electrons.

The charges of the protons and electrons cancel out in an atom

Q1 An atom of gallium has an atomic number of 31 and a mass number of 70. Give the number of electrons, protons and neutrons in the atom. *[3 marks]*

Q1 Video Solution

Elements

*An **element** is a substance made up of atoms that all have the **same** number of **protons** in their nucleus.*

Elements are Made Up of Atoms With the Same Atomic Number

1) The smallest part of an element that you can have is a single atom of that element.

2) The number of protons in the nucleus decides what type of atom it is.

> For example, an atom with one proton in its nucleus is hydrogen.
> An atom with two protons is helium.

3) If a substance only contains atoms with the same number of protons it's called an element.

4) So all the atoms of a particular element have the same number of protons.
And different elements have atoms with different numbers of protons.

5) There are about 100 different elements.
Each element is made up of only one type of atom — some examples are shown in the diagram below.

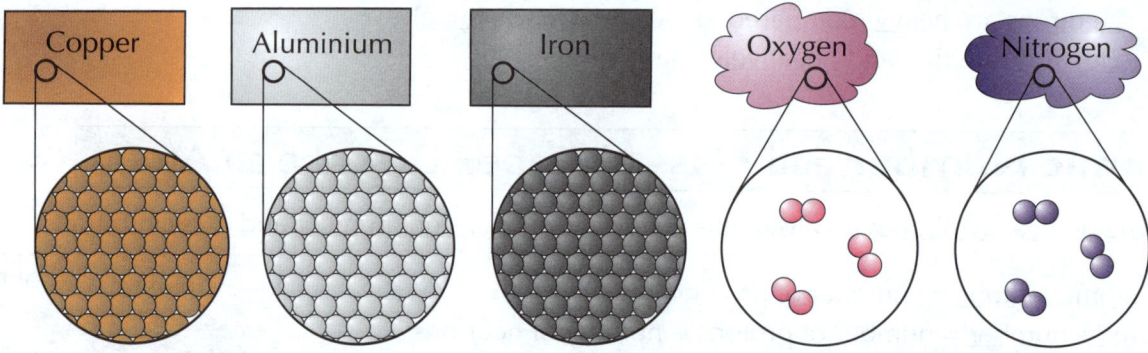

Atoms Can be Represented by Symbols

1) Atoms of each element can be represented by a one or two letter symbol.
2) You'll see these symbols on the periodic table (see page 180).
3) For example:

| C = carbon | O = oxygen | Na = sodium | Mg = magnesium | Fe = iron |

All atoms in an element have the same number of protons

Atoms and elements — make sure you know what they are and the differences between them. You don't need to learn all the symbols as you can use a periodic table, but it's handy to know the common ones.

Isotopes

What's inside different atoms of the **same element** can vary. Read on to find out how...

Isotopes are the Same Except for Extra Neutrons

1) Isotopes are:

   > Atoms with the same number of protons but a different number of neutrons

2) So they have the same atomic number but different mass numbers.

   The number of neutrons is the mass number minus the atomic number.

3) Carbon-12 and carbon-13 are isotopes of carbon.

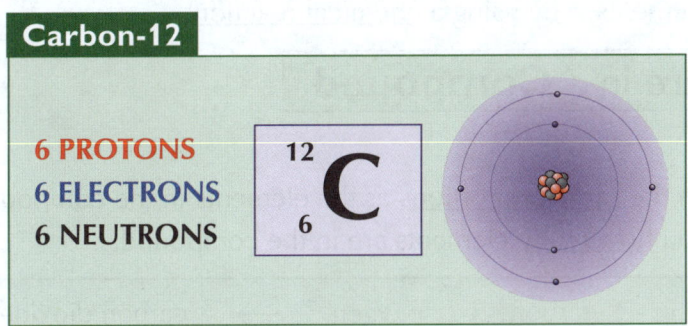

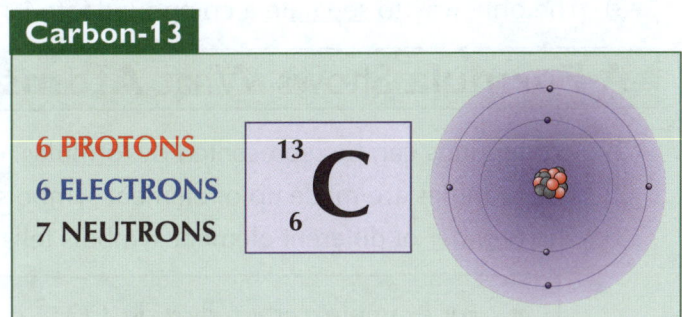

4) If an element has a number of isotopes, you can describe it using relative atomic mass (A_r) instead of mass number. This is an average mass.

5) A_r is worked out from the different masses and abundances (amounts) of each isotope.

   You can work out the 'sum of' two or more amounts by adding them together.

$$\text{relative atomic mass } (A_r) = \frac{\text{sum of (isotope abundance} \times \text{isotope mass number)}}{\text{sum of abundances of all the isotopes}}$$

6) You can use this formula to work out the relative atomic mass of an element:

EXAMPLE Copper has two stable isotopes. Cu-63 has an abundance of 69.2% and Cu-65 has an abundance of 30.8%. Calculate the relative atomic mass of copper to 1 decimal place.

abundance × mass number of Cu-63   abundance × mass number of Cu-65

$$\text{Relative atomic mass} = \frac{(69.2 \times 63) + (30.8 \times 65)}{69.2 + 30.8} = \frac{4359.6 + 2002}{100} = \frac{6361.6}{100} = 63.616 = \mathbf{63.6}$$

abundance of Cu-63 + abundance of Cu-65

Relative atomic mass is the average atomic mass of an element

Make sure you understand and know how to use the formula to find the A_r of an element.

Q1 What are isotopes? [1 mark]

Q2 Silicon, Si, has three stable isotopes. Si-28 has an abundance of 92.2%, Si-29 has an abundance of 4.7% and Si-30 has an abundance of 3.1%. Calculate silicon's relative atomic mass to 1 decimal place. [2 marks]

Topic C1 — Atomic Structure and the Periodic Table

Compounds

*It would be great if we only had to deal with elements. But unluckily for you, elements can mix and match to make lots of new substances called **compounds**. And this makes things a little bit more complicated...*

Atoms Join Together to Make Compounds

1) During a chemical reaction, at least one <u>new</u> substance is made. You can usually measure a <u>change in energy</u> such as a temperature change, as well.
2) When <u>two or more elements react</u>, they form <u>compounds</u>. <u>Compounds</u> are substances that contain atoms of different elements.
3) The atoms of each element are in <u>fixed proportions</u> (amounts) in the compound.
4) The atoms are held together by <u>chemical bonds</u>.
5) The only way to <u>separate</u> a compound into its elements is by using a <u>chemical reaction</u>.

There are different types of compound. For example, ionic compounds (see p.192-194) and covalent compounds (see p.196-198).

A Formula Shows What Atoms are in a Compound

1) Compounds can be represented by <u>formulas</u>.
2) The formulas are made up of element symbols in the <u>same proportions</u> as the elements in the compound.
3) The <u>number</u> of different element symbols tells you <u>how many</u> elements are in the compound.

- For example, carbon dioxide, CO_2, is a <u>compound</u> made from a <u>reaction</u> between carbon and oxygen.
- It contains <u>1 carbon atom</u> and <u>2 oxygen atoms</u>.

carbon + oxygen → carbon dioxide

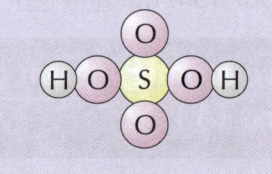

- Here's another example: the formula of <u>sulfuric acid</u> is H_2SO_4.
- So, each molecule contains <u>2 hydrogen atoms</u>, <u>1 sulfur atom</u> and <u>4 oxygen atoms</u>.

As an element, oxygen goes around in pairs of atoms (so it's O_2).

- There might be <u>brackets</u> in a formula, e.g. calcium hydroxide is $Ca(OH)_2$.
- The little number 2 outside the bracket means there's <u>two of everything</u> inside the brackets.
- So in $Ca(OH)_2$ there's <u>1 calcium atom</u>, <u>2 oxygen atoms</u> and <u>2 hydrogen atoms</u>.

4) Here are some examples of formulas which might come in handy:

Carbon dioxide — CO_2	Sodium chloride — NaCl	Calcium chloride — $CaCl_2$
Ammonia — NH_3	Carbon monoxide — CO	Sodium carbonate — Na_2CO_3
Water — H_2O	Hydrochloric acid — HCl	Sulfuric acid — H_2SO_4

The atoms in compounds are chemically joined

In your exams, you could be asked to give the name of a compound from its formula. Make sure you get lots of <u>practice</u> using the <u>periodic table</u> to find which elements the different <u>symbols</u> stand for — you'll save a lot of time if you <u>already know</u> where to look.

Q2 Video Solution

Q1 How many atoms are in one particle of Na_2CO_3? [1 mark]

Q2 A compound has the formula $Al_2(SO_4)_3$. Name the elements and state how many atoms of each element are represented in its formula. [1 mark]

Topic C1 — Atomic Structure and the Periodic Table

Chemical Equations

Chemical equations are used to show what is happening to substances involved in **chemical reactions**. They tell us what **atoms** are involved and how the substances change during a reaction.

Chemical Reactions are Shown Using Chemical Equations

1) One way to show a chemical reaction is to write a word equation.

2) Word equations show the names of the chemicals that are reacting and being produced.

> Here's an example — methane reacts with oxygen to make carbon dioxide and water:
>
> **methane + oxygen → carbon dioxide + water**
>
> The chemicals on the left-hand side of the equation are called the reactants (because they react with each other).
> The chemicals on the right-hand side are called the products (because they've been produced from the reactants).

Symbol Equations Show the Atoms on Both Sides

1) Chemical reactions can be shown using symbol equations.

2) Symbol equations just show the symbols or formulas of the reactants and products.

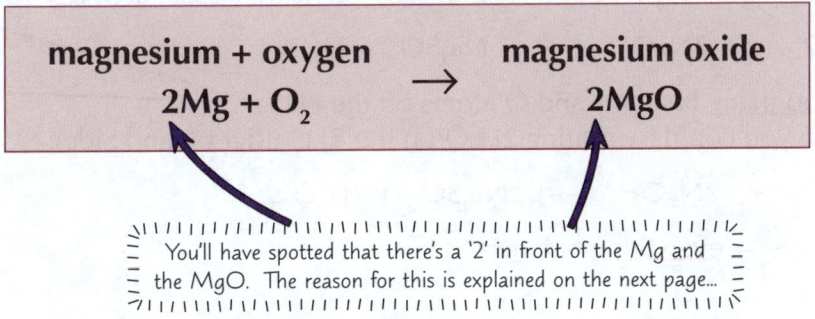

magnesium + oxygen → magnesium oxide
$2Mg + O_2$ → $2MgO$

You'll have spotted that there's a '2' in front of the Mg and the MgO. The reason for this is explained on the next page...

Symbol equations give you extra information

Writing symbol equations makes it easier to see how many atoms are involved in a reaction, and where they all end up. It also saves quite a lot of time, so it's no wonder scientists prefer them to word equations. You'll need to make sure you're comfortable using both for your exams though.

Topic C1 — Atomic Structure and the Periodic Table

Chemical Equations

Symbol Equations Need to be Balanced

1) There must always be the same number of atoms on both sides — they can't just disappear.
2) You balance the equation by putting numbers in front of the formulas.
Take this equation for reacting sulfuric acid with sodium hydroxide:

$$H_2SO_4 + NaOH \rightarrow Na_2SO_4 + H_2O$$

Left-hand side	Right-hand side
H = 3	H = 2
S = 1	S = 1
O = 5	O = 5
Na = 1	Na = 2

3) The formulas are all correct but the numbers of some atoms don't match up on both sides.
4) This equation needs balancing — see below for how to do this.

Here's How to Balance an Equation

The more you practise, the quicker you get, but all you do is this:

1) Find an element that doesn't balance and pencil in a number in front of one of the substances to try and sort it out.
2) See where it gets you. The equation may still not be balanced. Don't worry, just pencil in another number and see where that gets you.
3) Keep doing this until the equation is completely balanced.

You can't change formulas like H_2SO_4 to H_2SO_5. You can only put numbers in front of them.

EXAMPLE

In the equation above you'll notice we're short of Na atoms on the LHS (Left-Hand Side).

1) The only thing you can do about that is make it 2NaOH instead of just NaOH:

$$H_2SO_4 + 2NaOH \rightarrow Na_2SO_4 + H_2O$$

	LHS	RHS
H	= 4	= 2
S	= 1	= 1
O	= 6	= 5
Na	= 2	= 2

2) But that now gives too many H atoms and O atoms on the LHS. So to balance that up you could try putting $2H_2O$ on the RHS (Right-Hand Side):

$$H_2SO_4 + 2NaOH \rightarrow Na_2SO_4 + 2H_2O$$

3) And suddenly there it is — everything balances.

Getting good at balancing equations takes patience and practice

Remember, a number in front of a formula applies to the entire formula — so, $3CH_4$ means three lots of CH_4. The little numbers within or at the end of a formula only apply to the atom or brackets immediately before. So the 4 in CH_4 means there are 4 Hs, but there's just 1 C, not 4.

Q2 Video Solution

Q1 Balance the equation: $Fe + Cl_2 \rightarrow FeCl_3$ [1 mark]

Q2 Hydrogen and oxygen molecules are formed in a reaction where water splits apart.
For this reaction: a) State the word equation.
b) Give a balanced symbol equation. [3 marks]

Topic C1 — Atomic Structure and the Periodic Table

Warm-Up & Exam Questions

So, you reckon you know your elements from your compounds? Have a go at these questions and see how you do. If you get stuck on something, just flick back and give it another read through.

Warm-Up Questions

1) What does the mass number tell you about an atom?
2) What is the definition of an element?
3) True or false? A compound may contain atoms of only one element.
4) Name the compound that has the formula $CaCl_2$.
5) Balance this equation for the reaction of potassium (K) and water (H_2O):
   $K + H_2O \rightarrow KOH + H_2$

Exam Questions

1 This question is about atomic structure.

1.1 Which row of **Table 1** correctly shows the relative charges of a proton and a neutron?
Tick **one** box.

Table 1

Relative charge of a proton	Relative charge of a neutron	
+1	−1	☐
+1	0	☐
0	+1	☐
−1	0	☐

[1 mark]

1.2 Where are protons and neutrons found in an atom?
Tick **one** box.

☐ spread around the atom   ☐ in shells around the nucleus   ☐ in the nucleus   ☐ in the electrons

[1 mark]

1.3 An atom has 8 electrons. How many protons does the atom have?

[1 mark]

1.4 What is the relative mass of a proton?

[1 mark]

2 Table 2 gives some information about sodium.

Table 2

Element	Number of protons	Mass number
sodium	11	23

2.1 How many neutrons does sodium have?

[1 mark]

2.2 Give the atomic number of sodium.

[1 mark]

2.3 What is the chemical symbol for sodium?

[1 mark]

Topic C1 — Atomic Structure and the Periodic Table

Exam Questions

3 Methane (CH$_4$) reacts with oxygen (O$_2$) to make carbon dioxide (CO$_2$) and water (H$_2$O). *Grade 3-4*

3.1 Give the names of the reactants in this reaction.

[1 mark]

3.2 Give the names of the products in this reaction.

[1 mark]

3.3 Which molecule involved in the reaction is composed of only one element?

[1 mark]

3.4 Balance the symbol equation for the reaction below.

$$CH_4 + ...O_2 \rightarrow CO_2 + ...H_2O$$

[1 mark]

4 Sulfuric acid (H$_2$SO$_4$) reacts with ammonia (NH$_3$) to form ammonium sulfate, (NH$_4$)$_2$SO$_4$. *Grade 4-5*

4.1 Complete and balance the symbol equation below.

$$.................. + ... \rightarrow (NH_4)_2SO_4$$

[2 marks]

4.2 How many different elements are there in this reaction?

[1 mark]

4.3 How many hydrogen atoms are there in the formula of ammonium sulfate?

[1 mark]

5 The element boron has two isotopes, boron-10 and boron-11. Details about the boron-11 isotope are shown below. *Grade 4-5*

$$^{11}_{5}B$$

5.1 Complete the sentence below explaining what an isotope is. Use words from the box.

| protons | shells | neutrons | nuclei | electrons |

Isotopes are different forms of the same element, which have the

same number of .... but different numbers of ....

[2 marks]

5.2 How many protons are there in an atom of boron-11?

[1 mark]

5.3 Boron-10 has a mass number of 10 and a percentage abundance of 20%.
Boron-11 has a mass number of 11 and a percentage abundance of 80%.
Use the equation below to calculate the relative atomic mass of boron.

$$\text{relative atomic mass} = \frac{\text{sum of (isotope abundance} \times \text{isotope mass number)}}{\text{sum of abundances of all the isotopes}}$$

[3 marks]

Topic C1 — Atomic Structure and the Periodic Table

Mixtures

Mixtures in chemistry are just like mixtures in everyday life, lots of separate things all mixed together.

Mixtures are Easily Separated — Not Like Compounds

1) Mixtures contain at least two different elements or compounds.
2) There aren't any chemical bonds between the different parts of a mixture.

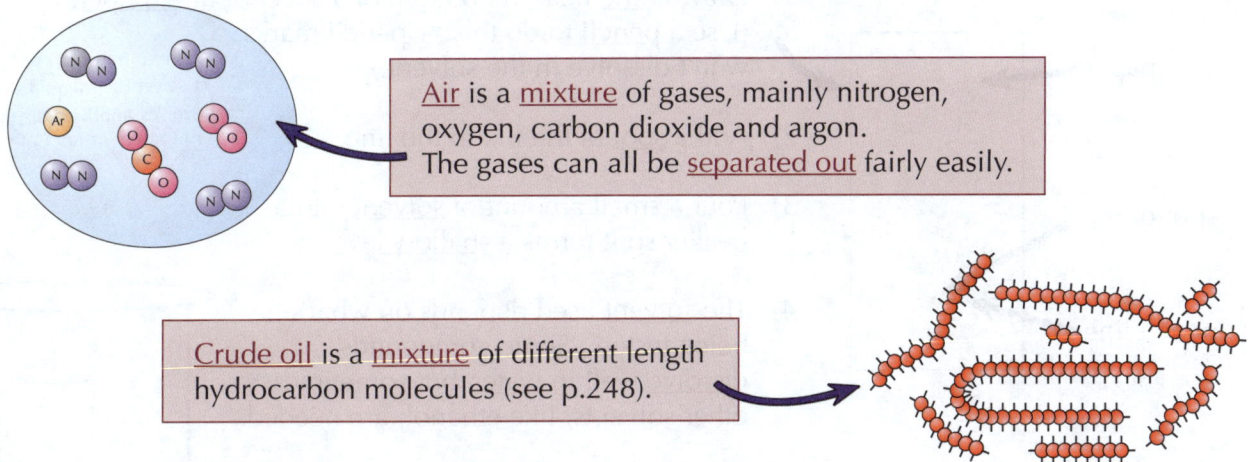

Air is a mixture of gases, mainly nitrogen, oxygen, carbon dioxide and argon. The gases can all be separated out fairly easily.

Crude oil is a mixture of different length hydrocarbon molecules (see p.248).

3) The different parts of a mixture can be separated out by methods such as:

- filtration (p.171)
- crystallisation (p.171)
- simple distillation (p.173)
- fractional distillation (p.174)
- chromatography (p.170)

4) The methods are all physical methods.
   This means they don't involve any chemical reactions, and don't form any new substances.

Each Part of a Mixture Keeps Its Own Properties

1) Properties describe what a substance is like and how it behaves, such as hardness or boiling point.
2) The properties of a mixture are just a mixture of the properties of the separate parts. The chemical properties of a substance aren't changed by it being part of a mixture.

For example, a mixture of iron powder and sulfur powder will show the properties of both iron and sulfur. It will contain grey magnetic bits of iron and bright yellow bits of sulfur.

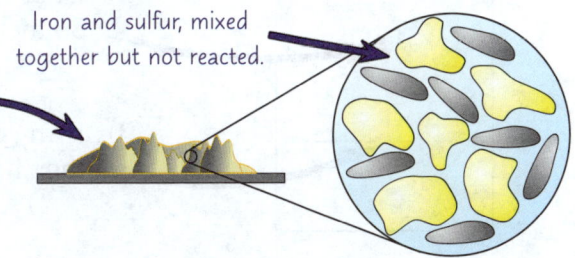

Iron and sulfur, mixed together but not reacted.

Mixtures can be separated without a chemical reaction

Remember that the different parts of mixtures aren't joined together chemically. This means their chemical properties are not changed and the compounds or elements can be separated by physical methods.

Topic C1 — Atomic Structure and the Periodic Table

Chromatography

Paper chromatography is a really useful technique to separate the compounds in a mixture.

You Need to Know How to Do Paper Chromatography

One method of separating substances in a mixture is chromatography. Chromatography can be used to separate different dyes in an ink. Here's how you can do it:

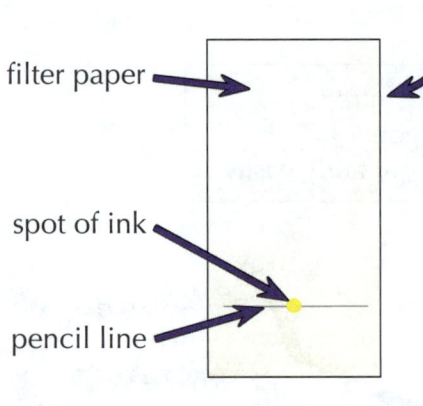

1) Draw a line near the bottom of a sheet of filter paper. (Use a pencil to do this — pencil marks won't dissolve in the solvent.)

A solvent is a liquid which dissolves another substance.

2) Add a spot of the ink to the line.

3) Pour a small amount of solvent into a beaker so it forms a shallow layer.

4) The solvent used depends on what's being tested. Some compounds dissolve well in water, but sometimes other solvents, like ethanol, are needed.

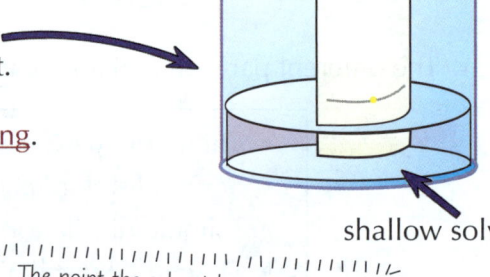

shallow solvent

5) Place the sheet in the beaker of solvent. Make sure the ink isn't touching the solvent — you don't want it to dissolve into it.

6) Place a lid on top of the container to stop the solvent evaporating.

7) The solvent seeps up the paper, carrying the ink with it.

8) When the solvent has nearly reached the top of the paper, take the paper out of the beaker and leave it to dry.

The point the solvent has reached as it moves up the paper is the solvent front.

9) The end result is a pattern of spots called a chromatogram.

Chromatography Separates the Parts of a Mixture

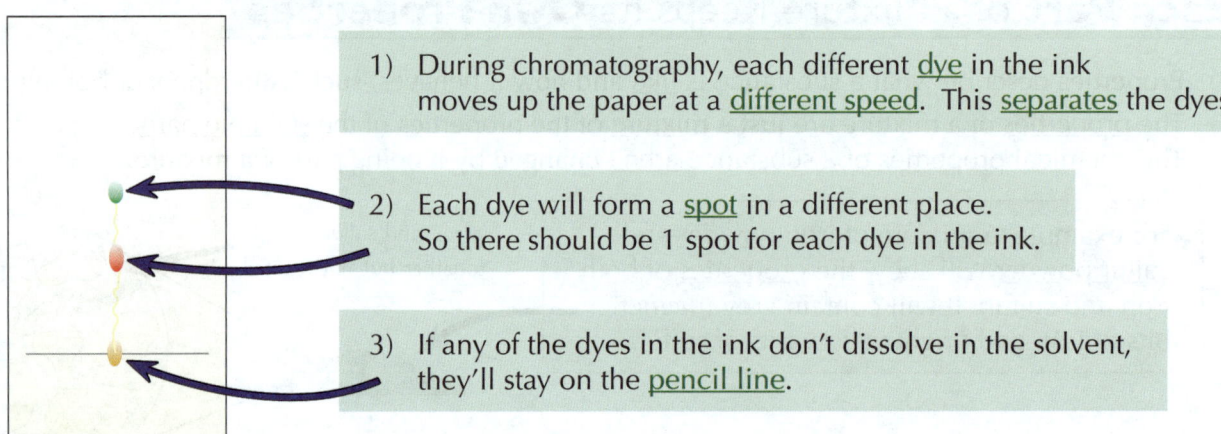

1) During chromatography, each different dye in the ink moves up the paper at a different speed. This separates the dyes.

2) Each dye will form a spot in a different place. So there should be 1 spot for each dye in the ink.

3) If any of the dyes in the ink don't dissolve in the solvent, they'll stay on the pencil line.

Chromatography separates the different dyes in inks

PRACTICAL TIP: Make sure you use a pencil to draw your baseline on the sheet of paper for your chromatogram. If you use a pen, all the components of the ink in the pen will get separated, along with the substance you're analysing, which will make your results very confusing.

Filtration and Crystallisation

*Filtration and crystallisation are **methods** of **separating mixtures**. Chemists use these methods all the time to separate **solids** from **liquids**, so it's worth making sure you know how to do them.*

Filtration Separates Insoluble Solids from Liquids

Filtration can be used to separate an insoluble solid from a liquid reaction mixture (insoluble solids can't be dissolved in the liquid). This can help make substances pure.

1) Put some filter paper in a funnel.
2) Pour the mixture into the filter paper.
3) Make sure the mixture doesn't go above the filter paper.
4) The liquid passes through the paper into the beaker. The solid is left behind in the filter paper.

Filter paper folded into a cone shape — the solid is left in the filter paper.

Two Ways to Separate Soluble Solids from Solutions

If a solid can be dissolved we say it's soluble. There are two methods you can use to separate a soluble salt from a solution — evaporation and crystallisation.

Evaporation

1) Slowly heat the solution in an evaporating dish. The solvent will evaporate.
2) Eventually, crystals will start to form.
3) Keep heating until all you have left are dry crystals.

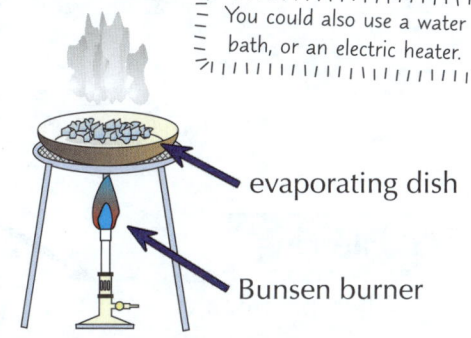

You could also use a water bath, or an electric heater.

evaporating dish

Bunsen burner

Evaporation is a really quick way of separating a soluble salt from a solution. But, you can only use it if the salt doesn't break down when it's heated. Otherwise, you'll have to use crystallisation.

Crystallisation

1) Gently heat the solution in an evaporating dish. Some of the solvent will evaporate.
2) Once some of the solvent has evaporated, or when you see crystals start to form, stop heating. Leave the solution to cool.
3) The salt should start to form crystals.
4) Filter the crystals out of the solution. Leave them in a warm place to dry.

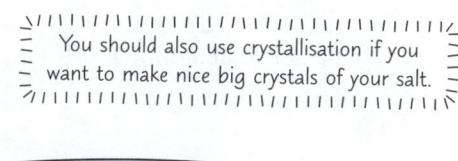

You should also use crystallisation if you want to make nice big crystals of your salt.

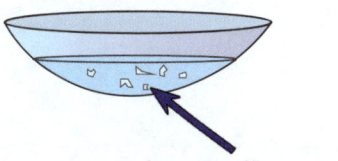

Salt crystallising out of solution.

Remember — soluble solids will dissolve in a solvent...

...but insoluble solids won't. The method you use to separate a solid from a solution depends on whether it's soluble or insoluble, so make sure you know the difference between the two.

Topic C1 — Atomic Structure and the Periodic Table

Filtration and Crystallisation

*Here's how you can put filtration and crystallisation to **good use**. Separating rock salt...*

Filtration and Crystallisation can be Used to Separate Rock Salt

1) Rock salt is a mixture of salt and sand. Salt dissolves in water and sand doesn't.
2) This difference in physical properties means we can separate them. Here's what to do...

1. Grinding

Grind the mixture to make sure the salt crystals are small, so will dissolve easily.

2. Dissolving

Put the mixture in water and stir. The salt will dissolve, but the sand won't.

You can heat the mixture to help dissolve the salt.

3. Filtering

Filter the mixture.

The grains of sand won't fit through the tiny holes in the filter paper, so they collect on the paper instead.

The salt passes through the filter paper as it's part of the solution.

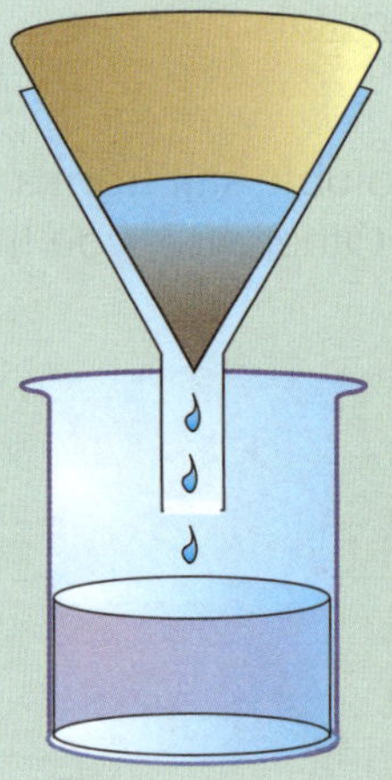

4. Evaporation

Evaporate the water from the salt so that it forms dry crystals.

You could also use crystallisation here if you wanted to make nice, big crystals (see previous page).

Separating rock salt requires filtration and evaporation

You may be asked how to separate another type of mixture containing insoluble and soluble solids — just apply the same method and think through what is happening in each stage.

Topic C1 — Atomic Structure and the Periodic Table

Simple Distillation

Distillation is used to separate mixtures which contain **liquids**. This first page looks at simple distillation.

Simple Distillation is Used to Separate Solutions

1) Simple distillation is used to separate a liquid from a solution.
2) First, the solution is heated.
3) The part of the solution that has the lowest boiling point evaporates first and turns into a gas.
4) The gas travels into the condenser.
5) In the condenser, the gas is cooled and condenses (turns back into a liquid).
6) The liquid drips out of the condenser and can be collected.
7) The rest of the solution is left behind in the flask.

> You can use simple distillation to get pure water from seawater. The water evaporates. It is then condensed and collected. This leaves the salt behind in the flask.

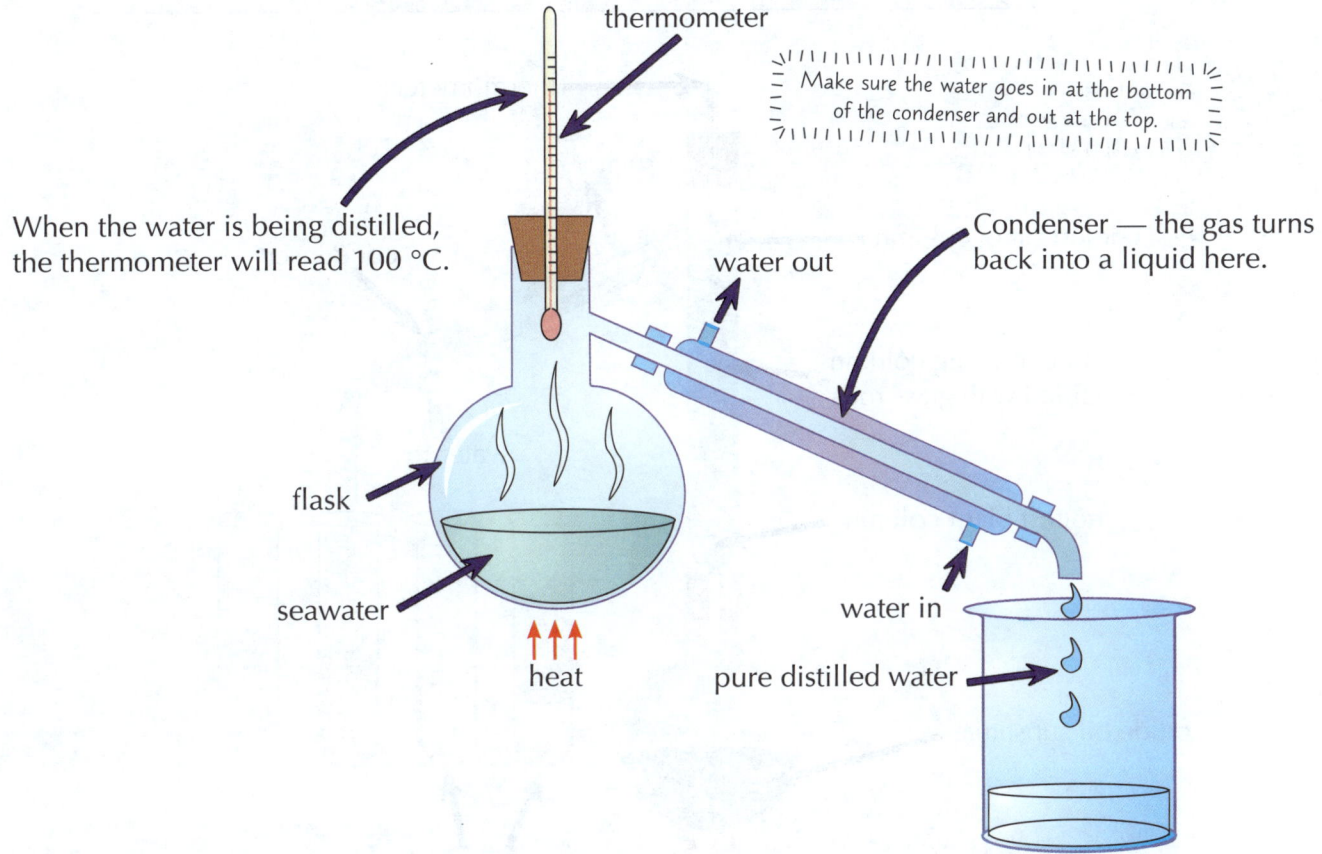

8) Simple distillation can't be used to separate mixtures of liquids with similar boiling points. So, you need to use another method instead — like fractional distillation (see next page).

Only gas should leave the flask during simple distillation
Make sure you don't put too much liquid in the flask, or heat it too vigorously — some of the unseparated solution could get into the condenser, meaning your product won't be pure.

Topic C1 — Atomic Structure and the Periodic Table

Fractional Distillation

*Another type of distillation is **fractional distillation**. This is trickier to carry out than simple distillation, but it can separate out **mixtures of liquids** even if their **boiling points** are close together.*

Fractional Distillation is Used to Separate a Mixture of Liquids

1) If you've got a mixture of liquids you can separate it using fractional distillation.
2) You put your mixture in a flask and stick a fractionating column on top, as shown below. Then you heat it.
3) The different liquids will have different boiling points.
4) So, they will evaporate at different temperatures.
5) The substance with the lowest boiling point evaporates first.
6) When the temperature on the thermometer matches the boiling point of this substance, it will reach the top of the column.
7) The substance will then enter the condenser, where it cools and condenses.
8) You can collect the liquid as it drips out of the condenser.
9) When the first liquid has been collected, you raise the temperature until the next one reaches the top.

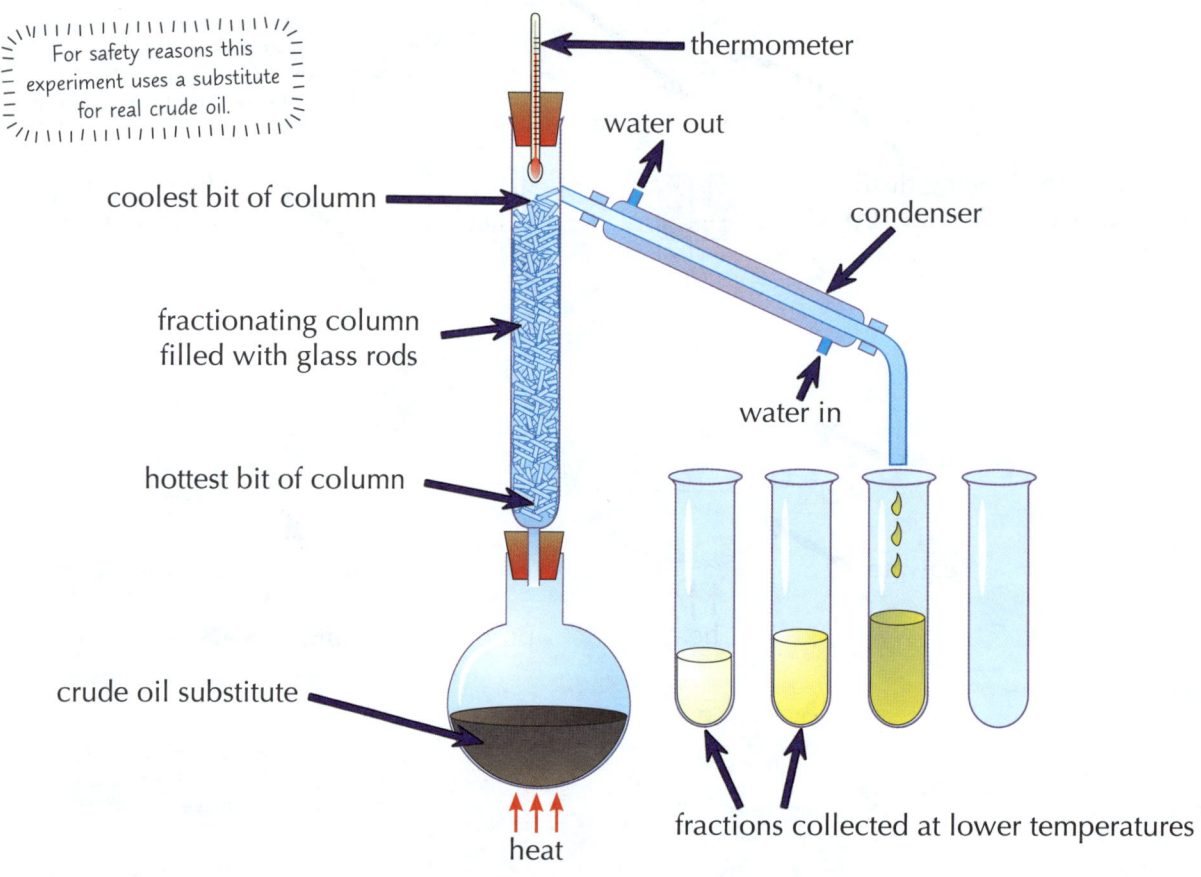

The diagram shows an experiment that can be used to show how fractional distillation of crude oil at a refinery works (see p.249).

Fractional distillation is used in the lab and industry

You've made it to the end of the pages on separation techniques, so make sure you understand what each of the methods can be used to separate and the apparatus set-up for each technique.

Q1 Propan-1-ol, methanol and ethanol have boiling points of 97 °C, 65 °C and 78 °C respectively. A student uses fractional distillation to separate a mixture of these compounds. State which liquid will be collected in the second fraction and explain why. [2 marks]

Q1 Video Solution

Topic C1 — Atomic Structure and the Periodic Table

Warm-Up & Exam Questions

So the last few pages have all been about mixtures and how to separate them. Here are some questions to get stuck into and make sure you know your filtration from your distillation...

Warm-Up Questions

1) True or False? The chemical properties of a substance are changed by it being part of a mixture.
2) Give the name of a method to separate a soluble solid from a solution.
3) Which part of the fractionating column is coolest during fractional distillation?

Exam Questions

1   A scientist is using paper chromatography to compare two different compounds. He draws a pencil line near the bottom of a sheet of filter paper and adds a spot of each of the different compounds to the line, leaving gaps between each spot.

1.1 Describe the next thing that the scientist should do with the filter paper.

[1 mark]

1.2 Why is pencil used to draw the line on the filter paper?

[1 mark]

2   A gardener uses a mixture of insoluble sharp sand and soluble ammonium sulfate fertiliser to treat his lawn.

A student plans to use the method below to separate the gardener's mixture into sharp sand and ammonium sulfate.

1. Mix the lawn sand with water and stir.
2. Filter the mixture using filter paper.
3. Pour the remaining solution into an evaporating dish and slowly heat it.

Explain the purpose of **each** of the steps in the student's method.

[3 marks]

3   Simple distillation can be used to separate mixtures of liquids. **Table 1** gives the boiling points of three liquids.

Figure 1

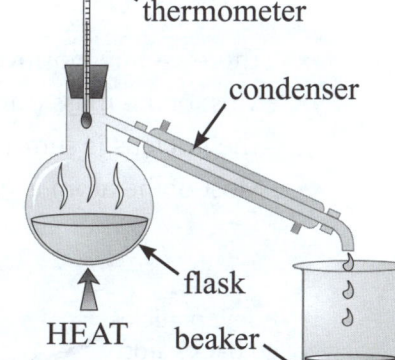

Table 1

Liquid	Boiling point (°C)
Methanoic acid	101
Propanone	56
Water	100

3.1 **Figure 1** shows the apparatus needed for simple distillation. Describe how the apparatus shown in **Figure 1** could be used to separate propanone from a solution of propanone and water. Propanone is highly flammable. Include any relevant safety precautions in your answer.

[6 marks]

3.2 Simple distillation can **not** be used to separate water from a solution of water and methanoic acid. Give a technique that **would** be suitable for separating water and methanoic acid.

[1 mark]

Topic C1 — Atomic Structure and the Periodic Table

The History of the Atom

You may have thought you were done with the **atom** after page 161. Unfortunately, you don't get away that easily. The next couple of pages are all about how **scientists** came to understand the atom as we do today.

Ideas About What Atoms Look Like Have Changed Over Time

1) Scientists used to think that atoms were solid spheres.

2) They then found atoms contain even smaller, negatively charged particles — electrons.

3) This led to a model called the 'plum pudding model' being created.

4) The plum pudding model showed the atom as a ball of positive charge with electrons scattered in this ball.

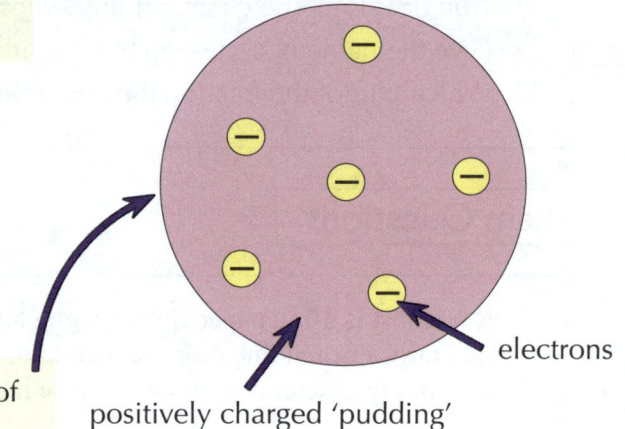

electrons

positively charged 'pudding'

Experiments Showed that the Plum Pudding Model Was Wrong

1) Later, scientists carried out alpha particle scattering experiments. They fired positively charged alpha particles at a very thin sheet of gold.

2) From the plum pudding model, they expected most of the particles to go straight through the sheet. They predicted that a few particles would change direction by a small amount.

3) But instead, some particles changed direction more than expected. A small number even went backwards.

4) This meant the plum pudding model couldn't be right.

5) So, scientists came up with the nuclear model of the atom:

- There's a tiny, positively charged nucleus at the centre of the atom.
- Most of the mass is in the nucleus.
- The nucleus is surrounded by 'cloud' of negative electrons.
- Most of the atom is empty space.

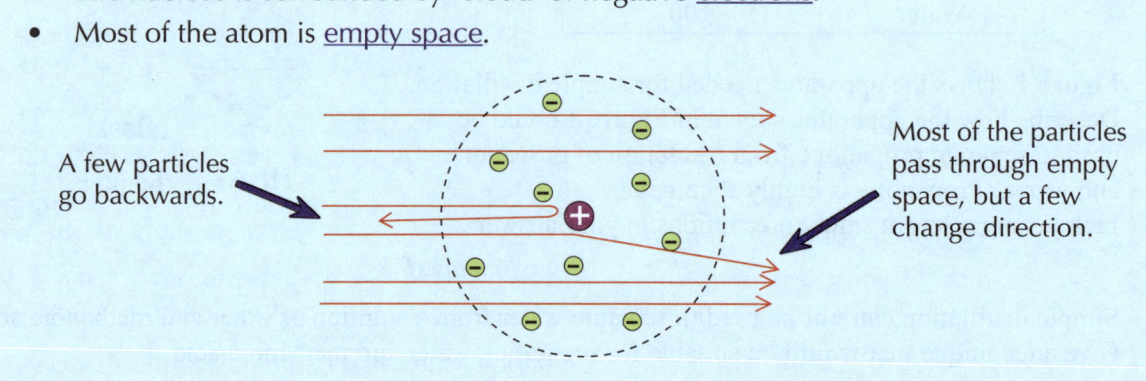

A few particles go backwards.

Most of the particles pass through empty space, but a few change direction.

Topic C1 — Atomic Structure and the Periodic Table

The History of the Atom

Bohr's **Nuclear Model** Explains a Lot

1) Niels Bohr changed the nuclear model of the atom.
   (See previous page for more information on the nuclear model).

2) He suggested that the electrons orbit (go around) the nucleus in shells (levels).

3) Each shell is a fixed distance from the nucleus.

4) Bohr's theory was supported by many experiments.
   Experiments later showed that Bohr's theory was correct.

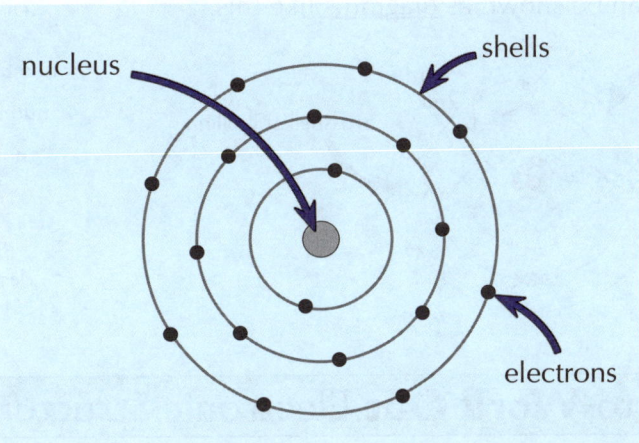

Later Experiments Found **Protons** and **Neutrons**

1) More experiments by scientists showed that the nucleus can be divided into smaller particles. Each particle has the same positive charge. These particles were named protons.

2) Experiments by James Chadwick showed that the nucleus also contained neutral particles — neutrons. This happened about 20 years after scientists agreed that atoms have nuclei.

3) This led to a model of the atom which was pretty close to the one we have today (see page 161).

 **Our understanding of what an atom looks like has changed**
The history of the atom is a fine example of the scientific method. It shows how our understanding of science can go through many stages, as scientists built upon other people's work with new evidence. And new evidence allows new predictions to be made, so the progress continues.

Topic C1 — Atomic Structure and the Periodic Table

Electronic Structure

Electrons don't just float around the nucleus randomly. They move in areas called **shells**.

Electron Shell **Rules**:

1) Electrons always move in shells (sometimes called energy levels).
2) The inner shells are always filled up first. These are the ones closest to the nucleus.
3) Only a certain number of electrons are allowed in each shell:

   1st shell: 2   2nd shell: 8   3rd shell: 8

4) Atoms are a lot more stable when they have full electron shells.
5) In most atoms, the outer shell is not full. These atoms will react to fill it.

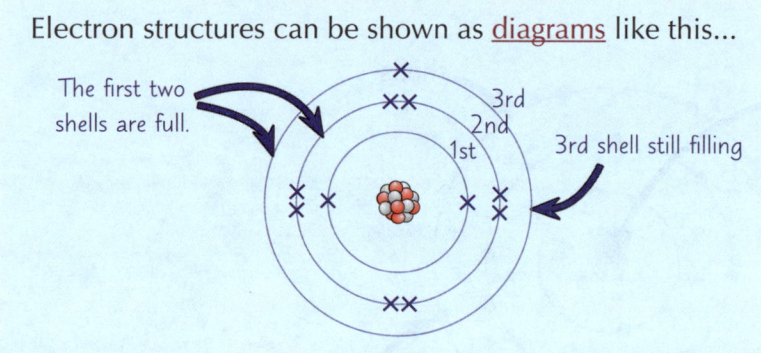

Electron structures can be shown as diagrams like this...

The first two shells are full.
3rd
2nd
1st
3rd shell still filling

...or as numbers like this: **2, 8, 1**

number of electrons in 1st shell
number of electrons in 2nd shell
number of electrons in 3rd shell

Both of these show the electron structure of sodium.

Follow the Rules to **Work Out** Electronic Structures

You can easily work out the electronic structures for the first 20 elements of the periodic table (things get a bit more complicated after that).

EXAMPLE   **What is the electronic structure of magnesium?**

1) From the periodic table, you can see that magnesium's atomic number is 12. This means it has 12 protons. So it must have 12 electrons.
2) Follow the 'Electron Shell Rules' above. The first shell can only take 2 electrons.   2...
3) The second shell can take up to 8 electrons.   2, 8...
4) So far we have a total of 10 electrons (2 + 8). So the third shell must also be partly filled with 2 electrons. This makes 12 electrons in total (2 + 8 + 2).

So the electronic structure for magnesium must be **2, 8, 2**.

Here are some more examples of electronic structures:

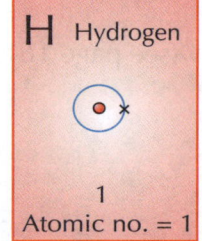
H Hydrogen
1
Atomic no. = 1

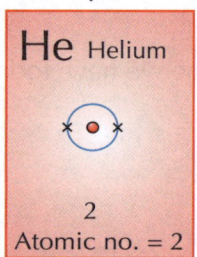
He Helium
2
Atomic no. = 2

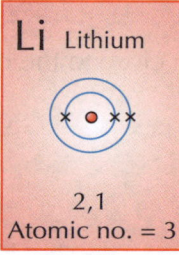
Li Lithium
2,1
Atomic no. = 3

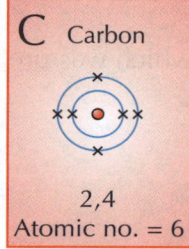
C Carbon
2,4
Atomic no. = 6

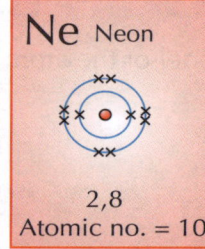
Ne Neon
2,8
Atomic no. = 10

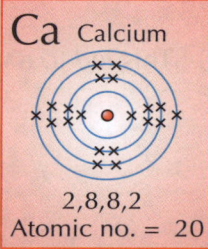
Ca Calcium
2,8,8,2
Atomic no. = 20

The first electron shell can only hold two electrons

Q1   Describe the order in which electron shells are filled up. **[1 mark]**

Q2   Give the electronic structure of argon (atomic number = 18). **[1 mark]**

Topic C1 — Atomic Structure and the Periodic Table

Development of the Periodic Table

We haven't always known as much about chemistry as we do now.
*Early chemists looked at **patterns** in the elements' properties to help them understand chemistry better.*

In the Early 1800s Elements Were Arranged By Atomic Weight

1) Until quite recently, scientists hadn't discovered protons, neutrons or electrons.
   So they had no idea of atomic number.

2) So scientists used atomic weight to arrange the elements into a periodic table.

3) These early periodic tables were not complete.

4) This is because not all of the elements had been found yet.

5) And putting the elements in order of atomic weight meant that
   some elements were also put in the wrong group (column).

Dmitri Mendeleev Left Gaps and Predicted New Elements

1) In 1869, a scientist called Mendeleev took all of the known elements and arranged them into a table.

```
         Mendeleev's Table of the Elements
H
Li  Be                              B  C  N  O  F
Na  Mg                              Al Si P  S  Cl
K   Ca  *  Ti V  Cr Mn Fe Co Ni Cu Zn *  *  As Se Br
Rb  Sr  Y  Zr Nb Mo *  Ru Rh Pd Ag Cd In Sn Sb Te I
Cs  Ba  *  *  Ta W  *  Os Ir Pt Au Hg Tl Pb Bi
```

2) He ordered them mainly by their atomic weight.

3) Sometimes he switched their positions or left gaps in the table.

4) This was so he could make sure that elements with similar properties stayed in the same groups.

5) Some of the gaps left space for elements that hadn't been found yet.
   Mendeleev used the position of the gaps to predict the properties of these elements.

6) New elements have been found since which fit into these gaps.
   This shows that Mendeleev's ideas were right.

> - Isotopes were discovered a while after Mendeleev made his Table of Elements.
> - Isotopes have different masses but share the same properties.
>   This means they have the same position on the periodic table.
> - So, Mendeleev was right to swap some elements around to keep properties together
>   — even if it meant they weren't in order of atomic weight.

By leaving gaps in the table Mendeleev had the right idea
Make sure you can describe what Mendeleev did to fix the problems with early periodic tables.

Topic C1 — Atomic Structure and the Periodic Table

The Modern Periodic Table

*Mendeleev got pretty close to producing something that you might **recognise** as a periodic table. The big breakthrough came when the **structure** of the **atom** was understood a bit better.*

The Periodic Table Helps you to See Patterns in Properties

1) There are about 100 elements.
2) In the periodic table the elements are laid out in order of increasing atomic number.
3) There are repeating (periodic) patterns in the properties of the elements. These periodic properties give the periodic table its name.
4) Metals are found to the left of the periodic table and non-metals are found to the right.

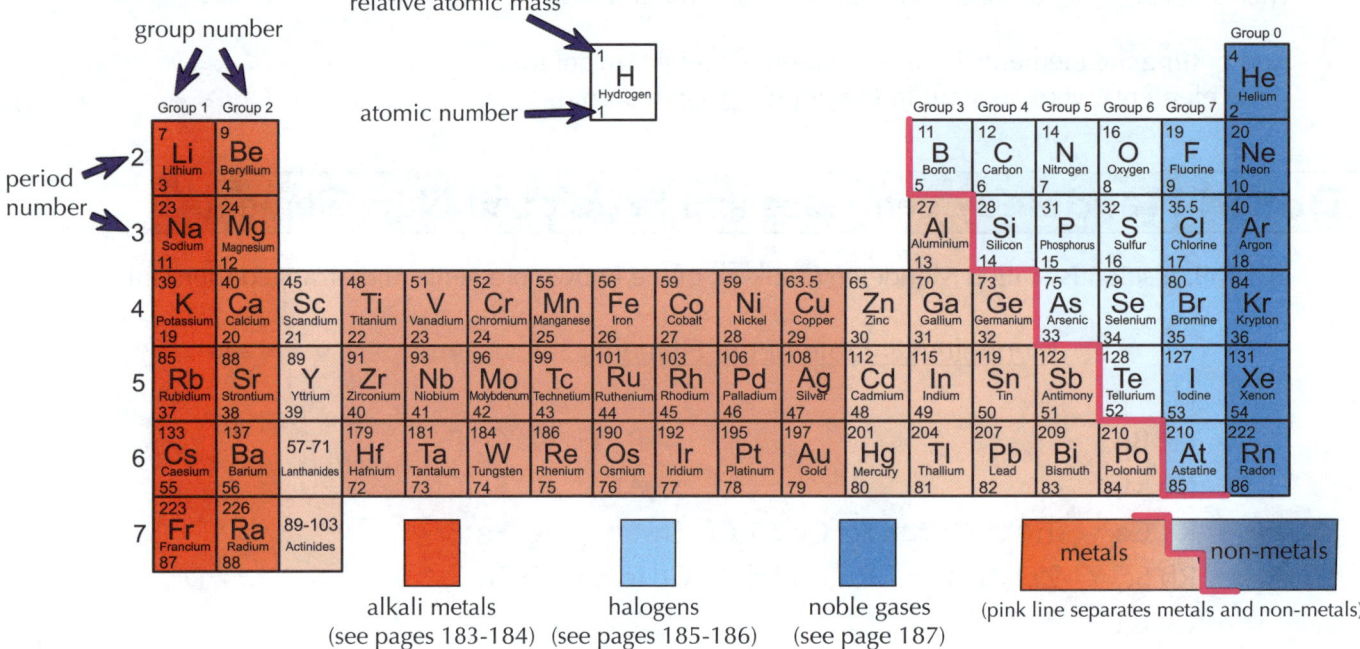

5) Elements with similar properties are arranged to form columns. These columns are called groups.
6) The group number of an element is the same as the number of outer shell electrons it has. (Except for Group 0 — helium has two electrons in its outer shell and the rest have eight.)
7) So, all the elements in a group have the same number of electrons in their outer shell. This means that elements in a group react in similar ways.

- The Group 1 elements are Li, Na, K, Rb, Cs and Fr.
- They all have one electron in their outer shells.
- They're all metals and they react in a similar way.

See pages 183-184 for more on Group 1 elements.

8) If you know the properties of one element, you can predict properties of other elements in that group.
9) You can also make predictions about trends in reactivity. E.g. in Group 1, the elements react more violently as you go down the group.
10) The rows in the periodic table are called periods. Each new period represents another shell of electrons.

The periodic table really helps with understanding chemistry

The periodic table is organised into groups and periods. This lets you see patterns in reactivity and properties. And this means you can make predictions on how reactions will occur.

Topic C1 — Atomic Structure and the Periodic Table

Warm-Up & Exam Questions

The last few pages have been tough with lots of information to learn. Luckily, here are some questions to get your head around to help you test your understanding.

Warm-Up Questions

1) Describe the 'plum pudding' model of the atom.
2) How many electrons can be held in the following:
   a) the first shell of an atom?
   b) the second shell of an atom?
3) Why did Mendeleev leave gaps in his table of the elements?

Exam Questions

1 The periodic table contains all the known elements arranged in order. *(Grade 3-4)*

1.1 How were elements generally ordered in early periodic tables?
Tick **one** box.

☐ alphabetically ☐ by atomic number ☐ by abundance ☐ by atomic weight

[1 mark]

1.2 How are the elements arranged in the modern periodic table?
Tick **one** box.

☐ alphabetically ☐ by atomic number ☐ by abundance ☐ by atomic weight

[1 mark]

1.3 Explain why elements in the same group of the periodic table have similar chemical properties.

[1 mark]

2 The atomic number of sulfur is 16. *(Grade 4-5)*

2.1 Complete **Figure 1** to show the electronic structure of sulfur.

Figure 1

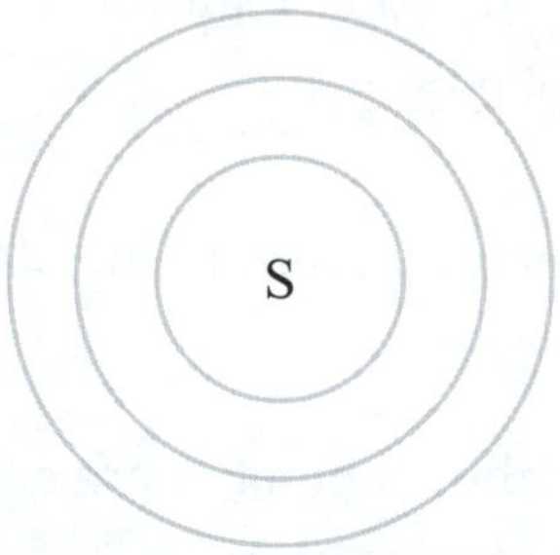

(S at centre with three concentric shells)

[1 mark]

2.2 Which group of the periodic table is sulfur in?

[1 mark]

2.3 Name another element that has the same number of electrons in its outer shell as sulfur.

[1 mark]

2.4 How many electrons does sulfur need to gain to have a full outer shell?

[1 mark]

Metals and Non-Metals

*Metals are used for all sorts of things so they're **really important** in modern life.*

Most Elements are Metals

1) Metals are elements which can form positive ions when they react.
2) They're towards the bottom and to the left of the periodic table.
3) Most elements in the periodic table are metals.
4) Non-metals are at the far right and top of the periodic table.
5) Non-metals don't usually form positive ions when they react.

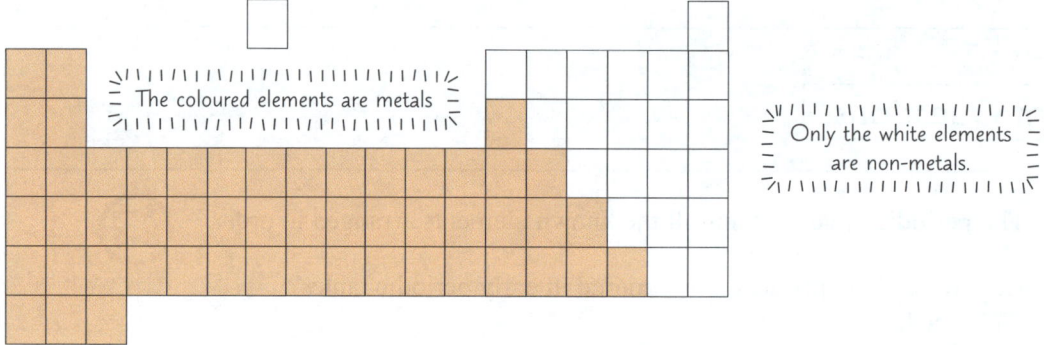

The Electronic Structure of Atoms Affects How They Will React

1) Atoms are more stable with a full outer shell. So, they react by losing, gaining or sharing electrons.
2) Metal elements are to the left and towards the bottom of the periodic table so they lose electrons quite easily. When this happens, they form positive ions, with a full outer shell.
3) Non-metals are to the right of the periodic table or towards the top, so it's easier for them to share or gain electrons to get a full outer shell.

Metals and Non-Metals Have Different Physical Properties

1) All metals have similar physical properties.

   - They're strong (hard to break), but can be bent or hammered into different shapes (malleable).
   - They're great at conducting heat and electricity.
   - They have high boiling and melting points.

2) Non-metals don't tend to show the same properties as metals.

   - They tend to be dull looking.
   - They're more brittle. This means they'll break more easily if you try to bend them.
   - They're not always solids at room temperature.
   - They don't usually conduct electricity.
   - They often have a lower density.

 **Metals have quite different properties from non-metals**
And you'll need to make sure you can remember them. Try covering up the lists of properties of metals and non-metals above and writing down as many as you can from memory.

Topic C1 — Atomic Structure and the Periodic Table

Group 1 Elements

*Group 1 elements are known as the **alkali metals**. As metals go, they're pretty **reactive**.*

The Group 1 Elements are Reactive, Soft Metals

1) The alkali metals are lithium, sodium, potassium, rubidium, caesium and francium.

2) They all have <u>one electron</u> in their outer shell. This makes them <u>very reactive</u>. It also gives them <u>similar properties</u>.

3) The alkali metals are all <u>soft</u>.

4) They all have <u>low density</u> (they're quite <u>light</u>).

There are Patterns in the Properties of Group 1 Metals

1) As you go down Group 1, the properties of the alkali metals change. For example:

- <u>Reactivity increases</u> — the outer electron is <u>more easily lost</u> as it gets <u>further</u> from the nucleus. This is because it's <u>less attracted</u> to the nucleus.

- <u>Melting</u> and <u>boiling</u> points get lower.

- <u>Relative atomic mass</u> goes up.

2) These patterns in the properties are called <u>trends</u>.

The properties of Group 1 metals change as you go down the group

In the exam you might be given a <u>trend</u> and then asked to <u>predict properties</u> of other Group 1 metals. For example, the <u>reactivity</u> of Group 1 metals <u>increases</u> as you go <u>down the group</u>, so you know that potassium will react more vigorously than sodium.

Topic C1 — Atomic Structure and the Periodic Table

Group 1 Elements

You met **Group 1 metals** on the previous page, so now it's time to learn about some of their **reactions**...

Alkali Metals Form **Ionic Compounds** with **Non-Metals**

The Group 1 elements <u>easily</u> lose their one outer electron to form a full outer shell.
So they form <u>1+ ions</u> easily.

Reaction With **Water**

1) The reactions are <u>vigorous</u> and produce a <u>metal hydroxide</u> and <u>hydrogen gas</u>:

   alkali metal + water → metal hydroxide + hydrogen

   For example:

   sodium + water → sodium hydroxide + hydrogen
   $$2Na_{(s)} + 2H_2O_{(l)} \rightarrow 2NaOH_{(aq)} + H_{2(g)}$$

   The little letters in brackets after each substance in the reaction show what state the substance is in — see p.207.

2) The <u>metal hydroxides</u> are compounds that <u>dissolve</u> in the <u>water</u>.
3) The <u>more reactive</u> (lower down in the group) an alkali metal is, the more <u>violent</u> the reaction.
4) Lithium, sodium and potassium <u>float</u> and <u>move</u> around the surface, <u>fizzing</u> furiously.
5) The reaction with potassium gives out enough energy to <u>ignite</u> the hydrogen (set it on fire).

Reaction With **Chlorine**

1) Group 1 metals react <u>vigorously</u> when heated in <u>chlorine gas</u> to form <u>white salts</u> called <u>metal chlorides</u>.

   alkali metal + chlorine → metal chloride

   For example:

   sodium + chlorine → sodium chloride
   $$2Na_{(s)} + Cl_{2(g)} \rightarrow 2NaCl_{(s)}$$

2) As you go down the group, the reaction gets <u>more vigorous</u>.

Reaction With **Oxygen**

1) Group 1 metals can react with <u>oxygen</u> to form a <u>metal oxide</u>.
2) Different types of <u>oxide</u> will form depending on the Group 1 metal.
3) Group 1 metals are <u>shiny</u> but when they react with oxygen in the air they turn a <u>dull grey</u> (they <u>tarnish</u>). This is because a layer of <u>metal oxide</u> is formed on the surface.

Alkali metals all react in a similar way...

...which is handy for you, because it means you don't need to learn the <u>reaction products</u> for each metal separately. You just need to remember the <u>types of product</u> that are formed during each reaction, then switch 'metal' for whichever <u>Group 1 element</u> you're interested in.

Q1 Video Solution

Q1 Write a word equation for the reaction between lithium and water. [1 mark]

Group 7 Elements

*The **Group 7** elements are known as the halogens. Like the **alkali metals**, **halogens** also show trends down the group. However, these trends are a bit different...*

The Halogens Show Patterns in their Properties

1) The halogens are fluorine, chlorine, bromine, iodine and astatine.

2) As elements, the halogens form molecules that contain two atoms. For example, chlorine (Cl_2) is a fairly reactive, poisonous green gas.

3) As you go down Group 7, the halogens:

- become less reactive — it's harder to gain an extra electron as the outer shell is further from the nucleus.

- have higher melting and boiling points.

- have higher relative atomic masses.

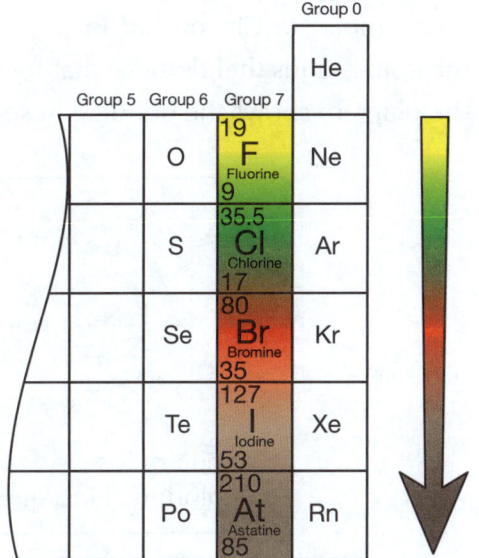

You can use these trends to predict properties of halogens. E.g. iodine will have a higher boiling point than chlorine as it's further down the group.

4) All the Group 7 elements react in similar ways. This is because they all have seven electrons in their outer shell.

Halogens can Form Molecular Compounds

1) When halogen atoms react with other non-metals, they share electrons and form covalent bonds (see page 196). This is so they can get a full outer shell.

2) These reactions form compounds with simple molecular structures (see p.197-198).

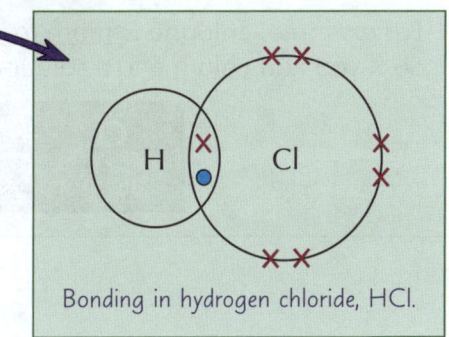

Bonding in hydrogen chloride, HCl.

Halogens all exist as molecules with two atoms

Just like alkali metals, you may be asked to predict the properties of a halogen from a given trend down the group. Make sure you understand why the halogens' electronic structures mean they react in similar ways.

Topic C1 — Atomic Structure and the Periodic Table

Group 7 Elements

You also need to know all about how the halogens **react** with **metals** and **halide salts**, so get reading this page...

Halogens Form Ionic Bonds with Metals

1) The halogens form 1– ions called halides:
   - fluoride, F^-
   - chloride, Cl^-
   - bromide, Br^-
   - iodide, I^-

2) Halides form when halogens bond with metals. For example Na^+Cl^- or $Ca^{2+}Br^-_2$.
3) The compounds (halide salts) that form have ionic structures.
4) The diagram shows the bonding in sodium chloride, NaCl.

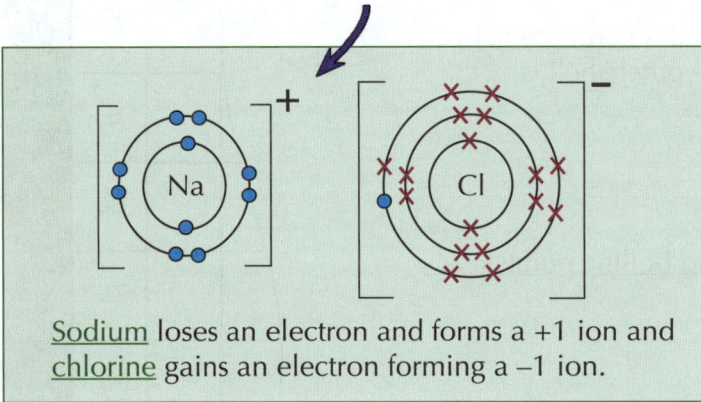

Sodium loses an electron and forms a +1 ion and chlorine gains an electron forming a –1 ion.

More Reactive Halogens Displace Less Reactive Halogens

1) A reaction can take place between a halogen and the halide salt of a less reactive halogen.
2) These are called displacement reactions.
3) When this happens, the less reactive halogen changes from a halide (1– ion) to a halogen. The more reactive halogen changes from a halogen into a halide ion and becomes part of the salt.

> For example, chlorine is more reactive than bromine.
> So if you add chlorine to a solution containing a bromide salt, bromine will be displaced.
>
> $Cl_{2(g)}$ + $2KBr_{(aq)}$ → $Br_{2(aq)}$ + $2KCl_{(aq)}$
> Pale green   Orange

Halogens higher up the group will displace the ones lower down

The halogens get less reactive as you go down the group. So a halogen will only be able to displace another halogen if it's higher up in Group 7. If it's lower down Group 7, no reaction will happen.

Q1 Give the balanced symbol equation for the displacement reaction between bromine and sodium iodide. [1 mark]

Q1 Video Solution

Topic C1 — Atomic Structure and the Periodic Table

Group 0 Elements

*The **noble gases don't react** with very much and you can't even see them — makes them a bit **dull** really.*

Group 0 Elements are All Unreactive, Colourless Gases

1) Group 0 elements are called the noble gases. They include the elements helium, neon and argon (plus a few others).
2) All elements in Group 0 are colourless gases at room temperature.
3) They all have eight outer shell electrons, apart from helium which has two. This means they have a stable full outer shell.

   Helium only has electrons in the first shell, which only needs 2 to be filled.

4) This stability makes them very unreactive (inert). This means they don't form molecules easily. So the elements are single atoms.
5) Because noble gases are unreactive, some reactions are carried out in an atmosphere that only contains a noble gas, instead of air.
6) This is done if the reactants could react with things in the air (e.g. oxygen or water) instead of taking part in the reaction you're trying to do. It's also done if the products react with things in the air.

There are Patterns in the Properties of the Noble Gases

1) As you go down Group 0, the relative atomic masses of the elements increase.
2) This means that as you go down the group, the elements have more electrons.
3) More electrons means stronger forces between atoms.
4) The stronger the forces, the higher the boiling point. So as you go down Group 0, the boiling points increase.
5) If you're given the boiling point of one noble gas you can predict the boiling point for another one. So make sure you know the pattern.

Noble Gas
helium
neon
argon
krypton
xenon
radon

Increasing boiling point ↓

EXAMPLE
Neon is a gas at 25 °C. Predict what state helium is at this temperature.

Helium has a lower boiling point than neon as it is further up the group.
So, helium must also be a gas at 25 °C.

EXAMPLE
Radon has a boiling point of –62 °C and krypton has a boiling point of –153 °C. Predict the boiling point of xenon.

Xenon comes in between radon and krypton in the group.
So, you can predict that its boiling point would be between their boiling points.
E.g. xenon has a boiling point of –100 °C.

The actual boiling point of xenon is –108 °C — which is between –62 °C and –153 °C. Just as predicted.

Just like other groups of elements, the noble gases follow patterns...

Although they're unreactive and hard to see, they're actually pretty useful. It took a while to discover them, but we now know all about them, including the trend in their boiling points.

Warm-Up & Exam Questions

These questions are all about the groups of the periodic table that you need to know about. Treat the exam questions like the real thing — don't look back through the book until you've finished.

Warm-Up Questions

1) Do metals form positive or negative ions?
2) Give two physical properties of most metals.
3) What is the product formed when lithium reacts with chlorine?
4) Do halide ions have a positive or a negative charge?
5) Which group of the periodic table are the noble gases in?

Exam Questions

1 **Table 1** shows some of the physical properties of four of the halogens.

Table 1

Halogen	Physical state at room temperature
Fluorine	gas
Chlorine	gas
Bromine	liquid
Iodine	solid

1.1 Which halogen in **Table 1** is the most reactive?

[1 mark]

1.2 Which halogen in **Table 1** has the highest melting point?

[1 mark]

2 **Figure 1** shows the periodic table.

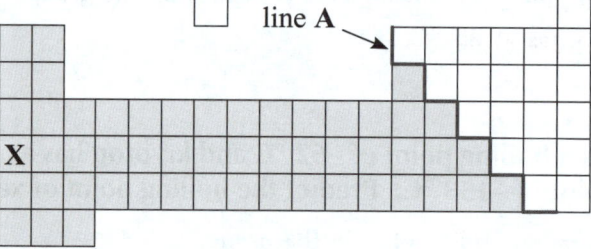

Figure 1

2.1 Element **X** is found in the first column of the periodic table.
What name is given to the elements found in the first column of the periodic table?
Tick **one** box.

☐ halogens ☐ alkali metals ☐ noble gases ☐ non-metals

[1 mark]

2.2 Element **Y** does not conduct electricity.
Predict whether element **Y** will be found to the left or the right of line **A** in **Figure 1**.
Explain your answer.

[2 marks]

Exam Questions

3 Group 1 elements include lithium, sodium and potassium.

3.1 Explain why the Group 1 elements react vigorously with water.

[1 mark]

3.2 The equation for the reaction of potassium with water is:

$$2K + 2H_2O \rightarrow 2KOH + H_2$$

Name the **two** products of the reaction.

[2 marks]

3.3 Potassium is more reactive than lithium and sodium.
Why do the Group 1 elements become more reactive as you go down Group 1?
Tick **one** box.

☐ The outer electron is closer to the nucleus and so more attracted to the nucleus.

☐ The outer electron is closer to the nucleus and so less attracted to the nucleus.

☐ The outer electron is further from the nucleus and so less attracted to the nucleus.

☐ The outer electron is further from the nucleus and so more attracted to the nucleus.

[1 mark]

4 Chlorine is a Group 7 element.
Chlorine's electron arrangement is shown in **Figure 2**.

Figure 2

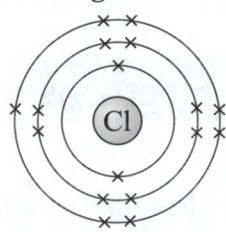

4.1 Chlorine is very reactive and forms compounds with metals.
What type of bonds form between chlorine and metals?

[1 mark]

When chlorine is bubbled through potassium iodide solution a reaction occurs.
The equation below shows the reaction.

$$Cl_{2(g)} + 2KI_{(aq)} \rightarrow I_{2(aq)} + 2KCl_{(aq)}$$

4.2 Identify the type of reaction that occurs.

[1 mark]

4.3 What does the reaction show about the relative reactivities of Cl_2 and I_2? Give a reason for your answer.

[2 marks]

4.4 None of the elements in Group 0 will react with potassium iodide or potassium bromide.
Using your knowledge of the electronic structure of the Group 0 elements,
explain why no reaction occurs.

[1 mark]

Ions

*Ions crop up all over the place in chemistry. You need to know **what** they are and **how** they form. Luckily for you, this page has got that covered. So crack on with it.*

Ions are Made When Electrons are Transferred

1) Ions are charged particles — for example Cl^- or Mg^{2+}.

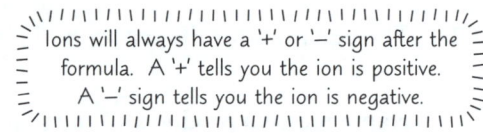

2) Ions are formed when atoms gain or lose electrons.

3) They do this to get a full outer shell — like a noble gas. This is because a full outer shell is very stable.

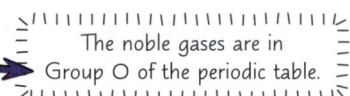

4) Metal atoms lose electrons from their outer shell to form positive ions.

5) Non-metal atoms gain electrons into their outer shell to form negative ions.

6) The number of electrons lost or gained is the same as the charge on the ion. For example:

> If 2 electrons are lost, the particle now has two more protons than electrons. So the charge is 2+.
>
> If 3 electrons are gained, the particle now has three more electrons than protons. So the charge is 3−.

Ionic Bonding — Transfer of Electrons

1) Metals and non-metals can react together.
2) When this happens, the metal atoms lose electrons to form positively charged ions.
3) The non-metal atoms gain these electrons to form negatively charged ions.
4) These oppositely charged ions are strongly attracted to one another by electrostatic forces.
5) This attraction is called an ionic bond. It holds the ions together to make an ionic compound.

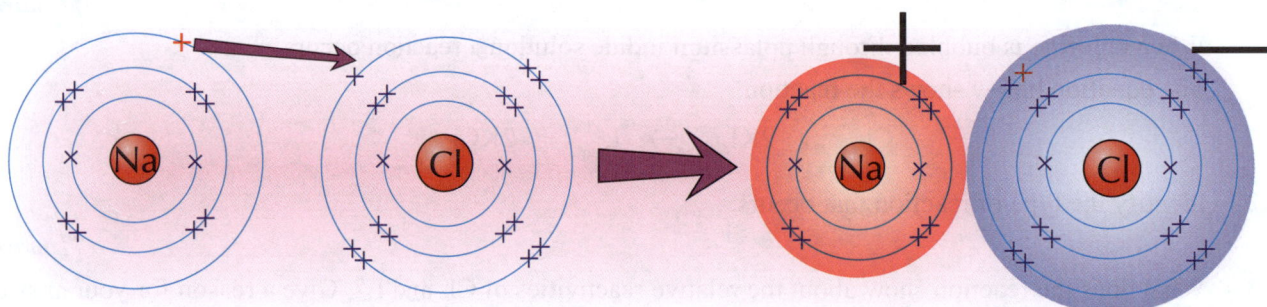

Metal atoms will form positively charged ions

Some atoms gain electrons and some atoms lose electrons. An ionic bond is just the attraction between the electrostatic charges of the newly formed ions.

Q1 Explain why simple ions often have noble gas electronic structures. [2 marks]

Ions

You need to be able to predict the ions that the atoms in Groups 1, 2, 6 and 7 will form.
Don't worry, though — the **periodic table** is here to help you.

You can **Work Out** What Ions are Formed by Groups **1, 2, 6 and 7**

1) Group 1 and 2 elements are metals. They lose electrons to form positive ions.
2) Group 6 and 7 elements are non-metals. They gain electrons to form negative ions.
3) Elements in the same group all have the same number of outer electrons. So they have to lose or gain the same number to get a full outer shell. And this means that they form ions with the same charges.
4) You don't have to remember what ions most elements form. You can just look at the periodic table.

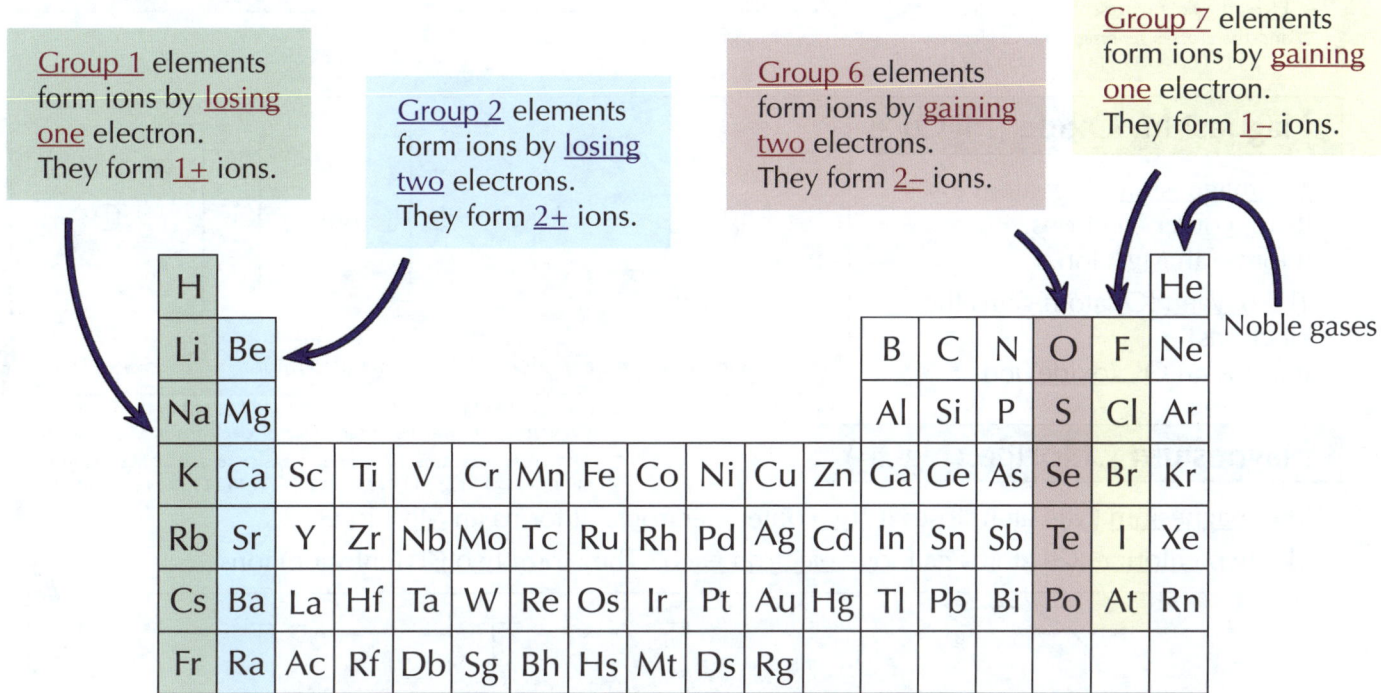

Group 1 elements form ions by losing one electron. They form 1+ ions.

Group 2 elements form ions by losing two electrons. They form 2+ ions.

Group 6 elements form ions by gaining two electrons. They form 2− ions.

Group 7 elements form ions by gaining one electron. They form 1− ions.

Noble gases

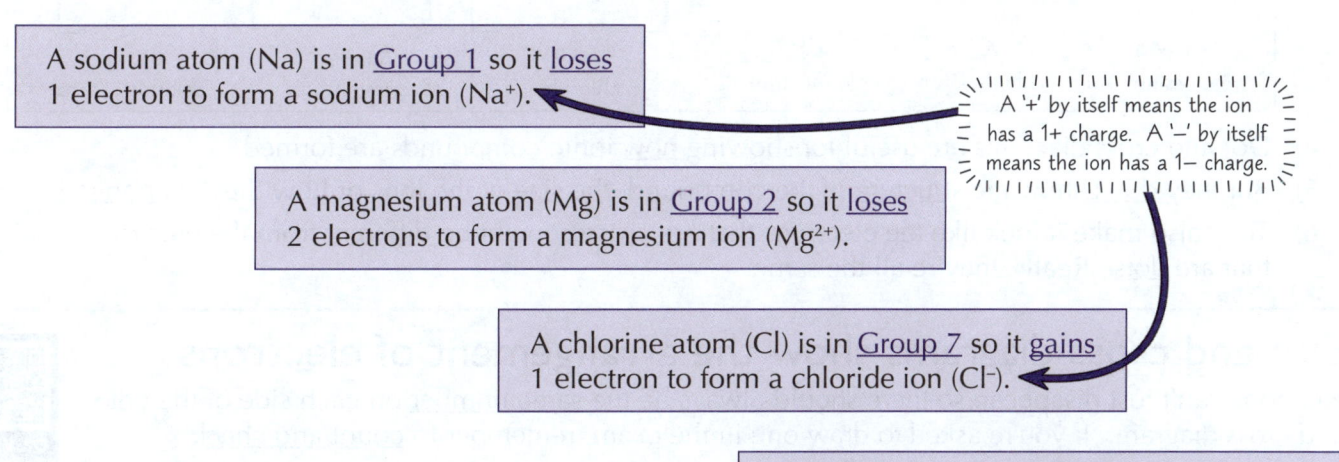

A sodium atom (Na) is in Group 1 so it loses 1 electron to form a sodium ion (Na$^+$).

A '+' by itself means the ion has a 1+ charge. A '−' by itself means the ion has a 1− charge.

A magnesium atom (Mg) is in Group 2 so it loses 2 electrons to form a magnesium ion (Mg^{2+}).

A chlorine atom (Cl) is in Group 7 so it gains 1 electron to form a chloride ion (Cl$^-$).

An oxygen atom (O) is in Group 6 so it gains 2 electrons to form an oxide ion (O^{2-}).

Topic C2 — Bonding, Structure and Properties of Matter

Ionic Bonding

Dot and Cross Diagrams Show How Ionic Compounds are Formed

1) Dot and cross diagrams show how electrons are arranged in an atom or ion.
2) Each electron is represented by a dot or a cross.
3) So these diagrams can show which atom the electrons in an ion originally came from.

Sodium Chloride (NaCl)

The sodium (Na) atom loses its outer electron. It forms an Na⁺ ion.

The chlorine (Cl) atom gains the electron.
It forms a Cl⁻ (chloride) ion.

Here, the dots represent the Na electrons and the crosses represent the Cl electrons.

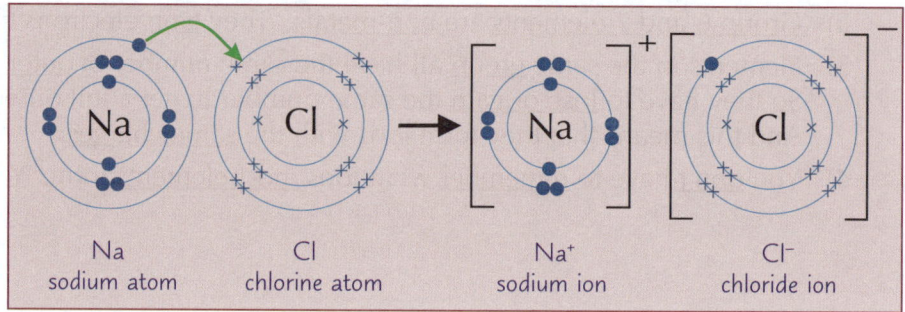

Magnesium Oxide (MgO)

The magnesium (Mg) atom loses its two outer electrons.
It forms an Mg²⁺ ion.

The oxygen (O) atom gains the electrons.
It forms an O²⁻ (oxide) ion.

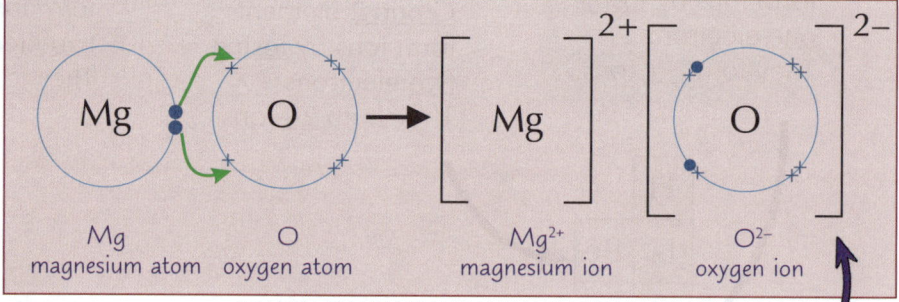

Magnesium Chloride (MgCl₂)

We've only shown the outer shells of electrons in these two dot and cross diagrams. It makes it easier to see what's going on.

The magnesium (Mg) atom loses its two outer electrons. It forms an Mg²⁺ ion.
The two chlorine (Cl) atoms gain one electron each. They form two Cl⁻ (chloride) ions.

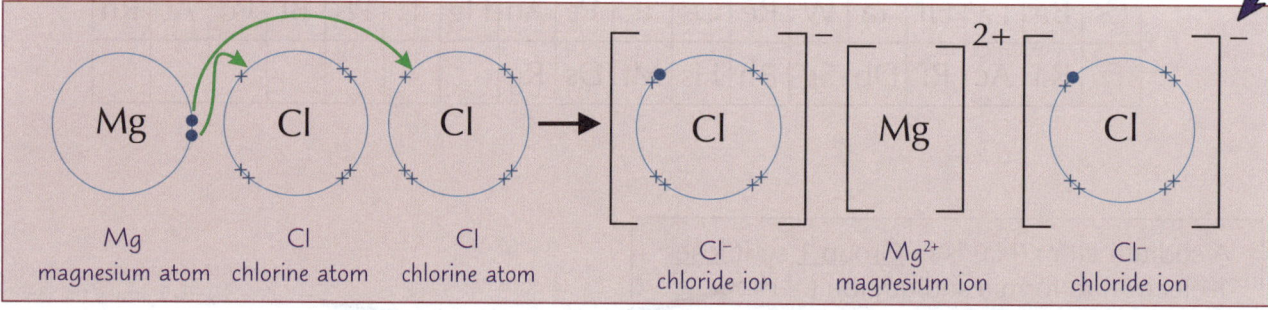

4) Dot and cross diagrams are useful for showing how ionic compounds are formed.
5) But they don't show the structure of the compound, the size of the ions or how they're arranged.
6) They also make it look like the electrons that are crosses might be different from the electrons that are dots. Really, they're all the same.

Dot and cross diagrams show the arrangement of electrons

Electrons can't just disappear, so there should always be the same number on each side of the dot and cross diagram. If you're asked to draw one in the exam, remember to count and check.

Q1 Describe what is represented on a dot and cross diagram by an arrow pointing from an electron shell of one atom to an electron shell of another atom. [1 mark]

Q2 Draw a dot and cross diagram to show how potassium (a Group 1 metal) and bromine (a Group 7 non-metal) form potassium bromide (KBr). [3 marks]

Q2 Video Solution

Topic C2 — Bonding, Structure and Properties of Matter

Ionic Compounds

Ionic compounds are just compounds that **only** contain **ionic bonds**. One of the main ionic compounds you need to know about is **sodium chloride**. And that's just **salt**. Nothing scary about that.

Ionic Compounds Have A Giant Ionic Lattice Structure

1) In ionic compounds, the ions are arranged in a pattern. This is called a giant ionic lattice.
2) There are strong electrostatic forces of attraction between oppositely charged ions.
3) These forces are called ionic bonds and they act in all directions.

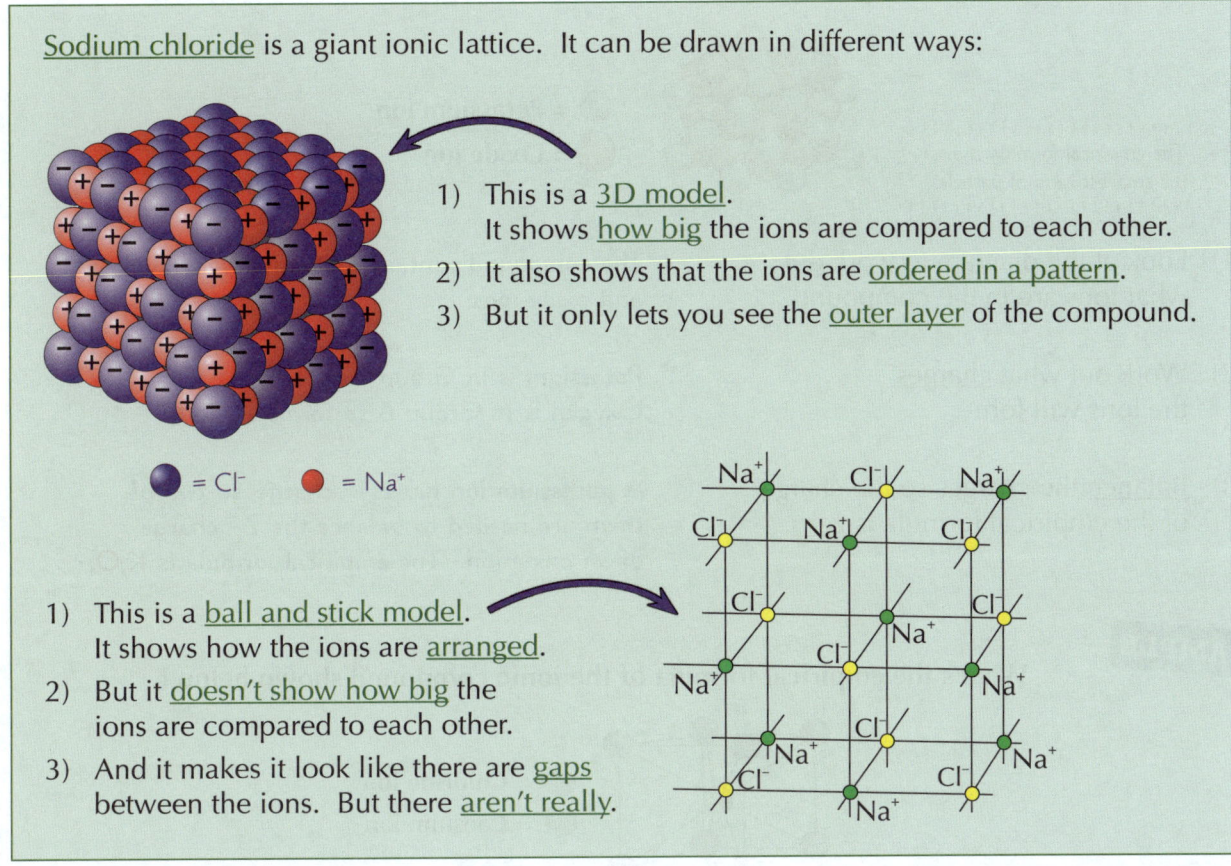

Sodium chloride is a giant ionic lattice. It can be drawn in different ways:

1) This is a 3D model.
   It shows how big the ions are compared to each other.
2) It also shows that the ions are ordered in a pattern.
3) But it only lets you see the outer layer of the compound.

● = Cl⁻   ● = Na⁺

1) This is a ball and stick model.
   It shows how the ions are arranged.
2) But it doesn't show how big the ions are compared to each other.
3) And it makes it look like there are gaps between the ions. But there aren't really.

Ionic Compounds All Have Similar Properties

1) They all have high melting points and high boiling points.
   This is because lots of energy is needed to break all the strong ionic bonds.
2) When they're solid, the ions are held in place, so the solid compounds can't conduct electricity.
3) When ionic compounds melt, the ions are free to move and they can conduct electricity.
4) Some ionic compounds dissolve in water.
   The ions can move in the solution, so they can conduct electricity.

Topic C2 — Bonding, Structure and Properties of Matter

Ionic Compounds

You might need to work out the **empirical formula** of an **ionic compound** from a **diagram** — here's how.

Use Charges to Find the Empirical Formula of an Ionic Compound

1) For a dot and cross diagram, just count up and write down how many ions there are of each element.
2) For a 3D diagram, use the diagram to work out what ions are in the compound.
3) Then balance the charges of the ions so that the overall charge on the compound is zero.

EXAMPLE What's the empirical formula of the ionic compound shown below?

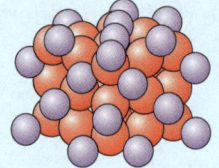

○ = Potassium ion
● = Oxide ion

The empirical formula shows the smallest ratio of particles.

1) Look at the diagram to work out what ions are in the compound.

    The compound contains potassium and oxide ions.

2) Work out what charges the ions will form.

    Potassium is in Group 1 so forms 1+ ions. Oxygen is in Group 6 so forms 2− ions.

3) Balance the charges so the charge of the empirical formula is zero.

    A potassium ion has a 1+ charge, so two of them are needed to balance the 2− charge of an oxide ion. The empirical formula is K_2O.

EXAMPLE What's the empirical formula of the ionic compound shown below?

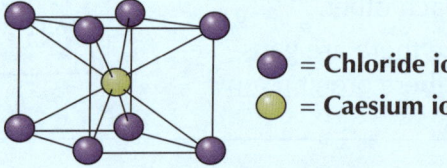

● = Chloride ion
● = Caesium ion

1) Look at the diagram to work out what ions are in the compound.

    The compound contains caesium ions and chloride ions.

2) Work out what charges the ions will form.

    Caesium is in Group 1 so forms 1+ ions. Chlorine is in Group 7 so forms 1− ions.

3) Balance the charges so the charge of the empirical formula is zero.

    A caesium ion has a 1+ charge, so you only need one to balance out the 1− charge of the chloride ion. The empirical formula is $CsCl$.

Ionic compounds have regular lattice structures

As long as you can find the charge of the ions in an ionic compound, you can work out the empirical formula. Try practising with different ionic compounds.

● = sulfide ion
● = magnesium ion

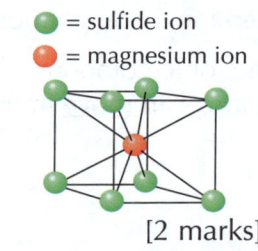

Q1 Video Solution

Q1 The structure of an ionic compound is shown on the right.
  a) Predict, with reasoning, whether the compound has a high or a low melting point. [2 marks]
  b) Explain why the compound can conduct electricity when molten. [1 mark]
  c) Use the diagram to find the empirical formula of the compound. [3 marks]

Topic C2 — Bonding, Structure and Properties of Matter

Warm-Up & Exam Questions

Congratulations, you got to the end of the pages on ions. Now test yourself with these questions...

Warm-Up Questions

1) What is an ion?
2) What is the charge on a ion formed from a Group 7 element?
3) Sodium chloride has a giant ionic structure. Does it have a high or a low boiling point?
4) True or false? Ionic compounds conduct electricity when dissolved in water.
5) What is the empirical formula of the compound containing Al^{3+} and OH^- ions only?
   A. $Al_3(OH)$ B. $Al(OH)_4$ C. $Al(OH)_3$ D. $Al(OH)_2$

Exam Questions

1 When lithium reacts with oxygen it forms the ionic compound Li_2O.

1.1 Complete **Figure 1** to show the electron arrangements and the charges on the ions when Li_2O is formed.

Figure 1

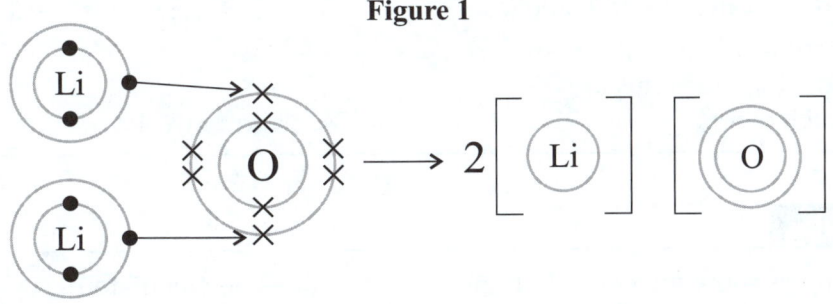

[2 marks]

1.2 What is the formula of the compound lithium forms with chlorine?
Tick **one** box.

☐ $LiCl_2$ ☐ $LiCl$ ☐ Li_2Cl ☐ $LiCl_3$

[1 mark]

2 Sodium and chlorine react to form sodium chloride, an ionic compound.

2.1 Explain how you can use the charges on the sodium and chloride ions to determine the empirical formula of sodium chloride.

[2 marks]

2.2 Explain why sodium chloride has a high melting point.

[2 marks]

2.3 Explain why solid sodium chloride does not conduct electricity, but molten sodium chloride does.

[2 marks]

2.4 Give **one** other way, besides melting, of making sodium chloride conduct electricity.

[1 mark]

Topic C2 — Bonding, Structure and Properties of Matter

Covalent Bonding

*Some elements form **covalent** bonds — they **share** electrons with each other in order to have full outer shells.*

Covalent Bonds — Sharing Electrons

1) When non-metal atoms bond together, they share pairs of electrons to make covalent bonds.
2) Covalent bonds are electrostatic forces and are very strong.
3) Atoms only share electrons in their outer shells.
4) Atoms get one extra shared electron for each single covalent bond that they form.
5) Each atom usually makes enough covalent bonds to fill up its outer shell. This makes them very stable.

There are Different Ways of Drawing Covalent Bonds

Nitrogen has five outer electrons. To form ammonia (NH_3) it forms three covalent bonds to get the extra 3 electrons it needs for a full outer shell. Here are different ways the bonding can be shown:

Dot and Cross Diagrams

1) Electrons shown in the overlap between two atoms are shared electrons.
2) Dot and cross diagrams show which atoms the electrons in a covalent bond come from.
3) But they don't show how the atoms are arranged, or how big the atoms are compared to each other.

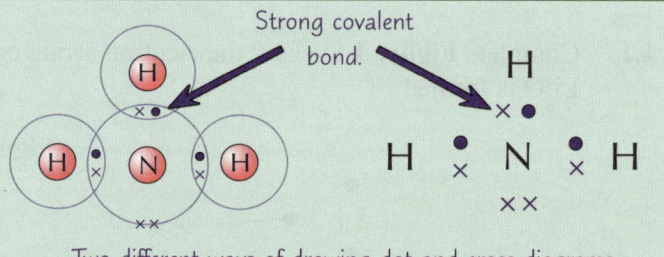

Two different ways of drawing dot and cross diagrams.

Displayed Formulas

1) Displayed formulas show the covalent bonds as single lines between atoms.
2) If it's a single covalent bond, there'll be one line. If it's a double covalent bond, there'll be two lines.
3) They're good for showing how atoms are connected in large molecules.
4) But they don't show the 3D structure of the molecule. They also don't show which atoms the electrons in the covalent bond have come from.

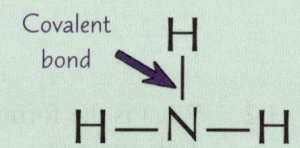

3D Models

1) 3D models show the atoms, the covalent bonds and how they're arranged.
2) But 3D models can be confusing for large molecules.
3) And they don't show where the electrons in the bonds have come from.

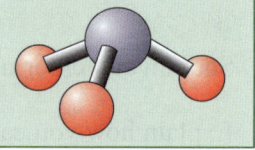

You can find the molecular formula of a simple molecular compound from these diagrams by counting up how many atoms of each element there are.

A molecular formula shows you how many atoms of each element are in a molecule.

EXAMPLE Find the molecular formula of ethane from the diagram of ethane.

In the diagram, there are two carbon atoms and six hydrogen atoms. So the molecular formula is C_2H_6.

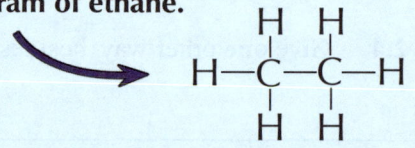

Topic C2 — Bonding, Structure and Properties of Matter

Covalent Bonding

*It's always handy to have an assortment of **examples** up your sleeve when you go into a chemistry exam. You've already seen ammonia, so use this page to familiarise yourself with a **few more** important molecules.*

Learn These Examples of Simple Molecular Substances

1) <u>Simple molecular substances</u> are made up of molecules that contain a <u>few atoms</u> joined together by <u>covalent bonds</u>.
2) Here are some <u>common examples</u> that you should know...

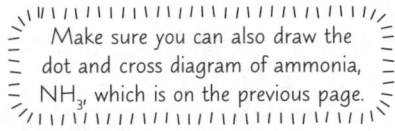

Make sure you can also draw the dot and cross diagram of ammonia, NH_3, which is on the previous page.

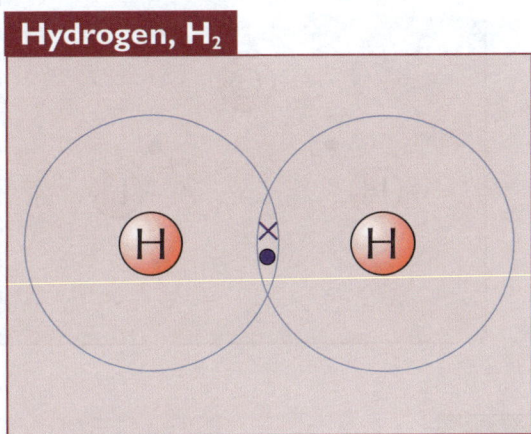

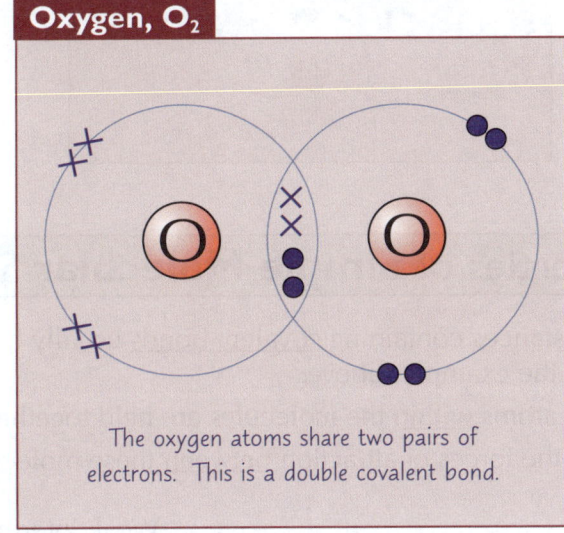

The oxygen atoms share two pairs of electrons. This is a double covalent bond.

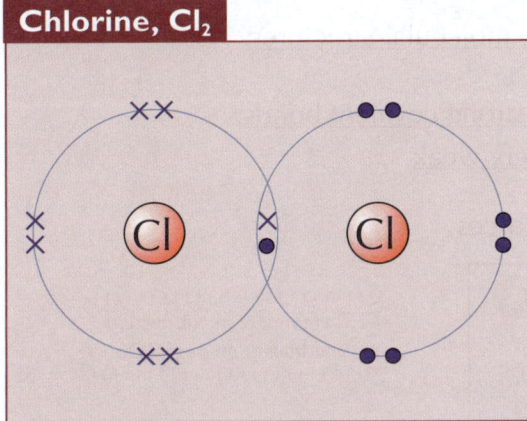

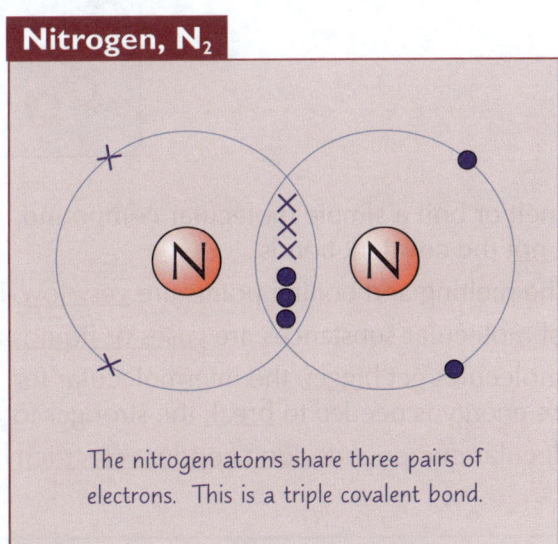

The nitrogen atoms share three pairs of electrons. This is a triple covalent bond.

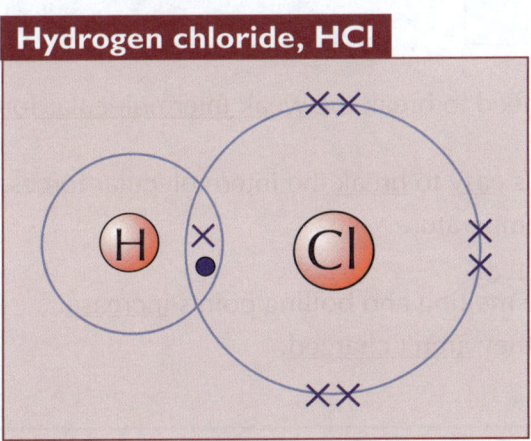

Covalent bonding involves sharing electrons

You might be asked to draw a <u>dot and cross diagram</u> for a simple molecule in the exam. The ones you need to know are shown on this page and the next page (oh, and don't forget to learn ammonia too).

Covalent Bonding

*Just in case you can't get enough of **simple molecular substances**, here's another **whole page** on them...*

Some More Examples of Simple Molecular Substances

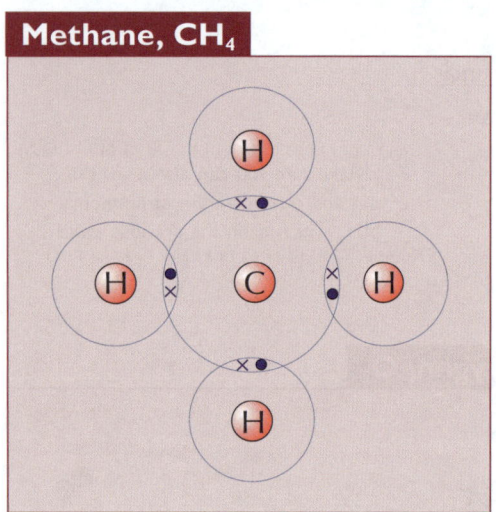

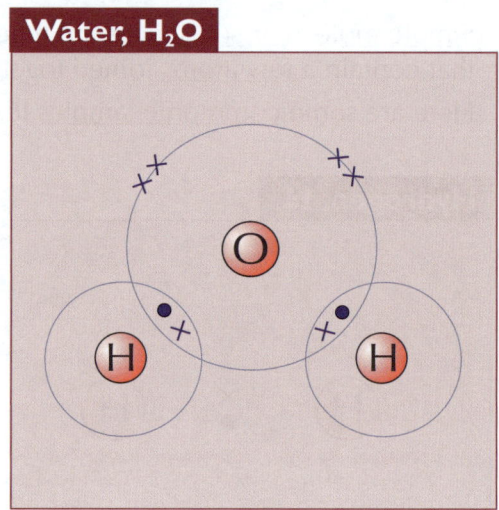

Properties of Simple Molecular Substances

1) Substances containing covalent bonds usually have simple molecular structures, like the examples above.
2) The atoms within the molecules are held together by very strong covalent bonds.
3) But the forces of attraction between these molecules are very weak.

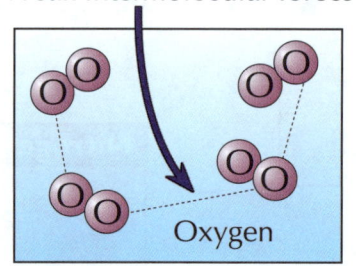

There's more about melting and boiling on page 206.

4) To melt or boil a simple molecular compound, you only need to break the weak intermolecular forces and not the covalent bonds.
5) So the melting and boiling points are very low, because it's easy to break the intermolecular forces.
6) Most molecular substances are gases or liquids at room temperature.
7) As molecules get bigger, the intermolecular forces get stronger. More energy is needed to break the stronger forces, so the melting and boiling points increase.
8) Molecular compounds don't conduct electricity because they aren't charged.

Simple molecular substances are easy to melt

Remember, it's the weak forces between molecules that are broken when a simple molecular substance melts — not the strong covalent bonds between the atoms in the molecules. It's easy to get the two mixed up, so get your head around it now before moving on.

Q1   Draw a dot and cross diagram to show the bonding in a molecule of ammonia (NH_3). [2 marks]

Q1 Video Solution

Topic C2 — Bonding, Structure and Properties of Matter

Warm-Up & Exam Questions

The questions on this page are all about covalent bonding. Go through them and if you have any problems, make sure you look back at the relevant pages again until you've got to grips with it all.

Warm-Up Questions

1) What is a covalent bond?
2) How many triple covalent bonds does a molecule of nitrogen have?
3) True or false? Most simple molecular substances are solids at room temperature.
4) Which forces are stronger in simple molecular substances — covalent bonds (between atoms) or intermolecular forces (between molecules)?

Exam Questions

1 The bonding in phosphorus trichloride (PCl_3) is shown in **Figure 1**. Only the outer electrons are shown.

Figure 1

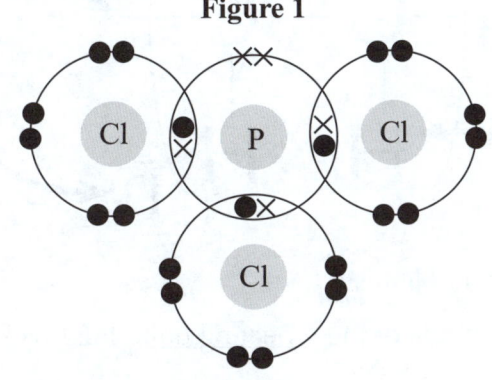

1.1 How many electrons are there in the outer shell of each atom in PCl_3?

[1 mark]

1.2 How many shared pairs of electrons are there in a molecule of PCl_3?

[1 mark]

1.3 How many double covalent bonds are there in a molecule of PCl_3?

☐ 0   ☐ 1   ☐ 2   ☐ 3

[1 mark]

2 Draw a dot and cross diagram for a molecule of hydrogen (H_2).

[1 mark]

3 Hydrogen chloride is a simple molecular substance.

3.1 How many single covalent bonds are there in a molecule of hydrogen chloride?

[1 mark]

3.2 Explain why hydrogen chloride has a low boiling point.

[1 mark]

3.3 Explain why hydrogen chloride gas has poor electrical conductivity.

[1 mark]

Topic C2 — Bonding, Structure and Properties of Matter

Polymers

*Simple molecular substances aren't the only covalent compounds you need to know about — **polymers** are also made up of **covalent bonds**. Their properties are a bit different to simple molecular compounds, though.*

Polymers Have Very Large Molecules

1) In a polymer, lots of small units are joined together to form a long molecule.

2) All the atoms in a polymer are joined by strong covalent bonds.

Instead of drawing out a whole polymer, you can draw a small part of it, called the repeating unit. The polymer is made up of this unit repeated over and over again.

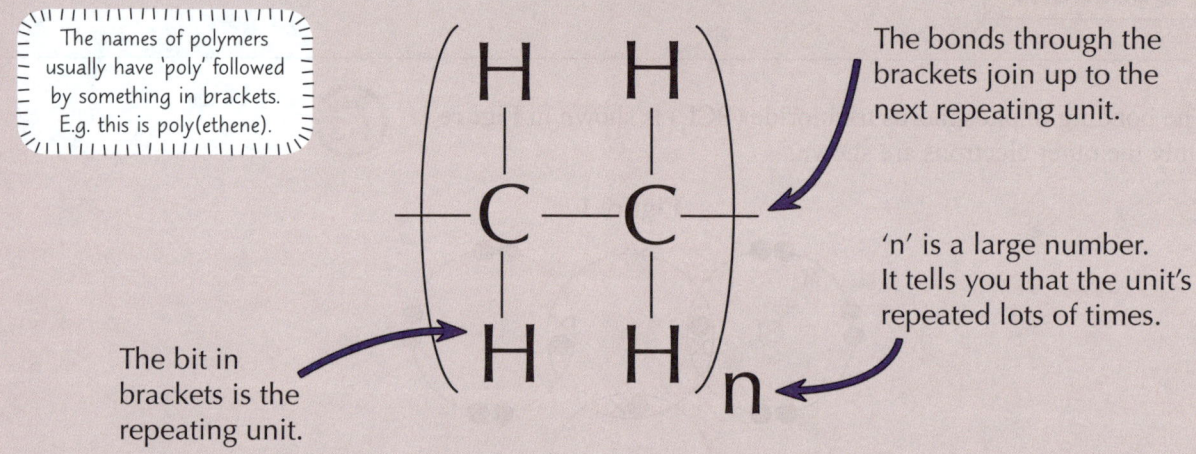

The names of polymers usually have 'poly' followed by something in brackets. E.g. this is poly(ethene).

The bit in brackets is the repeating unit.

The bonds through the brackets join up to the next repeating unit.

'n' is a large number. It tells you that the unit's repeated lots of times.

To find the molecular formula of a polymer...

- Write down the molecular formula of the repeating unit. Put brackets around it. Then put an 'n' outside.
- So for the polymer above, the molecular formula of the polymer is $(C_2H_4)_n$.

3) The intermolecular forces between polymer molecules are larger than between simple covalent molecules. This means more energy is needed to break them. So most polymers are solid at room temperature.

4) The intermolecular forces are still weaker than ionic or covalent bonds. This means they generally have lower melting and boiling points than ionic or giant covalent (see next page) compounds.

Polymers contain lots of strong covalent bonds

You've probably heard of a few polymers before — polyesters and polystyrene are used a lot in our day-to-day lives. Polymers are really useful materials, so you need to be able to recognise them.

Q1 The repeating unit of poly(chloroethene) is shown on the right. What's the molecular formula of poly(chloroethene)? [1 mark]

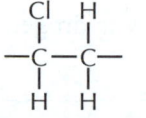

Topic C2 — Bonding, Structure and Properties of Matter

Giant Covalent Structures

*That's right — there's another class of **covalent structures** you need to know about. Here we go...*

Giant Covalent Structures Include Diamond, Graphite and Silica

1) In giant covalent structures, all the atoms are bonded to each other by strong covalent bonds.
2) They have very high melting and boiling points. This is because lots of energy is needed to break the covalent bonds between the atoms.
3) They don't contain charged particles, so they don't conduct electricity (except for a few weird exceptions such as graphite, see below).
4) The examples you should know about are diamond and graphite, which are both made from carbon atoms only, and silicon dioxide (silica).

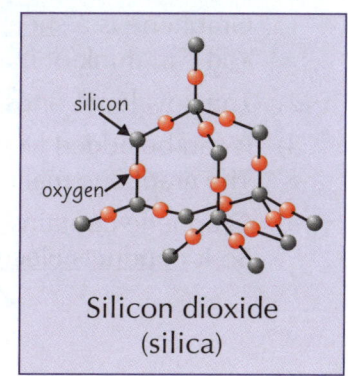
Silicon dioxide (silica)

Diamond is Very Hard

1) In diamond, each carbon atom forms four covalent bonds.
2) This makes diamond really hard.
3) It takes a lot of energy to break the covalent bonds. So diamond has a very high melting point.
4) Diamond doesn't conduct electricity because it has no free electrons or ions.

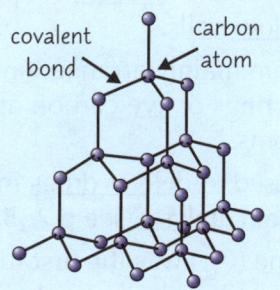

The structures are really much larger than we've shown on this page. The rest of the substance looks the same, so we've just drawn a section.

Graphite Contains Layers of Hexagons

1) Graphite contains layers of carbon atoms. The carbon atoms are arranged in hexagons (rings of six carbon atoms).
2) Each carbon atom forms three covalent bonds.
3) There aren't any covalent bonds between the layers. So the layers can move over each other. This makes graphite soft and slippery.
4) Lots of energy is needed to break the covalent bonds in the layers. So graphite has a high melting point.
5) Each carbon atom has one electron that's free to move (delocalised). So graphite conducts electricity and thermal energy (heat) — a bit like a metal does (see p.203).

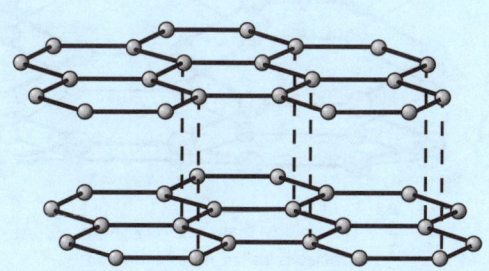

Giant covalent structures have high melting and boiling points

You do have to break the very strong covalent bonds between atoms in order to melt or boil a giant covalent structure. This gives them very high melting and boiling points.

Q1 Describe the structure and bonding of graphite. [4 marks]

Giant Covalent Structures

*Along with diamond and graphite, carbon atoms also make up **graphene** and **fullerenes**.*

Graphene is One Layer of Graphite

1) Graphene is a sheet of carbon atoms joined together in hexagons. You can think of it as one layer of graphite.
2) The covalent bonds make it very strong. It's also very light.
3) It can be added to other materials to make composites. The graphene makes the materials stronger but not much heavier.
4) Graphene contains electrons that are free to move. So it conducts electricity. This means it could be used in electronics.

Fullerenes Form Spheres and Tubes

1) Fullerenes are molecules of carbon, shaped like closed tubes or hollow balls.
2) The carbon atoms are mainly arranged in hexagons. They can also form rings of five carbon atoms or rings of seven carbon atoms.
3) Fullerenes can be used to deliver drugs into the body. They also make great catalysts (see p.235).
4) Buckminsterfullerene (C_{60}) was the first fullerene to be found. It's shaped like a hollow sphere (ball).

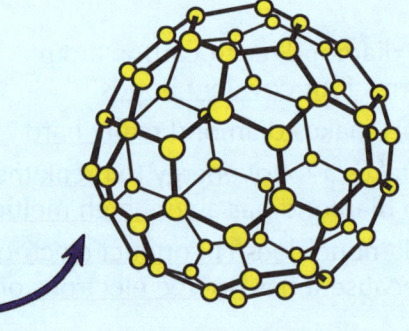

5) Fullerenes can form nanotubes — tiny carbon cylinders.
6) The ratio between the length and the diameter of nanotubes is very high (they're very long compared to their width).
7) Nanotubes have properties that make them useful in electronics. They can also be used to strengthen materials without adding much weight.

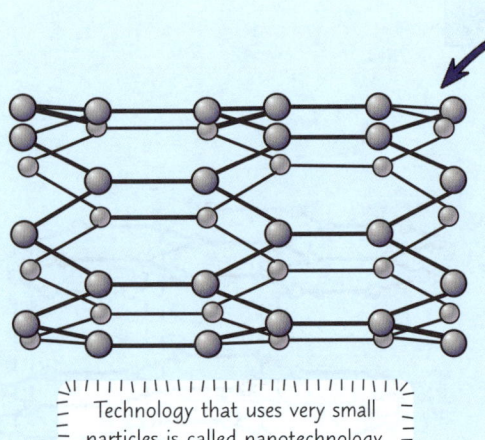

Technology that uses very small particles is called nanotechnology.

Carbon comes in many different forms

There's been quite a bit of information on the last two pages. Before you go on, make sure you can explain the properties of all these forms of carbon. Remember, just because they all contain carbon doesn't mean they're all the same — it all comes down to how the carbon atoms are arranged in the structure.

Topic C2 — Bonding, Structure and Properties of Matter

Metallic Bonding

Ever wondered what gives a metal its properties? Most of it comes down to bonding...

Metallic Bonding Involves Delocalised Electrons

1) <u>Metals</u> are <u>giant structures</u> of atoms.
   This means they contain <u>lots and lots</u> of metal atoms bonded together.
2) The electrons in the <u>outer shell</u> of the metal atoms are <u>free to move around</u> (delocalised).
3) There are strong forces of <u>electrostatic attraction</u> between the <u>positive metal ions</u> and the shared <u>negative electrons</u>.
4) These forces of attraction are known as <u>metallic bonds</u>.
   They <u>hold</u> the <u>atoms</u> together in a <u>regular pattern</u>.

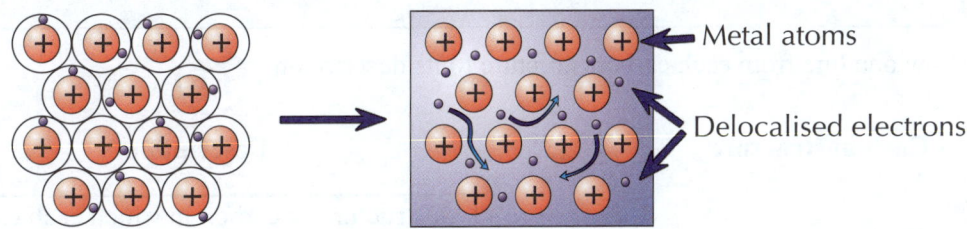

The Bonding in Metals Affects their Properties

1) Metallic bonds are very <u>strong</u>, so <u>lots of energy</u> is needed to break them.
2) This means that most substances with metallic bonds have very <u>high</u> melting and boiling points.
3) They're usually <u>solids</u> at room temperature.
4) The <u>delocalised electrons</u> in the metal are <u>free to move</u>.
5) These electrons can carry <u>electrical charge</u> and <u>thermal</u> (heat) energy through the whole structure.
6) This means metals are good <u>conductors</u> of <u>electricity</u> and <u>heat</u>.
7) The layers of atoms in a metal can <u>slide</u> over each other.
8) Because of this, metals can be <u>bent</u> or <u>formed</u> into different shapes.

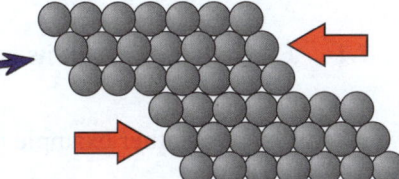

Metallic Bonding is Found in Alloys

1) <u>Pure metals</u> are often quite <u>soft</u>.
2) Most of the metals we use are <u>alloys</u>.
   An alloy is a <u>mixture</u> of <u>two or more metals</u> or a <u>metal and another element</u>.
3) Mixing another element with a pure metal causes the layers of metal atoms to <u>lose their shape</u>. This is because different elements have <u>different sized atoms</u>.
4) It becomes more <u>difficult</u> for the atoms to <u>slide</u> over each other.
5) This makes alloys <u>harder</u> and so <u>more useful</u> than pure metals.

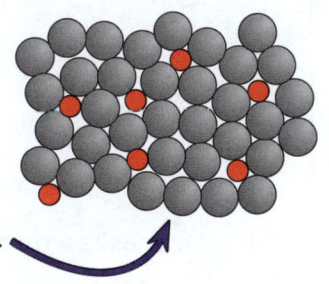

 **Metallic bonding is what makes metals, well... metals**
The bonding between atoms is the key to understanding the <u>structure</u> and <u>properties</u> of any substance — metals included. Take a moment to check you can <u>explain</u> how they are related.

Topic C2 — Bonding, Structure and Properties of Matter

Warm-Up & Exam Questions

Lots of information to learn on the previous few pages — here are some questions to test yourself on.

Warm-Up Questions

1) True or False? The repeating units in a polymer are held together with covalent bonds.
2) At room temperature, what state are most polymers in?
3) Describe the differences in the hardness and electrical conductivity of diamond and graphite.
4) Why can most metals be bent into different shapes?

Exam Questions

1 Draw one line from each carbon structure to its description.

Carbon structure | **Description**

Diamond

A structure of carbon in which each carbon atom forms four covalent bonds.

A carbon structure shaped like a closed tube or hollow ball.

Graphite

A layered structure in which each carbon atom forms three covalent bonds.

[2 marks]

2 Silicon carbide has a giant covalent structure and is a solid at room temperature.

2.1 Explain, in terms of its bonding and structure, why silicon carbide has a high melting point.

[2 marks]

2.2 Give **one** other example of a substance with a giant covalent structure.

[1 mark]

3 **Figure 1** shows the arrangement of atoms in two different materials.
A is a metal and B is an alloy.

Figure 1

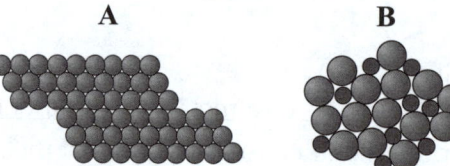

3.1 Define the term 'alloy'.

[1 mark]

3.2 Explain how delocalised electrons make metals good conductors of heat.

[2 marks]

3.3 Iron is a metal. Steel is an alloy of iron and carbon.
Explain why steel is harder than iron.

[4 marks]

Topic C2 — Bonding, Structure and Properties of Matter

States of Matter

*You can explain quite a bit of stuff in **chemistry** if you can get your head around this lot.*

The Three States of Matter — Solid, Liquid and Gas

1) Materials come in three different forms — solid, liquid and gas. These are the three states of matter.
2) Particle theory is a model where each particle is seen as a small, solid sphere (ball).
3) You can use particle theory to show what solids, liquids and gases are like. It can be used to explain how the particles in solids, liquids and gases behave.

The particles could be atoms, ions or molecules.

Solids

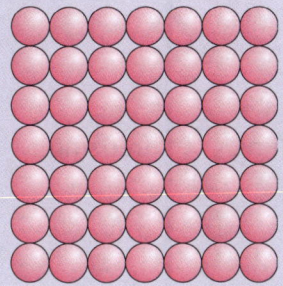

1) In solids, there are strong forces of attraction between particles.
2) The particles are held close together in fixed positions to form a pattern.
3) Solids have a fixed shape and volume.

Liquids

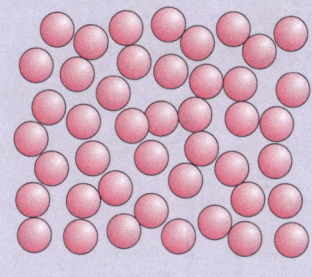

1) In liquids, there are weak forces of attraction between the particles.
2) They're randomly arranged and free to move past each other, but they tend to stick closely together.
3) Liquids have a fixed volume but don't keep a fixed shape. So they flow to fill the bottom of a container.

Gases

1) In gases, the forces of attraction between the particles are very weak.
2) The particles are free to move and are spaced far apart.
3) The particles in gases travel in straight lines.
4) Gases don't have a fixed shape or volume. They will always fill containers.

If you ever see something described as 'gaseous', it just means that it's a gas.

Particle theory explains solids, liquids and gases

Particle theory imagines all particles as small snooker ball-like spheres. It's not quite like that in real life, of course, but it's a good theory for understanding the properties and behaviour of the three states of matter.

States of Matter

Substances Can Change from One State to Another

Solid → Liquid → Gas

1) When a solid is heated, its particles gain energy and start to move about.

2) Some of the forces between the particles break.

3) At a temperature called the melting point, the particles have enough energy to break free from their positions. This is MELTING. The solid turns into a liquid.

4) When a liquid is heated, the particles get even more energy.

5) The forces holding the liquid together weaken and break.

6) At a temperature called the boiling point, the particles have enough energy to break the forces. This is BOILING. The liquid becomes a gas.

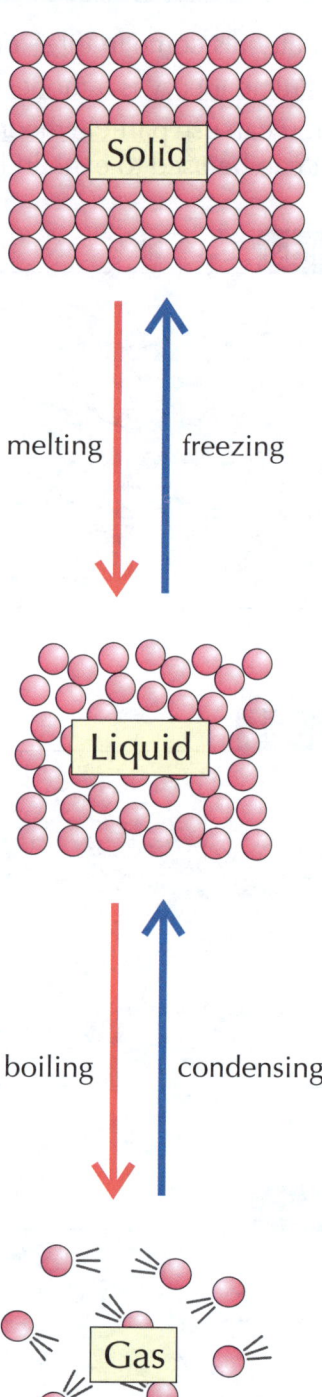

Gas → Liquid → Solid

1) As a gas cools, the particles have less energy.

2) Forces form between the particles.

3) At the boiling point, the forces between the particles are strong enough that the gas becomes a liquid. This is CONDENSING.

4) When a liquid cools, the particles have less energy, so move around less.

5) The forces between the particles become stronger.

6) At the melting point, the forces between the particles are so strong that they're held in place. The liquid becomes a solid. This is FREEZING.

The amount of energy needed for a substance to change state depends on how strong the forces between particles are. The stronger the forces, the more energy is needed to break them, and so the higher the melting and boiling points of the substance.

Topic C2 — Bonding, Structure and Properties of Matter

States of Matter

*The **state** a substance is in can be really important in working out **how it will react**. This page is all about how chemists **show** what the state of a substance is and how they **predict** what state a substance will be in.*

State Symbols Tell You the State of a Substance in an Equation

1) You saw on page 165 how a chemical reaction can be shown using a symbol equation.
2) Symbol equations can include state symbols next to each substance. They're always shown in brackets, and they're normally subscripts (slightly smaller and below the rest of the letters).
3) They tell you whether each substance is a solid, a liquid, a gas or dissolved in water:

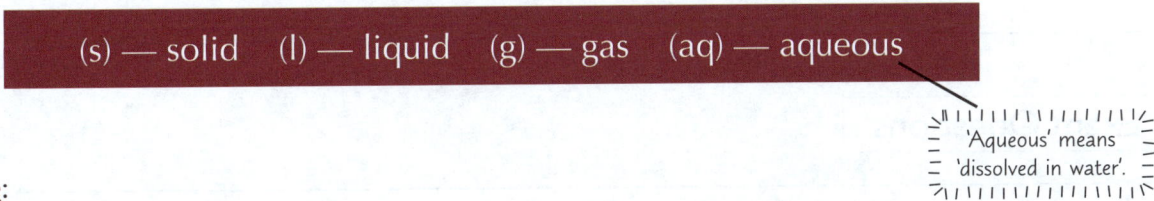

(s) — solid   (l) — liquid   (g) — gas   (aq) — aqueous

'Aqueous' means 'dissolved in water'.

For example:

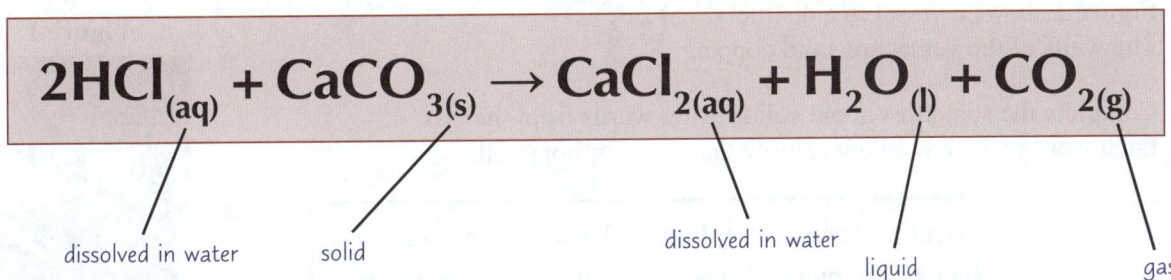

$$2HCl_{(aq)} + CaCO_{3(s)} \rightarrow CaCl_{2(aq)} + H_2O_{(l)} + CO_{2(g)}$$

dissolved in water — solid — dissolved in water — liquid — gas

You Have to be Able to Predict the State of a Substance

1) You can predict what state a substance is in at a certain temperature.
2) If the temperature's below the melting point of substance, it'll be a solid.
3) If it's above the boiling point, it'll be a gas.
4) If it's in between the two points, then it's a liquid.

The bulk properties such as the melting point of a material depend on how lots of atoms interact together. An atom on its own doesn't have these properties.

EXAMPLE
Which of the substances in the table is a liquid at room temperature (25 °C)?

	melting point	boiling point
oxygen	−219 °C	−183 °C
nitrogen	−210 °C	−196 °C
bromine	−7 °C	59 °C

Oxygen and nitrogen have boiling points below 25 °C, so will both be gases at room temperature.

The answer's **bromine**. It melts at −7 °C and boils at 59 °C. So, it'll be a liquid at room temperature.

Changing between states is reversible

Make sure you can describe what happens to particles, and the forces between them, as a substance is heated and cooled. Don't forget to learn the technical terms for each state change.

Q1 Ethanol melts at −114 °C and boils at 78 °C. Predict the state that ethanol is in at:
 a) −150 °C   b) 0 °C   c) 25 °C   d) 100 °C   [4 marks]

Q1 Video Solution

Topic C2 — Bonding, Structure and Properties of Matter

Warm-Up & Exam Questions

Reckon you know all there is to know about this section? Have a go at these questions and see how you get on. If you get stuck on something — just flick back and give it another read through.

Warm-Up Questions

1) What state of matter has a fixed arrangement of particles?
2) True or false? There are only weak forces of attraction between particles in a gas.
3) Describe what happens to the bonding in a substance when it condenses.
4) How you would show that sodium ions (Na^+) are dissolved in water in a symbol equation.
   A. $Na^{+(aq)}$ B. $Na^+_{(l)}$ C. $Na^+_{(aq)}$ D. $Na^+_{(s)}$

Exam Questions

1   **Figure 1** shows a vessel in a distillery. **Figure 1**
    The walls of the vessel are solid copper.

    Complete the sentences about solids using words from the box.
    Each word may be used once, more than once or not at all.

weak	move	colder	hotter	random
strong	expand	heavier	dissolve	regular

    In solids, there are .... forces of attraction between particles,

    which hold them in fixed positions in a .... arrangement.

    The particles don't .... from their positions, so solids keep their shape.

    The .... the solid becomes, the more the particles in the solid vibrate.

    [4 marks]

2   **Figure 2** shows a substance changing between solid, liquid and gas states.

    Figure 2

    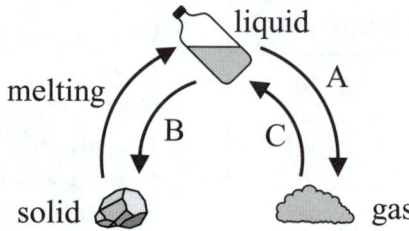

2.1 Give the letter of the arrow that represents **condensing**.

    [1 mark]

2.2 Give the name of the process represented by arrow **A**.

    [1 mark]

2.3 Use particle theory to explain why gases fill their containers.

    [2 marks]

Topic C2 — Bonding, Structure and Properties of Matter

Topic C3 — Quantitative Chemistry

Relative Formula Mass

*Calculating **relative formula mass** is important for lots of calculations in chemistry. It might sound a bit hard to begin with, but it gets easier with **practice**. We'd better get cracking...*

Compounds Have a Relative Formula Mass, M_r

To find the relative formula mass, M_r, of a compound, add together the relative atomic masses of all the atoms in the molecular formula.

> The relative atomic mass (A_r) of an element is on the periodic table. See page 163 for more.

EXAMPLES

a) Find the relative formula mass of $MgCl_2$.

1) Look up the relative atomic masses of all the elements in the compound on the periodic table.

> In the exams, you might be given the relative atomic masses you need in the question.

A_r of Mg = 24 A_r of Cl = 35.5

2) Add up all the relative atomic masses of the atoms in the compound.

> There are two chlorine atoms in $MgCl_2$, so the relative atomic mass of chlorine needs to be multiplied by 2.

Mg + (2 × Cl) = 24 + (2 × 35.5) = 24 + 71 = 95

M_r of $MgCl_2$ = **95**

b) Find the relative formula mass of $Ca(OH)_2$.

1) Look up the relative atomic masses of all the elements in the compound on the periodic table.

A_r of Ca = 40 A_r of O = 16 A_r of H = 1

2) Add up all the relative atomic masses of the atoms in the compound.

> The small number 2 after the bracket in the formula $Ca(OH)_2$ means that there's two of everything inside the brackets.

Ca + [(O + H) × 2] = 40 + [(16 + 1) × 2] = 40 + 34 = 74

M_r of $Ca(OH)_2$ = **74**

You Can Calculate the % Mass of an Element in a Compound

To work out the percentage mass of an element in a compound, you need to use this formula:

$$\text{Percentage mass of an element in a compound} = \frac{A_r \times \text{number of atoms of that element}}{M_r \text{ of the compound}} \times 100$$

EXAMPLE Find the percentage mass of sodium (Na) in sodium bromide (NaBr).

1) Look up the relative atomic masses of all the elements in the compound on the periodic table.

A_r of Na = 23 A_r of Br = 80

2) Add up the relative atomic masses of all the atoms in the compound to find the relative formula mass.

M_r of NaBr = 23 + 80 = 103

3) Use the formula to calculate the percentage mass.

$$\text{Percentage mass of sodium} = \frac{A_r \times \text{number of atoms of that element}}{M_r \text{ of the compound}} \times 100$$

$$= \frac{23 \times 1}{103} \times 100 = \mathbf{22\%}$$

Relative formula mass — add up all the relative atomic masses

You'll get a periodic table in the exams, which could be handy for these sorts of calculations — the relative atomic mass of an element is the bigger number next to the element's symbol.

Q2 Video Solution

Q1 Calculate the relative formula mass (M_r) of: a) CO_2 b) LiOH   c) H_2SO_4 [3 marks]

Q2 Calculate the percentage composition by mass of potassium in potassium hydroxide (KOH).   [2 marks]

Conservation of Mass

*You've probably realised by now that you can't **magic** stuff out of thin air. It can't magically **disappear**, either.*

In a Chemical Reaction, Mass Always Stays the Same

1) During a chemical reaction no atoms are lost and no atoms are made.
2) This means there are the same number and types of atoms on each side of a reaction equation.
3) Because of this, no mass is lost or gained — we say that mass is conserved (stays the same) in a reaction.

> For example: $2Li + F_2 \rightarrow 2LiF$
> In this reaction, there are 2 lithium atoms and 2 fluorine atoms on each side of the equation.

4) You can see that mass stays the same if you add up the relative formula masses of the substances on each side of a balanced symbol equation.
5) The total M_r of all the reactants will be the same as the total M_r of the products.

If you're not sure what the big numbers and the little numbers in reaction equations mean, see pages 164-166.

EXAMPLE Show that mass is conserved in this reaction: $2Li + F_2 \rightarrow 2LiF$.
Relative atomic masses (A_r): Li = 7, F = 19

1) Add up the relative formula masses on the left-hand side of the equation.

$$2 \times A_r(Li) + M_r(F_2) = (2 \times 7) + (2 \times 19)$$
$$= 14 + 38$$
$$= 52$$

2) Add up the relative formula masses on the right-hand side of the equation.

$$2 \times M_r(LiF) = 2 \times (7 + 19)$$
$$= 2 \times 26$$
$$= 52$$

The total M_r on the left-hand side of the equation is the same as the total M_r on the right-hand side, so mass is conserved.

You can Calculate the Mass of a Reactant or Product

1) You can use the idea of conservation of mass to work out the mass of a reactant or product in a reaction.
2) You need to know the masses of all the reactants and products except for one.
3) You can work out the total mass of everything on one side of the equation.
4) You can also work out the total mass of everything on the other side of the equation, except for the thing you don't know the mass of.
5) The mass of the thing you don't know is the difference between these two totals.

EXAMPLE 6 g of magnesium completely reacts with 4 g of oxygen in the following reaction:
$$2Mg + O_2 \rightarrow 2MgO$$
What mass of magnesium oxide is formed?

1) Find the total mass of reactants.   4 + 6 = 10 g
2) Magnesium oxide is the only product. So the mass of products you do know is 0 g.

Mass of magnesium oxide = 10 − 0 = **10 g**

The M_r of the reactants will always equal the M_r of the products

Don't forget that, in a reaction, the total mass of reactants is the same as the total mass of products.

Q1 Show that mass is conserved in the reaction: $H_2SO_{4(aq)} + 2NaOH_{(aq)} \rightarrow Na_2SO_{4(aq)} + 2H_2O_{(l)}$
$A_r(H) = 1$, $A_r(O) = 16$, $A_r(Na) = 23$, $A_r(S) = 32$
[5 marks]

Q1 Video Solution

Conservation of Mass

*Even though mass is always (always) conserved in a reaction, sometimes you can carry out an experiment where the mass of the **reaction container** changes. Time to find out why...*

If the Mass Seems to Change, There's Usually a Gas Involved

In some experiments, the mass of an unsealed reaction container might change during a reaction. This usually happens for one of two reasons...

1) If One of the Reactants is a Gas, the Mass Could Go Up

If the mass goes up, it's probably because one of the things that reacts is a gas that's found in air (e.g. oxygen) and all the things that are made are solids, liquids or in solution.

1) Before the reaction, the gas is floating around in the air. It's there, but it's not trapped in the reaction container. This means you can't measure its mass.
2) When the gas reacts, its atoms become part of the product, which is held inside the reaction container.
3) So the total mass of the stuff inside the reaction container goes up.

- When a metal reacts with oxygen in an unsealed container, the mass of the container goes up.
- This is because the mass of the oxygen atoms isn't measured when they're part of the gas, but it is when they're in the metal oxide.

$$\text{metal}_{(s)} + \text{oxygen}_{(g)} \rightarrow \text{metal oxide}_{(s)}$$

2) If One of the Products is a Gas, the Mass Could Go Down

If the mass goes down, it's probably because one of the products is a gas and all the things that react are solids, liquids or in solution.

1) Before the reaction, all the reactants are held in the reaction container.
2) If the container isn't sealed, then the gas can escape from the reaction container as it's formed.
3) It's no longer trapped in the reaction container, so you can't measure its mass.
4) This means the total mass of the stuff inside the reaction container goes down.

Remember from the particle model on page 205 that a gas will spread out to fill any container it's in. So if the reaction container isn't sealed, the gas will escape into the air.

- When a metal carbonate is heated, it can break down to form a metal oxide and carbon dioxide gas.
- When this happens, the mass of the reaction container will go down if it isn't sealed.
- But really, the mass of the metal oxide and the carbon dioxide formed will be the same as the mass of the metal carbonate.

Reactions where substances are heated and break down are called thermal decomposition reactions.

$$\text{metal carbonate}_{(s)} \rightarrow \text{metal oxide}_{(s)} + \text{carbon dioxide}_{(g)}$$

Mass is ALWAYS conserved in a reaction...

Everything has to be the same on both sides of a reaction equation. If it seems like it's not, the chances are there's a gas involved. So check those state symbols for the little $_{(g)}$.

Q1 Video Solution

Q1 During a reaction, the mass in a reaction container increases. What does this tell you about the states of matter of the products and reactants? [2 marks]

Q2 A scientist carries out the reaction below. He does it in an unsealed reaction container: $2Na_{(s)} + 2HCl_{(aq)} \rightarrow 2NaCl_{(aq)} + H_{2(g)}$. Predict how the mass of the reaction container will change as the reaction takes place. [1 mark]

Topic C3 — Quantitative Chemistry

Concentrations of Solutions

Lots of reactions take place between substances that are dissolved in a **solution**. And sometimes it's useful to find out the **mass** of a substance that's dissolved in a **solution**. Hold onto your hats and concentrate...

Concentration is a Measure of How Crowded Things Are

1) The amount of a substance (e.g. the mass) in a certain volume of a solution is called its concentration.
2) The more substance that's dissolved in a certain volume, the more concentrated the solution.

Concentration can be Measured in g/dm³

1) You can find the concentration of a solution if you know the mass of the substance dissolved and the volume of the solution.

   The thing that dissolves the solid is called a 'solvent'. The solid that dissolves in the solvent is called a 'solute'.

2) The units will be units of mass/units of volume. For example, g/dm³.
3) Here's how to calculate the concentration of a solution in grams per decimetre cubed (g/dm³):

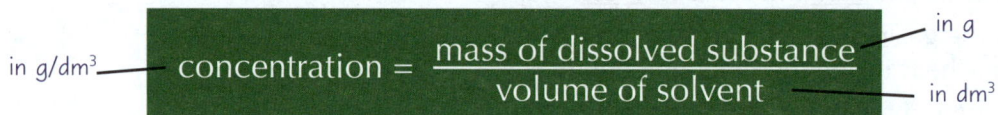

in g/dm³ — concentration = mass of dissolved substance / volume of solvent — in g, in dm³

EXAMPLES

a) 30 g of sodium chloride is dissolved in 0.2 dm³ of water. What's the concentration of this solution in g/dm³?

1 dm³ = 1000 cm³

Put the numbers into the formula to calculate the concentration.

concentration = $\frac{30}{0.2}$ = **150 g/dm³**

b) 15 g of salt is dissolved in 500 cm³ of water. What's the concentration of this solution in g/dm³?

1) The units of the volume need to be dm³ so that the units of concentration are in g/dm³. So make the units of volume dm³ by dividing by 1000:

   500 ÷ 1000 = 0.5 dm³

2) Now you've got the mass and the volume in the right units, just stick them in the formula:

   concentration = $\frac{15}{0.5}$ = **30 g/dm³**

4) You can rearrange the equation above to find the mass of substance dissolved in a certain volume of solution if you know its concentration.
5) Here's a formula triangle to help with rearranging the equation:
6) To use the formula triangle, just cover up the thing you want to find with your finger and write down what's left showing.

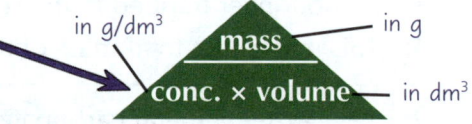

in g/dm³ — mass — in g; conc. × volume — in dm³

EXAMPLE

A solution of magnesium chloride has a concentration of 24 g/dm³. What mass of magnesium chloride is there in 0.40 dm³ of this solution?

1) You want to find the mass. So cover up 'mass' in the formula triangle. This leaves 'concentration × volume'.

   mass = conc. × volume

2) Use this equation to calculate the mass:

   mass = 24 × 0.40 = **9.6 g**

When you measure something like the mass of a substance or the volume of a solution, there's always some uncertainty to the measurement. You can calculate this uncertainty from a range — see pages 9 and 15 for more.

That formula triangle is very useful...

Make sure you know the equation for concentration. And make sure you can rearrange it to find the mass of solute or the volume of the solution. Practising using the formula triangle should help.

Q1 Calculate the concentration, in g/dm³, of a solution that contains 0.6 g of salt in 15 cm³ of solvent.   [2 marks]

Q1 Video Solution

Topic C3 — Quantitative Chemistry

Warm-Up & Exam Questions

There's no getting around it — there's quite a bit of maths in Topic 3. But the best way to get good at maths questions is lots of practice. So here are some questions to help you do just that...

Warm-Up Questions

1) What is meant by the relative formula mass of a compound?
2) The A_r of hydrogen (H) is 1 and the A_r of oxygen (O) is 16. Calculate the M_r of water (H_2O).
3) True or false? Conservation of mass means that no mass is lost or gained during a reaction.
4) What does it usually mean if the mass of a container increases during a reaction?
5) A student is calculating the concentration of a solution.
   The mass of the dissolved substance is in grams and the volume of the solvent is in dm^3.
   What units should the student use for the concentration?

Exam Questions

1 Which of the following is the M_r of calcium chloride ($CaCl_2$)? *Grade 1-3*
Tick **one** box.
Relative atomic masses (A_r): Cl = 35.5, Ca = 40

☐ 54 ☐ 75.5 ☐ 111 ☐ 71

[1 mark]

2 Which of the following compounds has a relative formula mass of 62? *Grade 3-4*
Tick **one** box.
Relative atomic masses (A_r): F = 19, Na = 23, Mg = 24, Cl = 35.5, K = 39, Br = 80

☐ sodium chloride, NaCl ☐ potassium bromide, KBr

☐ magnesium fluoride, MgF_2 ☐ sodium bromide, NaBr

[1 mark]

3 Zinc carbonate decomposes when it is heated. The equation for this reaction is: *Grade 3-4*

$ZnCO_{3(s)}$ → $ZnO_{(s)}$ + $CO_{2(g)}$
zinc carbonate → zinc oxide + carbon dioxide

In one reaction, a sample of zinc carbonate completely decomposed.
The decomposition produced 48.6 kg of zinc oxide and 26.4 kg of carbon dioxide.
Calculate the mass of zinc carbonate that decomposed.

[1 mark]

4 The formula of the compound zinc cyanide is $Zn(CN)_2$. *Grade 4-5*
Calculate the relative formula mass of zinc cyanide.

[2 marks]

Exam Questions

5 A solution of calcium hydroxide, Ca(OH)₂, can be known as limewater. *Grade 4-5*

5.1 Calculate the relative formula mass of calcium hydroxide, Ca(OH)₂.
Relative atomic masses (A_r): H = 1, O = 16, Ca = 40

[1 mark]

Calcium hydroxide reacts with nitric acid. The word equation for this reaction is:

calcium hydroxide + nitric acid → calcium nitrate + water

5.2 In one experiment, 18.5 g of calcium hydroxide reacted completely with 31.5 g of nitric acid. 9 g of water was produced in the reaction. What mass of calcium nitrate was formed?

[1 mark]

6 A student carries out the following reaction in an unsealed container: *Grade 4-5*

$$2HCl_{(aq)} + MgCO_{3(s)} \rightarrow MgCl_{2(aq)} + H_2O_{(l)} + CO_{2(g)}$$

6.1 Calculate the relative formula mass of $MgCO_3$.
Relative atomic masses (A_r): C = 12, O = 16, Mg = 24

[1 mark]

6.2 How will the mass of the reaction vessel and its contents change during the reaction?
Tick **one** box.

☐ The mass will decrease ☐ The mass will stay the same

☐ The mass will increase ☐ It is impossible to tell from the information given

[1 mark]

7 A teacher makes up a solution of copper sulfate to use in an experiment. *Grade 4-5*
She dissolves copper sulfate powder in 1500 cm³ of water.
The concentration of copper sulfate in the solution is 12 g/dm³.

7.1 Convert 1500 cm³ to dm³.

[1 mark]

7.2 Calculate the mass of copper sulfate the teacher used.

[2 marks]

8 The compound potassium hydroxide has the formula KOH. *Grade 4-5*
Potassium hydroxide dissolves easily in water.

8.1 A scientist dissolved 40 g of potassium hydroxide in 0.25 dm³ of water.
Calculate the concentration of the solution formed in g/dm³.

[2 marks]

8.2 Calculate the percentage by mass of potassium in potassium hydroxide.
Give your answer to 3 significant figures.
Relative atomic masses (A_r): H = 1, O = 16, K = 39

[3 marks]

Topic C3 — Quantitative Chemistry

Topic C4 — Chemical Changes

Acids, Bases and Their Reactions

Acids and *bases* crop up everywhere in chemistry — so here's the lowdown on the basics of pH...

The pH Scale Goes From 0 to 14

1) The pH scale is a measure of how acidic or alkaline a solution is.
2) The lower the pH of a solution, the more acidic it is.
3) The higher the pH of a solution, the more alkaline it is.
4) A neutral substance (e.g. pure water) has pH 7.

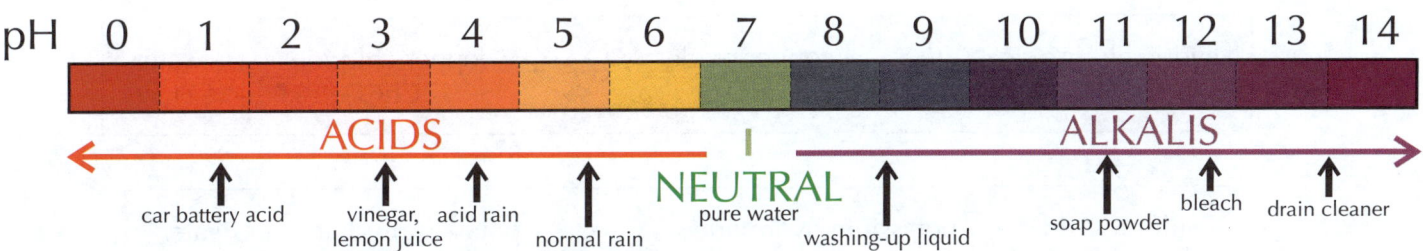

Acids and Bases Neutralise Each Other

1) When acids dissolve in water, they form solutions with a pH of less than 7. Acids form H^+ ions in water.
2) Bases have pHs greater than 7.
3) Alkalis are bases that dissolve in water to form solutions with a pH greater than 7. Alkalis form OH^- ions in water. For example, soluble metal hydroxides are alkalis.

The reaction between acids and bases is called neutralisation:

$$Acid + Base \rightarrow Salt + Water$$

Neutralisation between acids and alkalis can be shown using H^+ and OH^- ions like this:

$$H^+_{(aq)} + OH^-_{(aq)} \rightarrow H_2O_{(l)}$$

Hydrogen (H^+) ions react with hydroxide (OH^-) ions to produce water.

The products of neutralisation reactions have a pH of 7. This means they're neutral.

4) You can use a chemical called an indicator to tell when a neutralisation reaction is over.
5) An indicator is a dye that changes colour depending on whether it's above or below a certain pH.
6) The indicator is added to the acid or alkali you're neutralising. Then the other substance is gradually added. The indicator will change colour when the neutralisation reaction is over.
7) Universal indicator is an indicator that gives the colours shown on the pH scale above. It turns green when the pH of a solution is neutral.

There's more on indicators on page 394.

Interesting fact — your skin is slightly acidic (pH 5.5)

When you mix an acid with an alkali, hydrogen ions from the acid react with hydroxide ions from the alkali to make water. The leftover bits of the acid and alkali make a salt.

Acids, Bases and Their Reactions

There's more about neutralisation reactions coming up on this page...

Metal Oxides and Metal Hydroxides are Bases

1) Metal oxides and metal hydroxides react with acids in neutralisation reactions to form a salt and water.
2) The salt that forms depends upon the acid and the metal ion in the oxide or hydroxide.
3) HCl reacts to form chlorides, H_2SO_4 reacts to form sulfates and HNO_3 reacts to form nitrates.

> hydrochloric acid + copper oxide → copper chloride + water
> $2HCl$ + CuO → $CuCl_2$ + H_2O

> sulfuric acid + potassium hydroxide → potassium sulfate + water
> H_2SO_4 + $2KOH$ → K_2SO_4 + $2H_2O$

> nitric acid + sodium hydroxide → sodium nitrate + water
> HNO_3 + $NaOH$ → $NaNO_3$ + H_2O

Acids and Metal Carbonates Produce Carbon Dioxide

Metal carbonates are also bases. They react with acids to produce a salt, water and carbon dioxide.

> sulfuric acid + calcium carbonate → calcium sulfate + water + carbon dioxide
> H_2SO_4 + $CaCO_3$ → $CaSO_4$ + H_2O + CO_2

> hydrochloric acid + sodium carbonate → sodium chloride + water + carbon dioxide
> $2HCl$ + Na_2CO_3 → $2NaCl$ + H_2O + CO_2

Acid + Base → Salt + Water (and sometimes carbon dioxide)

There's a whole lot of chemical symbols and equations in this section, so now is a good time to make sure you're comfortable using them. Look back at pages 164-166 if you need more help.

Q1 Calcium carbonate is added to hydrochloric acid. Write the word equation and the balanced symbol equation for the reaction that occurs. [3 marks]

Q1 Video Solution

Topic C4 — Chemical Changes

Acids, Bases and Their Reactions

*Talking about **acids** and **bases** is all well and good, but you need to know how to measure their **pH** too. You also need to know the correct method for making **soluble salts** from an acid and an insoluble base.*

You Can Measure the pH of a Solution

1) As you saw on page 215, an indicator is a dye that changes colour depending on whether it's above or below a certain pH.
2) Wide range indicators are substances that gradually change colour as pH changes.
3) They're useful for estimating the pH of a solution.
4) For example, Universal indicator is a wide range indicator. See page 215 for a reminder of its colours.
5) A pH probe attached to a pH meter can also be used to measure pH electronically.
6) The probe is put in the solution and the pH is shown as a number. This means it's more accurate than an indicator.

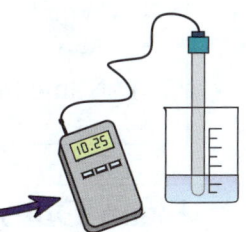

Remember, acids have a pH of less than 7. Bases have a pH of more than 7.

You can Make Soluble Salts Using an Insoluble Base

1) If you react an acid with an insoluble base or a metal, you can make a soluble salt.
2) First, pick the acid that contains the same negative ion as the salt you want to make. For example, to make copper chloride, you'd choose hydrochloric acid.
3) Then pick an insoluble base with the same positive ion as the salt you want to make. You could use an insoluble metal oxide, hydroxide, or carbonate.
4) So to make copper chloride, you'd choose copper oxide, copper hydroxide or copper carbonate. Here's the equation for making copper chloride from hydrochloric acid and copper oxide:

$$CuO_{(s)} + 2HCl_{(aq)} \rightarrow CuCl_{2\,(aq)} + H_2O_{(l)}$$

5) Here's the method for making a soluble salt using an acid and an insoluble base.

> 1) Gently warm the dilute acid using a Bunsen burner, then turn off the Bunsen burner.
> 2) Add the insoluble base to the acid until no more reacts (you'll see the solid at the bottom of the flask).
> 3) Filter out the solid that hasn't reacted to get the salt solution (see p.171).
> 4) To get pure, solid crystals of the salt, you need to crystallise it (see p.171).
> 5) To do this, gently heat the solution using a water bath or an electric heater. Some of the water will evaporate. Stop heating the solution and leave it to cool.
> 6) Crystals of the salt should form, which can be filtered out of the solution and then dried.

You might get to carry these experiments out in class

Remember, you could be asked to describe how you would make a pure, dry sample of a given soluble salt, so make sure you understand the method and that you could suggest a suitable acid and base to use.

Topic C4 — Chemical Changes

Warm-Up & Exam Questions

So you think you know everything there is to know about acids? Time to put yourself to the test.

Warm-Up Questions

1) What range of values can pH take?
2) What term is used to describe a solution with a pH of 7?
3) True or false? Universal indicator will turn blue if it is added to a substance with a pH of 7.
4) Which of the following substances is a base?
   A. $CaCO_3$ B. HCl C. HNO_3 D. CO_2
5) Name the two substances formed when sulfuric acid reacts with copper hydroxide.

Exam Questions

1 A student had a sample of acid in a test tube. He gradually added some base to the acid. *Grade 3-4*

1.1 Which negative ion is produced when a soluble base is added to an aqueous solution? Tick **one** box.

☐ H^- ☐ CO_3^{2-} ☐ O^{2-} ☐ OH^-

[1 mark]

1.2 What type of reaction took place in the student's experiment? Tick **one** box.

☐ thermal decomposition   ☐ redox   ☐ combustion   ☐ neutralisation

[1 mark]

2 All metal hydroxides are bases. They can react with acids to form a salt and water. *Grade 3-4*

2.1 Sodium hydroxide is a soluble base. What name is given to bases that dissolve in water? Tick **one** box.

☐ Alkalis   ☐ Acids   ☐ Indicators   ☐ Oxides

[1 mark]

2.2 Nitric acid reacts with magnesium hydroxide to produce magnesium nitrate and water.
Complete and balance the symbol equation for this reaction. The formula of a nitrate ion is NO_3^-.

$2HNO_3 + Mg(OH)_2 \rightarrow$ .... + ...

[2 marks]

PRACTICAL

3 A student is making a sample of a salt by reacting calcium carbonate with hydrochloric acid. *Grade 4-5*

3.1 Name the salt formed in this reaction.

[1 mark]

The student adds solid calcium carbonate to the acid until the reaction is complete.
This is shown by a change in colour of an indicator in the solution.
The student then crystallises the solution to obtain the salt.

3.2 Suggest why the student's method will **not** produce a pure sample of the salt.

[1 mark]

3.3* Describe a method the student could use to make a sample of the soluble salt, that does **not** involve measuring the pH of the solution. Include in your answer how the student could use filtration and crystallisation to make pure and dry crystals of the soluble salt.

[6 marks]

Topic C4 — Chemical Changes

Metals and Their Reactivity

You can place **metals** in order of reactivity. This can be really useful for **predicting** their **reactions**.

The Reactivity Series — How Easily a Metal Reacts

1) The reactivity series lists metals in order of how reactive they are (their reactivity).
2) Metals react to form positive ions.
3) So for metals, their reactivity depends on how easily they lose electrons and form positive ions.
4) The higher up the reactivity series a metal is, the more easily it forms positive ions.

Make sure you learn this list.

Carbon and hydrogen are non-metals but are often included in the reactivity series.

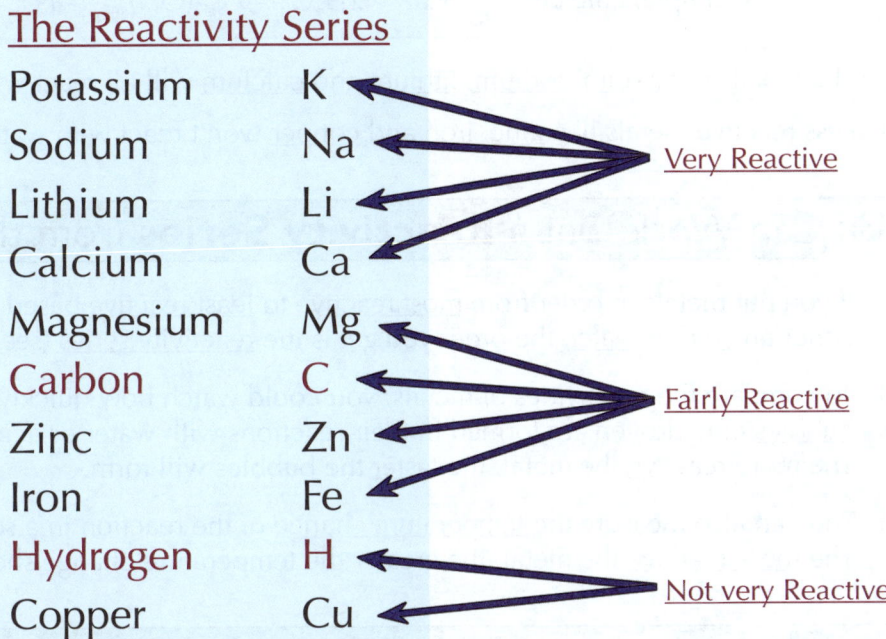

The Reactivity Series

Potassium	K	Very Reactive
Sodium	Na	
Lithium	Li	
Calcium	Ca	
Magnesium	Mg	Fairly Reactive
Carbon	C	
Zinc	Zn	
Iron	Fe	
Hydrogen	H	Not very Reactive
Copper	Cu	

Metals React With Acids

1) Some metals react with acids to produce a salt and hydrogen gas.

HCl reacts to form chloride salts, H_2SO_4 reacts to form sulfate salts.

Acid + Metal → Salt + Hydrogen

hydrochloric acid + magnesium → magnesium chloride + hydrogen $2HCl + Mg \rightarrow MgCl_2 + H_2$
sulfuric acid + zinc → zinc sulfate + hydrogen $H_2SO_4 + Zn \rightarrow ZnSO_4 + H_2$
hydrochloric acid + iron → iron chloride + hydrogen $2HCl + Fe \rightarrow FeCl_2 + H_2$

2) **Very reactive** metals like potassium, sodium, lithium and calcium react explosively with acids.
3) **Less reactive** metals such as magnesium, zinc and iron react less violently with acids.
4) In general, copper won't react with cold, dilute acids.

You need to make sure you know the reactivity series

The reactivity series is useful for predicting the reactions of metals, but you might not be given it in the exam. You need to learn it. If you're struggling, try making up a mnemonic to help you remember — or just use this one: **P**eople **S**ay **L**ong **C**alls **M**ake **C**artoon **Z**ebras **I**nto **H**appy **C**ows.

Topic C4 — Chemical Changes

Metals and Their Reactivity

*As well as reacting with acids, metals can also react with **water** and the **salts** of other metals.*

Some Metals React with **Water**

1) Many metals will react with water.

$$\text{Metal + Water} \rightarrow \text{Metal Hydroxide + Hydrogen}$$

For example, calcium: $Ca_{(s)} + 2H_2O_{(l)} \rightarrow Ca(OH)_{2(aq)} + H_{2(g)}$

2) The metals potassium, sodium, lithium and calcium will all react with water.
3) Less reactive metals like zinc, iron and copper won't react with water.

You Can Work Out a **Reactivity Series** from the Reactions of Metals

1) If you put metals in order from most reactive to least reactive based on their reactions with either an acid or water, the order you get is the reactivity series (see the previous page).
2) To compare the reactivities of metals, you could watch how quickly bubbles of hydrogen are formed in their reactions with water or acid. The more reactive the metal, the faster the bubbles will form.
3) You can also measure the temperature change of the reaction in a set time period. The more reactive the metal, the greater the temperature change should be.

For these experiments to be fair, the mass and surface area of the metals should be the same each time.

More Reactive Metals can Displace Less Reactive Metals from Salts

Displacement reactions involve one metal kicking another one out of a compound. Here's the rule:

A more reactive metal will displace a less reactive metal from its compound.

- For example, iron is more reactive than copper. So if you add solid iron to copper sulfate solution, you get a displacement reaction.
- The iron kicks the copper out of copper sulfate. You end up with iron sulfate solution and copper solid.

$$\text{iron + copper sulfate} \rightarrow \text{iron sulfate + copper}$$
$$Fe_{(s)} + CuSO_{4(aq)} \rightarrow FeSO_{4(aq)} + Cu_{(s)}$$

Metals at the top of the reactivity series are highly reactive

There are plenty of chemical equations to get your head around here. If the question doesn't ask you for a name, you can save time in the exam if you just write a chemical symbol instead of a full name — but always check that you're using the right one, otherwise you could lose marks.

Q1 Give the balanced equation, including state symbols, for the reaction of sodium and water. [3 marks]

Q1 Video Solution

Topic C4 — Chemical Changes

Extracting Metals

*A **few metals** are found in the ground in their **pure forms**. No such luck with the rest though — they need to be **extracted** from other compounds before you can use them. Here's how it works...*

Metals Often Have to be Separated from their Oxides

1) Lots of common metals, like iron and aluminium, react with oxygen to form oxides.
2) This process is an example of oxidation.
3) These oxides are often the ores that the metals are removed (extracted) from.
4) A reaction that separates a metal from its oxide is called a reduction reaction.

An ore is a type of rock that contains metal compounds. Most metals are found in the earth as ores.

Oxidation = Gain of Oxygen

For example, magnesium is oxidised to make magnesium oxide.

$$2Mg + O_2 \rightarrow 2MgO$$

Reduction = Loss of Oxygen

For example, copper oxide is reduced to copper.

$$2CuO + C \rightarrow 2Cu + CO_2$$

Some Metals can be Extracted by Reduction with Carbon

1) Some metals can be extracted from their ores using a reaction with carbon.
2) In this reaction, the ore is reduced as oxygen is removed from it. Carbon gains oxygen, so it is oxidised.
3) For example:

iron(III) oxide + carbon → iron + carbon dioxide
$$2Fe_2O_3 + 3C \rightarrow 4Fe + 3CO_2$$

Iron has lost oxygen. — Carbon has gained oxygen.

4) The reactivity series can tell you if a metal can be extracted with carbon.

- Metals above carbon in the reactivity series are extracted using electrolysis (p.223-224). This is expensive as it takes lots of energy to melt the ore and to produce the electricity.
- Electrolysis is also used to extract metals that react with carbon.
- Metals below carbon in the reactivity series can be extracted by reduction using carbon. For example, iron oxide is reduced in a blast furnace to make iron (see above).
- This is because carbon can only take the oxygen away from metals which are less reactive than carbon itself is.

Make sure you can explain how and why different metals are extracted in different ways.

5) Some metals are so unreactive they are found in the earth as the metal itself. For example, gold.

Carbon can't reduce things above it in the reactivity series

Make sure you understand the difference between reduction and oxidation and make sure that you can spot which substance has been reduced and which substance has been oxidised in a reaction.

Q1   Write a balanced equation for the reduction of lead oxide, PbO, by carbon, C.   [2 marks]

Q2   A mining company tried to extract calcium from its ore by reduction with carbon. The process did not work. Explain why.   [1 mark]

Q1 Video Solution

Topic C4 — Chemical Changes

Warm-Up & Exam Questions

Hoping to test your knowledge with some testing chemistry questions? You're in luck...

Warm-Up Questions

1) True or false? The reactivity of a metal is determined by how easily it forms a positive ion.
2) Magnesium is reacted with dilute hydrochloric acid. What is the name of the salt formed?
3) Which two substances are formed when a metal reacts with water?

Exam Questions

1 Which of the following metals would displace zinc from a solution of zinc chloride but **would not** displace calcium from a solution of calcium chloride?

☐ Magnesium   ☐ Sodium   ☐ Copper   ☐ Iron

[1 mark]

2 **Figure 1** shows part of the reactivity series of metals. Carbon and hydrogen have also been included in this reactivity series.

Figure 1

Potassium   K
Sodium   Na
Calcium   Ca
Magnesium   Mg
CARBON   C
Zinc   Zn
Iron   Fe
HYDROGEN   H
Copper   Cu

2.1 Name **one** metal from **Figure 1** that is more reactive than magnesium.

[1 mark]

2.2 Name **one** metal from **Figure 1** which would **not** displace hydrogen from sulfuric acid.

[1 mark]

2.3 A student places a small piece of zinc into dilute acid. The mixture produces bubbles of hydrogen gas fairly slowly.

Predict whether the reaction would be more or less violent if iron was used instead of zinc. Explain your answer.

[2 marks]

2.4 Name **one** metal from **Figure 1** that could **not** be extracted from its ore by reduction with carbon.

[1 mark]

3 A student placed pieces of copper, zinc and an unknown metal, **X**, in zinc sulfate solution and copper sulfate solution and left them for an hour. The student's results are shown in **Table 1**.

3.1 Explain how you can tell that metal **X** is more reactive than copper.

[1 mark]

3.2 Explain why there was no reaction between copper and zinc sulfate.

[1 mark]

3.3 Suggest the name of the metal, **X**.

[1 mark]

Table 1

	zinc	copper	metal **X**
reaction with zinc sulfate	no reaction	no reaction	no reaction
reaction with copper sulfate	reaction	no reaction	reaction

Topic C4 — Chemical Changes

Electrolysis

*Electrolysis uses **electricity** to cause a reaction. You need to know how it works, so here we go...*

Electrolysis Means 'Splitting Up with Electricity'

1) An electrolyte is just a liquid or solution that can conduct electricity. For example, an ionic compound that's either dissolved in water, or melted so it's a liquid.

2) An electrode is a solid that is put in the electrolyte and conducts electricity.

3) In electrolysis, an electric current is passed through an electrolyte. The ions move towards the electrodes, where they react. The compound then breaks down.

4) Positive ions in the electrolyte move towards the cathode (negative electrode). Here, they gain electrons.

5) Negative ions in the electrolyte move towards the anode (positive electrode). Here, they lose electrons.

6) The ions form the uncharged element. The ions are said to be discharged from the electrolyte.

7) A flow of charge is created through the electrolyte as the ions travel to the electrodes.

Electrolysis of Molten Ionic Solids Forms Elements

1) Molten ionic compounds can be electrolysed because the ions can move freely and conduct electricity.

2) Molten ionic liquids are always broken up into their elements.

3) The metal forms at the cathode. The non-metal is formed at the anode.

> For example, when molten lead bromide is electrolysed, lead forms at the cathode and bromine forms at the anode.

See the next page for another example of how electrolysis of molten ionic compounds works.

4) The electrodes should be inert (unreactive) so they don't react with the electrolyte.

Positive ions move towards the negative electrode

It's easy to get in a muddle with electrolysis, but if you can remember that the cathode is the negative electrode and so it attracts the positive metal ions then you're halfway there. And the anode is just the opposite — it's positive, so it attracts the negative non-metal ions.

Topic C4 — Chemical Changes

Electrolysis

You saw on page 221 how carbon can be used to **extract** some metals from their **ores**. Now it's time to look at how **electrolysis** can also be used for extracting metals.

Metals can be Extracted From Their Ores Using Electrolysis

1) Aluminium is extracted from an ore that contains aluminium oxide, Al_2O_3.
2) Aluminium oxide has a very high melting point so it's mixed with a substance called cryolite. This lowers the melting point.
3) The positive Al^{3+} ions are attracted to the negative electrode where they form aluminium atoms.
4) The negative O^{2-} ions are attracted to the positive electrode where they react to form O_2 molecules.

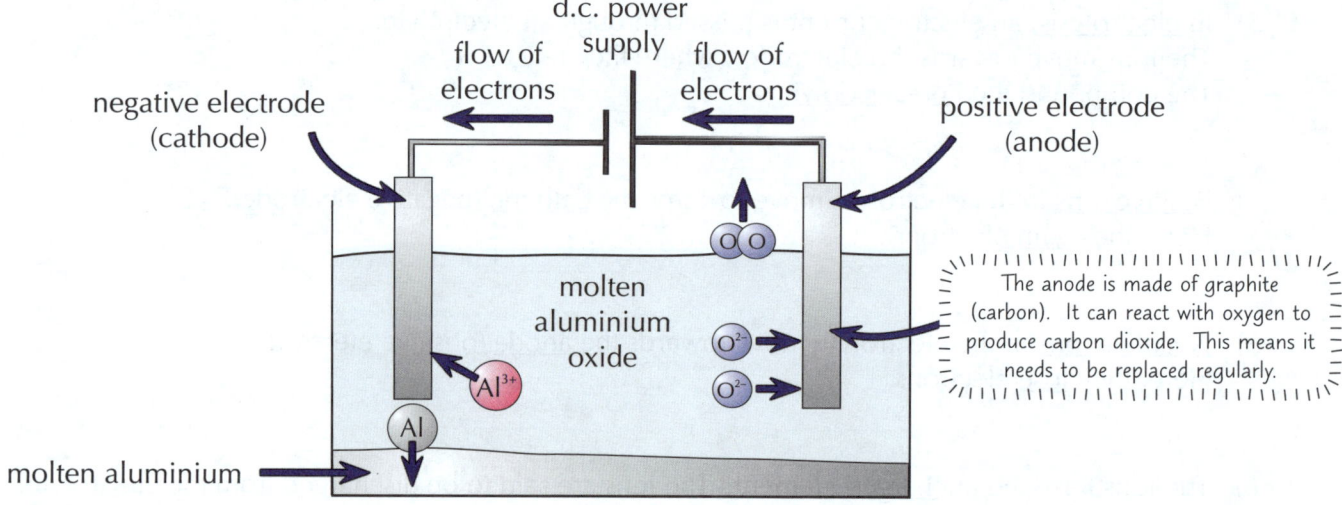

At the Negative Electrode

1) Metals form positive ions.
2) They're attracted to the negative electrode.
3) Aluminium is produced.

At the Positive Electrode

1) Non-metals form negative ions.
2) They're attracted to the positive electrode.
3) Oxygen is produced.

Overall Equation:

aluminium oxide → aluminium + oxygen
$$2Al_2O_{3(l)} \rightarrow 4Al_{(l)} + 3O_{2(g)}$$

Electrolysis is used to extract reactive metals from their ores

Extracting aluminium by electrolysis is handy, but it does have downsides. In industry, the mixture of aluminium oxide and cryolite is heated to around 960 °C. These temperatures and the electrical current used in electrolysis require large amounts of energy. So the whole process is expensive.

Q1 A student carries out electrolysis on molten calcium chloride. What is produced at:
   a) the anode? b) the cathode? [2 marks]

Topic C4 — Chemical Changes

Electrolysis of Aqueous Solutions

When you electrolyse a salt that's dissolved in water, you also have to think about the ions from the **water**.

You Can **Predict** what Forms when a **Salt Solution** is **Electrolysed**

1) Water can break down into H^+ and OH^- ions.

$$H_2O_{(l)} \rightleftharpoons H^+_{(aq)} + OH^-_{(aq)}$$

The $\rightleftharpoons$ symbol in this reaction shows that it's reversible. For more about reversible reactions see p.242-243.

2) So in solutions that contain water, there will be the ions from the ionic compound as well as hydrogen ions (H^+) and hydroxide ions (OH^-) from the water.

3) H^+ ions and metal ions will move to the cathode.

4) If the metal's more reactive than hydrogen, hydrogen gas will form.

5) If the metal is less reactive than hydrogen, a solid layer of the pure metal will form.

6) If the salt contains halide ions (Cl^-, Br^-, I^-), chlorine, bromine or iodine will form at the anode.

7) If no halide ions are present, then the OH^- ions lose electrons and oxygen will form at the anode.

Example 1: Electrolysis of **Copper Sulfate Solution**

A solution of copper(II) sulfate ($CuSO_4$) contains four different ions: Cu^{2+}, SO_4^{2-}, H^+ and OH^-.

At the cathode:
- Copper metal is less reactive than hydrogen.
- So copper metal is produced.

At the anode:
- There aren't any halide ions present.
- So oxygen and water are produced.

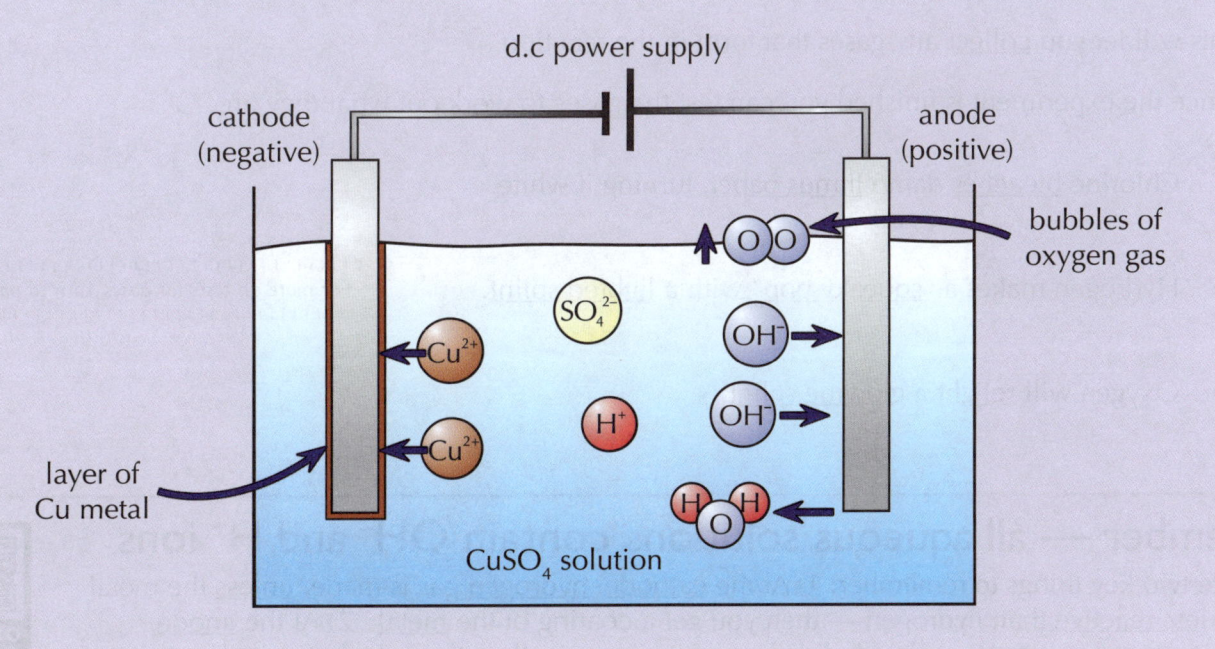

Topic C4 — Chemical Changes

Electrolysis of Aqueous Solutions

Example 2: Electrolysis of **Sodium Chloride Solution**

A solution of sodium chloride (NaCl) contains four different ions: Na^+, Cl^-, OH^- and H^+.

At the cathode:
- Sodium metal is more reactive than hydrogen.
- So hydrogen gas is produced.

At the anode:
- Chloride ions are in the solution.
- So chlorine gas is produced.

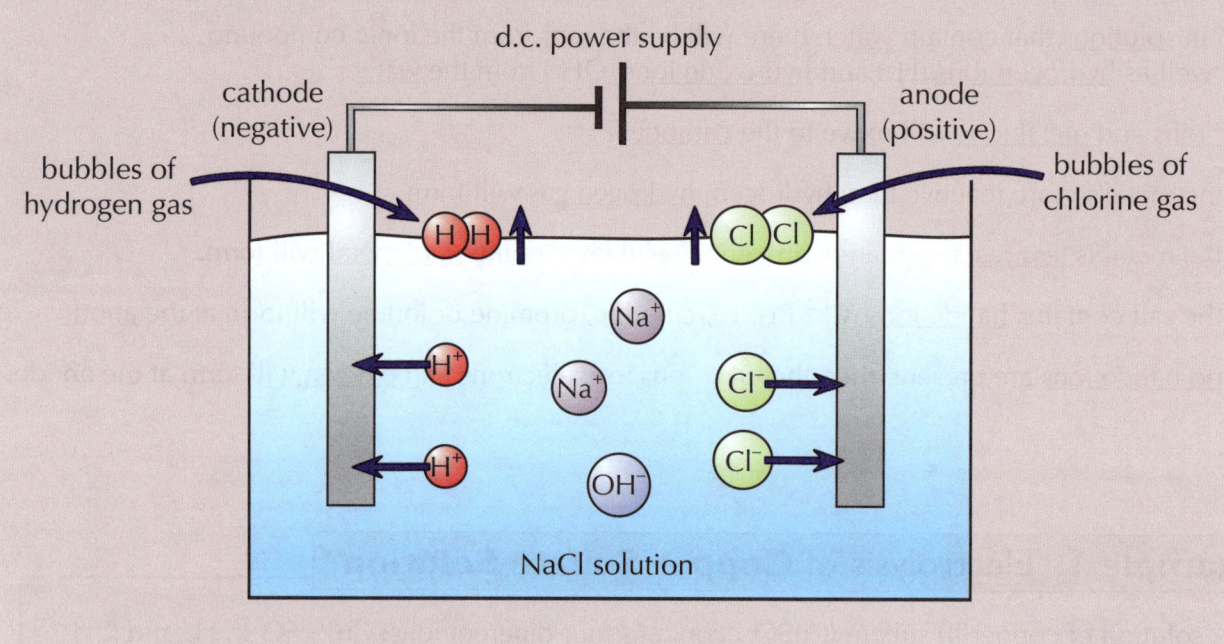

You can do **Electrolysis** in the **Lab**

PRACTICAL

1) You can set up an electrolysis experiment in the lab like the set-up on page 396.
2) This will let you collect any gases that form in the reaction.
3) Once the experiment is finished you can test the gases to work out what they are.

- Chlorine bleaches damp litmus paper, turning it white.
- Hydrogen makes a "squeaky pop" with a lighted splint.
- Oxygen will relight a glowing splint.

For more on tests for gases, turn to page 259

Remember — all aqueous solutions contain OH^- and H^+ ions

There are two key things to remember: 1) At the cathode, hydrogen gas is made, unless the metal ions are less reactive than hydrogen — then you get a coating of the metal. 2) At the anode, oxygen and water are made, unless halide ions are present — then the halogen is made.

Q1   An aqueous solution of copper bromide, $CuBr_2$, is electrolysed using inert electrodes.
State what is produced at: a) the anode,   b) the cathode.   [2 marks]

Q1 Video Solution

Topic C4 — Chemical Changes

Warm-Up & Exam Questions

Time to test your knowledge. Try and get through the following questions. If there's anything you're not quite sure about, have a look at the pages again until you can answer all the questions.

Warm-Up Questions

1) Are metals deposited at the anode or the cathode during electrolysis?
2) True or False? The electrolysis of molten zinc chloride produces zinc and chlorine gas.
3) What is formed at the cathode during the electrolysis of an aqueous solution of potassium hydroxide?
   A. Potassium metal
   B. Hydrogen gas
   C. Oxygen and water
   D. Chlorine gas

Exam Questions

1 Sodium chloride, NaCl, is an ionic compound.
Use words from the box to complete the sentences below.

Grade 1-3

| negative | gaseous | sodium | aqueous | hydrogen | hydroxide |

When molten sodium chloride is electrolysed, the .... chloride ions move towards the anode and form chlorine gas. At the cathode, the .... ions gain electrons.

[2 marks]

2 Figure 1 shows the electrolysis of molten lead bromide, $PbBr_2$.

Grade 4-5

Figure 1

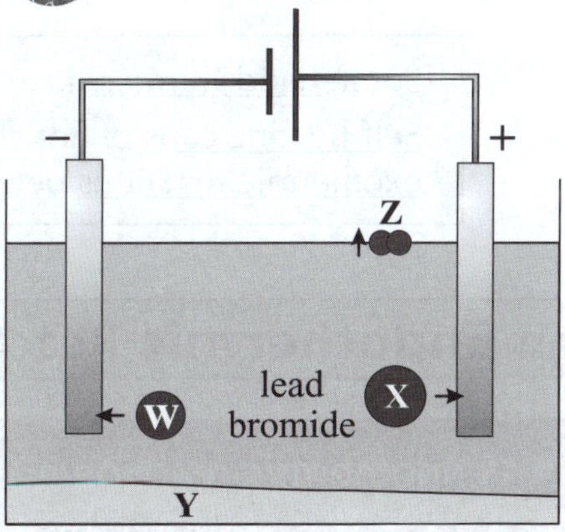

2.1 Which substance is represented in Figure 1 by the letter **W**?

☐ Br_2 ☐ Br^-
☐ Molten lead ☐ Pb^{2+}

[1 mark]

2.2 Which substance is represented in Figure 1 by the letter **X**?

☐ Br_2 ☐ Br^-
☐ Molten lead ☐ Pb^{2+}

[1 mark]

2.3 Which substance is represented in Figure 1 by the letter **Y**?

☐ Br_2 ☐ Br^- ☐ Molten lead ☐ Pb^{2+}

[1 mark]

2.4 Which substance is represented in Figure 1 by the letter **Z**?

☐ Br_2 ☐ Br^- ☐ Molten lead ☐ Pb^{2+}

[1 mark]

Topic C4 — Chemical Changes

Topic C5 — Energy Changes

Exothermic and Endothermic Reactions

Whenever chemical reactions occur, there are changes in energy. This means that when chemicals get together, things either heat up or cool down. I'll give you a heads up — this page is a good 'un.

Energy is Moved Around in Chemical Reactions

1) Chemicals store a certain amount of energy — and different chemicals store different amounts.
2) Sometimes, the products of a reaction store more energy than the reactants. This means that the products have taken in energy from the surroundings during the reaction.
3) But if the products store less energy, then the extra energy was transferred (given out) to the surroundings during the reaction.
4) The amount of energy transferred is the difference between the energy of the products and the energy of the reactants.
5) The overall amount of energy doesn't change. This is because energy stays the same (is conserved) in reactions — it can't be made or destroyed, only moved around.
6) This means the amount of energy in the universe always stays the same.

In an Exothermic Reaction, Energy is Given Out

> An exothermic reaction is one which gives out energy to the surroundings. This is shown by a rise in temperature of the surroundings.

1) Examples of exothermic reactions include:

   - Burning fuels — also called combustion.
   - Neutralisation reactions (acid + alkali).
   - Many oxidation reactions.

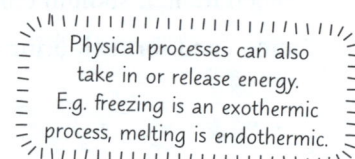
Physical processes can also take in or release energy. E.g. freezing is an exothermic process, melting is endothermic.

2) Exothermic reactions have lots of everyday uses. For example:

   Some hand warmers use an exothermic reaction to release energy. Self-heating cans of hot chocolate and coffee also use exothermic reactions between chemicals in their bases.

In an Endothermic Reaction, Energy is Taken In

> An endothermic reaction is one which takes in energy from the surroundings. This is shown by a fall in temperature of the surroundings.

1) Examples of endothermic reactions include:

   - The reaction between citric acid and sodium hydrogencarbonate.
   - Thermal decomposition (when a substance breaks down when it's heated).

2) Exothermic reactions have lots of everyday uses. For example:

   Endothermic reactions are used in some sports injury packs. The chemical reaction allows the pack to become instantly cooler without having to put it in the freezer.

Measuring Energy Changes

*Sometimes it's not enough to just know if a reaction is **endothermic** or **exothermic**. You may also need to measure **how much** the **temperature** changes during a reaction.*

Energy Transfer can be Measured

You can use an experiment to investigate the temperature change of a chemical reaction.

1) You can do this by taking the temperature of the reactants and mixing them in a polystyrene cup.
2) If the temperature of the solution rises during the reaction, record the highest temperature that it reaches.
   If the temperature of the solution falls during the reaction, record the lowest temperature that it reaches.
3) To find the temperature change, take away this temperature from the temperature of the reactants.
4) If the temperature goes up, the reaction's exothermic. If the temperature goes down, the reaction's endothermic.
5) The biggest problem with temperature measurements is the amount of energy lost to the surroundings.
6) You can reduce this by putting a lid on the polystyrene cup and putting the cup into a beaker of cotton wool.

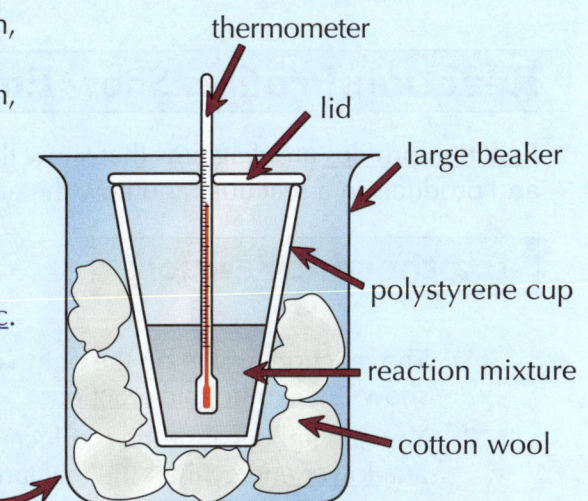

This method works for a number of different reactions. For example:

- neutralisation reactions.
- reactions between metals and acids.
- reactions between acids and carbonates.
- displacement reactions of metals.

The Temperature Change Depends on Different Variables

1) You can also use this method to investigate what effect different variables have on the temperature change — e.g. the mass or concentration of the reactants used.
2) Here's how you could test the effect of acid concentration on the temperature change of a neutralisation reaction between hydrochloric acid (HCl) and sodium hydroxide (NaOH):

   1) Put 25 cm³ each of 10 g/dm³ hydrochloric acid and sodium hydroxide in separate beakers.
   2) Place the beakers in a water bath set to 25 °C until they are both at the same temperature (25 °C).
   3) Add the HCl followed by the NaOH to a polystyrene cup with a lid — as in the diagram above.
   4) Take the temperature of the mixture every 30 seconds, and record the highest temperature.
   5) Use your results to work out the temperature change of the reaction.
   6) Repeat these steps using 20 g/dm³ and then 30 g/dm³ of hydrochloric acid. Then compare your results to see how acid concentration affects the temperature change of the reaction.

 **To get a reasonably accurate reading, insulate your reaction**
It's really important the reaction mixture is well insulated in the method shown above. Without insulation, heat might escape and the temperature reading of the solution would be affected.

Topic C5 — Energy Changes

Reaction Profiles

Reaction profiles are handy little diagrams which show you the **changes in energy** during a reaction.

Activation Energy is Needed to Start a Reaction

1) The activation energy is the minimum amount of energy the reactants need to have to react when they collide with each other.
2) The greater the activation energy, the more energy needed to start the reaction. This energy has to be given, e.g. by heating the reaction mixture.

There's more on activation energy and collision theory on pages 234-235.

Reaction Profiles Show Energy Changes

Reaction profiles are diagrams that show the difference between the energies of the reactants and products in a reaction, and how the energy changes over the course of the reaction.

Exothermic Reactions

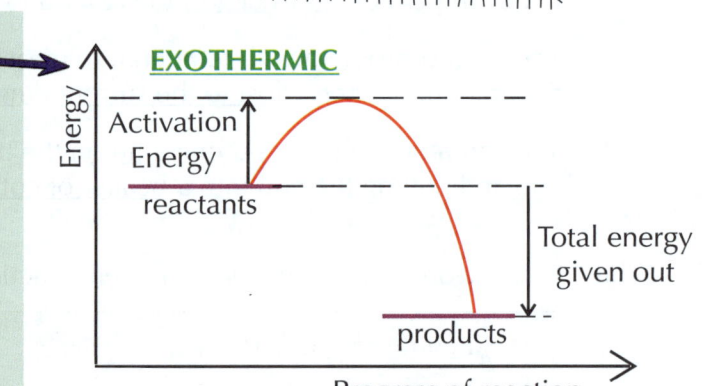

1) The reaction profile on the right shows an exothermic reaction.
2) You can tell because the products are at a lower energy than the reactants.
3) The difference in height between the reactants and the products shows the overall energy change in the reaction (the energy given out).
4) The rise in energy at the start shows the energy needed to start the reaction. This is the activation energy.

Reaction profiles are sometimes called energy level diagrams.

Endothermic Reactions

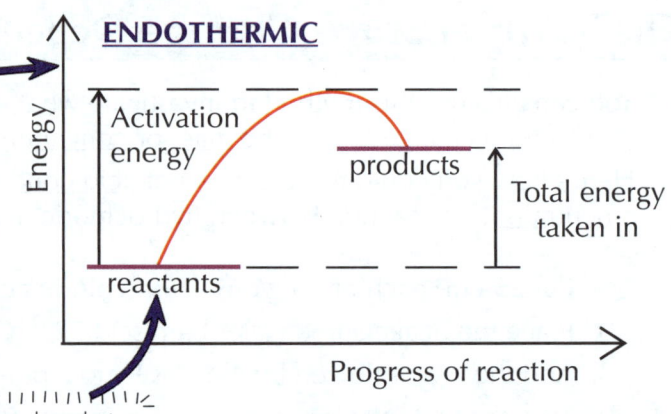

1) In this reaction profile, the products are at a higher energy than the reactants. So the reaction is endothermic.
2) The difference in height shows the overall energy change during the reaction (the energy taken in).
3) The rise in energy at the start is the activation energy.

You can also write the formulas for the reactants and products on the reaction profile instead of 'reactants' and 'products'.

A reaction profile shows the energy levels during a reaction

The diagrams above might seem a bit confusing at first — remember, it's the energy in the chemicals themselves, not in their surroundings, which is being shown in the diagrams.
To help you remember them, why not draw them out and label them for yourself.

Q1 Video Solution

Q1 Here is the equation for the exothermic reaction between methane and oxygen:
$CH_{4(g)} + 2O_{2(g)} \rightarrow CO_{2(g)} + 2H_2O_{(g)}$. Draw a reaction profile for this reaction. [3 marks]

Topic C5 — Energy Changes

Warm-Up & Exam Questions

A whole bunch of reactions, an experiment, some funny diagrams — there's a lot to get your head around on the last few pages. Here are some questions so that you can check how you're getting on.

Warm-Up Questions

1) True or False? The amount of energy in the universe always stays the same.
2) Give one example of an everyday use of an exothermic reaction.
3) How will the temperature of the surroundings change during an endothermic reaction?
4) Is the reaction between citric acid and sodium hydrogencarbonate exothermic or endothermic?
5) What is shown by the difference in height between the reactants and products in the reaction profile for an exothermic reaction?

Exam Questions

1 Which of the following types of reaction is most likely to be endothermic?
Tick **one** box.

☐ combustion ☐ neutralisation ☐ thermal decomposition ☐ oxidation

[1 mark]

2 A student measured the temperature change during a reaction between sodium hydroxide and hydrochloric acid using the apparatus shown in **Figure 1**.

2.1 Before mixing the reagents, the student measured the temperature of each of them. Suggest why.

[1 mark]

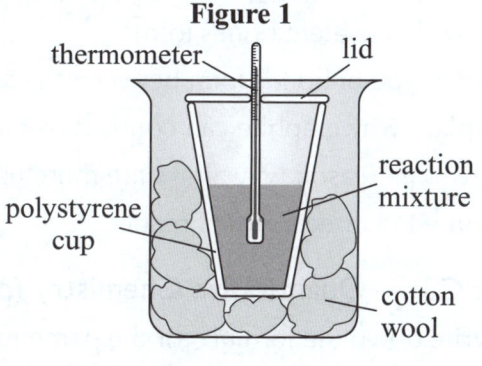

2.2 State the purpose of the cotton wool and the lid in the experimental set-up in **Figure 1**.

[1 mark]

2.3 How could the student test the repeatability of the temperature change that they recorded during the experiment?

[2 marks]

3 The diagrams in **Figure 2** represent the energy changes in four different chemical reactions.

Figure 2

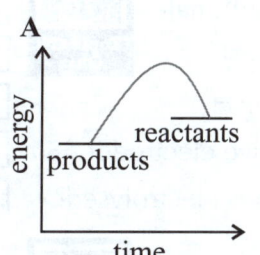 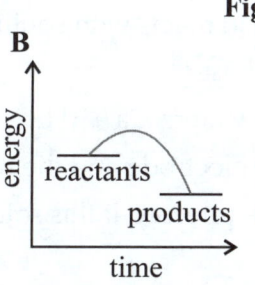 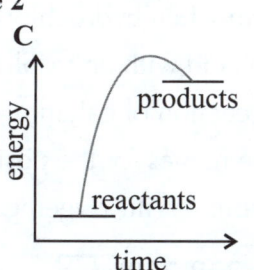 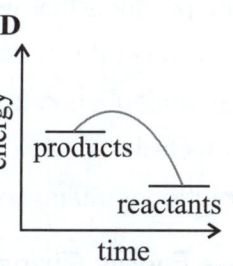

Write the letter of **one** diagram, **A**, **B**, **C** or **D**, which correctly shows an endothermic reaction.

[1 mark]

Topic C5 — Energy Changes

Revision Summary for Topics C1-5

That's Topics C1-5 done — time to check how much you actually remember.
- Try these questions and tick off each one when you get it right.
- When you're completely happy with a topic, tick it off.

For even more practice, try the Retrieval Quizzes for Topics C1-5 — just scan the QR codes!

Topic C1 — Atomic Structure and the Periodic Table (p.161-187)

1) Draw an atom. Label the nucleus and the electrons.
2) What is the smallest part of an element that you can have?
3) Balance these equations: a) $Mg + O_2 \rightarrow MgO$ b) $H_2SO_4 + NaOH \rightarrow Na_2SO_4 + H_2O$
4) What method could you use to separate dyes in an ink?
5) Which method of separation is useful to separate an insoluble solid from a liquid?
6) What is the maximum number of electrons that can be in the third shell of an atom?
7) What is the electronic structure of sodium?
8) What does the group number of an element tell you about its electrons?
9) State two trends as you go down Group 1 of the periodic table.
10) How do the boiling points of halogens change as you go down the group?
11) Halogens form ions when they react with metals. What is the charge of the halogen ions?
12) Why are the Group 0 elements single atoms?

Topic C2 — Bonding, Structure and Properties of Matter (p.190-207)

13) What type of bond is formed when a metal reacts with a non-metal?
14) Draw dot and cross diagrams to show the formation of:
    a) sodium chloride   b) magnesium oxide   c) magnesium chloride
15) List three properties of ionic compounds.
16) How do covalent bonds form?
17) What type of bonds form between the atoms in a polymer?
18) Explain why graphite can conduct electricity.
19) Give one reason why alloys are more useful than metals.
20) Name the three states of matter.

Topic C3 — Quantitative Chemistry (p.209-212)

21) Write down the formula for the percentage mass of an element in a compound.
22) What do the following terms mean: a) solvent   b) solute?

Topic C4 — Chemical Changes (p.215-226)

23) Name the products that will form when hydrochloric acid reacts with sodium carbonate.
24) What do you get if you react an acid with an insoluble base?
25) Complete the equation for the reaction of calcium with water: $Ca + ?H_2O \rightarrow ? + H_2$
26) During electrolysis, what are the names for the positive electrode and the negative electrode?
27) A salt solution contains halide ions. Will oxygen gas be released if this solution is electrolysed?

Topic C5 — Energy Changes (p.228-230)

28) In an exothermic reaction, is energy transferred to or from the surroundings?
29) Define what is meant by an endothermic reaction.
30) Sketch a reaction profile for an exothermic reaction.

Topic C6 — The Rate and Extent of Chemical Change

Rates of Reaction

*Rates of reaction are pretty **important**. In **industry**, the **faster** you make **chemicals**, the **faster** you make **money**.*

The Speed of a Reaction is Called its Rate

1) The rate of a chemical reaction is how fast the reactants are changed into products.
2) Some reactions are very slow, for example, rusting. Others, like burning, are fast.
3) Graphs can show you how the rate (speed) of a reaction changes.
4) The steeper the line on the graph, the faster the rate of reaction.
5) Over time the line becomes less steep as the reactants are used up.

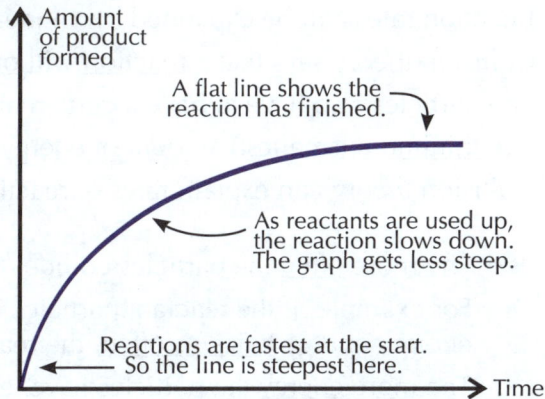

Reaction Rates Can Change when the Reaction Conditions Change

1) Faster reactions have steeper lines to begin with and become flat more quickly.
2) The graph below shows how the speed of a particular reaction changes under different conditions.

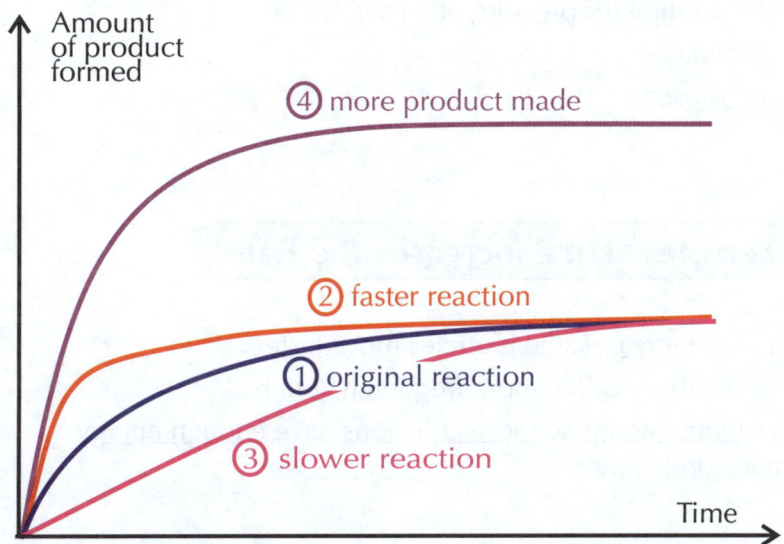

- Line 1 shows the original reaction.
- Line 2 shows the same reaction taking place faster. It starts more steeply than line 1. It also goes flat sooner.
- Line 3 shows the same reaction taking place more slowly. It isn't as steep at the start as line 1 and goes flat later.
- Lines 1, 2 and 3 all meet at the same level. This shows that they all produce the same amount of product. They just take different times to produce it.
- Line 4 shows more product is formed. This can only happen if there were more reactants at the start.

Factors Affecting Rates of Reaction

*I'd ask you to **guess** what these two pages are about, but the **title** pretty much says it all really. Read on...*

Particles Must **Collide** with **Enough Energy** in Order to **React**

1) Reaction rates can be explained by an idea called collision theory.
2) Collision theory says that a reaction will only take place when particles collide (crash into each other).
3) The particles also have to have a certain amount of energy when they collide, otherwise they won't react.
4) The minimum (smallest) amount of energy they need is called the activation energy.
5) Collision theory can explain rates of reactions in a bit more detail too...

- The more often the particles collide, the faster the reaction will happen.
- For example, if the reactant particles in a certain reaction collide with enough energy twice as often, the reaction will happen twice as fast.
- The more energy the particles have, the faster the reaction will be.
- This is because there's more chance that they'll have at least the activation energy.

How often the particles collide is sometimes called the 'collision frequency'.

The **Rate of Reaction** Depends on **Four Things**

1) Temperature.
2) Concentration of a solution (or pressure of a gas).
3) Surface area of a solid.
4) Whether a catalyst is used.

Increasing the **Temperature** Increases the Rate

1) When temperature increases, the particles move faster.
2) If they move faster, they collide more frequently (often).
3) They also have more energy, so more collisions have enough energy to make the reaction happen.

If particles collide twice as often with enough energy to react, the rate of reaction will be twice as fast. If they collide three times as often it'll be three times as fast.

Increasing the **Concentration** or **Pressure** Increases the Rate

1) If a solution is more concentrated, it has more particles in the same volume.
2) And when the pressure of a gas is increased, it means the same number of particles are now in a smaller space.
3) So collisions between the reactant particles are more frequent.

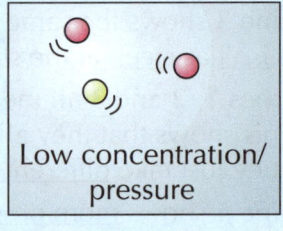

Low concentration/ pressure

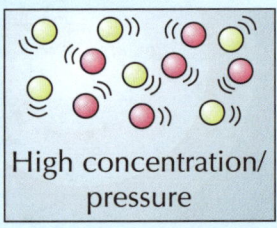

High concentration/ pressure

Topic C6 — The Rate and Extent of Chemical Change

Factors Affecting Rates of Reaction

Increasing the **Surface Area** Increases the Rate

1) If one of the reactants is a solid, then breaking it up into smaller pieces will increase its surface area to volume ratio.
2) This means the same amount of solid has a bigger surface area.
3) So more of the solid's particles are available to particles of the other reactant. And collisions will be more frequent.

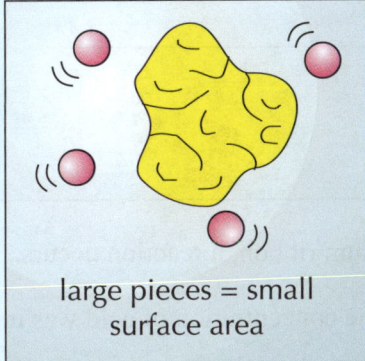

large pieces = small surface area

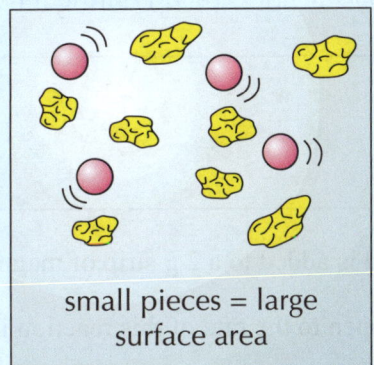

small pieces = large surface area

Using a **Catalyst** Increases the Rate

1) A catalyst is a substance that speeds up a reaction. None of it gets used up in the reaction.
2) This means it's not part of the reaction equation.
3) Different catalysts are needed for different reactions.
4) Catalysts work by providing a different pathway for a reaction. The new pathway has a lower activation energy, so less energy is needed for the reaction to happen.
5) Enzymes are biological catalysts — they catalyse reactions in living things.

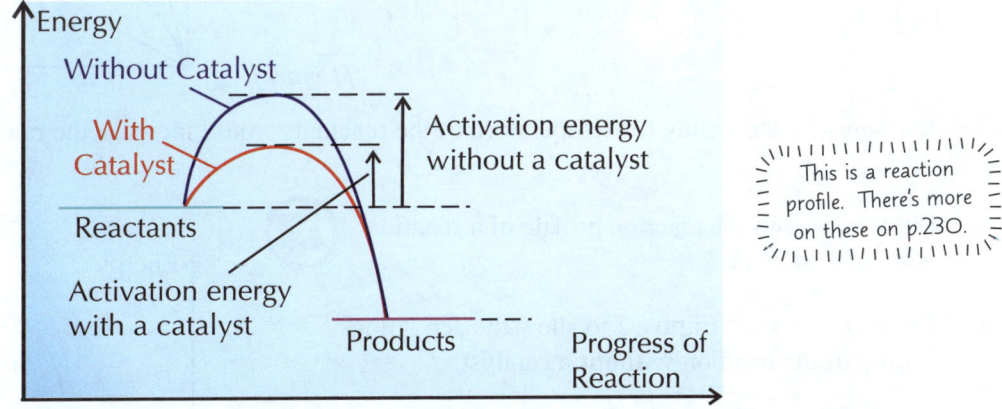

This is a reaction profile. There's more on these on p.230.

It's easier to learn stuff when you know the reasons for it

Once you've learnt everything off these two pages, the rates of reaction stuff should start making a lot more sense to you. The idea's fairly simple — the more often particles bump into each other, and the harder they hit when they do, the faster the reaction happens.

Topic C6 — The Rate and Extent of Chemical Change

Warm-Up & Exam Questions

It's easy to think that you've understood something when you've just read through it. These questions should test whether you really understand the previous chunk of pages, and get you set for the next bit.

Warm-Up Questions

1) Give an example of a reaction that happens very slowly, and a reaction that is very fast.
2) According to collision theory, what must happen in order for two particles to react?
3) Explain why breaking a solid reactant into smaller pieces can affect the rate of a reaction.
4) True or False? A catalyst is not used up during a reaction.

Exam Questions

1 When hydrochloric acid is added to a 2 g strip of magnesium ribbon, a reaction occurs. *(Grade 1-3)*

1.1 Predict what would happen to the rate of this reaction if the concentration of acid was increased.
[1 mark]

1.2 Predict what would happen to the rate of this reaction if 2 g of powdered magnesium was used instead of the strip of magnesium ribbon.
[1 mark]

2 **Figure 1** shows how the speed of a reaction changed under three different conditions. *(Grade 4-5)*

Figure 1
(Graph: Amount of product formed vs Time, showing three curves A, B, C where A plateaus highest, B middle, C lowest)

2.1 Which line shows the reaction that produced the most product? Tick **one** box.

☐ A   ☐ B   ☐ C
[1 mark]

2.2 Which line shows the slowest reaction? Tick **one** box.

☐ A   ☐ B   ☐ C
[1 mark]

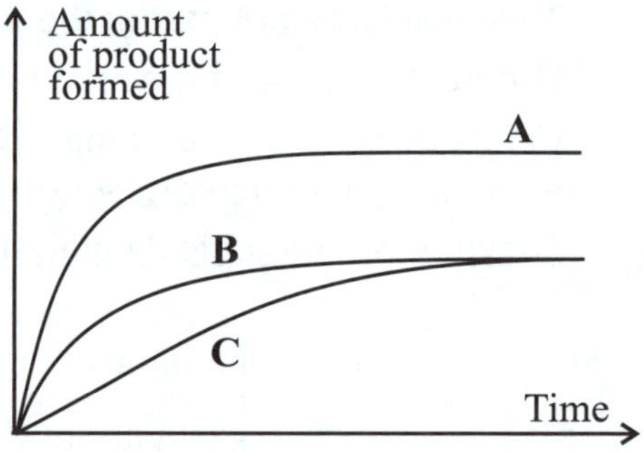

2.3 Explain why increasing the temperature of the reactants would increase the rate of the slowest reaction.
[4 marks]

3 **Figure 2** shows the reaction profile of a reaction without a catalyst. *(Grade 4-5)*

Figure 2
(Reaction profile graph: Energy vs Progress of Reaction, showing Reactants, a hump, and Products at a lower energy)

3.1 Draw an arrow on **Figure 2** to show the activation energy of the reaction **without** a catalyst.
[1 mark]

3.2 On **Figure 2**, draw a curve to show what the reaction profile would look like **with** a catalyst.
[1 mark]

Topic C6 — The Rate and Extent of Chemical Change

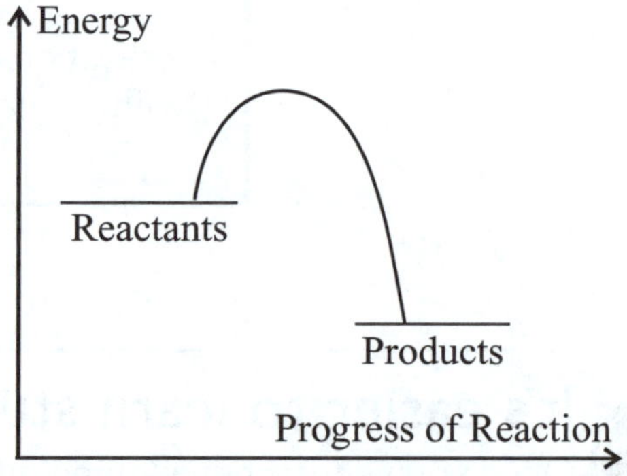

Measuring Rates of Reaction

*All this **talk** about reaction rate is fine and dandy, but it's no good if you can't **measure** it. You can investigate how **concentration** affects the rate of a reaction by **measuring** the **volume of gas** given off.*

Marble Chips and Hydrochloric Acid React to Produce a Gas

1) Measure out a set volume of dilute hydrochloric acid using a measuring cylinder.
2) Carefully pour it into a conical flask.
3) Measure out a set mass of marble chips.
4) Add the marble chips to the flask and quickly attach a delivery tube and gas syringe to the flask. You need to do this quickly before any gas escapes.
5) Start the stopwatch straight away.
6) Carbon dioxide gas will start to collect in the gas syringe. Take readings of the volume of gas at regular intervals (e.g. every 10 seconds) and write them into a table.

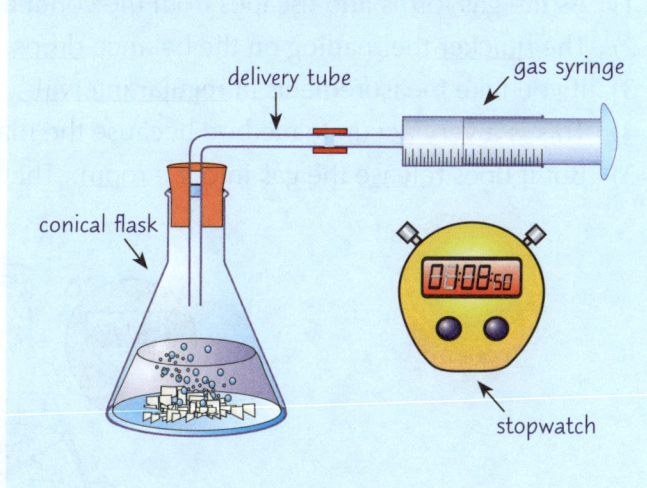

Gas syringes usually give volumes to the nearest cm³, so they're quite accurate. Be careful though — if the reaction is too vigorous, you could blow the plunger out of the syringe.

Repeat the Experiment Using Different Concentrations of Acid

1) To investigate the effect of concentration on the rate of reaction, you'll need to repeat the experiment using different concentrations of acid. For example, you might use three different concentrations.
2) To make your experiment a fair test, you should only change the concentration of acid.
3) All the other variables (that's everything else) need to be kept the same. For example, keep the volume of acid, the mass of marble chips and the temperature the same each time.

You Then Need to Interpret Your Results

1) The more gas given off in a set amount of time, the faster the reaction.
2) You can use the results in your table to draw a graph. This makes it easier to see how concentration has affected the rate. (See pages 240 and 241 for more on drawing graphs and calculating rates.)

Measuring the volume of gas means you can find the rate

As exciting as everything on this page is, don't get carried away and forget about safety precautions — working with acids and gases can be dangerous. Always wear safety goggles, and make sure you don't let harmful gases escape into the room.

Topic C6 — The Rate and Extent of Chemical Change

 **Measuring Rates of Reaction**

Measuring the volume of gas given off isn't the only way of investigating the rate of a reaction. There are two more methods you need to be familiar with.

You Can Use a Mass Balance to Measure Amount of Gas Produced

1) As the gas forms and escapes from the container, the mass of the reaction mixture falls.
2) The quicker the reading on the balance drops, the faster the reaction.
3) If you take measurements at regular intervals, you can plot a graph.
4) This is a very accurate method because the mass balance is very accurate.
5) But it does release the gas into the room. That's not good if the gas is toxic (poisonous).

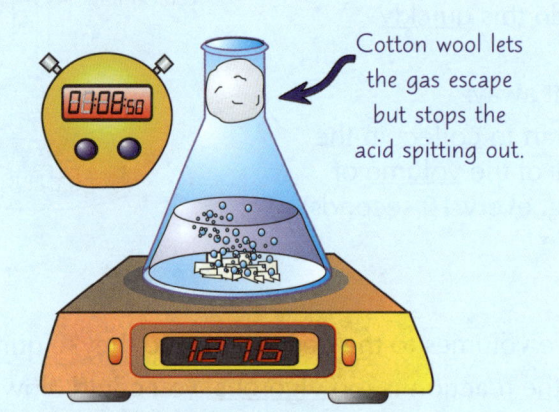
Cotton wool lets the gas escape but stops the acid spitting out.

You Can Time How Long it Takes For a Solid Product to Form

1) Some reactions start with a transparent (see-through) solution and produce a solid product.
2) The solid product (called a precipitate) will make the solution go cloudy. Another way of saying this is to say that its turbidity increases.
3) You can look at a mark through the solution and measure how long it takes for it to disappear. The faster the mark disappears, the quicker the reaction.
4) The results are subjective (there isn't just one right answer). This is because people might not agree over the exact point when the mark 'disappears'.

 **Each of these methods has pros and cons**

For example, the mass balance method is only accurate if the flask isn't too hot. Otherwise, the loss in mass that you see might be partly due to evaporation of liquid as well as the loss of gas formed during the reaction. And of course, it only works if a gas is given off during the reaction.

Topic C6 — The Rate and Extent of Chemical Change

Measuring Rates of Reaction **PRACTICAL**

*Here's how to use the last method from the previous page. It's **less accurate** than using a mass balance, but you need to know how to do it in case you want to investigate a reaction that **doesn't produce a gas**.*

Sodium Thiosulfate and Hydrochloric Acid Produce a Precipitate

- Sodium thiosulfate and hydrochloric acid (HCl) are both clear solutions. They react together to form a yellow precipitate of sulfur.
- The yellow precipitate causes the solution to turn cloudy.
- The time taken for this to happen can be measured to calculate the rate of reaction.

1) Start by adding a set volume of dilute sodium thiosulfate to a conical flask.
2) Place the flask on a piece of paper with a black cross drawn on it.
3) Add some dilute hydrochloric acid to the flask and start the stopwatch.
4) Watch the black cross disappear through the cloudy sulfur and time how long it takes to go.

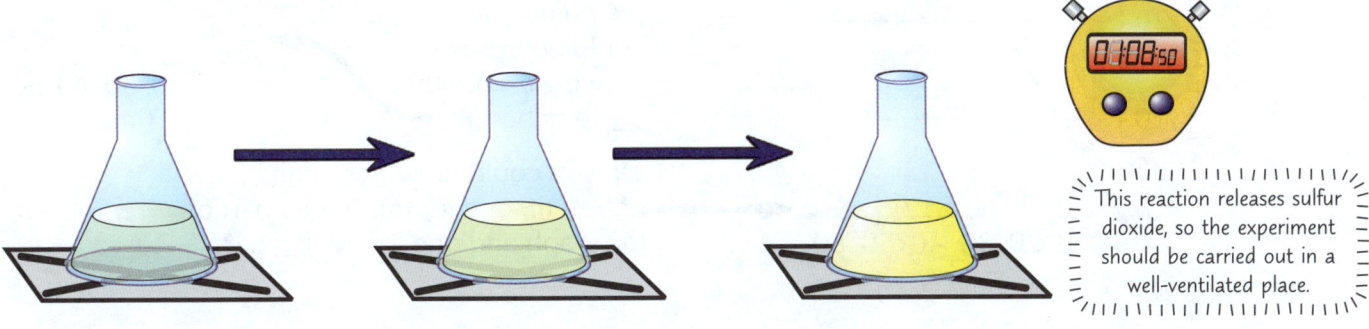

This reaction releases sulfur dioxide, so the experiment should be carried out in a well-ventilated place.

5) Repeat the reaction using different concentrations of one reactant, e.g. the hydrochloric acid. (Only change the concentration of one reactant at a time though.)
6) Make sure you control all the other variables (keep everything else the same). For example, the volumes of solutions, the temperature and the size of the flask all need to be kept the same.
7) You'll end up with a set of results that show how long it takes for the cross to disappear at different concentrations of acid. Like this:

Concentration of HCl (g/dm^3)	20	35	50	65	80
Time taken for mark to disappear (s)	193	184	178	171	164

8) The higher the concentration, the faster the reaction, so the less time it takes for the mark to disappear.

Make sure you use the right method for your experiment

The method shown on this page only works if there's a really obvious change in the solution. If there's only a small change in colour, it might not be possible to observe and time the change.

Topic C6 — The Rate and Extent of Chemical Change

Graphs of Reaction Rate Experiments

You might remember a bit about graphs on reaction rate from page 233 — well this page shows you how to draw and interpret them for real experiments.

You Can Draw a Graph of Your Results

The type of graph you can draw depends on what experiment you did.
Here's how to draw a graph to show the volume of gas given off during a reaction.

EXAMPLE

Draw a graph of the results in the table.

Time (s)	0	10	20	30	40	50	60
Volume of gas (cm³)	0	9.5	15	18.5	20	20	20

1) Put time on the x-axis and volume of gas on the y-axis.
   (The x-axis goes along the page, the y-axis goes up it.)
2) Carefully draw a small cross to show how much gas had been produced at each time interval.
3) Draw a line of best fit through the points. You could do this by drawing a smooth curve of best fit.
4) Or, you could draw two straight lines of best fit, one for the sloped part of the graph and one for the flat part.

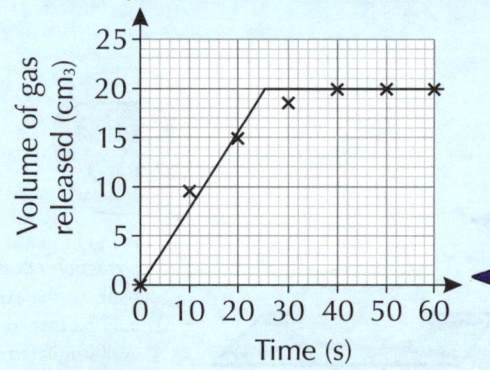

Tangents Help You Compare Reaction Rates at Different Points

If your graph is a curve, it's not always easy to see how the rate changes during a reaction. Tangents can help make this clearer. A tangent is a straight line that touches the curve at one point and doesn't cross it.

EXAMPLE

The graph below shows the volume of gas produced during a chemical reaction.
Is the rate fastest at 20 seconds or 30 seconds?

1) Position a ruler on the graph at a point where you want to know the rate — here it's 30 seconds.
2) Adjust the ruler until the space between the ruler and the curve is equal on both sides of the point.
3) Draw a line along the ruler to make the tangent. Extend the line right across the graph.
4) Do the same thing at 20 seconds.
5) Compare how steep the tangents are.
6) The tangent at 20 seconds (the blue line) is steeper than the tangent at 30 seconds (the red line).
   This means that the reaction is faster at 20 seconds.

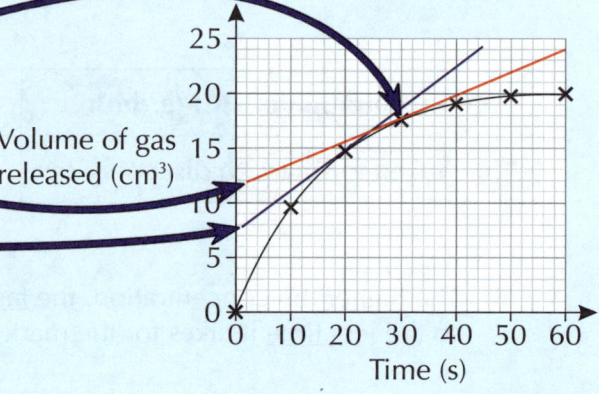

You can use tangents to measure how the rate changes

When you draw a tangent, don't forget to make it just touch the curve without crossing it.

Q1 Draw a graph of the results in this table.
Include a curved line of best fit. [3 marks]

Time (s)	0	10	20	30	40	50
Volume of CO₂ (cm³)	0	18	29	33	34	35

Q1 Video Solution

Working Out Reaction Rates

You need to be able to do some **calculations** to work out **reaction rates**.

Here's How to **Work Out** the **Rate** of a Reaction

$$\text{Mean Rate of Reaction} = \frac{\text{Amount of reactant used or amount of product formed}}{\text{Time}}$$

This equation is for <u>mean</u> rate of reaction. So it lets you work out the <u>average rate</u> over an <u>amount of time</u>.

EXAMPLE A reaction takes 120 seconds. 3.0 g of product are made. Find the mean rate of reaction.

Mean Rate = amount of product formed ÷ time
= 3.0 g ÷ 120 s = **0.025 g/s**

Gases can be measured in cm^3, so if the product you measured was a gas the rate could be measured in cm^3/s rather than in g/s.

You Can Calculate the **Mean Reaction Rate** from a **Graph**

1) To find the <u>mean rate</u> for the <u>whole reaction</u>, start by working out when the reaction <u>finished</u>. This is when the line goes <u>flat</u>.
2) Then work out how much <u>product</u> was <u>formed</u> (or how much <u>reactant</u> was <u>used up</u>).
3) Then <u>divide this</u> by the <u>total time taken</u> for the reaction to finish.

EXAMPLE The graph shows the volume of gas released by a reaction, measured at regular intervals. Find the mean rate of the reaction.

1) Work out when the reaction <u>finished</u>.
   The line goes flat at 50 s.
2) Work out how much <u>product</u> was <u>formed</u>.
   20 cm^3 of gas was formed.
3) <u>Divide</u> this by the <u>time taken</u> for the reaction to finish.
   mean rate = 20 cm^3 ÷ 50 s = **0.40 cm^3/s**

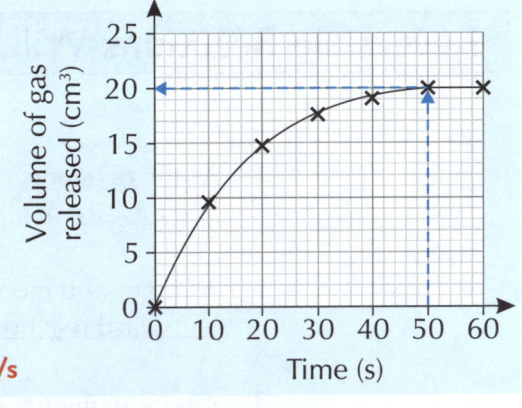

4) You can also use the graph to find the <u>mean rate</u> of reaction between <u>two points</u> in time:

EXAMPLE Find the mean rate of reaction between 20 s and 40 s.

1) Work out how much <u>gas</u> was produced <u>between</u> 20 s and 40 s.
   At 20 s, 15 cm^3 had been produced.
   At 40 s, 19 cm^3 had been produced.
   Volume released between 20 and 40 s was:
   19 cm^3 − 15 cm^3 = 4 cm^3
2) Work out the <u>time difference</u> between 20 s and 40 s.
   40 s − 20 s = 20 s
3) <u>Divide</u> the amount of gas produced by the time taken.
   mean rate = 4 cm^3 ÷ 20 s = **0.2 cm^3/s**

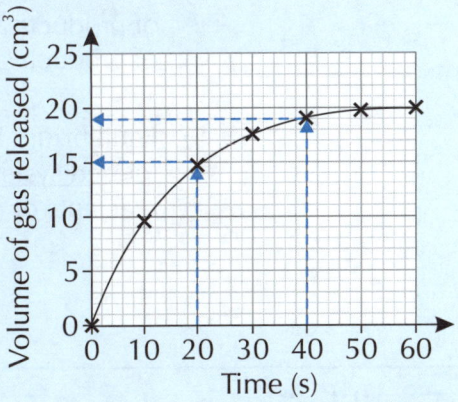

The mean reaction rate is the average reaction rate over time

Q1 A reaction takes 200 s. 6.0 g of reactant are used up.
What is the mean rate of the reaction? [2 marks]

Topic C6 — The Rate and Extent of Chemical Change

Reversible Reactions

*Reversible reactions are what they sound like — **reactions** that can be **reversed**. So they can go **backwards**.*

Reversible Reactions Go Both Ways

1) This equation shows a reversible reaction.

2) The products (C and D) react to form the reactants (A and B) again.

3) You can tell it's a reversible reaction because of the ⇌ symbol.

4) The reaction of A and B is called the forward reaction. The reaction of C and D is the backward reaction.

Reversible Reactions Will Reach Equilibrium

1) As the reactants react, their concentrations fall. The forward reaction slows down.

2) As more and more products are made the backward reaction will speed up.

3) After a while the forward reaction and backward reaction will be going at exactly the same rate. The system is at equilibrium.

4) Equilibrium doesn't mean that there are the same amounts of products and reactants. It just means that the amounts of products and reactants aren't changing any more.

5) Equilibrium is only reached if the reaction takes place in a 'closed system'. A closed system just means that none of the reactants or products can escape and nothing else can get in.

Equilibrium — lots of activity, but not to any great effect

The idea of equilibrium is something that you need to get to grips with, as things will get more complicated on the next page. Have another read and make sure you've got the basics sorted before moving on. Remember, equilibrium means that things are staying nice and stable — but that doesn't mean that the reactions aren't happening. It just means that the amounts of reactants and products aren't changing.

Topic C6 — The Rate and Extent of Chemical Change

Reversible Reactions

*In a reversible reaction, both the forward and the backward reactions are happening **at the same time**. Some conditions can be **more favourable** to one direction than the other, making that reaction happen faster.*

Reversible Reactions Have an **Overall Direction**

1) Once a reaction is at equilibrium, there could be <u>more</u> of the <u>products</u> than reactants. When this happens, we say the equilibrium <u>favours</u> the <u>forwards direction</u>.

2) If there are <u>more reactants</u> than products then the equilibrium <u>favours</u> the <u>backwards direction</u>.

3) You can <u>change</u> the <u>direction</u> by <u>changing</u> the <u>conditions</u> (the temperature, pressure or concentration).

> - <u>Ammonium chloride</u> breaks down to form <u>ammonia</u> and <u>hydrogen chloride</u>.
> - The hydrogen chloride can then react with the ammonia to make ammonium chloride again.
> - The reaction equation is:
>
> $$\text{ammonium chloride} \underset{\text{cool}}{\overset{\text{heat}}{\rightleftharpoons}} \text{ammonia} + \text{hydrogen chloride}$$
>
> - If you <u>heat</u> this reaction, it will go in the <u>forwards</u> direction. You'll get more ammonia and hydrogen chloride.
> - If you <u>cool</u> it, it will go in the <u>backwards</u> direction. You'll get more ammonium chloride.

Reversible Reactions Can Be **Endothermic** and **Exothermic**

1) If the reaction is <u>endothermic</u> (takes in heat) in one direction, it will be <u>exothermic</u> (give out heat) in the other.

    See page 228 for more on endothermic and exothermic reactions.

2) The amount of energy <u>taken in</u> by the endothermic reaction is the <u>same</u> as the amount <u>given out</u> during the exothermic reaction.

3) A good example is the <u>thermal decomposition</u> of hydrated copper sulfate:

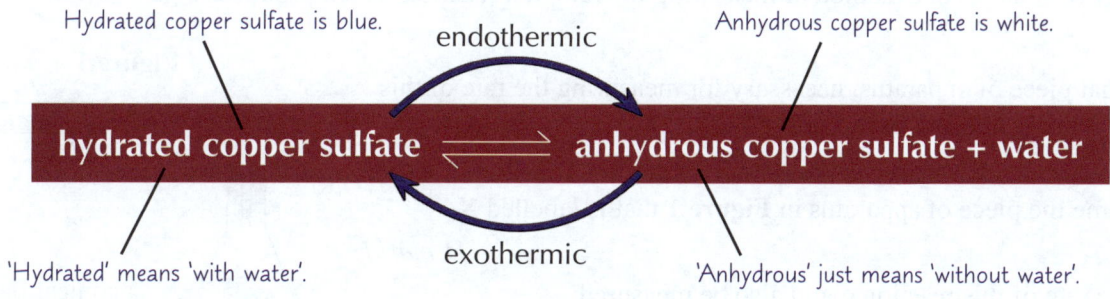

Hydrated copper sulfate is blue. — endothermic — Anhydrous copper sulfate is white.

hydrated copper sulfate ⇌ anhydrous copper sulfate + water

'Hydrated' means 'with water'. — exothermic — 'Anhydrous' just means 'without water'.

More of the products = reaction favours the forwards direction

This whole exothermic/endothermic thing is a fairly simple idea — don't be put off by the long words. Remember, "<u>exo-</u>" = <u>external</u>, "<u>-thermic</u>" = <u>heat</u>, so an exothermic reaction is one that <u>gives out</u> heat. And "<u>endo-</u>" = erm... the other one. OK, there's no easy way to remember that one. Tough.

Topic C6 — The Rate and Extent of Chemical Change

Warm-Up & Exam Questions

Not long now till this section's over, but first there are some questions for you to tackle.

Warm-Up Questions

1) Name one piece of apparatus that could be used to measure the rate of a reaction that gives off a gas.
2) Reaction A forms more product than Reaction B over 30 seconds. Which reaction, A or B, has a higher rate?
3) True or false? A steep tangent on a rate graph means that the rate is slow.
4) If there are more reactants than products in a reversible reaction at equilibrium, does the equilibrium favour the forwards or backwards direction?

Exam Questions

1   In the reaction below, substances A and B react to form substances C and D.

$$2A + B \rightleftharpoons 2C + D$$

1.1   What can you deduce about this reaction from the symbol $\rightleftharpoons$?

[1 mark]

1.2   When the reaction takes place in a closed system, equilibrium is reached. Complete the sentences. Use words from the box.

| equal | stay the same | forwards | rise | fall |
| reversible | changing | constant | backwards | |

As A and B react, their concentrations will .... and the rate of the forwards reaction will decrease. As C and D are made, the .... reaction speeds up. After a while, the amounts of products and reactants will be .... The system is at equilibrium.

[3 marks]

PRACTICAL

2   **Figure 1** shows one method of measuring the rate of a reaction which produces a gas.

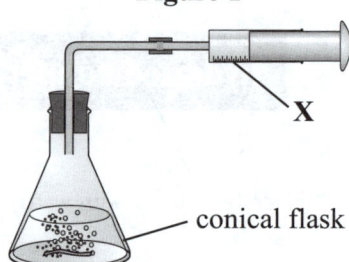

Figure 1

2.1   What piece of apparatus, necessary for measuring the rate of this reaction, is missing from **Figure 1**?

[1 mark]

2.2   Name the piece of apparatus in **Figure 1** that is labelled **X**.

[1 mark]

The rate of this reaction could also be measured using a mass balance.

2.3   Give **one** advantage of using a mass balance to measure the rate of a reaction.

[1 mark]

2.4   Give **one** possible safety concern when using a mass balance to measure the rate of this reaction.

[1 mark]

Topic C6 — The Rate and Extent of Chemical Change

Exam Questions

3 Hydrogen peroxide is a liquid that decomposes into water and oxygen.

Samples of three catalysts with the same surface area were added to some hydrogen peroxide solution. The same volume and concentration of hydrogen peroxide was used each time. The volume of oxygen produced over time was measured and recorded, and is shown in **Figure 2**.

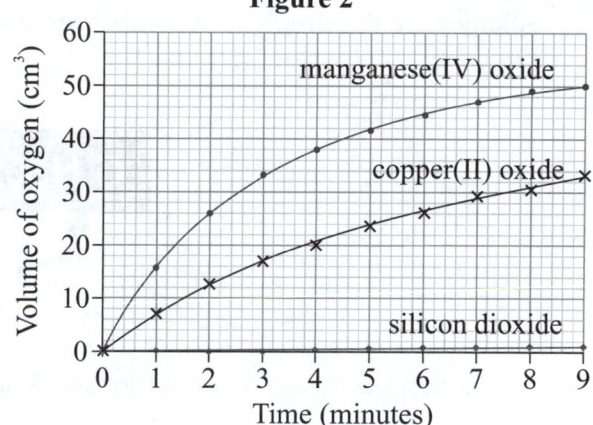

3.1 How much oxygen was produced in the first 3 minutes with copper(II) oxide?
[1 mark]

3.2 Identify, with a reason, the most effective catalyst.
[2 marks]

3.3 Describe how you could use tangents to compare the rates of the reactions at 2 minutes.
[3 marks]

PRACTICAL

4* A reaction between two colourless solutions, **X** and **Y**, produces a yellow precipitate.

Describe how you could use the change in colour to investigate how the rate of this reaction changes with the concentration of Y. Describe the method and equipment you would use, and the measurements you would make.
[4 marks]

5 **Figure 3** shows how the volume of hydrogen gas produced in a reaction varies with time.

The mean rate of a reaction is calculated using the equation:

$$\text{Mean rate of reaction} = \frac{\text{Amount of product formed}}{\text{Time}}$$

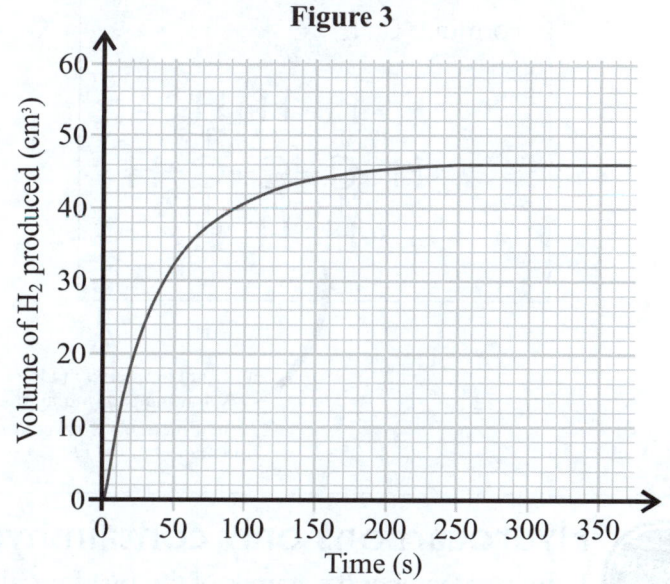

5.1 Use the equation to calculate the mean rate for the whole reaction. Include appropriate units in your answer.
[3 marks]

5.2 Use the equation to calculate the mean rate of reaction between 40 and 150 seconds.
[3 marks]

Topic C7 — Organic Chemistry

Hydrocarbons

*Organic chemistry is about compounds that contain **carbon**. **Hydrocarbons** are the simplest organic compounds. As you're about to discover, their **properties** are affected by their **structure**.*

Alkanes Only Have C–C and C–H Single Bonds

1) <u>Hydrocarbons</u> are compounds formed from carbon and hydrogen atoms <u>only</u>.

2) <u>Alkanes</u> are the simplest type of hydrocarbon. They have the general formula:

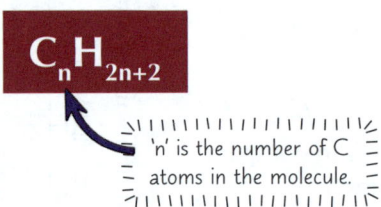

$$C_nH_{2n+2}$$

'n' is the number of C atoms in the molecule.

3) In alkanes, each carbon atom forms four <u>single covalent bonds</u>.

4) The first four alkanes are <u>methane</u>, <u>ethane</u>, <u>propane</u> and <u>butane</u>.

Methane
Formula: CH_4
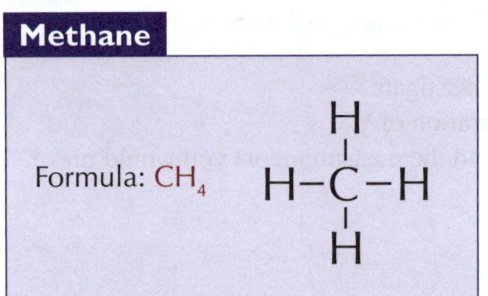

Ethane
Formula: C_2H_6
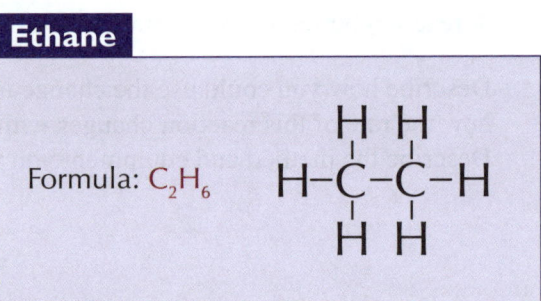

Propane
Formula: C_3H_8
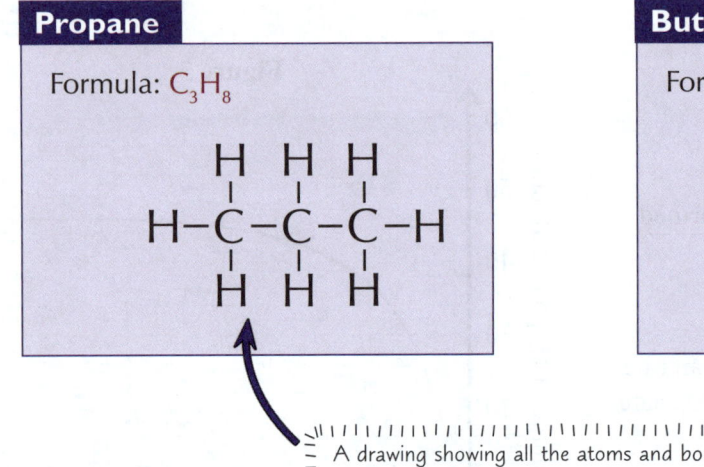

Butane
Formula: C_4H_{10}
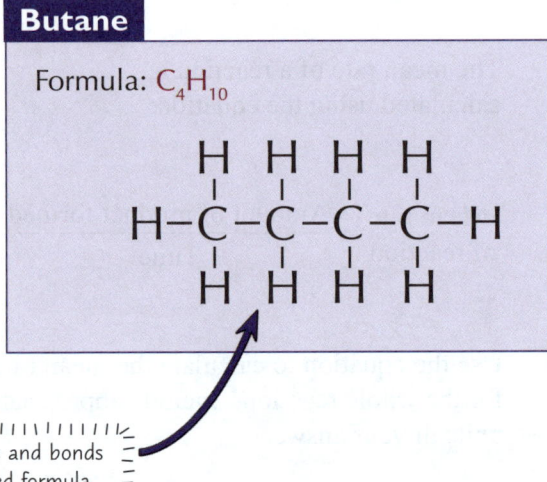

A drawing showing all the atoms and bonds in a molecule is called a displayed formula.

Hydrocarbons only contain hydrogen and carbon

REVISION TIP — To help remember the names of the <u>first four alkanes</u> just remember: <u>M</u>ice <u>E</u>at <u>P</u>eanut <u>B</u>utter. Practise drawing the structures and get to grips with their formulas using the general formula.

Hydrocarbons

Hydrocarbon **Properties Change** as the Chain Gets **Longer**

1) As the length of the carbon chain changes, the properties of the hydrocarbon change.
2) The shorter the hydrocarbon chain...

   ...the more runny a hydrocarbon is — that is, the less viscous (gloopy) it is.

   ...the lower its boiling point will be.

   ...the more flammable (easier to ignite) the hydrocarbon is.

Complete Combustion Occurs When There's Plenty of **Oxygen**

1) The complete combustion of a hydrocarbon in oxygen releases lots of energy.
2) This makes them useful as fuels.
   The properties of the different hydrocarbons affect exactly how they're used for fuels.
3) The only waste products of complete combustion are carbon dioxide and water vapour.

   hydrocarbon + oxygen → carbon dioxide + water   (+ energy)

4) During combustion, both carbon and hydrogen from the hydrocarbon are oxidised.
   Oxidation is the gain of oxygen.
5) You need to be able to give a balanced symbol equation for the complete combustion of a simple hydrocarbon when you're given its molecular formula. Here's an example:

   See p.166 for more on balancing equations.

> **EXAMPLE** Write a balanced equation for the complete combustion of methane (CH_4).
>
> 1) On the left hand side, there's one carbon atom, so only one molecule of CO_2 is needed to balance this.
>
> $CH_4 + ?O_2 \rightarrow CO_2 + ?H_2O$
>
> 2) On the left hand side, there are four hydrogen atoms, so two water molecules are needed to balance them.
>
> $CH_4 + ?O_2 \rightarrow CO_2 + 2H_2O$
>
> 3) There are four oxygen atoms on the right hand side of the equation. Two oxygen molecules are needed on the left to balance them.
>
> $CH_4 + 2O_2 \rightarrow CO_2 + 2H_2O$

Alkanes are useful fuels as they release energy when burnt

Shorter hydrocarbons are less viscous, more volatile and easier to ignite than longer hydrocarbons.
Q1 Write a balanced symbol equation for the complete combustion of ethane, C_2H_6. [2 marks]

Topic C7 — Organic Chemistry

Crude Oil

*Crude oil has fuelled **modern life** — it would be a very different world if we hadn't discovered oil.*

Crude Oil is Made Over a Long Period of Time

1) Crude oil is a fossil fuel found in rocks.
2) Fossil fuels are natural substances. They can be used as a source of energy.
3) Crude oil formed mainly from the remains of plankton, as well as other plants and animals. These died millions of years ago and were buried in mud.

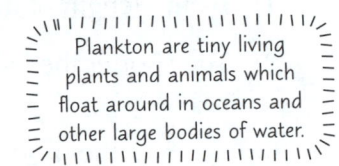
Plankton are tiny living plants and animals which float around in oceans and other large bodies of water.

- Fossil fuels like coal, oil and gas are called non-renewable fuels.
- This is because they take so long to make that they're being used up much faster than they're being formed.
- They're finite resources (see p.270) — one day they'll run out.

Crude Oil has Various Important Uses in Modern Life

1) Oil provides the fuel for most modern transport — cars, trains, planes, the lot.

2) Diesel oil, kerosene, heavy fuel oil and LPG (liquefied petroleum gas) all come from crude oil.

3) Petrochemicals are compounds that come from crude oil. The petrochemical industry uses some of the compounds from crude oil as a feedstock to make new compounds for use in things like...

- polymers (e.g. plastics)
- solvents
- lubricants
- detergents

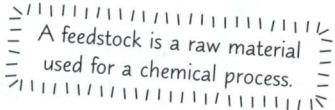
A feedstock is a raw material used for a chemical process.

4) All the products you get from crude oil are examples of organic compounds. Organic compounds are compounds containing carbon atoms.

5) Most of the organic compounds in crude oil are hydrocarbons (see pages 246-247). These hydrocarbons are a range of different sizes.

6) You can get a large variety of products from crude oil. This is because carbon atoms can bond together to form different groups called homologous series.

7) These groups contain similar compounds which have many properties in common. Alkanes and alkenes are both examples of different homologous series.

Crude oil isn't only used for fuel

Crude oil is a fossil fuel, but the compounds found in it are useful for manufacturing all kinds of things. Make sure you can name some examples of the different fuels that come from crude oil, and of product types which are made from crude oil feedstocks (e.g. detergents).

Topic C7 — Organic Chemistry

Fractional Distillation

Crude oil can be used to make loads of useful things, such as fuels. But you can't just put crude oil in your car. First, the different hydrocarbons have to be separated. That's where *fractional distillation* comes in.

Fractional Distillation is Used to Separate Hydrocarbon Fractions

Crude oil is a mixture of lots of different hydrocarbons, most of which are alkanes. The different compounds in crude oil are separated by fractional distillation. Here's how it works:

1) The oil is heated until most of it has evaporated (turned into gas). The gases enter a fractionating column (and the liquid bit is drained off).
2) In the column it's hot at the bottom and gets cooler as you go up.

Hydrocarbons are molecules containing only hydrogen and carbon.

The shorter hydrocarbons have low boiling points. This means that they're still gases at low temperatures. So they don't condense and turn back into liquids until they move up near the top of the column, where they cool down a lot.

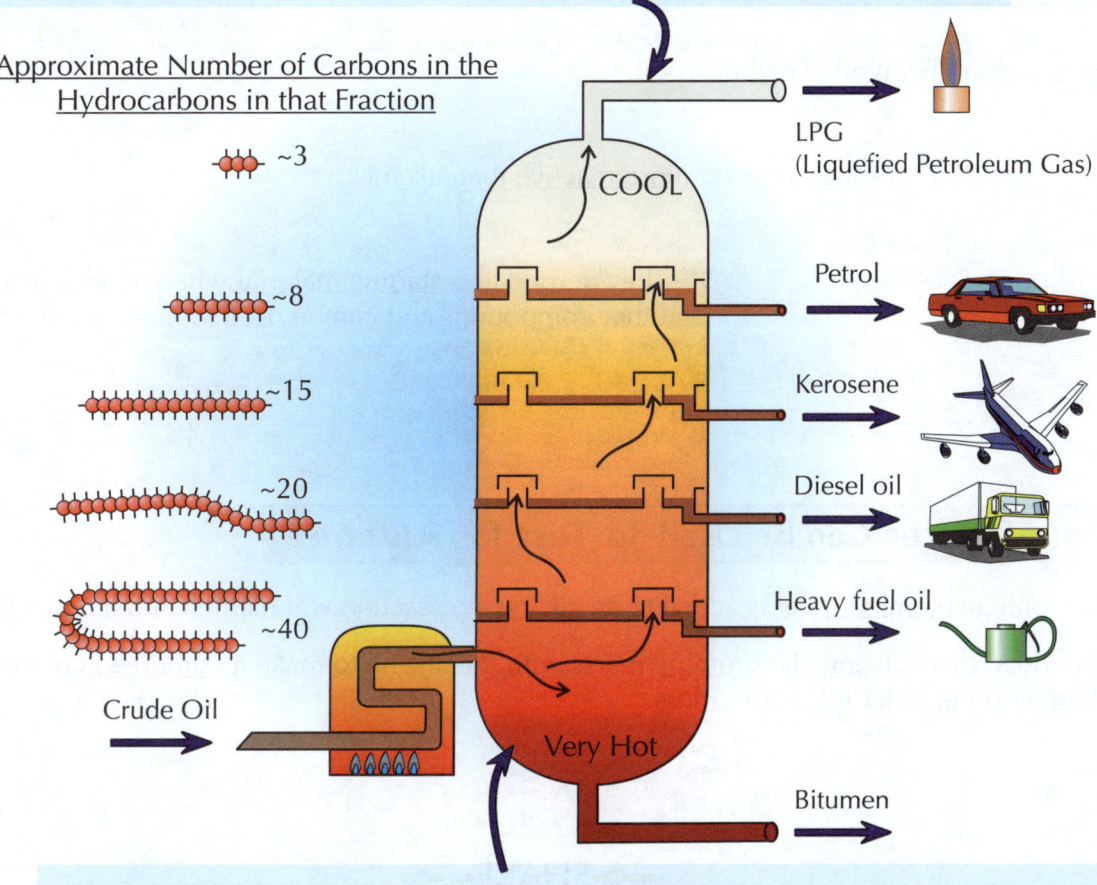

The longer hydrocarbons have high boiling points. This means that they'll only stay a gas if it's very hot. As they move up the fractionating column, it gets cooler. So they condense back into liquids and drain out of the column early on, when they're near the bottom.

3) You end up with the crude oil mixture separated into different fractions (parts), e.g. petrol and diesel oil.
4) Each fraction contains a mixture of hydrocarbons. All of the hydrocarbons in one fraction contain a similar number of carbon atoms. This means they'll have similar boiling points.

Hydrocarbon fractions drain off according to their boiling points

Q1 Petrol drains further up a fractionating column than diesel.
Use the diagram of the fractionating column to explain
why the boiling point of petrol is lower than that of diesel. [2 marks]

Topic C7 — Organic Chemistry

Cracking

*Crude oil fractions from fractional distillation can be split into **smaller molecules**. This is called **cracking**. It's super important, otherwise we might not have enough fuel for cars and planes and things.*

Cracking Means Splitting Up Long-Chain Hydrocarbons

1) There is a high demand for fuels with small molecules.
2) This is because short-chain hydrocarbons tend to be more useful than long-chain hydrocarbons.
3) So, lots of longer alkane molecules are turned into smaller, more useful ones. This is done by a process called cracking.
4) Some of the products of cracking are useful as fuels, e.g. petrol for cars.

Cracking Makes Alkanes and Alkenes

1) Alkenes are another type of hydrocarbon.

2) Alkenes are a lot more reactive than alkanes.

3) They're used as a starting material when making lots of other compounds and can be used to make polymers.

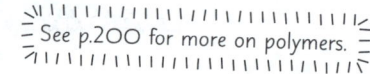

See p.200 for more on polymers.

Bromine Water Can Be Used To Test for Alkenes

1) When orange bromine water is added to an alkane, no reaction will happen and it'll stay bright orange.
2) If it's added to an alkene, the bromine reacts with the alkene to make a colourless compound. So the bromine water turns colourless.

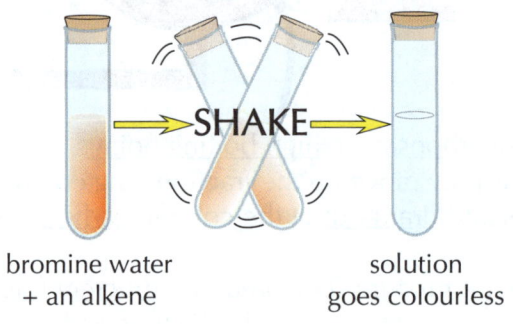

bromine water + an alkene → SHAKE → solution goes colourless

Cracking produces more of the hydrocarbons we need

There's not much you can do with longer-chain hydrocarbons. But shorter-chain ones are useful as fuels and to make other compounds like polymers. Cracking means we can make use of all those long chain hydrocarbons that we otherwise wouldn't need, which is why it's so important to the petrochemical industry.

Cracking

There are **Different Methods** of **Cracking**

1) Cracking is a thermal decomposition reaction.
2) This means the molecules are broken down by heating them.
3) This can by done by catalytic cracking or by steam cracking.

Catalytic Cracking

Catalysts speed up reactions without getting used up (see p.235).

1) Long-chain hydrocarbons are heated to turn them into a gas (vapour).
2) Then the vapour is passed over a hot powdered aluminium oxide catalyst.
3) The long-chain molecules split apart on the surface of the specks of catalyst.

Steam Cracking

1) Long-chain hydrocarbons are heated to turn them into a gas.
2) The hydrocarbon vapour is mixed with steam.
3) They are then heated to a very high temperature which splits them into smaller molecules.

You Can Write **Equations** for Cracking Reactions

You might be asked to work out the formula of the products or reactants involved in a cracking reaction. You can do this by balancing the number of carbons and hydrogens on each side of the reaction.

EXAMPLE
Decane can be cracked to form octane and one other product. The equation for the cracking of decane, $C_{10}H_{22}$, is shown below. Complete the equation.

$$C_{10}H_{22} \rightarrow C_8H_{18} + \text{.....................}$$

1) There needs to be the same number of carbon and hydrogen atoms on each side of the equation.

2) The number of carbon atoms in the missing product equals the number of carbons in $C_{10}H_{22}$ minus the number of carbons in C_8H_{18}.

   number of C atoms = 10 − 8 = 2

3) The number of hydrogen atoms in the missing product equals the number of hydrogens in $C_{10}H_{22}$ minus the number of hydrogens in C_8H_{18}.

   number of H atoms = 22 − 18 = 4

4) Put these numbers into the formula of the missing product.

   $C_{10}H_{22} \rightarrow C_8H_{18} + C_2H_4$

Cracking always involves heating

In catalytic cracking the hydrocarbons are heated with a catalyst and in steam cracking they're heated with steam. Simple really. Which gives you more time to practise balancing your equations.

Q1   Pentane, C_5H_{12}, can be cracked into ethene and one other hydrocarbon. Give the balanced symbol equation for the cracking reaction.   [1 mark]

Topic C7 — Organic Chemistry

Warm-Up & Exam Questions

Hydrocarbons contain only hydrogen and carbon atoms.
This page contains only Warm-Up and Exam Questions. Time to get thinking.

Warm-Up Questions

1) Name the first four alkanes.
2) What type of bonds hold the atoms in alkanes together?
3) Why are alkanes often used as fuels?
4) What size hydrocarbon molecules are cracked?

Exam Questions

1 Which alkane is shown in **Figure 1**?
Tick **one** box.

Figure 1

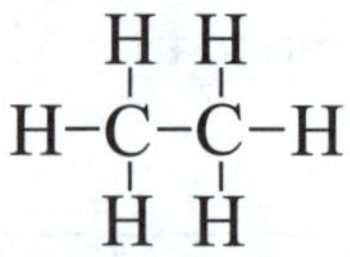

H H
| |
H–C–C–H
| |
H H

☐ propane   ☐ ethane   ☐ methane

[1 mark]

2 Which statement correctly describes how the properties of hydrocarbons change as the length of the carbon chain increases?
Tick **one** box.

☐ Flammability increases and viscosity increases.

☐ Flammability increases and viscosity decreases.

☐ Flammability decreases and viscosity increases.

☐ Flammability decreases and viscosity decreases.

[1 mark]

3 Alkanes are a homologous series of hydrocarbons made up of chains of carbon atoms surrounded by hydrogen atoms.

Pentane is an alkane with the formula C_5H_{12}.

3.1 Name the alkane that has one fewer carbon atoms than pentane.

[1 mark]

3.2 Give the **formula** of the alkane that has one fewer carbon atoms than pentane.

[1 mark]

3.3 Alkanes are a product made from crude oil. They are often used as fuels.
Give **two** other uses of compounds made from crude oil.

[2 marks]

Topic C7 — Organic Chemistry

Exam Questions

4 Which of the following statements best describes what occurs during a combustion reaction?
Tick **one** box.

☐ Oxygen reacts with a fuel and energy is released.

☐ A fuel takes in energy and decomposes.

☐ Carbon dioxide reacts with a fuel and energy is released.

☐ A fuel reacts with water vapour in clouds to produce acid rain.

[1 mark]

5 When nonane is cracked, heptane (an alkane) and ethene (an alkene) can be produced.
Figure 2 shows the displayed formulas of nonane, heptane and ethene.

Figure 2

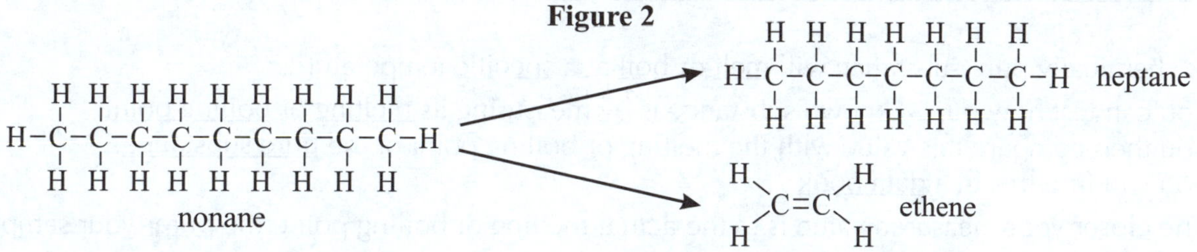

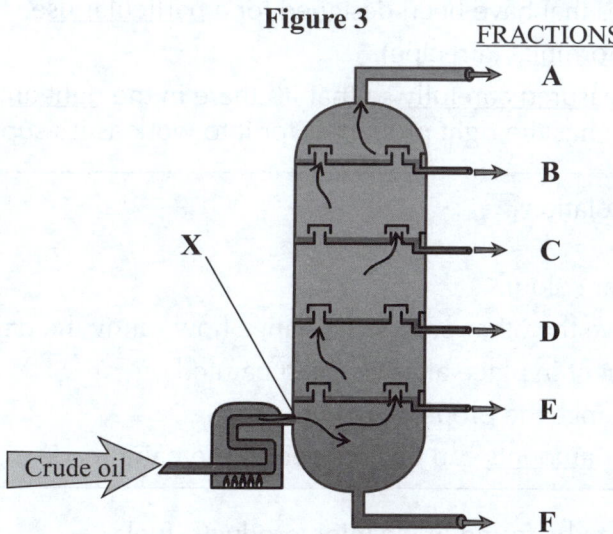

5.1 Complete the chemical equation for the cracking of nonane to make heptane and ethene.

$C_9H_{20} \rightarrow$ .... + ...

[1 mark]

5.2 Samples of heptane and ethene were added to two separate test tubes of bromine water and shaken.
Describe what you would expect to happen in each case.

[2 marks]

6 Crude oil can be separated into a number of different compounds in a fractional distillation column. **Figure 3** shows a fractional distillation column.

Figure 3

FRACTIONS: A, B, C, D, E, F

Crude oil enters at point X.

6.1 Which letter, **A-F**, represents the liquefied petroleum gas (LPG) fraction?

[1 mark]

6.2 The crude oil is heated until it becomes a gas and then enters the column at point **X** on **Figure 3**.
The gas moves through the column and different fractions exit at different points.
Explain why fractions with longer hydrocarbon chains exit near the bottom of the column.

[3 marks]

Topic C7 — Organic Chemistry

Topic C8 — Chemical Analysis

Purity and Formulations

*In a perfect world, every compound a chemist made would be **pure**. Unfortunately, in the real world it **doesn't** always work out like that. Luckily, there are ways to find out **how pure** a substance is.*

Purity Has a Different Meaning in Chemistry to Everyday

1) Usually when you say that a substance is pure you mean that nothing has been added to it. So it's in its natural state. For example: pure milk or beeswax.
2) In chemistry, a pure substance is something that only contains one compound or element all the way through. It's not mixed with anything else.

The Boiling or Melting Point Tells You How Pure a Substance Is

1) A chemically pure substance will melt or boil at a specific temperature.
2) You can test how pure a known substance is by measuring its melting or boiling point. You then compare this value with the melting or boiling point of the pure substance. You can find this in a data book.
3) The closer your measured value is to the actual melting or boiling point, the purer your sample is.
4) Impurities in your sample will lower the melting point. They may also cause the sample to melt across a wider range of temperatures.
5) Impurities in your sample will increase the boiling point. They may also cause the sample to boil across a range of temperatures.

A Formulation is a Mixture with Exact Amounts of its Parts

1) Formulations are useful mixtures that have been designed for a particular use.
2) They are made by following a 'formula' (a recipe).
3) Each part of a formulation is measured carefully so that it's there in the right amount. This makes sure the formulation has the right properties for it to work as it's supposed to.

Take a look at p.169 for more on mixtures.

> For example, paints are formulations.
> They are made up of:
> - Pigment — gives the paint colour.
> - Solvent — used to dissolve the other parts and change how runny the paint is.
> - Binder — holds the pigment in place after it's been painted on.
> - Additives — added to change the properties of the paint.
>
> The chemicals used and their amounts can be changed so the paint made is right for the job.

4) In everyday life, formulations can be found in cleaning products, fuels, medicines, cosmetics, fertilisers, metal alloys and even food and drink.

Make sure you can use data to identify pure and impure substances...

Knowing if a product is pure is really important for making things such as medicines or food. Extra stuff in it by mistake could change the properties of the product, and even make it dangerous.

Paper Chromatography

*You met chromatography on page 170. Now it's time to see **how it works**.*

Chromatography uses Two Phases

1) Chromatography is a method used to separate the substances in a mixture.

2) It can then be used to identify the substances.

3) The type of chromatography you need to know about is paper chromatography. Like all types of chromatography, it has two 'phases'.
   - A mobile phase — where the molecules can move. In paper chromatography, this is a solvent (e.g. water or ethanol).
   - A stationary phase — where the molecules can't move. In paper chromatography, this is the paper.

4) During paper chromatography the solvent moves up the paper. As the solvent moves, it carries the substances in the mixture with it.

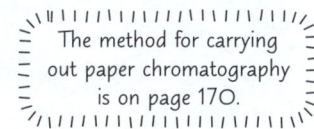

The method for carrying out paper chromatography is on page 170.

5) In a chromatography experiment, the amount of time a chemical spends dissolved in the solvent or stuck on the paper is called its 'distribution'.

6) The more soluble a chemical is, the more time it spends dissolved in the solvent. This means that the chemical will move further up the paper.

7) Different chemicals may be dissolved in the solvent for different amounts of time. So the different chemicals will move different distances up the paper.

8) This means they separate into different spots.

Chromatography revision — it's a phase you have to get through...
Chromatography works because each of the chemicals in a mixture spends different amounts of time dissolved in the mobile phase and stuck to the stationary phase. It's great — all you need is some paper and a bit of solvent. There's more about using it to identify chemicals coming up on the next few pages.

Topic C8 — Chemical Analysis

PRACTICAL: Interpreting Chromatograms

*Now that you know a bit of the **theory** behind how paper chromatography works, here's how you can use a **chromatogram** to analyse a particular substance and find out **what's in it**.*

The Result of Chromatography is Shown on a Chromatogram

1) Chromatograms show the result of chromatography experiments.

2) The solvent front is the furthest point reached by the solvent during a chromatography experiment.

3) Chemicals move different distances up the paper. So different spots show different chemicals.

4) The number of spots on a chromatogram is the smallest possible number of chemicals in the mixture.

5) Sometimes more than one chemical may travel the same distance up the paper. This means that these chemicals will only form one spot between them.

6) If you repeat the experiment with a different solvent, you'll get a different chromatogram. The spots may have travelled different distances compared to the solvent front. There might also be a different number of spots on the chromatogram.

7) If you only get one spot in lots of different solvents, there's only one chemical in the substance. This means the substance is pure.

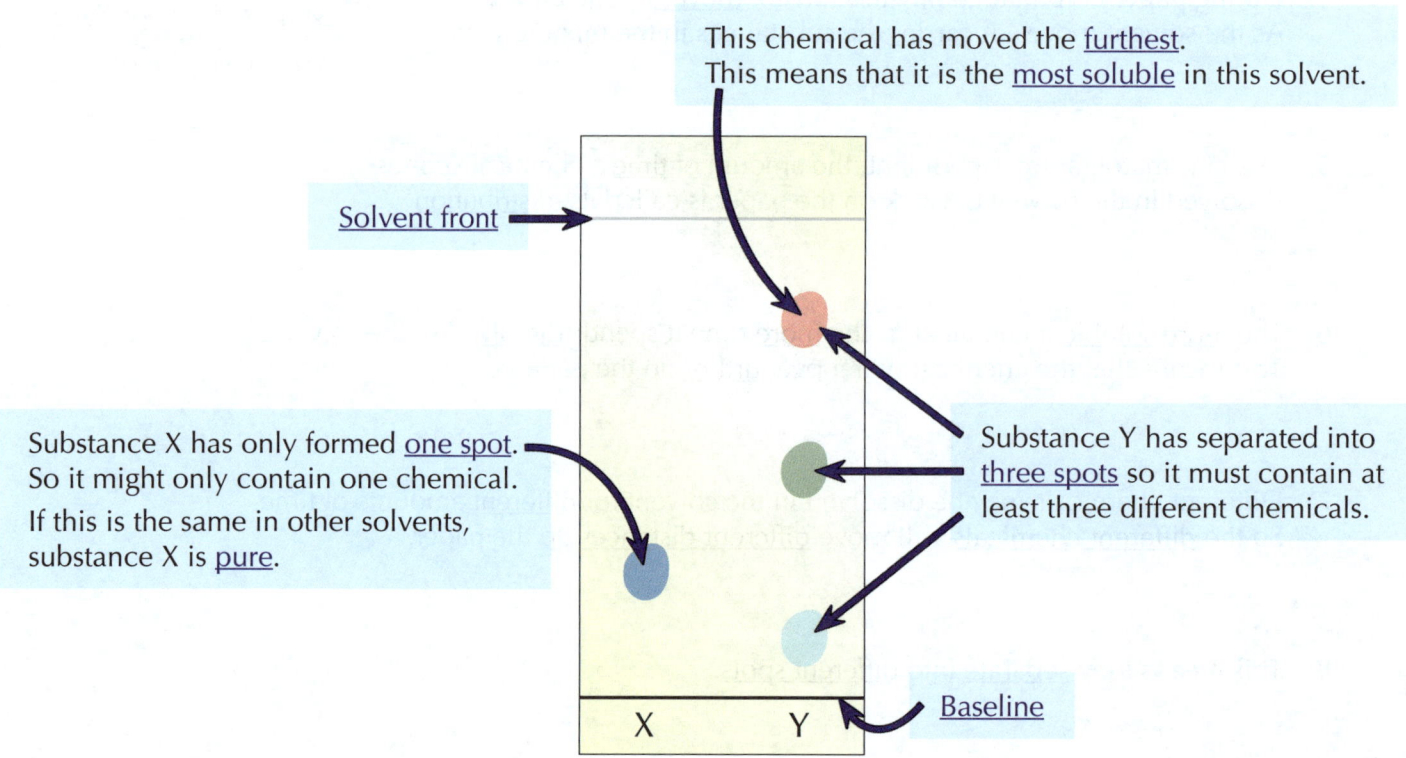

A pure substance will only ever produce one spot

PRACTICAL TIP — When you're doing paper chromatography, you might end up with a spot left sitting on the baseline, even after your solvent has run all the way up the paper. Any substance that stays on the baseline is insoluble in that solvent. If this happens, you could try the experiment using a different solvent, and see if the mystery substance dissolves in it.

Topic C8 — Chemical Analysis

Interpreting Chromatograms — PRACTICAL

*If you were sad the last page on **chromatography** was finished — fear not. There's more to come on this page.*

You can Calculate the R_f Value for Each Chemical

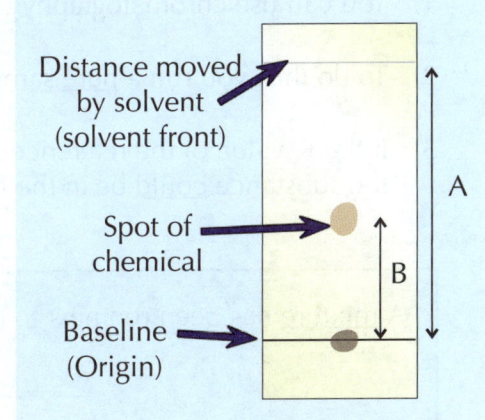

1) An R_f value is the ratio between the distance travelled by the dissolved substance and the distance travelled by the solvent.

2) The further a substance moves through the stationary phase, the larger the R_f value.

3) You can calculate R_f values using the formula:

$$R_f = \frac{\text{distance moved by substance (B)}}{\text{distance moved by solvent (A)}}$$

This is the distance from the baseline to the centre of the spot.

EXAMPLE

A chromatography experiment looking at the colours in a dye produces the chromatogram shown on the right. Calculate the R_f value for the red spot.

1) Measure the distance moved by the red spot (B). This is the distance from the baseline to the centre of the spot.   **29 mm**

2) Measure the distance moved by the solvent (A).   **41 mm**

3) Calculate the R_f value.

$$R_f = \frac{\text{distance moved by substance (B)}}{\text{distance moved by solvent (A)}} = \frac{29}{41} = 0.70731... = 0.71$$

Give your answer to the smallest number of significant figures in the calculation.

4) The R_f value of a chemical will change if you change the solvent.

You need to learn the formula for R_f

R_f values always lie between 0 and 1, as the solvent always travels further than any of the substances in the mixture. If you work out an R_f value to be outside this range, you know you've gone wrong somewhere (e.g. you may have written the fraction in the formula upside-down).

Q1 A spot on a chromatogram moved 6.3 cm from the baseline. The solvent front moved 8.4 cm. Calculate the R_f value. [1 mark]

Topic C8 — Chemical Analysis

PRACTICAL — Interpreting Chromatograms

*Time for one last page on chromatography — this one's all about using **references** to identify substances.*

You Can Identify Substances in Mixtures Using Chromatography

1) You can use chromatography to see if a mixture contains a certain substance.

2) To do this, you run a pure sample of that substance (a reference) next to the mixture.

3) If the R_f value of the reference compound matches one of the spots in the mixture, the substance could be in the mixture. For example:

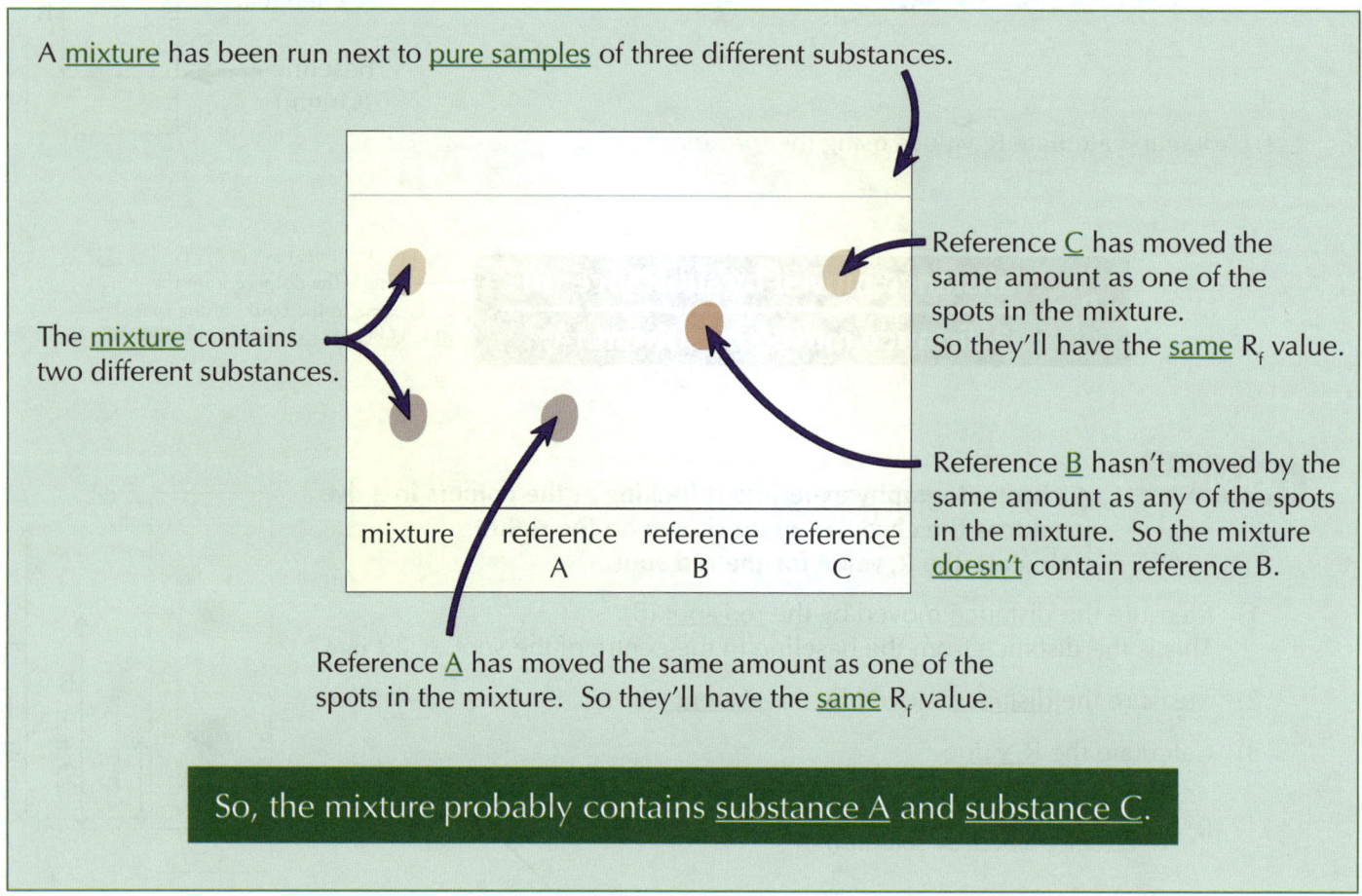

A mixture has been run next to pure samples of three different substances.

The mixture contains two different substances.

Reference C has moved the same amount as one of the spots in the mixture. So they'll have the same R_f value.

Reference B hasn't moved by the same amount as any of the spots in the mixture. So the mixture doesn't contain reference B.

Reference A has moved the same amount as one of the spots in the mixture. So they'll have the same R_f value.

So, the mixture probably contains substance A and substance C.

4) If the R_f values match in one solvent, you can check to see if the chemicals are the same by repeating with a different solvent. If they match again, it's likely that they're the same.

Reference spots are pure samples of known substances

Chromatography can be a bit tricky to get your head around sometimes, especially when you're given a chromatogram that contains quite a few different spots. Make sure you're really comfortable with how to use a chromatogram to work out what compounds are present in a mixture. It'll be worth it come the exam.

Topic C8 — Chemical Analysis

Tests for Gases

Yep, that's right, you need to revise tests for your test. There are a few ways to **test for gases**, but you only need to know these **four**. And two of them use fire. Pretty cool.

There are Tests for Four Common Gases

1) Chlorine

Chlorine bleaches damp litmus paper, turning it white.

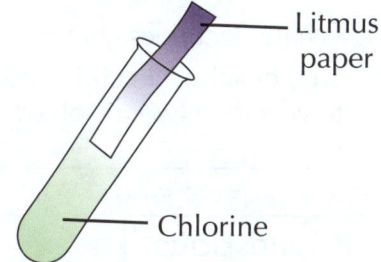

2) Oxygen

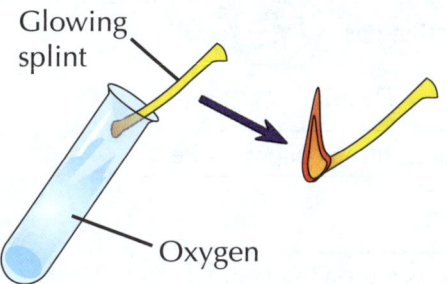

If you put a glowing splint inside a test tube containing oxygen, the oxygen will relight the glowing splint.

3) Carbon Dioxide

You can test for carbon dioxide by bubbling it through a solution of calcium hydroxide.
If the gas is carbon dioxide, the solution will turn cloudy.
You can also do this test by shaking the gas with the solution.

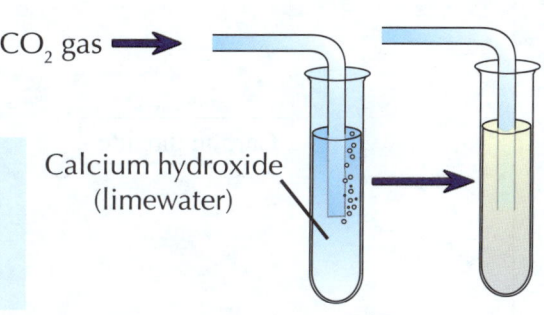

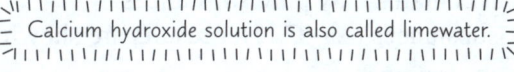
Calcium hydroxide solution is also called limewater.

4) Hydrogen

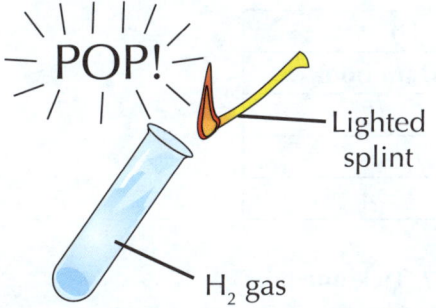

If you hold a lit splint at the open end of a test tube containing hydrogen, you'll get a "squeaky pop".

These are all really useful tests to know...
The method you use to collect a gas will depend on whether it's lighter or heavier than air. If it's heavier (like chlorine), you have the test tube the right way up and the gas will sink to the bottom. If it's lighter (like hydrogen), you have the test tube upside-down and the gas will rise to fill it.

Topic C8 — Chemical Analysis

Warm-Up & Exam Questions

Look, a chromatography question — those things are fun. Get your detective hat on and get stuck in...

Warm-Up Questions

1) What effect will impurities in a substance have on its boiling point?
2) Give an example of a formulation used in everyday life.
3) What effect does chromatography have on a mixture?
4) True of False? In chromatography, an R_f value represents the ratio between the distance travelled by two dissolved substances.

Exam Questions

1   Draw **one** line from each gas to the correct result of the test for that gas.

Gas	Result
Hydrogen	Turns damp litmus paper white
	Relights a glowing splint
Chlorine	Produces a "squeaky pop" with a lit splint
Carbon dioxide	Turns a solution of calcium hydroxide cloudy

[3 marks]

2   A scientist is preparing a formulation that contains only pure substances. She measures the melting point of three substances, **A – C**. Her results are shown in **Table 1**.

Table 1

Substance	Melting point (°C)	
	Experimental	Data book
A	14	17
B	49	49
C	21 – 24	24

Which of the substances could the scientist use in her formulation? Tick **one** box.

☐ Substance B only ☐ Substance A and C

☐ Substance C only ☐ Substance A, B and C

[1 mark]

Topic C8 — Chemical Analysis

Exam Questions

3 Different groups of seaweed contain different types of a pigment called chlorophyll.
Table 2 shows which types of chlorophyll each group of seaweed contains.
Figure 1 shows the results of a chromatography experiment to analyse an unknown seaweed.

Table 2

Group of seaweed	Type of chlorophyll		
	a	b	c
Red	✓		
Brown	✓		✓
Green	✓	✓	

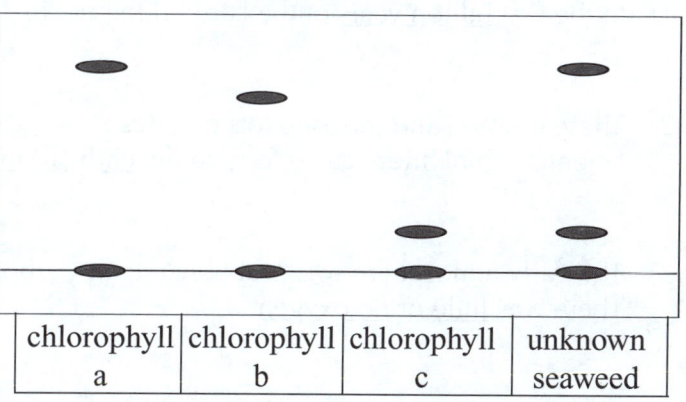

Figure 1

Use **Table 2** and **Figure 1** to identify which group the unknown seaweed belongs to.

[1 mark]

4 A scientist used chromatography to study the compounds present in five food colourings. Four of the colourings were unknown (**A – D**). The other colouring was sunrise yellow. The results are shown in **Figure 2**.

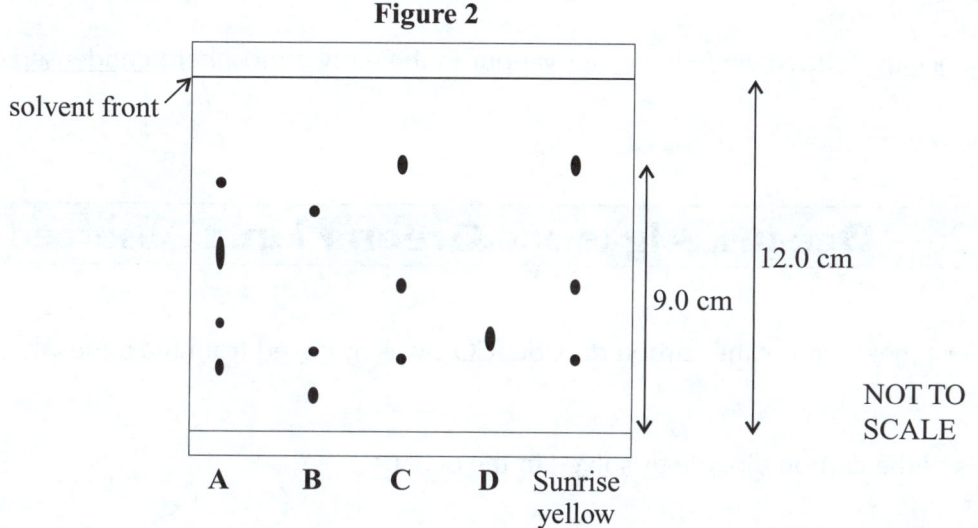

Figure 2

NOT TO SCALE

4.1 Which food colouring contained the compound that had the strongest attraction to the stationary phase?

[1 mark]

4.2 Which of the food colourings, **A-D**, could be the same as sunrise yellow?

[1 mark]

4.3 The R_f value of a spot can be calculated using the following formula:

$$R_f = \frac{\text{distance moved by substance}}{\text{distance moved by solvent}}$$

Calculate the R_f value for the spot of chemical in sunrise yellow which is furthest up the chromatogram.

[2 marks]

Topic C8 — Chemical Analysis

Topic C9 — Chemistry of the Atmosphere

The Evolution of the Atmosphere

*Theories for how the Earth's atmosphere **evolved** have changed a lot over the years. It's hard to gather evidence from such a **long time period** and from **so long ago** (4.6 billion years). Here's one idea we've got:*

Phase 1 — Volcanoes Gave Out Gases

1) In the first billion years of the Earth's lifetime, its surface was covered in volcanoes.

2) These erupted and released lots of gases. Scientists think these gases formed the early atmosphere.

3) The early atmosphere was probably mostly carbon dioxide (CO_2). There was little or no oxygen.

4) This is quite like the atmospheres of Mars and Venus today.

5) Volcanoes also released nitrogen (this built up in the atmosphere over time), water vapour and small amounts of methane and ammonia.

6) The oceans formed when the water vapour in the early atmosphere condensed (turned to liquid).

Phase 2 — Oceans, Algae and Green Plants Absorbed CO_2

1) Over time, much of the carbon dioxide (CO_2) was removed from the atmosphere.

2) Lots of the carbon dioxide dissolved in the oceans.

3) The dissolved carbon dioxide formed carbonates that precipitated as small, solid particles (sediments).

> Precipitation is the formation of an insoluble solid from a solution.

4) When green plants and algae evolved, they took in some carbon dioxide during photosynthesis (see next page).

Before volcanic activity, the Earth didn't even have an atmosphere

One way scientists can get information about what Earth's atmosphere was like in the past is from Antarctic ice cores. Each year a layer of ice forms with tiny bubbles of air trapped in it. The deeper you go in the ice, the older the air. So analysing bubbles from different layers shows you how the atmosphere has changed.

The Evolution of the Atmosphere

Some Carbon Became Trapped in Fossil Fuels and Rocks

Some of the carbon that organisms took in from the atmosphere and oceans became locked up in rocks and fossil fuels after the organisms died.

1) When sea organisms die, they fall to the seabed and get buried. Over millions of years, they're squashed down. This forms sedimentary rocks (e.g. coal and limestone), oil and gas. The carbon gets trapped within them.

2) Things like coal, crude oil and natural gas that are made this way are called 'fossil fuels'.

3) Crude oil and natural gas are formed from the remains of plankton that settled on the seabed.

4) Coal is made from thick layers of plants that died and then settled on the seabed.

5) Limestone is mostly made of calcium carbonate from the shells and skeletons of marine organisms.

Phase 3 — Green Plants and Algae Produced Oxygen

1) Algae evolved about 2.7 billion years ago. Then green plants evolved over the next billion years or so.

2) Green plants and algae produce oxygen in a reaction called photosynthesis:

$$\text{carbon dioxide} + \text{water} \xrightarrow{\text{light}} \text{glucose} + \text{oxygen}$$

$$6CO_2 + 6H_2O \rightarrow C_6H_{12}O_6 + 6O_2$$

3) Over time, oxygen levels built up in the atmosphere. This meant that animals could evolve.

4) The proportions of gases in the atmosphere have been similar for about the last 200 million years. It is made up of about:

- 80% (⁴/₅) nitrogen,
- 20% (¹/₅) oxygen,
- small amounts (less than 1%) of other gases (mainly carbon dioxide, noble gases and water vapour).

Not too much CO_2 and enough O_2 — perfect for life on Earth
You need to know the rough proportions of the gases in Earth's atmosphere, so don't just skip over the numbers — you could get asked about them in the exam. Make sure you know that nitrogen is 80%, oxygen is 20% and everything else makes up less than 1% of our atmosphere.

Topic C9 — Chemistry of the Atmosphere

Climate Change and Greenhouse Gases

*Greenhouse gases are important but can also cause **problems** — it's all about keeping a delicate **balance**.*

Carbon Dioxide is a Greenhouse Gas

1) Greenhouse gases include carbon dioxide, methane and water vapour.
2) Greenhouse gases keep the Earth warm enough to support life. Here's how they work:

   1) The Sun gives out short wavelength radiation.

   2) This radiation is reflected back by the Earth as long wavelength radiation. This is thermal (heat) radiation. It's then absorbed by greenhouse gases.

   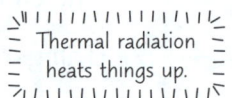
   Thermal radiation heats things up.

   3) Greenhouse gases then give out this radiation in all directions.

   4) Some radiation heads back towards the Earth and warms up the surface. This is the greenhouse effect.

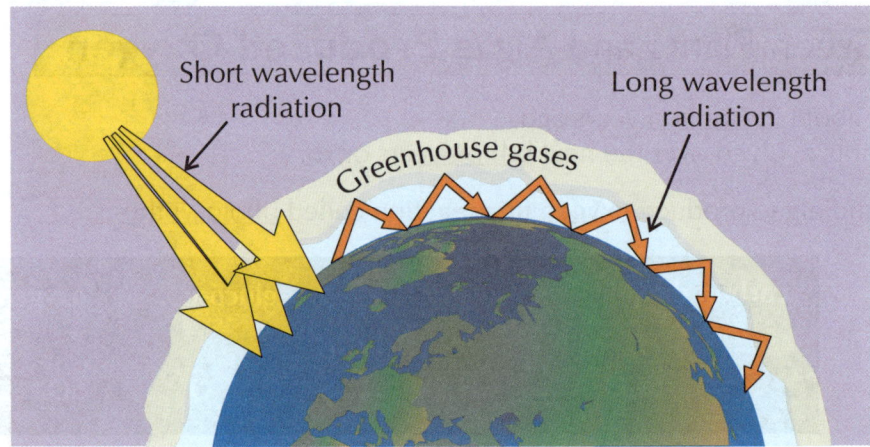

3) Some forms of human activity increase the amount of greenhouse gases in the atmosphere. For example:

   1) Deforestation: fewer trees means that less carbon dioxide is taken in for photosynthesis.
   2) Burning fossil fuels: releases carbon dioxide.
   3) Agriculture: more farm animals produce more methane when they digest their food.
   4) Creating waste: more landfill sites and more waste from farming means more carbon dioxide and methane is released when the waste breaks down.

Greenhouse gases aren't all bad — we need them to survive

Without greenhouse gases our planet would be incredibly cold — the greenhouse effect warms the Earth enough for it to support living things. Without it we wouldn't be here. But the overall balance of gases in the atmosphere matters, as you're about to find out...

Q1 Describe the greenhouse effect and how it affects global temperature. [4 marks]

Topic C9 — Chemistry of the Atmosphere

Climate Change and Greenhouse Gases

Increasing Carbon Dioxide is Linked to Climate Change

1) Recently, the average temperature of the Earth's surface has been going up.

2) Scientists agree that this has been caused by human activities that release greenhouse gases, including carbon dioxide.

3) They believe this is leading to climate change.

4) Evidence for this has been peer-reviewed (see page 2). This means that the information is reliable.

5) However, the Earth's climate is very complex. So, it's very hard to make a model that isn't oversimplified.

See page 3 for more on science in the media.

6) This has led to people forming their own theories and opinions, particularly in the media. These stories aren't based on good evidence — they may be biased or only give some of the information.

If something's biased, it favours one point of view in a way that's not backed up by facts.

Climate Change Could Have Dangerous Consequences

1) Higher global temperatures are causing ice in the Arctic and Antarctic to melt — causing sea levels to rise. If sea levels keep rising, this will lead to more flooding.

2) Changes in rainfall are causing some regions to get too much or too little water.

3) Storms may become more frequent and severe.

4) Changes in temperature and rainfall are having an effect on the production of food in certain places.

People can get quite hot under the collar talking about all this...

That's because climate change could have a massive impact on many people's lives. So it's very important to recognise when results might be biased or changed in favour of one particular view. Data should always be checked and peer-reviewed before any final conclusions are made.

Topic C9 — Chemistry of the Atmosphere

Carbon Footprints

*Scientists believe that greenhouse gas emissions from **human activities** are causing **climate change**. Knowing what things release lots of carbon dioxide could help **stop** climate change from happening.*

Carbon Footprints are Tricky to Measure

1) Carbon footprints are a measure of the amount of carbon dioxide and other greenhouse gases released over the full life cycle of something. That can be almost anything:
   - a service (e.g. the school bus).
   - an event (e.g. the Olympics).
   - a product (e.g. a toastie maker).
2) Measuring the total carbon footprint of something can be really hard or even impossible.
3) But a rough calculation can give a good idea of what things release the most greenhouse gases. So, people can then avoid using them in the future.

There are Ways of Reducing Carbon Footprints

1) You can reduce a carbon footprint by reducing the amount of greenhouse gases given out by a process.
2) Here are some things that can be done:

> - Using renewable energy sources (sources that won't run out) or nuclear energy instead of fossil fuels.
> - Using processes that use less energy or produces less waste (decomposing waste releases methane).
> - Governments could tax companies or individuals based on the amount of greenhouse gases they emit. This could encourage people to use processes which use less fuel and are less polluting.
> - Governments can also put a limit on emissions of all greenhouse gases that companies make. They can then sell licences for emissions up to that cap.
> - There's also technology that captures carbon dioxide before it's released into the atmosphere. This carbon dioxide is then stored deep underground.

But Making Reductions is Still Difficult

1) Reducing greenhouse gas emissions isn't simple.
2) Alternative technologies that release less carbon dioxide still need a lot of work.
3) Many governments are worried that making these changes will affect the economies of communities. This could be bad for people's well-being — especially those in developing countries.
4) This makes it hard for countries to agree to reduce emissions.
5) Individuals in developed countries also need to make changes to their lifestyles. But this is tricky when some don't want to and others don't understand why the changes are important or how to make them.

Topic C9 — Chemistry of the Atmosphere

Air Pollution

Increasing carbon dioxide is causing climate change. But CO_2 isn't the only gas released when fossil fuels burn — you can also get other nasty gases like **oxides of nitrogen**, **sulfur dioxide** and **carbon monoxide**.

Combustion of Fossil Fuels Releases Gases and Particles

1) Fossil fuels, such as crude oil and coal, contain hydrocarbons (see page 246).
2) Hydrocarbons can combust (burn in oxygen). There are two types of combustion:

   Complete combustion — when there's plenty of oxygen around all of the fuel burns.

   Incomplete combustion — when there's not enough oxygen around some of the fuel does not burn.

   There's more about complete combustion on p.247.

3) Both types of combustion release carbon dioxide and water vapour into the atmosphere.
4) During incomplete combustion, solid particles (called particulates) of soot (carbon) and unburnt fuel are also released. Carbon monoxide gas is also produced.
5) Particulates in the air and carbon monoxide can cause all sorts of problems:

Particulates

1) If particulates are breathed in, they can get stuck in the lungs and cause damage. This can lead to respiratory (breathing) problems.
2) They're also bad for the environment — they reflect sunlight back into space. This means that less light reaches the Earth — causing global dimming.

Carbon Monoxide

1) Carbon monoxide (CO) is really dangerous because it can stop your blood from carrying enough oxygen around the body.
2) A lack of oxygen in the blood can lead to fainting, a coma or even death.
3) Carbon monoxide doesn't have any colour or smell, so it's very hard to detect. This makes it even more dangerous.

Sulfur Dioxide and Oxides of Nitrogen Can be Released

1) Other pollutants are also released from burning fossil fuels.
2) Sulfur dioxide (SO_2) is released during the combustion of fossil fuels that contain sulfur impurities.
3) Nitrogen oxides form in a reaction between nitrogen and oxygen in the air. This reaction is caused by the heat of the burning fossil fuels.
4) These gases mix with clouds and cause acid rain.
5) Acid rain kills plants. It also damages buildings, statues and metals.
6) Sulfur dioxide and nitrogen oxides also cause respiratory problems if they're breathed in.

You can test for sulfur impurities in a fuel by bubbling the gases from combustion through a solution containing Universal indicator. If the fuel contains sulfur, the Universal indicator will turn red.

Fossil fuels are bad news — but we need them for many things...

...so a big reduction in their use is probably hard to achieve. Make sure you know the different pollutants that are given out when fuels burn, and the differences between complete and incomplete combustion.

Topic C9 — Chemistry of the Atmosphere

Warm-Up & Exam Questions

There's lots of important information in this section, from the Earth's atmosphere to climate change and pollution. Answer these questions to see what you can remember and what you need to go over again.

Warm-Up Questions

1) Where do scientists think the gases that made up Earth's early atmosphere came from?
   A. Space   B. The ocean   C. Volcanoes   D. Burning fossil fuels
2) Give an example of a sedimentary rock.
3) Give an example of a fossil fuel.
4) True or false? Nuclear energy has a lower carbon footprint than burning fossil fuels.
5) Give one problem caused by sulfur dioxide in the atmosphere.

Exam Questions

1 Green plants and algae had a significant effect on Earth's early atmosphere.

1.1 Use words from the box to complete the sentences below.

| oxygen   carbon   carbon dioxide   particulates   nitrogen |

When green plants first evolved, the Earth's atmosphere was mostly .... gas .

These plants produced .... by the process of photosynthesis.

Some of the .... from dead plants eventually became 'locked up' in fossil fuels.

[3 marks]

1.2 Complete the balanced symbol equation for photosynthesis.

$$6 \,\rule{1cm}{0.4pt}\, + \,\rule{0.5cm}{0.4pt}\, H_2O \rightarrow C_6H_{12}O_6 + 6 \,\rule{1cm}{0.4pt}\,$$

[2 marks]

1.3 Which line in **Table 1** shows the correct percentages of each gas in the Earth's atmosphere today?

Table 1

	Percentage in atmosphere (%)			
	Carbon dioxide	Oxygen	Nitrogen	Water vapour
A	20	70	10	less than 1
B	10	20	70	less than 1
C	less than 1	20	80	less than 1
D	less than 1	10	80	10

☐ A   ☐ B   ☐ C   ☐ D

[1 mark]

Topic C9 — Chemistry of the Atmosphere

Exam Questions

2 Fossil fuels contain hydrocarbons. *(Grade 3-4)*

2.1 Which of the following is a fossil fuel?
Tick **one** box.

☐ Limestone ☐ Natural gas ☐ Carbon dioxide ☐ Water

[1 mark]

When hydrocarbons burn without enough oxygen present, solid particles and carbon monoxide can be produced.

2.2 Name the type of reaction that occurs when hydrocarbons burn in oxygen.

[1 mark]

2.3 What name is given to the solid particles that can be produced when hydrocarbons burn without enough oxygen present?

[1 mark]

2.4 Describe **two** problems caused by these solid particles in the air.

[2 marks]

2.5 Give **two** reasons why carbon monoxide is difficult to detect.

[1 mark]

3 This question is about greenhouse gases and climate change. *(Grade 4-5)*

3.1 Which of the following is an example of a greenhouse gas?
Tick **one** box.

☐ Oxygen ☐ Nitrogen ☐ Methane ☐ Argon

[1 mark]

3.2 Greenhouse gases absorb thermal radiation that has been reflected by the Earth.
Explain how this leads to global warming.

[3 marks]

3.3 Give **two** human activities that have increased the amount of greenhouse gases in Earth's atmosphere.

[2 marks]

3.4 Give **one** possible consequence of climate change.

[1 mark]

3.5 Describe **one** action that governments could take to try to reduce the amount of greenhouse gases in the atmosphere.

[1 mark]

4* Earth's early atmosphere contained large amounts of carbon dioxide. *(Grade 4-5)*

Describe how the oceans, plants and algae removed some of this carbon dioxide, and how this eventually led to the formation of coal.

[4 marks]

Topic C9 — Chemistry of the Atmosphere

Topic C10 — Using Resources

Finite and Renewable Resources

*There are lots of different resources that humans use for things like **electricity**, **heating**, **travelling**, **building materials** and **food**. Some of these resources can be replaced, some can't.*

Natural Resources Come From the Earth, Sea and Air

1) Natural resources form by themselves — they're not made by humans. They include anything that comes from the earth, sea or air, e.g. cotton and oil.
2) Some natural products can be replaced or improved by man-made products or processes. For example:

> - Rubber is a natural product that comes from the sap of a tree. But we can now make polymers (see p.200) to replace some natural rubber to make things like tyres.
> - Wool is a natural product that comes from animals such as sheep. But scientists have developed synthetic (man-made) fibres that we can use instead of wool to make things like jumpers and blankets.

3) Agriculture (farming) helps to increase our supply of natural resources to provide food, timber, clothing and fuel. It also provides conditions which can make natural resources better for our needs.

> E.g. the development of fertilisers means we can increase the amount of crops grown in a given area.

Some Natural Resources will Run Out

1) Renewable resources can be remade at least as fast as we use them.
2) This means that they can be replaced fairly quickly.

> - For example, timber is a renewable resource. Trees can be planted following a harvest and only take a few years to regrow.
> - Other examples of renewable resources include fresh water and food.

3) Finite (non-renewable) resources are remade very slowly (or not at all). So we use them up quicker than we can replace them. This means that they'll eventually run out.
4) Finite resources include fossil fuels and nuclear fuels, as well as minerals and metals found in the ground.
5) We can process many finite resources to provide fuels and materials necessary for modern life. For example:

> - Fractional distillation (see p.249) is used to produce usable products such as petrol from crude oil.
> - Metal ores are reduced to produce pure metals (see p.221).

Natural resources can be renewable or finite

It's important that we're always able to get the resources we need to survive. Make sure you know the difference between a finite and a renewable resource and a few examples of each.

Resources and Sustainability

Tables, Charts and Graphs can Tell You About Different Resources

You can <u>interpret</u> information about resources from information that's given to you.

EXAMPLE
The table below shows how long it takes for three resources to form. The resources are coal, wood and cotton. Work out which resource is coal.

	Time it takes to form
Resource 1	10 years
Resource 2	120-180 days
Resource 3	10^6 years

Wood and cotton are both renewable resources.

Coal is a finite resource.

Finite resources take a very long time to form.

Resource 3 takes a much longer amount of time to form compared to Resources 1 and 2.

Coal is Resource 3.

10^6 is a quick way of showing 1 000 000. This is because $10^6 = 10 \times 10 \times 10 \times 10 \times 10 \times 10 = 1 000 000$.

We Need to Consider the Future When Choosing Resources

1) <u>Sustainable development</u> means thinking about the needs of <u>people today</u> without damaging the lives of <u>people in the future</u>.

2) Using, extracting and processing resources can be unsustainable. This could be because:

   - Some resources are <u>non-renewable</u> — they'll run out one day. For example, the <u>raw materials</u> used to make metals, building materials, many plastics and things made from clay and glass are <u>limited</u>.
   - Extraction processes can use lots of <u>energy</u> and produce lots of <u>waste</u>.
   - Turning resources into useful materials, like <u>glass</u> or <u>bricks</u>, often uses <u>energy</u> made from <u>finite resources</u>.

   Extraction processes separate the materials you want from the other things that they're mixed with.

3) One way to be more sustainable is to use <u>fewer</u> finite resources. This reduces both the use of finite resources and anything needed to produce them.

4) We can do this by <u>reusing</u> and <u>recycling</u> materials when we're finished with them. During recycling, <u>waste</u> is <u>processed</u> so that it can be used to make <u>new products</u>.

   There's more on reusing and recycling on the next page.

5) We can't stop using finite resources completely. But scientists can <u>develop</u> processes that use <u>less</u> and <u>reduce</u> damage to the environment.

We need to be responsible with resources that won't last forever

Sustainable development is talked about a lot these days, however it's often quite <u>hard</u> to achieve in practice. It can be a <u>difficult</u> topic, too — making the changes we need to isn't always easy and can have a big impact on many people's lives. Plus people <u>can't always agree</u> on the best thing to do.

Topic C10 — Using Resources

Reuse and Recycling

*Supplies of many materials used in the modern world are **limited**. Once they're finished with, it's usually far better to **recycle** them than to use new finite resources, which will run out.*

Recycling Metals is Important

1) Mining and extracting metals takes lots of energy. Most of this energy comes from burning fossil fuels.

2) It's usually better to recycle metals instead of making new metals.

> Benefits of recycling:
> - It often uses much less energy than the amount needed to make a new metal.
> - It helps save some of the finite amount of each metal in the earth.
> - It cuts down on the amount of waste getting sent to landfill.

3) Metals can be recycled by melting them and then moulding (recasting) them into the shape of a new product.

4) Sometimes, different metals won't need to be completely separated before recycling. The amount of separation depends on what the final product will be.

> For example, waste steel and iron can be kept together. This is because they can both be added to iron in a blast furnace. This means that less iron ore will be needed.

A blast furnace is used to extract iron from its ore at a high temperature using carbon.

Glass can be Reused or Recycled

1) Reusing or recycling glass can help sustainability.

2) This reduces the amount of energy used for making new glass.

3) It also means that less glass is thrown away, so less waste is produced.

> - Glass bottles can often be reused without reshaping.
> - Some glass products can't be reused so they're recycled for a different use instead.
> - The glass is crushed and melted. It's then reshaped to make other glass products like jars.

Recycling is key to sustainability — it's useful in lots of ways...

Remember that recycling doesn't just reduce the use of raw materials, it also reduces the amount of energy used, the amount of damage to the environment and the amount of waste that is produced.

Topic C10 — Using Resources

Life Cycle Assessments

*If a company wants to manufacture a new product, they carry out a **life cycle assessment (LCA)**.*

Life Cycle Assessments Show Total Environmental Costs

1) A life cycle assessment (LCA) looks at every stage of a product's life to assess the impact (effect) it would have on the environment.
2) Here are the four different stages:

1) Getting the Raw Materials

1) Lots of raw materials need to be extracted (separated from other materials) before we can use them for a product.
2) Extracting raw materials can damage the local environment, e.g. mining metals.
3) Extraction uses lots of energy. This can result in pollution.
4) Raw materials often need to be processed (e.g. by changing their shape and properties) to turn them into useful materials.
5) This often needs large amounts of energy. E.g. extracting metals from ores (see p.221 and 224) or fractional distillation of crude oil (see p.249).

2) Manufacture and Packaging

1) Making products and their packaging can use a lot of energy and other resources.
2) It can cause lots of pollution too.
3) Chemical reactions are sometimes used to make products. These reactions also produce waste products which need to be got rid of.
4) Some of this waste can be turned into other useful chemicals. This reduces the amount that ends up polluting the environment.

3) Using the Product

See page 264 for more on greenhouse gases.

1) The use of a product can also damage the environment.
2) For example, burning fuels releases greenhouse gases and other harmful substances. Fertilisers can drain into streams and rivers, causing harm to plants and animals.
3) It's also important to think about how long a product is used for or how many uses it gets. Products that need lots of energy to produce but are used for ages mean less waste in the long run.

4) Product Disposal

1) Products are often thrown away in landfill sites.
2) This takes up space and pollutes land and water. For example, paint can wash off a product and get into rivers.
3) Energy is used to transport waste to landfill. This can release pollutants into the atmosphere, such as carbon monoxide (see p.267) and carbon dioxide.
4) Products might be incinerated (burnt), which causes air pollution.

LCAs look at the effect a product has on the environment

LCAs can show us how to improve products, so we can make them less damaging to the environment. However, LCAs can also be very time consuming and expensive, because there is so much to think about.

Topic C10 — Using Resources

Life Cycle Assessments

You can compare the life cycle assessments of **similar products** to see which has the **smallest effect** on the environment. You might want to do this to see which material will affect the environment the **least**.

You Can **Compare** Life Cycle Assessments for **Plastic** and **Paper Bags**

1) You may be asked to compare life cycle assessment (LCA) information about paper and plastic bags.
2) You can then decide which type of bag is the least harmful to the environment.

Life Cycle Assessment Stage	Plastic Bag	Paper Bag
Raw Materials	Crude oil	Wood
Manufacturing and Packaging	Plastics are made from compounds extracted from crude oil by fractional distillation and processed by cracking and polymerisation. Waste is reduced as the other fractions of crude oil have other uses.	Pulped wood is processed using lots of energy. Lots of waste is made.
Using the Product	Can be reused several times. Can be used for other things as well as shopping. For example, as bin liners.	Usually only used once.
Product Disposal	Recyclable but many types aren't biodegradable. Take up space in landfill and pollute land.	Can be recycled. Biodegradable and non-toxic.

3) LCAs have shown that even though plastic bags aren't usually biodegradable, they may be less harmful to the environment.
4) This is because they take less energy to make and have a longer lifespan than paper bags.

If something's biodegradable, it can be broken down naturally by microorganisms (tiny living things like bacteria).

There are **Problems** with Life Cycle Assessments

1) It's quite easy to measure things like the use of energy or resources, and the production of some types of waste. So we can give all of these measurements a number in an LCA.
2) But measuring some effects is much harder.

- For example, plastic bags that litter the environment don't look very nice.
- But measuring how unattractive something looks isn't easy. The person measuring it has to use their own judgement.

3) So, producing an LCA can involve the feelings of the person carrying out the assessment as well as facts. So the results could change depending on who does the assessment. This means LCAs can be biased.
4) Selective LCAs only show some of the impacts that a product has on the environment. So these can also be biased because they can be written to purposely support the claims of a company. This would give the company positive advertising.

If something is biased, that means it favours one point of view in a way that isn't backed up by facts.

LCAs aren't all they are cracked up to be...

In the exam, you may have to compare LCAs for plastic and paper bags and decide which one is the most environmentally friendly. Each bag has good and bad points, but don't forget that not all environmental impacts can be measured in an LCA. Also, the results of an LCA can be biased.

Topic C10 — Using Resources

Warm-Up & Exam Questions

That's some more revision done and dusted — now it's time to test yourself on how much you've taken in. Have a go at the questions on this page to see if you need to look back at some topics.

Warm-Up Questions

1) Give one example of a natural resource which has been replaced by a man-made alternative.
2) True or false? One way to be more sustainable is to use fewer finite resources.
3) Give two positive effects of recycling metals.
4) What is the purpose of a life cycle assessment?

Exam Questions

1   Natural resources are formed without human input and are used for construction, fuel and food.

1.1   What is a renewable natural resource?
Tick **one** box.

☐ A natural resource that can't be remade by humans.

☐ A natural resource that can be remade at least as fast as we use it.

☐ A natural resource that is only remade very slowly.

[1 mark]

1.2   Aluminium is used to make soft drink cans.
Extracting aluminium uses a large amount of energy.
Suggest **one** way that the use of soft drink cans can be made more sustainable.

[1 mark]

2   Table 1 shows part of a life cycle assessment for two types of bag.

Table 1

	Raw materials	Manufacturing	Reusability	Disposal
Plastic bag	Crude oil	Waste from manufacturing process can be used to make other products.	Can be reused many times, in different ways.	Recyclable, but often not biodegradable. Many end up in landfill sites or the oceans.
Paper bag	Wood	High energy process that generates lots of waste.	Usually not reusable.	Can be recycled. Biodegradable and non-toxic.

2.1* Use the information in **Table 1** and your own knowledge to evaluate which type of bag is better for the environment.

[6 marks]

2.2   Give **two** reasons why a life cycle assessment may be biased.

[2 marks]

Potable Water and Water Treatment

We all need safe drinking water. The **way** that water's made safe depends on **local conditions**.

Potable Water is Water You Can Drink

1) Potable water is water that's safe for humans to drink. We need it to live.
2) Some water is naturally potable, but most water needs to be treated before it's safe to drink.
3) Potable water isn't pure. Pure water only contains H_2O molecules but potable water can contain lots of other dissolved substances. *See p.254 for more on purity.*
4) For water to be safe to drink, it must:
   - not have high levels of dissolved salts,
   - have a pH between 6.5 and 8.5,
   - not have any bad things in it (like bacteria or other microbes).

How Potable Water is Produced Depends on Where You Are

1) Fresh water is water that doesn't have much dissolved in it. Rainwater is a type of fresh water.
2) When it rains, water can either collect as surface water or as ground water.

   Surface water: collects in lakes, rivers and reservoirs (places for storing liquids).
   Ground water: collects in rocks that trap water underground.

3) When producing potable water, companies need to choose a suitable source of freshwater. In the UK, the source of fresh water used depends on location.
4) Surface water tends to dry up first. So in warm areas, such as the south-east, most of the water supply comes from ground water. This is because it is underground so doesn't dry up.

Most Fresh Water Needs to be Treated to Make it Safe

1) Fresh water contains low levels of dissolved substances.
2) It still needs to be treated to make it safe before we use it.
3) This process includes:

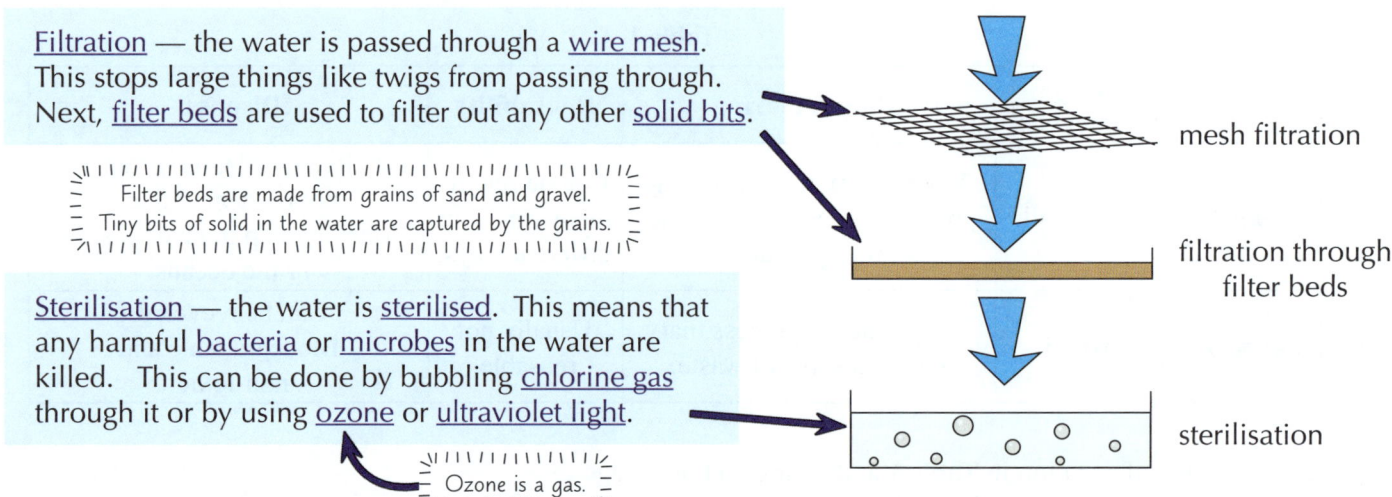

Filtration — the water is passed through a wire mesh. This stops large things like twigs from passing through. Next, filter beds are used to filter out any other solid bits.

Filter beds are made from grains of sand and gravel. Tiny bits of solid in the water are captured by the grains.

Sterilisation — the water is sterilised. This means that any harmful bacteria or microbes in the water are killed. This can be done by bubbling chlorine gas through it or by using ozone or ultraviolet light.

Ozone is a gas.

- mesh filtration
- filtration through filter beds
- sterilisation

Waste water must be treated before it's safe to drink

A lot of the water that we drink comes from rain. Thankfully in the UK, we have an endless supply...

Q1 Describe the steps used to treat fresh water to make it potable. [2 marks]

Topic C10 — Using Resources

Potable Water and Water Treatment

*Some countries have **limited supplies** of fresh water. They need to use **other** water sources, like the sea.*

Potable Water Can Be Made From Seawater

1) In some very dry countries there's not enough surface or ground water.
   So instead they use seawater to provide potable water.

2) Seawater contains salts which need to be removed before we can drink it.

3) These salts are removed by a process called desalination.

Distillation Can Be Used to Purify Seawater   **PRACTICAL**

1) Distillation can be used to remove the salt from seawater (desalination).

2) You can test and purify a sample of water in the lab using distillation:

Seawater can also be purified by a process called reverse osmosis (see next page).

   1) Use a pH meter to test the pH of a sample of water (see p.217).
      If the pH is too high or too low, you'll need to neutralise it (make it neutral).
      You do this by adding some acid (if the sample's alkaline) or some alkali
      (if the sample's acidic) until the pH is 7.

   2) Set up the equipment as shown in the diagram below.

Neutral solutions are neither acidic nor alkaline. They have a pH of 7. E.g. pure water is neutral. There's more on this on page 215.

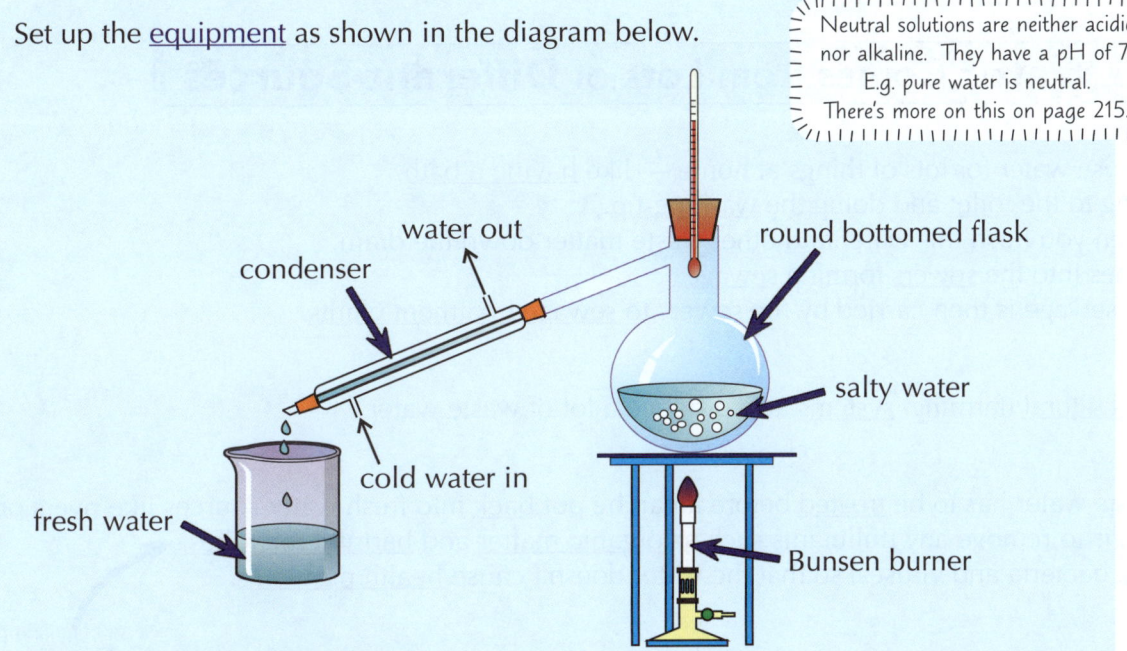

   3) Heat the water in the flask using a Bunsen burner.

   4) As the water heats up, it becomes a gas (it evaporates).
      The gas then enters the condenser as steam.
      Cold water is pumped around the condenser to cool the steam inside of it.
      This drop in temperature makes the steam condense back into liquid water.

   5) Collect the water running out of the condenser in a beaker.

   6) Retest the pH of the water with a pH meter to check it's neutral.

   7) After the water has been distilled (all of the water has evaporated from the flask),
      see whether there are any crystals in the round bottomed flask.
      If there are crystals it means that there were salts in the water before you distilled it.

Topic C10 — Using Resources

Potable Water and Water Treatment

*You saw on the previous page how **distillation** can be used to purify seawater. There's another method you need to know about — **reverse osmosis**.*

Seawater can be Purified Using Reverse Osmosis

1) Seawater can be purified by processes that use thin layers of material called membranes. Membranes have really tiny holes that only let certain things pass through them.

2) One of these processes is called reverse osmosis. The salty water is passed through a membrane. The membrane lets water molecules pass through but traps the salts. This separates them from the water.

There's more on purifying seawater using distillation on the previous page.

3) Both distillation and reverse osmosis need loads of energy to work which makes them expensive. This is why they're not used when there are other sources of fresh water available.

Waste Water Comes from Lots of Different Sources

1) We use water for lots of things at home — like having a bath, going to the toilet and doing the washing-up. When you flush this water and other waste matter down the drain, it goes into the sewers forming sewage. The sewage is then carried by the sewers to sewage treatment plants.

2) Agricultural (farming) systems also produce a lot of waste water.

3) Waste water has to be treated before it can be put back into fresh water sources like rivers or lakes. This is to remove any pollutants such as organic matter and harmful microbes (e.g. bacteria and viruses) so that the water doesn't cause health problems.

4) Waste water from industrial processes also has to be collected and treated.

5) Industrial waste water can contain organic matter and harmful chemicals. So it needs further treatment before it's safe to put back into the environment.

Organic matter contains carbon compounds that come from the remains and waste of organisms.

The water that comes out of our taps has been treated

Location is a really important factor in determining how water is treated. For example, in the UK, there is a lot of fresh water available so this is filtered and then sterilised. However, in very dry countries a more expensive process may have to be used, such as the distillation or reverse osmosis of sea water.

Topic C10 — Using Resources

Potable Water and Water Treatment

*Dealing with waste water is really important to make sure we don't **pollute** our environment.*

Sewage **Treatment** Happens in **Several Stages**

1) Some of the processes involved in treating waste water at sewage treatment plants are shown below.

1) Screening
The sewage is screened to remove any large bits of material (like twigs or plastic bags) and any grit (small bits of stone and sand).

2) Sedimentation
Then it goes through sedimentation. The heavier solids sink to the bottom to produce sludge. The lighter effluent (liquid waste) floats on the top.

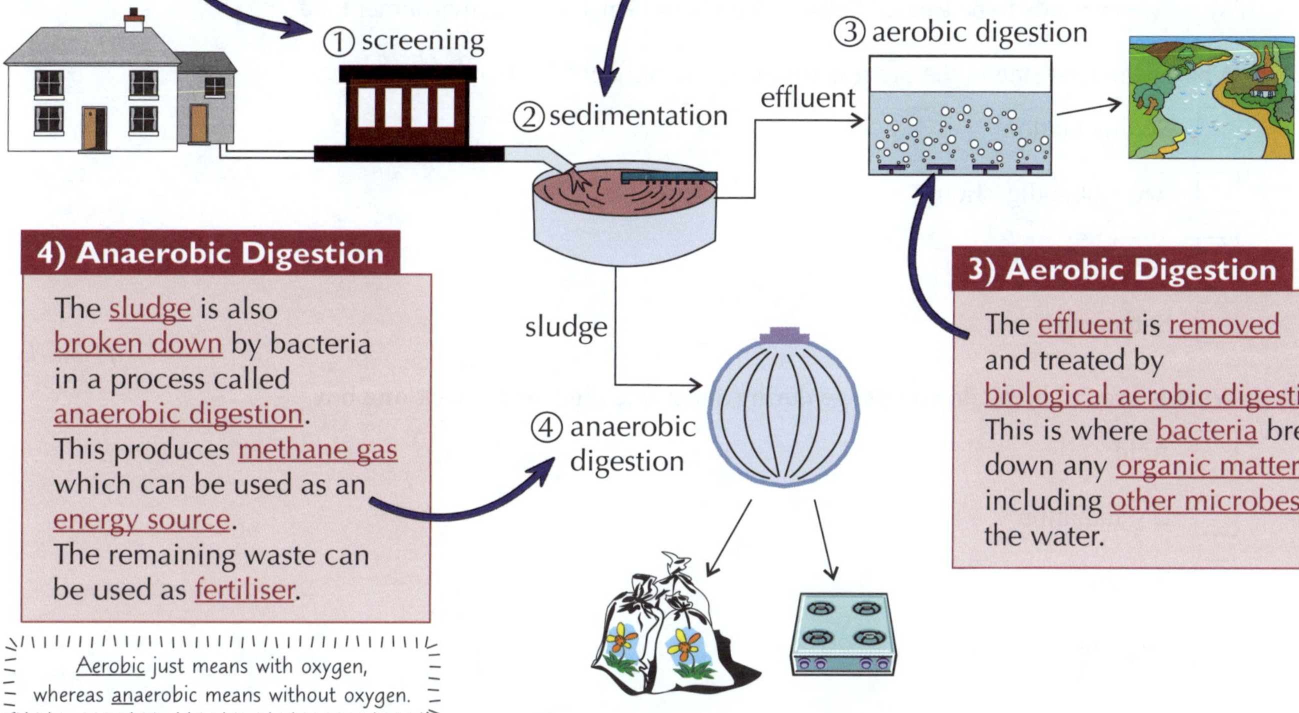

4) Anaerobic Digestion
The sludge is also broken down by bacteria in a process called anaerobic digestion. This produces methane gas which can be used as an energy source. The remaining waste can be used as fertiliser.

3) Aerobic Digestion
The effluent is removed and treated by biological aerobic digestion. This is where bacteria break down any organic matter — including other microbes in the water.

Aerobic just means with oxygen, whereas anaerobic means without oxygen.

2) Waste water containing toxic substances needs extra stages of treatment. This may include adding chemicals, UV radiation or using membranes.

3) Sewage treatment has more stages than treating fresh water but uses less energy than the desalination of salt water. So it could be used as an option in areas where there's not much fresh water. However, people don't like the idea of drinking water that used to be sewage.

Waste water can be recycled to produce potable water

To learn the stages of water treatment, cover the diagram, write out each step and then check it.

Q1 Name and describe the first two stages of waste water treatment at a sewage treatment plant. [2 marks]

Q1 Video Solution

Warm-Up & Exam Questions

Now you know all there is to possibly know about water it's time to test yourself with some questions.

Warm-Up Questions

1) Describe one feature of potable water that makes it safe for drinking.
2) True or False? The first step in treating fresh water involves mesh filtration.
3) What two things are produced following the sedimentation stage of water treatment?
4) True or False? Sewage treatment has more stages than the treatment of fresh water.

Exam Questions

1 Waste water needs to be treated before it can be returned to the environment. *(Grade 1-3)*

1.1 What is the first step in the process of sewage treatment? Tick **one** box.

☐ Sedimentation
☐ Anaerobic digestion
☐ Aerobic digestion
☐ Screening

[1 mark]

1.2 What is used to break down sewage during anaerobic digestion? Tick **one** box.

☐ Acid
☐ UV radiation
☐ Bacteria
☐ Oxygen

[1 mark]

2 Fresh water can be treated to produce potable water. *(Grade 3-4)*

2.1 What is potable water?
Tick **one** box.

☐ Water that only contains H_2O molecules.
☐ Water with a pH above 9.
☐ Water that comes from the surface.
☐ Water that is safe to drink.

[1 mark]

2.2 Why is fresh water filtered during the production of potable water?

[1 mark]

2.3 Suggest **one** thing that could be used to sterilise the water during the sterilisation stage of the process.

[1 mark]

Topic C10 — Using Resources

Exam Questions

PRACTICAL

3 A student wants to carry out distillation of a sample of seawater.
   Figure 2 shows the apparatus that the student is going to use.

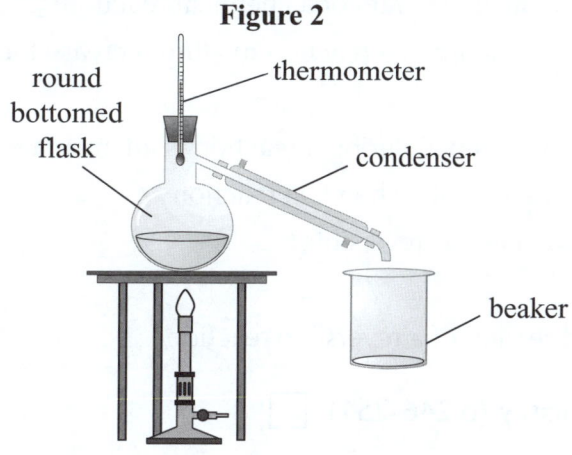

3.1 Use an **X** to label the point on **Figure 2** where the pure water is collected.

[1 mark]

3.2 Describe how the processes of evaporation and condensation are used to obtain a sample of pure water from seawater.

[2 marks]

3.3 The sample of seawater has a high pH.
Suggest **one** thing that the student could do to change the pH of the water so that it is safe to drink.

[1 mark]

3.4 How will the student know when all the water has been distilled?

[1 mark]

When the distillation was complete, the student noticed crystals in the round bottomed flask.

3.5 Explain why crystals formed in the flask at the end of the distillation.

[1 mark]

4 This question is about waste water.

4.1 Explain why waste water has to be treated before it can be returned to the environment.

[2 marks]

4.2 During the sedimentation step of sewage treatment, effluent and sludge are separated from each other. Describe how effluent and sludge are then treated before being returned to the environment.

[3 marks]

A factory makes cleaning products.

4.3 The factory's waste water needs more treatment than household waste water before it can be returned to the environment. Explain why.

[1 mark]

5 Membranes are thin layers of material that contain very small holes.

5.1 Explain how reverse osmosis uses membranes to purify seawater.

[2 marks]

5.2 Why is reverse osmosis an expensive process?

[1 mark]

Topic C10 — Using Resources

Revision Summary for Topics C6-10

That wraps up Topics C6-10 — time to see if you've got this topic in the bag.
- Try these questions and tick off each one when you get it right.
- When you're completely happy with a topic, tick it off.

For even more practice, try the Retrieval Quizzes for Topics C6-10 — just scan the QR codes!

Topic C6 — The Rate and Extent of Chemical Change (p.233-243)
1) What are the four factors that affect the rate of a chemical reaction?
2) Why does increasing the temperature of a reaction mixture increase the reaction rate?
3) What is a catalyst?
4) Explain why measuring a mass change during a reaction is an accurate method of measuring rate.
5) Give two possible units for the rate of a chemical reaction.
6) What symbol shows that a reaction is reversible?
7) What is a closed system?
8) How can you change the direction of a reversible reaction?

Topic C7 — Organic Chemistry (p.246-251)
9) What two elements do hydrocarbons contain?
10) Draw the displayed formula of butane.
11) What did crude oil form from?
12) Where do the shortest carbon chains condense in a fractional distillation column?
13) How is catalytic cracking carried out?

Topic C8 — Chemical Analysis (p.254-259)
14) In chemistry, what does it mean if a substance is pure?
15) Describe how you could use the boiling point of a substance to test whether it is pure.
16) What is the solvent front on a chromatogram?
17) Give the formula for working out the R_f value of a substance.
18) How can you test if a gas in a test tube is oxygen?

Topic C9 — Chemistry of the Atmosphere (p.262-267)
19) Name three of the gases that scientists think were present in the early atmosphere.
20) Give one way that the levels of carbon dioxide in the early atmosphere were reduced.
21) How do greenhouse gases help to support life on Earth?
22) Give two things that can be done to reduce carbon footprints.
23) Describe how acid rain forms.

Topic C10 — Using Resources (p.270-279)
24) What is sustainable development?
25) What happens to glass when it is sent to be recycled into a new product?
26) What are the four stages of a life cycle assessment (LCA)?
27) Why might companies want to use selective life cycle assessments?
28) Name two processes you could use to purify sea water.
29) Name two different sources of waste water.
30) What happens during the screening step of sewage treatment?

Topic P1 — Energy

Energy Stores

Energy is *never used up*. It's just the way that it's *stored* that *changes*.
There are eight *energy stores* you need to know about, so let's get cracking...

Energy is Transferred Between Energy Stores

Energy is stored in the different energy stores of an object.
You need to know the following energy stores:

1) KINETIC — anything moving has energy in its kinetic energy store.

2) THERMAL — all objects have energy in this store. The hotter the object, the more energy in the store.

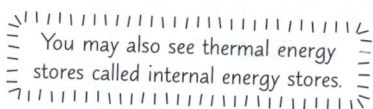
You may also see thermal energy stores called internal energy stores.

3) CHEMICAL — anything that can release energy by a chemical reaction has energy in this store, e.g. food.

4) GRAVITATIONAL POTENTIAL — any object raised above ground level has energy in this store.

5) ELASTIC POTENTIAL — anything stretched has energy in this store, like springs and rubber bands.

6) ELECTROSTATIC — e.g. two charges that attract or repel each other have energy in this store.

7) MAGNETIC — e.g. two magnets that attract or repel each other have energy in this store.

8) NUCLEAR — the nucleus of an atom releases energy from this store in nuclear reactions.

You'll learn more about nuclear reactions on pages 338-339.

No matter what store it's in, it's all energy...

In the exam, make sure you refer to energy in terms of the store it's in. For example, if you're describing energy in a hot object, say it 'has energy in its thermal energy store'.

Energy Transfer

*Now you know about the different energy stores, it's time to find out how energy is **transferred** between them.*

Energy can be Transferred in Four Ways

MECHANICALLY
1) This happens when a force does work (p.349) on an object.
2) For example, a force pushing an object along the floor.

ELECTRICALLY
1) This happens when a moving charge does work (p.321).
2) For example, when a current flows through a light bulb.

BY HEATING
1) When energy is transferred from a hotter object to a colder object.
2) For example, when a pan of water is heated on a hob.

BY RADIATION
1) This happens when energy is transferred by e.g. sound or light.
2) For example, when energy from the Sun travels to Earth by light.

When a System Changes, Energy is Transferred

1) A system is just the single object or a group of objects that you're interested in.
2) When a system changes, the way energy is stored changes in one of the ways above.
3) Closed systems are systems where no matter (stuff) or energy can enter or leave.
4) When a closed system changes, there is no net (overall) change in the total energy of the system.
5) For example...

> A cold spoon sealed in a flask of hot soup is a closed system.
> Energy is transferred from the thermal energy store of the soup to the thermal energy store of the spoon by heating.
> But no energy leaves the system. The total energy stays the same.

Heating a Material Transfers Energy to its Thermal Energy Store

1) As a material is heated, energy is transferred to its thermal energy store.
2) This causes its temperature to increase.
3) You need to be able to describe the changes in how energy is stored when an object is heated.
4) For example, if you boil water in an electric kettle:

> Energy is transferred electrically to the thermal energy store of the kettle's heating element.
> Energy is then transferred by heating to the water's thermal energy store.
> So the temperature of the water increases.

Don't talk about 'electrical energy stores'...

It's easy to accidentally refer to electrical energy stores, rather than talk about how energy is transferred electrically. Make sure you don't get it mixed up the exam.

Topic P1 — Energy

Mechanical Energy Transfer

*It's time to look at **mechanical energy transfers** a bit more. They happen when a **force does work**.*

Forces Cause Mechanical Energy Transfers

1) If a force moves an object, then work is done.
2) Work done is the same as energy transferred.
3) So energy is transferred mechanically when a force moves an object.
4) For any given situation, you'll have to describe the changes in how energy is stored.
5) Here are a few examples:

There's more on work done on p.349.

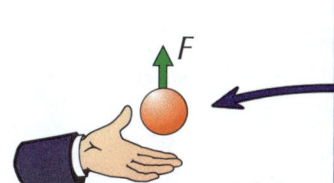

Example 1 — A ball thrown into the air
A boy throws a ball upwards. The boy exerts a force on the ball.
Energy is transferred mechanically from the chemical energy store of the boy's arm to the kinetic energy stores of the ball and arm.

Example 2 — A ball dropped from a height
The ball is accelerated by the constant force of gravity.
Energy is transferred mechanically from the ball's gravitational potential energy store to its kinetic energy store.

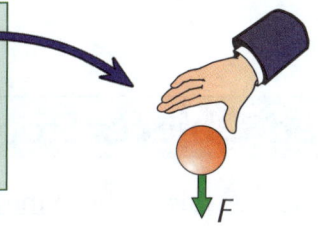

Example 3 — A car slowing down
Friction acts between the car's brakes and its wheels.
Energy is transferred mechanically from the wheels' kinetic energy stores to the thermal energy store of the surroundings.

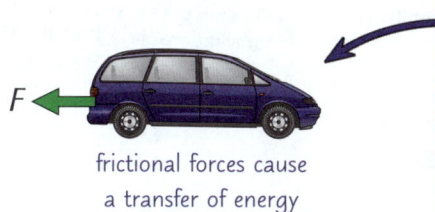

frictional forces cause a transfer of energy

Example 4 — A car hitting a wall
When the car and the wall touch, there is a normal contact force (p.347) on both of them.
Energy is transferred mechanically from the car's kinetic energy store to lots of other energy stores.
Some energy is transferred to the elastic potential and thermal energy stores of the wall and the car.
Some energy might also be transferred away by sound waves.

Energy is transferred between the different stores of objects...

Energy stores pop up everywhere in physics. You need to be able to describe how energy is transferred, and which stores it gets transferred between, for any scenario.

Q1   Describe the energy transfers that occur when the wind causes a windmill to spin.   [3 marks]

Q1 Video Solution

Topic P1 — Energy

Kinetic and Potential Energy Stores

Kinetic, *gravitational potential* and *elastic potential* are three important energy stores. Here's how to **calculate** the amount of energy in each of them.

Movement Means Energy in an Object's Kinetic Energy Store

1) Energy is transferred to the kinetic energy store when an object speeds up.
2) Energy is transferred away from this store when an object slows down.
3) There's a slightly tricky formula for finding the energy in an object's kinetic energy store:

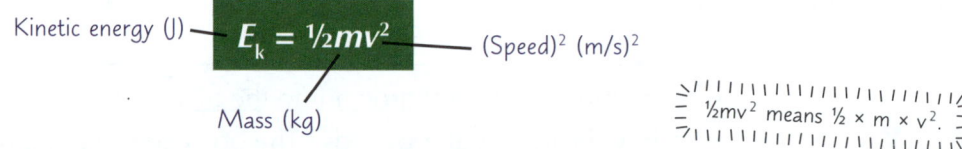

Kinetic energy (J) — $E_k = \frac{1}{2}mv^2$ — (Speed)² (m/s)²
Mass (kg)

½mv² means ½ × m × v².

EXAMPLE

A car of mass 2500 kg is travelling at 20 m/s. Calculate the energy in its kinetic energy store.

$E_k = \frac{1}{2} \times 2500 \times 20^2 = 500\,000$ J

Raised Objects Store Energy in Gravitational Potential Energy Stores

1) All objects raised above the ground gain energy in their gravitational potential energy (g.p.e.) store.
2) You can find the energy in an object's g.p.e. store using:

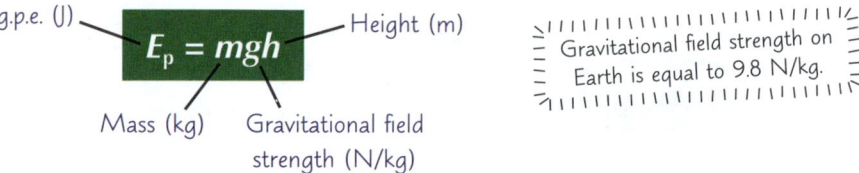

g.p.e. (J) — $E_p = mgh$ — Height (m)
Mass (kg)   Gravitational field strength (N/kg)

Gravitational field strength on Earth is equal to 9.8 N/kg.

Stretching can Transfer Energy to Elastic Potential Energy Stores

1) Stretching or squashing an object can transfer energy to its elastic potential energy store.
2) The energy in the elastic potential energy store of a stretched spring can be found using:

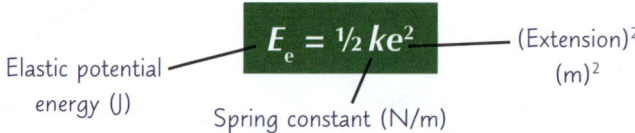

Elastic potential energy (J) — $E_e = \frac{1}{2}ke^2$ — (Extension)² (m)²
Spring constant (N/m)

3) This equation only works if the limit of proportionality has not been passed (p.352).

Greater height = more energy in gravitational potential stores...

Make sure you know what all the variables are in the equations above, and the units they're in.

Q1 A 2.0 kg object is dropped from a height of 10 m.
    Calculate the speed of the object after it has fallen 5.0 m, assuming there is no air resistance.
    Give your answer to 2 significant figures. *g* = 9.8 N/kg. [5 marks]

Q1 Video Solution

Conservation of Energy

*Repeat after me: **energy** is **NEVER** destroyed. Make sure you learn that fact, it's really important.*

Energy is Always Wasted in any Energy Transfer

1) When energy is transferred between stores, some energy is transferred to the store you want it in.
2) This energy is usefully transferred.
3) But in any energy transfer, some energy is always dissipated.
4) This means the energy is transferred to useless stores.
5) These useless energy stores are usually thermal energy stores.
6) This energy is often described as 'wasted' energy.

> When you use a mobile phone, energy is transferred from the chemical energy store of the battery.
> Some energy is usefully transferred.
> But some is dissipated to the thermal energy store of the phone.

You Need to Know the Conservation of Energy Principle

> Energy can be transferred usefully, stored or dissipated, but can never be created or destroyed.

1) This means that whenever a system changes, all the energy is simply moved between stores.
2) It never disappears.
3) This is true for every energy transfer.
4) Even when energy is dissipated (or wasted), it isn't gone.
5) It's just been transferred to an energy store that we didn't want.
6) Ways of reducing unwanted energy transfers are given on p.294.

You Can Calculate Energy Transfers Using Conservation of Energy

You can make calculations when energy is transferred between two stores. For example:

1) You saw on p.285 that a falling object transfers energy from its g.p.e. store to its kinetic energy store.
2) The conservation of energy principle (above) says that energy can't be destroyed.
3) So for a falling object when there's no air resistance:

> Energy lost from the g.p.e. store = Energy gained in the kinetic energy store

---

Energy is always conserved, but it can be wasted...

The energy transferred away from a falling object's gravitational potential energy store is only equal to the energy gained in its kinetic energy store if there's no air resistance. In reality, some energy would be wasted. For example, some energy would be transferred mechanically to the thermal energy stores of the air.

Specific Heat Capacity

Specific heat capacity is really just a sciencey way of saying **how hard** it is to **heat** something up...

Different Materials Have Different Specific Heat Capacities

1) Some materials need more energy to increase their temperature than others.
2) These materials also transfer more energy when they cool down again.
3) They can 'store' a lot of energy.
4) The amount of energy stored or released as a material changes temperature depends on the specific heat capacity of the material:

> Specific heat capacity is the amount of energy needed to raise the temperature of 1 kg of a material by 1 °C.

There's a Helpful Formula Involving Specific Heat Capacity

Below is the equation that links energy transferred to specific heat capacity (the Δ's just mean "change in").

Change in thermal energy (J) — Temperature change (°C)

$$\Delta E = mc\Delta\theta$$

Mass (kg) — Specific heat capacity (J/kg°C)

EXAMPLE

A hot block of metal cools from 55 °C to 25 °C. The block has a mass of 0.50 kg and is made from a material that has a specific heat capacity of 320 J/kg°C. Calculate the energy transferred from the block as it cooled.

1) First, calculate the change in the block's temperature. 55 °C − 25 °C = 30 °C
2) The numbers are in the correct units. So put them into the equation. $\Delta E = mc\Delta\theta$
   = 0.50 × 320 × 30
3) The unit for energy is joules (J). = 4800 J

If you're **not** working out the energy, you'll have to rearrange the equation, so a formula triangle will come in dead handy. To use them, cover up the thing you want to find and write down what's left showing. You write the bits of the formula in the triangle like this:

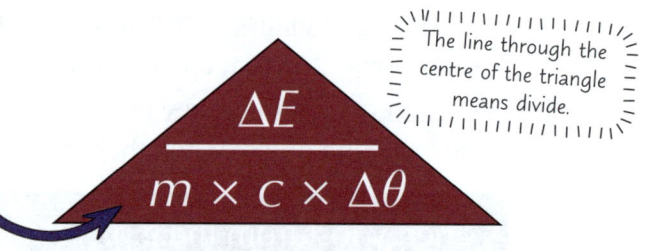

The line through the centre of the triangle means divide.

Some substances can store more energy than others...

Learn the definition of specific heat capacity and make sure you know how to use the formula above.

Q1 Find the final temperature of 5 kg of water, at an initial temperature of 5 °C, after 50 kJ of energy has been transferred to it. The specific heat capacity of water is 4200 J/kg°C. [3 marks]

Q1 Video Solution

Investigating Specific Heat Capacity

*Time for a **practical**. Woohoo I hear you shout! Maybe not, but you do have to know it I'm afraid.*

Investigate the Specific Heat Capacity of a Solid Block

1) Measure the mass of the block.

2) Wrap it in an insulating layer (e.g. thick newspaper) to reduce energy losses.

3) Set up the apparatus as shown.

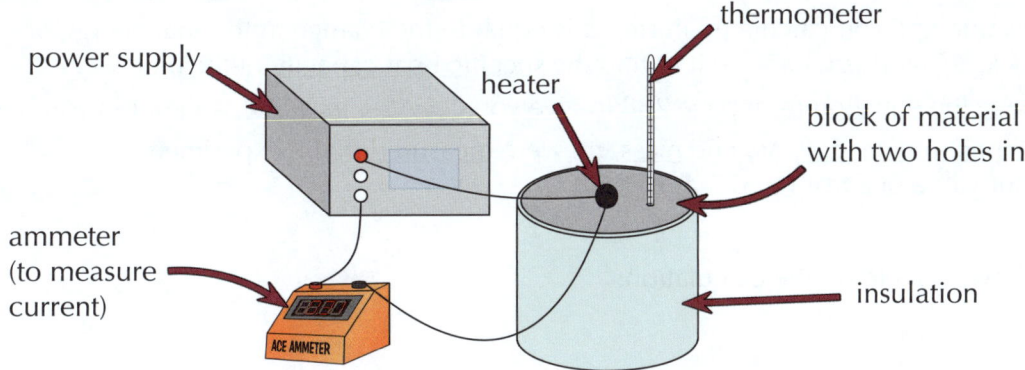

4) Measure the starting temperature of the block.

5) Turn on the power supply and start a stopwatch.

6) Record the potential difference, V, of the power supply and the current, I. They shouldn't change at all.

7) After 10 minutes, take a reading of the block's temperature.

8) Turn off the heater and work out the temperature change.

9) Calculate the power of the heater using $P = VI$ (p.322).

10) You can use this to calculate the specific heat capacity of the material the block is made from (see the next page).

Think about how you could improve your experiments...

If the hole in your material is bigger than your thermometer, you could put a small amount of water in the hole with the thermometer. This helps the thermometer to measure the temperature of the block more accurately, as water is a better thermal conductor than air.

Topic P1 — Energy

PRACTICAL: Investigating Specific Heat Capacity

You're not quite done yet — now that the fiddly bit is over, it's time to whip your calculator out to find the **specific heat capacity** of the **material** from your results.

Calculating the Specific Heat Capacity

To calculate the specific heat capacity from your results from the previous page, you need ideas about work done and energy transferred:

1) When you turn on the power, the current in the circuit does work on the heater.
2) Energy is transferred electrically from the power supply to the heater's thermal energy store.
3) The energy transferred to the heater is given by $E = Pt$ (p.293). (P is the power of the heater and t is how long the heater is on for.)
4) This energy is then transferred to the material's thermal energy store by heating.
5) So the value of E you calculated in step 3 is equal to the change in thermal energy of the block, ΔE, and you can use it to find the specific heat capacity of the block, c.
6) Rearrange the equation from page 288 to give you $c = \Delta E \div (m \times \Delta\theta)$, and put in your results.
7) The temperature change, $\Delta\theta$, and mass, m, were measured in the experiment. Use your value of E from step 3 as ΔE.

This example shows how to do the calculation:

EXAMPLE

A 1.0 kg block of material is heated using a 10 V power supply.
The starting temperature of the block is 20 °C.
The current through the heater is recorded as 10 A.
After 60 seconds, the final temperature of the block is 26 °C.
Calculate the specific heat capacity of the material of the block.

1) Calculate the power of the heater. $P = V \times I = 10 \times 10 = 100$ W
2) Calculate the energy transferred. $E = P \times t = 100 \times 60 = 6000$ J
3) Find the change in temperature. $\Delta\theta = 26 - 20 = 6$ °C
4) Calculate the specific heat capacity. $c = \Delta E \div (m \times \Delta\theta) = 6000 \div (1.0 \times 6) = $ **1000 J/kg °C**

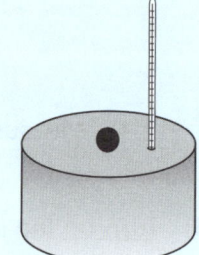

You Can Find the Specific Heat Capacities of Liquids Too

1) You can repeat the experiment shown on the previous page with different materials to see how their specific heat capacities compare.
2) For a liquid, place the heater and thermometer into an insulated beaker with a known mass of the liquid.
3) Then carry out the rest of the experiment in exactly the same way as above.

Insulation reduces the energy transferred to the surroundings...

If you repeat the experiment with a liquid, don't let it start boiling as this will affect your results.

Q1   A student uses an 80 W heater to heat a 2 kg block of metal for 200 s. The temperature of the block increases by 20 °C. Calculate the specific heat capacity of the metal. Assume that all the energy from the heater is transferred to the metal block.   [5 marks]

Q1 Video Solution

Warm-Up & Exam Questions

These questions give you chance to use your knowledge about energy transfers and specific heat capacity.

Warm-Up Questions

1) Give two ways in which energy can be transferred.
2) Describe how energy is transferred between stores as someone throws a ball upwards.
3) True or false? The energy in an object's gravitational potential energy store is proportional to its mass.
4) Which has more energy in its kinetic energy store: a person walking at 3 miles per hour, or a lorry travelling at 60 miles per hour?
5) State the principle of the conservation of energy.

Exam Questions

1 An electric heater is placed in a hole in a metal block and switched on. *(Grade 1-3)*
Energy is transferred from the thermal energy store of the heater.
Where is the energy transferred to? Tick **one** box.

☐ The thermal energy store of the metal block

☐ The chemical energy store of the heater

☐ The kinetic energy store of the metal block

☐ The kinetic energy store of the heater

[1 mark]

2 This question is about energy transfers. *(Grade 3-4)*

Draw one line from each scenario to the energy store that energy is transferred away from.

A skydiver falling from an aeroplane.	elastic potential
A substance undergoing a nuclear reaction.	gravitational potential
A piece of burning coal.	chemical
	nuclear

[3 marks]

Topic P1 — Energy

Exam Questions

3 Figure 1 shows a toy car that is launched by pulling it backwards onto a spring. The spring compresses, and then the car is released.

Figure 1

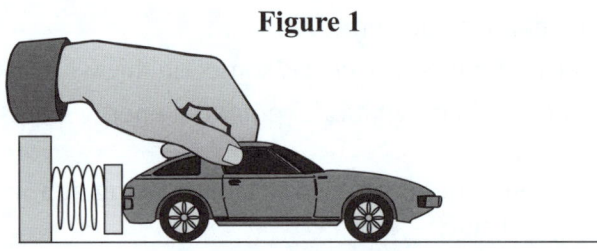

3.1 Name the store that energy is transferred from as the car is released.

[1 mark]

Immediately after the launch, the car moves away at a speed of 0.9 m/s.
The car has a mass of 0.20 kg.

3.2 Write down the equation that links the energy in an object's kinetic energy store, mass and speed.

[1 mark]

3.3 Calculate the energy in the kinetic energy store of the car immediately after it is released.

[2 marks]

4 A motor lifts a load of mass 20 kg.
The load gains 137.2 J of energy in its gravitational potential energy store.

4.1 Write down the equation that links the energy in an object's gravitational potential energy store, mass, gravitational field strength and height.

[1 mark]

4.2 Calculate the height through which the motor lifts the load.
Assume the gravitational field strength = 9.8 N/kg

[3 marks]

4.3 The motor releases the load and the load falls.
Ignoring air resistance, describe the changes in the way energy is stored that take place as the load falls.

[2 marks]

4.4 Describe how your answer to **4.3** would differ if air resistance was not ignored.

[1 mark]

5 36 000 J of energy is transferred to change the temperature of a 0.5 kg concrete block by 80 °C.

Calculate the specific heat capacity of the concrete block in J/kg °C.
Use the correct equation from the Physics Equation Sheet on page 538.

[3 marks]

Power

*The **more powerful** a device is, the **more energy** it will transfer in a certain amount of **time**.*

Power is the 'Rate of Doing Work'

1) Power is the rate of energy transfer.
2) You can also say it's the rate of doing work.
3) This just means that power is how fast energy is transferred or how fast work is done.
4) Power is measured in watts.
5) One watt = 1 joule of energy transferred per second.
6) You can calculate power using these equations:

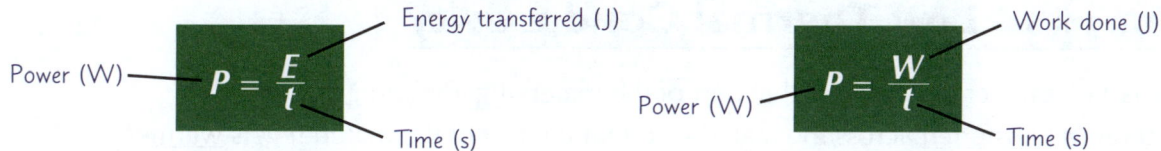

Power (W) — $P = \dfrac{E}{t}$ — Energy transferred (J), Time (s)

Power (W) — $P = \dfrac{W}{t}$ — Work done (J), Time (s)

> Take two cars that are the same in every way apart from the power of their engines. Both cars race the same distance along a straight race track to a finish line. The car with the more powerful engine will reach the finish line faster than the other car. This is because it will transfer the same amount of energy but over a shorter time.

EXAMPLE

a) It takes 8000 J of work to lift a stuntman to the top of a building.
A motor takes 50 s to make the lift.
Calculate the power of the motor.

1) The numbers are in the correct units.
2) Put the numbers into the equation for power in terms of work.

$P = W \div t$
$= 8000 \div 50$
$= 160$ W

b) A second motor has a power of 200 W.
It lifts the stuntman for 30 s.
Calculate the energy transferred by the motor.

1) Rearrange the power equation for energy transferred.
2) Put the numbers in.
3) Remember energy is in joules.

$P = E \div t$ so $E = P \times t$
$= 200 \times 30$
$= 6000$ J

A large power doesn't always mean a large force...

A powerful device is not necessarily one which can exert a strong force (although it usually ends up that way). A powerful device is one which transfers a lot of energy in a short space of time.

Q1 A motor transfers 4.8 kJ of energy in 2 minutes. Calculate its power output. [3 marks]

Q1 Video Solution

Topic P1 — Energy

Reducing Unwanted Energy Transfers

There are a few ways you can **reduce** the amount of energy running off to a **completely useless** store.

Lubrication Reduces Frictional Forces

1) Friction acts between all objects that rub together.
2) This causes some energy in the system to be dissipated.
3) Lubricants can be used to reduce the friction between the objects.
4) For example, oil in car engines reduces friction between all of the moving parts.
5) This reduces the amount of dissipated energy.

Insulation Has a Low Thermal Conductivity

1) When part of a material is heated, that part of the material gains energy.
2) This energy is transferred across the material so that the rest of the material gets warmer.
3) For example, if you heated one end of a metal rod, the other end would eventually get warmer. This is known as conduction.
4) Thermal conductivity is a measure of how quickly energy is transferred by conduction through a material.
5) Materials with a high thermal conductivity transfer lots of energy in a short time.
6) Materials with a low thermal conductivity are called thermal insulators.
7) Thermal insulators can reduce unwanted transfers by heating, e.g. in the home.

Insulation is Important for Keeping Buildings Warm

You can keep your home cosy and warm by reducing the rate of cooling. How quickly a building cools depends on:

1) How thick its walls are. The thicker the walls are, the slower a building will cool.
2) The thermal conductivity of its walls. Building walls from a material with a low thermal conductivity reduces the rate of cooling.
3) How much thermal insulation there is, e.g. loft insulation reduces energy losses through the roof.

Energy Transfers Involve Some Wasted Energy

1) You saw on page 287 that some energy is always wasted when energy is transferred.
2) The less energy that is wasted, the more efficient the energy transfer is.
3) The efficiency of an energy transfer is a measure of the amount of energy that ends up in useful energy stores.
4) But as some energy is always wasted, nothing is 100% efficient.

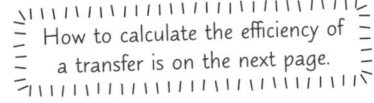
How to calculate the efficiency of a transfer is on the next page.

Having a well-insulated home can reduce your heating bills...

When people talk of energy loss, it's not that the energy has disappeared. It still exists (see page 287), just not in the store we'd like it to be. For example, in a car, you want the energy to transfer to the kinetic energy store of the wheels, and not to the thermal energy stores of the moving components.

Topic P1 — Energy

Efficiency

Devices have **energy transferred** to them, but only transfer **some** of that energy to **useful energy stores**. Wouldn't it be great if we could tell **how much** it **usefully transfers**? That's where **efficiency** comes in.

There Are **Two Efficiency Equations**

1) The efficiency for any energy transfer can be <u>worked out</u> using this equation:

$$\text{Efficiency} = \frac{\text{Useful output energy transfer}}{\text{Total input energy transfer}}$$

This gives efficiency as a decimal, but you can turn it into a percentage — see below.

EXAMPLE

36 000 J of energy is transferred to a television. It transfers 28 800 J of this energy usefully. Calculate the efficiency of the television. Give your answer as a percentage.

1) Put the numbers you're given <u>into the equation</u>.

    efficiency = useful output energy transfer ÷ total input energy transfer
    = 28 800 ÷ 36 000
    = 0.8

2) To change a <u>decimal</u> to a <u>percentage</u>, <u>multiply</u> your answer <u>by 100</u>.

    0.8 × 100 = 80, so efficiency = **80%**

2) You might not know the <u>energy</u> input and output of a device.
3) But you can use its <u>power input</u> and <u>output</u> to calculate its <u>efficiency</u>:

$$\text{Efficiency} = \frac{\text{Useful power output}}{\text{Total power input}}$$

EXAMPLE

A blender is 70% efficient. It has a total input power of 600 W. Calculate the useful power output.

1) Change the <u>efficiency</u> from a <u>percentage</u> to a <u>decimal</u>. To do this, <u>divide</u> the percentage <u>by 100</u>.

    efficiency = 70% ÷ 100 = 0.7

2) <u>Rearrange</u> the equation for <u>useful power output</u>.

3) <u>Stick in</u> the numbers and find the useful power output.

    useful power output = efficiency × total power input
    useful power output = 0.7 × 600 = **420 W**

The less energy a device wastes, the higher its efficiency...

Make sure you can <u>use</u> and <u>rearrange</u> the equations for efficiency, then have a go at these questions.

Q1 A motor in a remote-controlled car transfers 300 J of energy into the car's energy stores. 225 J are transferred to the car's kinetic energy stores. Calculate the efficiency of the motor. **[2 marks]**

Q2 A machine has a useful power output of 900 W and a total power input of 1200 W. In a given time, 72 kJ of energy is transferred to the machine. Calculate the amount of energy usefully transferred by the machine in this time. **[4 marks]**

Topic P1 — Energy

Warm-Up & Exam Questions

Don't let your energy dissipate. These questions will let you see how efficient your revision has been.

Warm-Up Questions

1) What is power?
2) What is 1 watt equivalent to: 1 J, 1 Js or 1 J/s?
3) What should you consider when choosing a material to build house walls from to minimise the rate of cooling?
4) Which has a greater rate of energy transfer: a material with a high thermal conductivity or a material with a low thermal conductivity?
5) Why is the efficiency of an appliance always less than 100%?

Exam Questions

1 The motor of an electric scooter moves the scooter along a flat, horizontal course in 20 seconds. During this time the motor does 1 kJ of work. *(Grade 3-4)*

1.1 Write down the equation that links power, work done and time.

[1 mark]

1.2 Calculate the power of the motor.

[2 marks]

1.3 Suggest **one** way in which unwanted energy transfers in the scooter could be reduced.

[1 mark]

1.4 The scooter's motor is replaced with a more powerful, but otherwise identical, motor. It moves along the same course. How will the performance of the scooter differ from before? Tick **one** box.

☐ It will be faster. ☐ It will be slower. ☐ It will use less energy. ☐ It will use more energy.

[1 mark]

2 An electric fan transfers 7250 J of energy. 2 kJ of this is wasted energy. *(Grade 4-5)*

2.1 Suggest **one** way in which energy is wasted by the fan.

[1 mark]

The useful output energy of the fan is given by the equation below:

useful output energy transfer = total input energy transfer − wasted output energy transfer

2.2 Calculate the energy that is usefully transferred by the fan.

[2 marks]

2.3 The efficiency of a device is given by the equation below:

$$\text{efficiency} = \frac{\text{useful output energy transfer}}{\text{total input energy transfer}}$$

Calculate the efficiency of the fan. Give your answer to **two** significant figures.

[2 marks]

Topic P1 — Energy

Energy Resources and their Uses

There are lots of **energy resources** available on Earth. They are either **renewable** or **non-renewable** resources.

Non-Renewable Energy Resources Will Run Out One Day

1) Non-renewable energy resources are fossil fuels and nuclear fuel.
2) The three main fossil fuels are coal, oil and (natural) gas.
3) We can't replace non-renewable energy resources as quickly as we're using them.

Renewable Energy Resources Will Never Run Out

1) Renewable energy resources can be replenished (replaced) as quickly as they are being used.
2) The renewable energy resources you need to know are:

   1) The Sun (Solar)   3) Water waves   5) Biofuels   7) Geothermal
   2) Wind   4) Hydro-electricity 6) Tides

Energy Resources can be Used for Transport...

1) Transport uses both renewable and non-renewable energy resources. For example:

NON-RENEWABLE ENERGY RESOURCES	RENEWABLE ENERGY RESOURCES
Petrol or diesel is used in most vehicles. They're both created from oil. Coal is used in steam trains to boil water to produce steam.	Vehicles can run on pure biofuels (p.298) or a mix of a biofuel and petrol or diesel.

2) Electricity can also be used for transport — e.g. electric cars and some trains.
3) The electricity can be generated using renewable or non-renewable energy resources (p.298-302).

...And for Heating

Energy resources are also needed for heating things, like your home.

NON-RENEWABLE ENERGY RESOURCES	RENEWABLE ENERGY RESOURCES
Natural gas is burnt to heat water in a boiler. This hot water is then pumped into radiators. Gas fires burn natural gas to heat rooms. Coal is burnt in open fireplaces. Electric heaters use electricity which can be generated from non-renewable energy resources.	Biofuel boilers work in the same way as gas boilers. A geothermal heat pump uses geothermal energy resources (p.300) to heat buildings. Solar water heaters use the Sun to heat water which is then pumped into radiators in the building. Electric heaters can use electricity generated from renewable resources.

Fossil fuels are produced over millions of years — quite a wait...

Make sure you know the difference between renewable and non-renewable energy resources.

Biofuels

Biofuels are a promising renewable energy resource, as they can be **used** and **stored** in much the same way as **fossil fuels**. However, they need a lot of **space** to grow...

You Need to Be Able to Compare Resources

1) You're about to learn all about the main energy resources.
2) You need to be able to:

   - describe their effects on the environment (e.g. pollution).
   - compare their reliability (whether they can be trusted to provide energy when we need it).

3) So make sure you're paying attention.

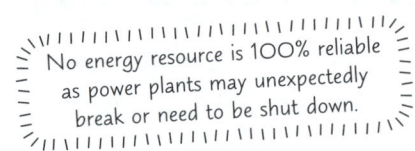
No energy resource is 100% reliable as power plants may unexpectedly break or need to be shut down.

Biofuels are Made from Plants and Waste

1) Biofuels are fuels created from plant products or animal dung.
2) They can be burnt to produce electricity or used to run cars in the same way as fossil fuels.

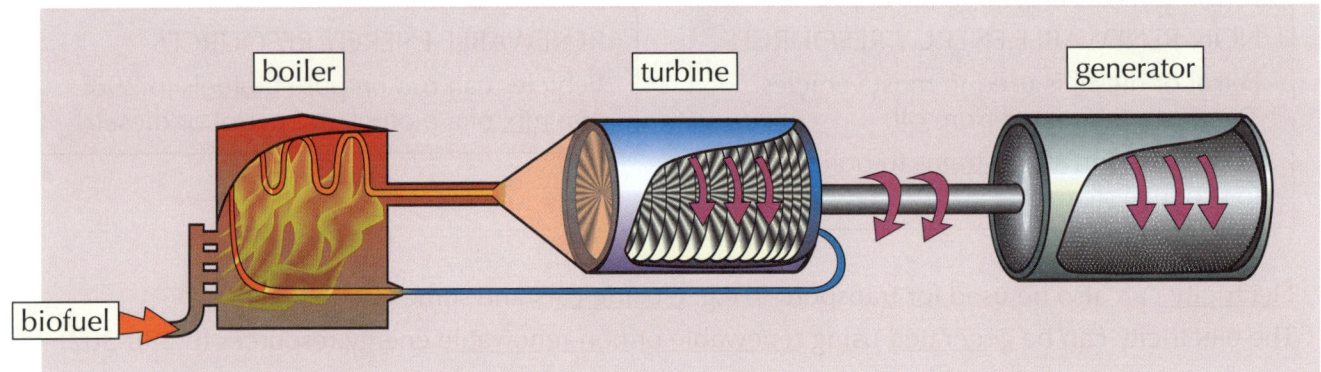

3) They produce carbon dioxide when they're burnt.
4) But the plants used to make biofuels will have absorbed a lot of carbon dioxide while they were growing.
5) In some places, large areas of forest have been cleared to make room to grow biofuels.
6) This leads to lots of animals losing their natural habitats.
7) Crops can be grown throughout the year.
8) Extra biofuels can be constantly produced and stored for when they are needed.
9) So biofuels are fairly reliable.

In theory, biofuels are carbon neutral...

You've probably heard that releasing carbon dioxide (CO_2) into the atmosphere is bad as it leads to global warming. When you burn biofuels, you do release CO_2 into the atmosphere. But, when the plants grew, they absorbed this CO_2 from the atmosphere for photosynthesis. This is why biofuels are often said to be 'carbon neutral'.

Topic P1 — Energy

Wind Power and Solar Power

*Coming up are two of the most visible renewable energy resources — **wind** and **solar power**.*

Wind Power — Lots of Wind Turbines

1) Wind turbines are usually put up in open spaces.
2) When the wind turns the blades, electricity is produced.
3) They produce no pollution once they're built.
4) And they do no permanent (lasting) damage to the landscape. If you remove the turbines, the area goes back to normal.
5) However, they're not as reliable as other energy resources.
6) They don't produce electricity when the wind stops.
7) Turbines are also stopped if the wind is too strong. This stops them getting damaged.
8) It's also impossible to increase supply when there's extra demand (p.323) for electricity.

Solar Cells — Expensive but No Environmental Damage

Solar cells generate electricity directly from sunlight.

1) They create no pollution once they're built.
2) But quite a lot of energy is used to build them.
3) Solar power only generates electricity during the day.
4) In sunny countries solar power is a very reliable source of energy.
5) It's still fairly reliable in cloudy countries like Britain.
6) Like wind, you can't increase the power output when there is extra demand.

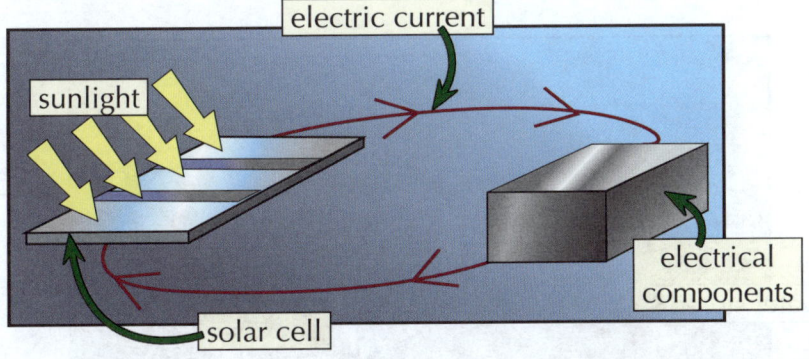

People love the idea of wind power — just not in their back yard...

It's easy to think that renewables are the answer to the world's energy problems. However, they have their downsides, and we definitely couldn't rely on them totally at present. Make sure you know the pros and cons for wind and solar power because there are more renewables coming up on the next page.

Topic P1 — Energy

Geothermal and Hydro-electric Power

*Here are some more examples of **renewable energy resources** — **geothermal** and **hydro-electric**. These ones are a bit more **reliable** than wind and solar — read on to find out why.*

Geothermal Power uses Underground Thermal Energy Stores

1) Geothermal power uses energy from the thermal energy stores of hot rocks below the Earth's surface.
2) It can be used to generate electricity or to heat buildings.
3) Geothermal power is very reliable because the hot rocks are always hot.
4) Most geothermal power stations only have a small impact on the environment.

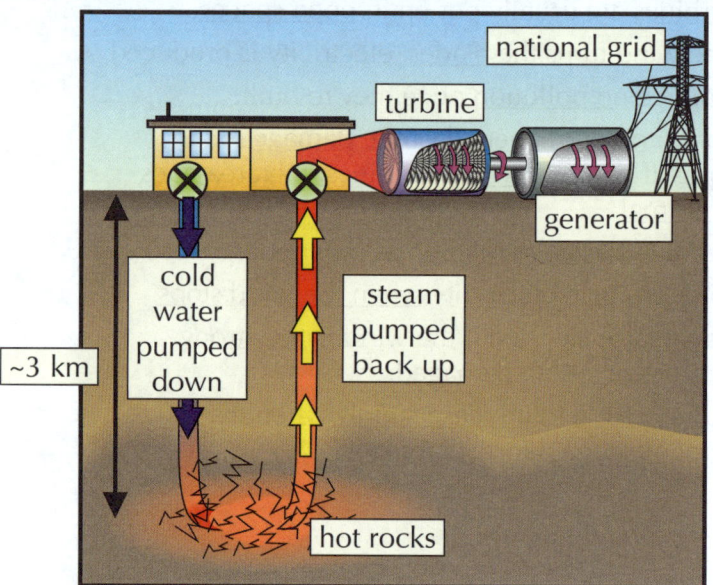

Hydro-electric Power Uses Falling Water

1) Hydro-electric power usually involves building a big dam across a valley.
2) The valley is usually flooded.
3) Water is allowed to flow out through turbines, which generates electricity.
4) There is no pollution when it's running.
5) But there is a big impact on the environment due to the flooding of the valley.
6) Plants rot and release greenhouse gases which lead to global warming (see p.302).
7) Animals and plants also lose their habitats (where they live).
8) There's no problem with reliability in countries that get rain regularly.
9) And it can respond straight away when there's extra demand (p.323) for electricity.

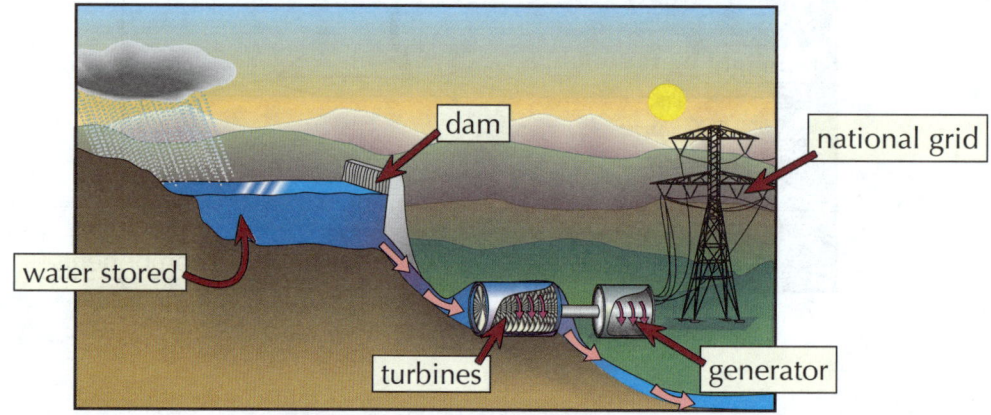

Falling water loses energy from its gravitational potential energy store...

You don't need to be able to describe how these power sources work in detail, but you should be able to comment on their reliability and describe the impact that they have on the environment.

Topic P1 — Energy

Wave Power and Tidal Barrages

*Good ol' **water**. Not only can we drink it, we can also use it to **generate electricity**. It's easy to get confused between **wave** and **tidal** power as they both involve the seaside — but don't. They are completely different.*

Wave Power — Lots of Little Wave-Powered Turbines

1) Turbines around the coast are turned by water waves and electricity is generated.
2) There is no pollution.
3) But they disturb the seabed and the habitats of animals.
4) They are fairly unreliable, as waves tend to die down when the wind drops.

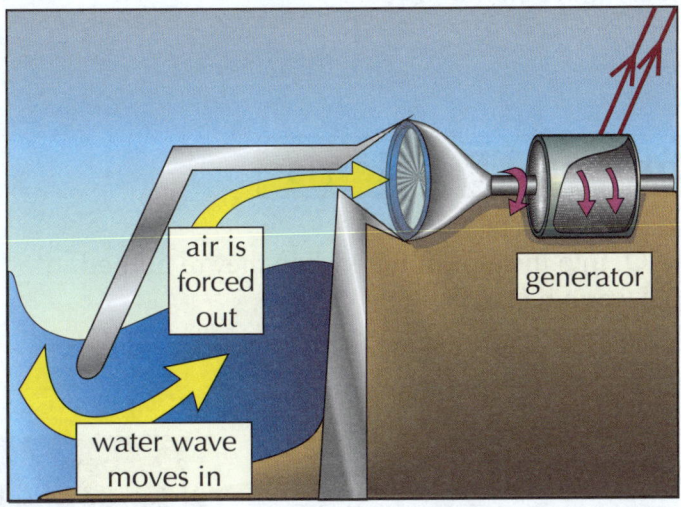

Tidal Barrages Use the Tides of the Sea

1) Tidal barrages are big dams (with turbines in them) built across rivers.
2) Water passing through the turbines generates electricity.
3) The amount of energy generated changes with the tides.
4) But tidal barrages are very reliable, as we can predict the tides (we know what they're going to do).
5) There is no pollution.
6) But they do change the habitat of the wildlife, e.g. birds and sea creatures.
7) And often fish are killed as they swim through the turbines.

Wave and tidal — power from the motion of the ocean...

The first large-scale tidal barrages started being built in the 1960s, so tidal power isn't a new thing. Wave power is still pretty experimental though. Make sure you can tell the two of them apart.

Non-Renewables

Renewable resources may sound like **great news** for the **environment**. But when it comes down to it, they **don't** currently meet all our needs — so we still need those nasty, polluting **non-renewables**.

Non-Renewables are Reliable...

1) Fossil fuels and nuclear energy are reliable.
2) There's enough fossil and nuclear fuels to meet current demand.
3) We always have some in stock so power plants can respond quickly to changes in demand.
4) However, these fuels are slowly running out. Some fossil fuels may run out within a hundred years.

...But Create Environmental Problems

1) Coal, oil and gas release CO_2 into the atmosphere when they're burned. All this CO_2 leads to global warming.
2) Burning coal and oil also releases sulfur dioxide, which causes acid rain.
3) Acid rain makes lakes and rivers acidic, which can kill animals and plants. It can also damage trees and soils.
4) Coal mining makes a mess of the landscape.
5) And it destroys the habitats of local animals and plants.
6) Oil spills cause big environmental problems and harm sea creatures.
7) Nuclear power is clean but the nuclear waste is very dangerous and difficult to get rid of.
8) Nuclear power also carries the risk of a big accident that could release a lot of radiation, like the Fukushima disaster in Japan.

Global warming is where greenhouse gases cause the Earth to warm up.

Radiation can be very dangerous to humans — see p.342-343 for more.

Currently we Still Need Non-Renewables

1) Our use of electricity increased a lot in the 1900s.
2) This was because the population and the number of things that used electricity increased.
3) But electricity use in the UK has been falling slowly since around the year 2000.
4) This is because we're trying harder to be energy efficient and save energy.
5) At the moment, we use non-renewables for most of our electricity, transport and heating.

We're just going to have to deal with non-renewables for now...

Non-renewables cause a lot of long-term problems, but they also have a few short-term advantages. Make sure you can balance up the advantages and disadvantages of fossil and nuclear fuels.

Topic P1 — Energy

Limitations on the Use of Renewables

*Although we still rely on **non-renewables** for **a lot** of our energy needs at the moment, the balance is **shifting**.*

People Want to use **More Renewable** Energy Resources

1) We now know that non-renewables are very bad for the environment and will run out one day (p.302).
2) This makes many people want to use renewable energy resources as they are better for the environment.
3) Many people also think it's better to move to renewables before non-renewables run out.
4) Pressure from other countries and the public has meant that governments have begun to introduce targets for using renewable energy resources.
5) This puts pressure on energy providers to build new renewable power plants. If they don't, they may lose business and money in the future.
6) Car companies have also had to change to become more environmentally-friendly.
7) Cars that can run on electricity are already on the market. The electricity can be generated using renewable energy resources.

The Use of **Renewables** is **Limited** by Lots of **Factors**

1) There's a lot of scientific evidence supporting renewables.
2) But scientists can only give advice. They don't have the power to make people, companies or governments change their ways (see p.4).
3) Moving to renewables can be limited by money.

- Building new renewable power plants costs money.
- Some renewable resources are less reliable than other resources, so a mixture of different resources would need to be used.
- This costs even more money.
- Cars that run on electricity are more expensive than petrol cars.

4) Moving to renewables can also be affected by politics, people and ethics (if something is right or wrong).

- The cost of switching to renewable power will have to be paid through energy bills or taxes.
- Governments often don't want to suggest raising taxes as this may make them unpopular.
- Some people don't want to or can't afford to pay. There are arguments about whether it's ethical (right or wrong) to make them pay.
- Many people also don't want to live near to a power plant (like a wind farm or hydro-electric dam).
- And some think it's not ethical to make people put up with new power plants built near to them.

Going green is on-trend this season...

So with some people wanting to help the environment, others not wanting to be inconvenienced, and greener alternatives being expensive to set up, the energy resources we use are changing. Just not particularly quickly.

Warm-Up & Exam Questions

This is the last set of warm-up and exam questions on Topic 1. They're not *too* horrendous, I promise.

Warm-Up Questions

1) Name two fossil fuels.
2) Explain why solar power is usually less reliable than fossil fuels.
3) Describe one way that renewable energy resources can be used to power vehicles.
4) Give two ways in which using coal as an energy resource causes environmental problems.
5) True or false? Geothermal power is a reliable energy resource.

Exam Questions

1 Which of the following energy resources is a renewable energy resource? Tick **one** box. *(Grade 1-3)*

☐ coal ☐ nuclear ☐ wind ☐ oil

[1 mark]

2 Which of the following is a possible disadvantage of using fossil fuels to generate electricity? Tick **one** box. *(Grade 3-4)*

☐ They cannot generate electricity at night.

☐ They don't release very much energy.

☐ The waste produced is dangerous to dispose of.

☐ They can produce sulfur dioxide and cause acid rain.

[1 mark]

3 Natural gas is a non-renewable energy resource that can be used to generate electricity. *(Grade 4-5)*

3.1 Name **one** other use of natural gas.

[1 mark]

3.2 Give **two** reasons why we still use non-renewable energy resources.

[2 marks]

4 The government of a country needs to generate more electricity to support a growing population. The government has considered using wind, hydro-electric power and tides to generate electricity. *(Grade 4-5)*

4.1* Compare the environmental impact and reliability of wind power and hydro-electric power.

[6 marks]

4.2 The government choose to generate electricity using tidal barrages. Give **one** environmental advantage of generating electricity using tidal barrages.

[1 mark]

Topic P1 — Energy

Topic P2 — Electricity

Current and Circuit Symbols

Circuit diagrams are a useful way of showing how the **components** of a circuit are **arranged**. You need to be able to **recognise** the **symbols** used to represent different components. Read on for the complete run down...

Learn these Circuit Diagram Symbols

1) You need to be able to understand circuit diagrams and draw them using the correct symbols.

2) These are the symbols you need to know:

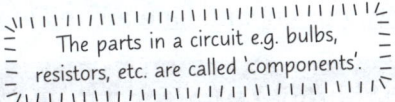
The parts in a circuit e.g. bulbs, resistors, etc. are called 'components'.

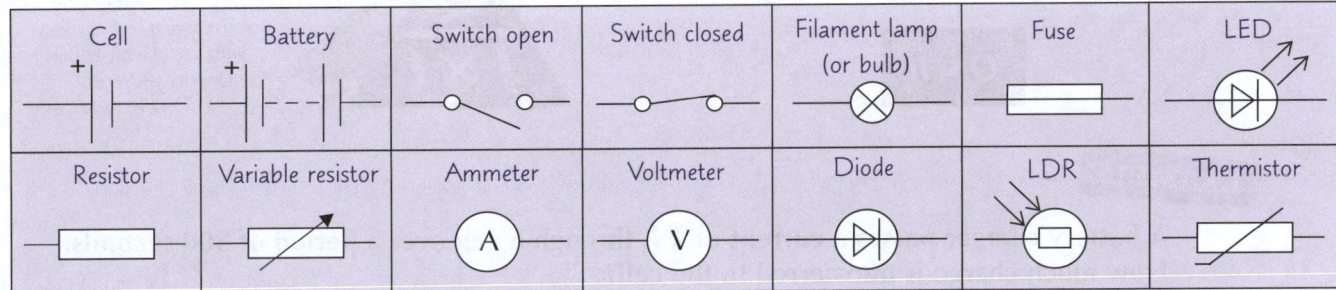

3) Follow these rules to draw a circuit diagram:

- Make sure all the wires in your circuit are straight lines.
- Make sure that the circuit is closed. This means you can follow a wire from one end of the cell or battery, through any components, to the other end of the cell or battery.

Potential Difference Causes Charge to Flow Round a Circuit

1) Electric current is a flow of electrical charge.
   - Current is measured in amperes, A.
   - Charge is measured in coulombs, C.
   - In a single, closed loop the current is the same everywhere in the circuit (see p.315).
   - The size of the current tells you how fast the charge is flowing. This is known as the rate of flow of charge.

2) The potential difference is the 'driving force' that pushes charge around the circuit.
   - Electrical charge will only flow round a complete (closed) circuit if something is providing a potential difference, e.g. a battery.
   - Potential difference is measured in volts (V).

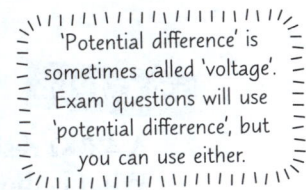
'Potential difference' is sometimes called 'voltage'. Exam questions will use 'potential difference', but you can use either.

3) Resistance is anything that slows down the flow of charge.
   - Resistance is measured in ohms (Ω).

Charge won't flow without a push...

Left to its own devices, charge won't move anywhere. To get a current flowing, charge needs a 'push'. This push comes in the form of a potential difference, supplied by a source like a cell.

Charge and Resistance Calculations

*Prepare yourself to meet two of the most **important equations** in electricity. They're all about **charge**, **current** and **potential difference**... Now if that doesn't tempt you to read this page, I don't know what will.*

Total Charge Through a Circuit Depends on Current and Time

Charge flow, current and time are related by this handy equation:

Charge flow (C) = Current (A) × Time (s)

Q = It

Use this formula triangle to rearrange the equation. Just cover up the thing you're trying to find, and what's left visible is the formula you're after.

EXAMPLE

A battery charger passes a current of 2 A through a cell over a period of 300 seconds. How much charge is transferred to the cell?

Just substitute the values into the equation above and calculate the charge.

$Q = It = 2 × 300$
$= 600$ C

Remember charge is measured in coulombs.

There's a Formula Linking Potential Difference and Current

1) The current flowing through a component depends on the potential difference across it and the resistance of the component.

Resistance measures how much the current is slowed down.

The greater the resistance across a component, the smaller the current that flows (for a given potential difference across the component).

2) The formula linking potential difference (pd) and current is:

Potential Difference (V) = Current (A) × Resistance (Ω)

V = IR

EXAMPLE

A 4.0 Ω resistor in a circuit has a potential difference of 6.0 V across it. What is the current through the resistor?

1) Cover the *I* in the formula triangle to find that $I = V ÷ R$. $I = V ÷ R$
2) Substitute in the values you have, and work out the current. $I = 6.0 ÷ 4.0 = 1.5$ A

Increasing resistance reduces current...

There's more coming up on resistance, so make sure you're happy with this page before moving on.

Q1 An appliance is connected to a 230 V source.
Calculate the resistance of the appliance if a current of 5.0 A is flowing through it. [3 marks]

Q1 Video Solution

Topic P2 — Electricity

Ohmic Conductors

*A circuit **component** can be either **ohmic** or **non-ohmic**. It all depends on whether or not its **resistance** stays the same when the **current** is increased. Read on to find out all the details.*

Ohmic Conductors Have a Constant Resistance

1) The resistance of an ohmic conductor doesn't change with current.
2) Ohmic conductors only have a fixed resistance if their temperature doesn't change.
3) Wires and resistors are examples of ohmic conductors.
4) $V = IR$ (see p.306), so if resistance is constant, increasing potential difference will lead to an increase in current.
5) So,

> For an ohmic conductor at a fixed temperature, the current flowing through it is directly proportional to the potential difference across it.

6) This means that if you multiply the potential difference by a certain amount, the current will be multiplied by the same amount. For example, if the potential difference doubles, the current doubles too.

Some Components Have a Changing Resistance

The resistance of some components does change with current. For example, a filament lamp or a diode.

Filament Lamps

1) Filament lamps contain a wire (the filament), which is designed to heat up and 'glow' as the current increases.
2) So as the current increases, the temperature of the filament increases.
3) Resistance increases with temperature, so the resistance increases with current.

A higher current makes a filament glow brighter. So a higher pd means a brighter lamp.

Diodes

1) For diodes, the resistance depends on the direction of the current.
2) A diode will let current flow in one direction. It has a very high resistance in the opposite direction, which makes it hard for a current to flow that way.

Resistance changes when temperature changes...

Remember that ohmic conductors will only have a constant resistance at a constant temperature. In general, resistance increases with temperature (though there are some exceptions, like thermistors — see p.312). So if the temperature is changing, the resistance of your component will be changing too.

Investigating Resistance

Resistance depends on various things. Here's an experiment you can do to investigate one of them — how the resistance varies with the length of the conductor.

You Can Investigate the Factors Affecting Resistance

The resistance of a circuit can depend on a number of factors, like whether components are in series or parallel, p.314 and p.316, or the length of wire used in the circuit. You can investigate the effect of wire length using the circuit below.

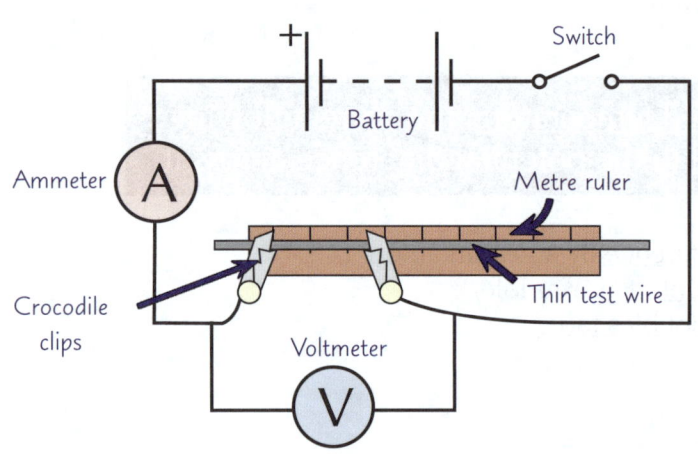

The Ammeter

1) Measures the current (in amps) flowing through the test wire.
2) The ammeter must always be placed in series with whatever you're investigating.

The Voltmeter

1) Measures the potential difference (or pd) across the test wire (in volts).
2) The voltmeter must always be placed in parallel around whatever you're investigating (p.316) — NOT around any other bit of the circuit, e.g. the battery.

Measure Potential Difference and Current for Different Lengths

1) Attach a crocodile clip to the wire level with 0 cm on the ruler.
2) Attach the second crocodile clip to the wire a short distance from the first clip.
3) Write down the length of the wire between the clips.
4) Close the switch, then record the current through the wire and the pd across it.
5) Use $R = V \div I$ (from the equation $V = IR$ on p.306) to calculate the resistance of the wire.
6) Open the switch and move the second crocodile clip along the wire.
7) Repeat steps 3 to 6 for a range of wire lengths.

Plot a Graph of your Results

1) Plot a graph of resistance against wire length.
2) Draw a line of best fit through your points.
3) Your graph should be a straight line through the origin (where length and resistance are both zero).
4) This means resistance is directly proportional to length — the longer the wire, the greater the resistance.

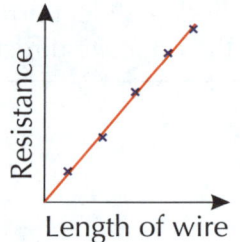

 **Be careful with the temperature of the wire...**

If a large current flows through a wire, it can cause it to heat up (which will increase the resistance). So use a low pd to stop it getting too hot and turn off the circuit between readings to let it cool.

I-V Characteristics

I-V characteristics tell you how the **current** through a component changes as the **potential difference** across it is increased. You need to know the I-V characteristics of **ohmic conductors**, **filament lamps** and **diodes**.

I-V Characteristics Show How Current Changes With Pd

1) An 'I-V characteristic' graph shows how the current (I) flowing through a component changes as the potential difference (V) across it changes.
2) Components with straight line I-V characteristics are called linear components (e.g. a fixed resistor).
3) Components with curved I-V characteristics are non-linear components (e.g. a filament lamp or a diode).
4) To find the resistance at any point on the I-V characteristic, first read off the values of I and V at that point. Then use $R = V \div I$ (from $V = IR$ on page 306).

Three Very Important I-V Characteristics

Make sure you know these graphs really well — you might be asked to sketch one in the exam.

Ohmic Conductor (e.g. resistor at a constant temperature)

1) Current is directly proportional to potential difference.
2) So the graph of current against potential difference is a straight line.

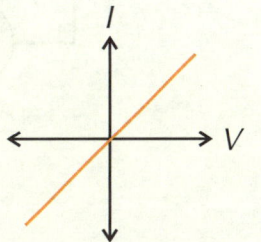

Filament Lamp

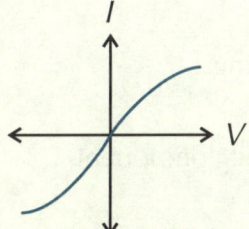

1) Temperature increases as current increases.
2) So resistance increases.
3) This makes it harder for current to flow.
4) So the graph gets less steep.

Diode

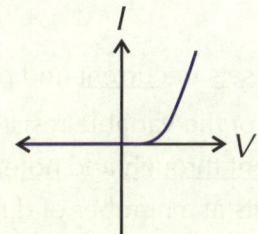

1) Current only flows in one direction.
2) The diode has very high resistance in the reverse direction.

You may be asked to interpret an I-V characteristic...

Make sure you take care when reading values off the graph. Pay close attention to the axes, and make sure you've converted all values to the correct units before you do any calculations.

Q1 Explain the shape of the filament lamp I-V characteristic above, for the quadrant where I and V are positive. [3 marks]

Q1 Video Solution

Topic P2 — Electricity

Investigating *I-V* Characteristics

On the previous page you met **I-V characteristics**. Now, here's all you need to know to set up a **practical** in order to collect data to draw some I-V characteristics of your very own.

You Can **Investigate** *I-V* Characteristics

You should do this experiment for different components, including a filament lamp, a diode and a resistor at a fixed temperature.

1) Set up the test circuit shown below.

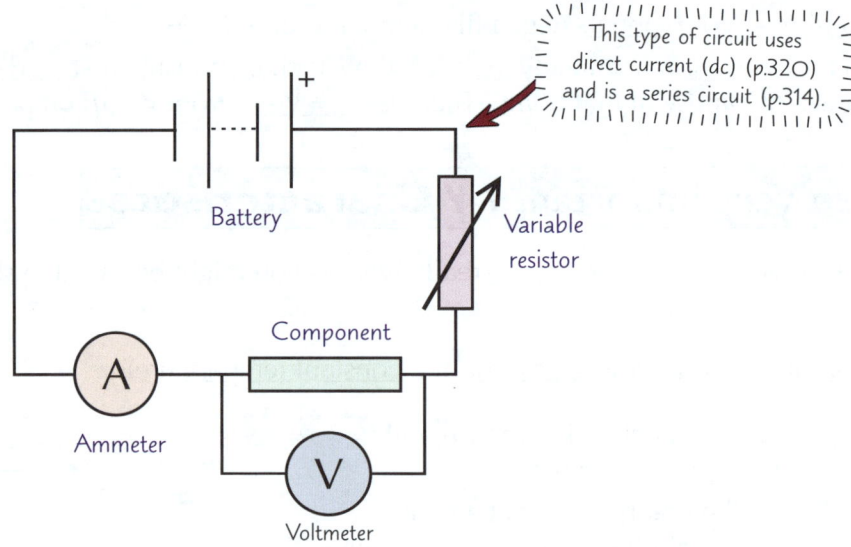

This type of circuit uses direct current (dc) (p.320) and is a series circuit (p.314).

2) The variable resistor is used to change the current in the circuit. This changes the potential difference across the component.

3) Now you need to get sets of current and potential difference readings:
   • Set the resistance of the variable resistor.
   • Measure the current through and potential difference across the component.
   • Take measurements at a number of different resistances.

4) Swap over the wires connected to the battery to reverse the direction of the current. The ammeter should now display negative readings.

5) Repeat step 3 to get results for negative values of current.

6) Plot a graph with current on the y-axis and potential difference on the x-axis.

Ammeters must be connected in series, voltmeters in parallel...

With any circuit, ammeters and voltmeters need to be connected correctly. Ammeters must be in series and voltmeters must be parallel to the component you're investigating (see p.308).

Topic P2 — Electricity

Warm-Up & Exam Questions

Phew — circuits aren't the easiest thing in the world, are they? Make sure you've understood the last few pages by trying these questions. If you get stuck, just go back and re-read the relevant page.

Warm-Up Questions

1) Draw the symbol for a light-emitting diode (LED).
2) What is an electric current?
3) What are the units of resistance?
4) Give an example of an ohmic conductor.
5) How should a voltmeter be connected in a circuit to measure the pd across a component?
6) What is an *I-V* characteristic?

Exam Questions

1   **Figure 1** shows a circuit diagram. The circuit's switch is closed for 2 minutes. While the switch is closed the ammeter reads 0.30 A.   *Grade 3-4*

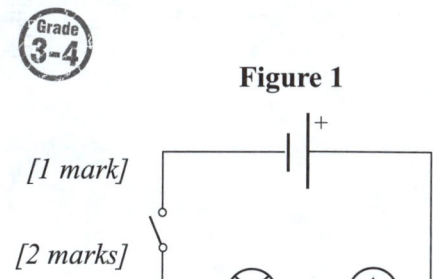

Figure 1

1.1 Write down the equation that links current, charge and time.

[1 mark]

1.2 Calculate the total charge that flows through the filament lamp.

[2 marks]

1.3 Use words from the box to complete the sentence.

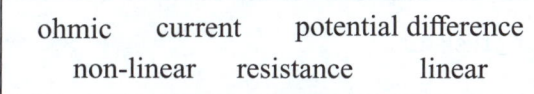

ohmic   current   potential difference
non-linear   resistance   linear

A filament lamp is a _____ component. As the temperature of the filament increases, the _____ of the lamp increases.

[2 marks]

2   A student wants to produce a graph of current against potential difference for component X. **Figure 2** shows an incomplete diagram of the circuit he is going to use.   *Grade 4-5*

Figure 2

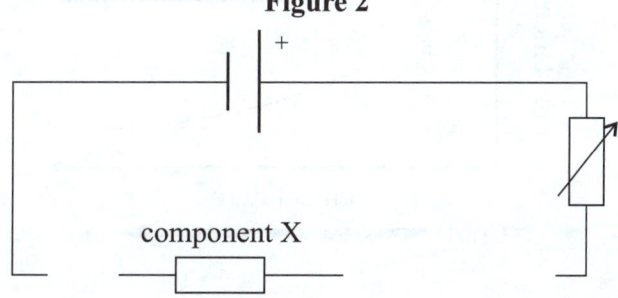

component X

2.1 Complete the circuit by adding an ammeter and a voltmeter.

[2 marks]

2.2 Predict the shape of the graph that the student will produce. Explain your answer.

[2 marks]

Circuit Devices

You might consider yourself a bit of an expert in **circuit components** — you're enlightened about bulbs, you're switched on to switches... Just make sure you know these ones as well — they're a bit trickier.

A Light-Dependent Resistor or "LDR"

1) The resistance of an LDR changes as the intensity of light changes.
2) In bright light, the resistance is low.
3) In darkness, the resistance is high.
4) LDRs have lots of uses including turning on lights when it gets dark.
5) This can be used in automatic night lights, or outdoor lighting.
6) They're also used in burglar detectors.

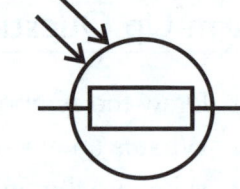

This is the circuit symbol for a light-dependent resistor.

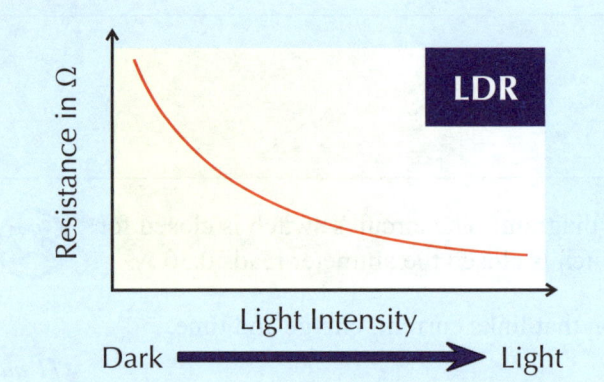

Thermistor Resistance Decreases as Temperature Increases

1) A thermistor is a temperature dependent resistor.
2) In hot conditions, the resistance drops.
3) In cool conditions, the resistance goes up.
4) Thermistors are used in car engines and central heating thermostats.
5) Thermostats turn the heating on when it's cool and off when it's warm.

This is the circuit symbol for a thermistor.

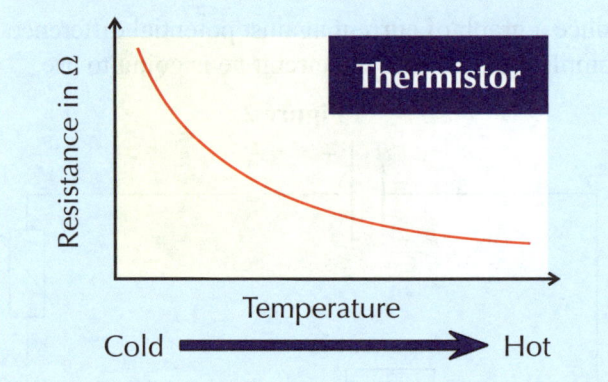

Thermistors and LDRs have many applications...

And they're not just limited to the examples on this page. Oh no. For example, LDRs are used in digital cameras to control how long the shutter should stay open for. If the light level is low, changes in the resistance cause the shutter to stay open for longer than if the light level was higher. How interesting.

Sensing Circuits

*Now you've learnt about what **LDRs** and **thermistors** do, it's time to take a look at how they're put to use.*

You Can Use Thermistors in Sensing Circuits

Sensing circuits can be used to automatically change the pd across components depending on changes in the environment.

1) The circuit on the right is a sensing circuit used to control a fan in a room.
2) The potential difference of the power supply is shared out between the thermistor and the fixed resistor (see p.314).
3) How much pd each one gets depends on their resistances.
4) The larger a component's resistance, the more of the pd it takes.
5) This circuit means that the pd across the fan goes up as the room gets hotter. Here's why:

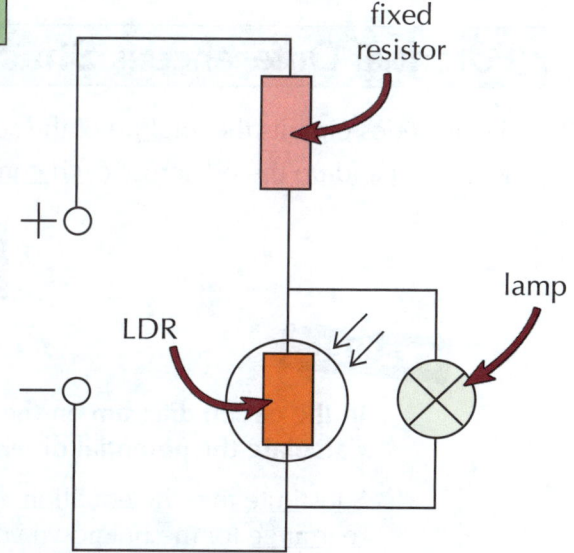

- As the room gets hotter, the resistance of the thermistor decreases.
- The thermistor takes a smaller share of the pd from the power supply.
- So the pd across the fixed resistor rises.
- The pd across the fixed resistor is equal to the pd across the fan (you'll see why on p.316).
- So the pd across the fan rises too, making the fan go faster.

6) If you connected the fan across the thermistor instead, the circuit would do the opposite.
7) The fan would slow down as the room got hotter.

You Can Also Use LDRs in Sensing Circuits

1) If you used an LDR instead of a thermistor in the circuit above it would be a light sensing circuit.
2) For example, the bulb in the sensing circuit on the right is connected across the LDR. This means it gets brighter as the room gets darker.

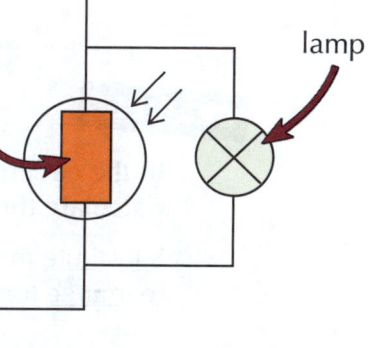

Sensing circuits react to changes in the surroundings...

Sensing circuits are a useful application of thermistors and LDRs, but can be tricky to make sense of. They rely on the properties of series and parallel circuits, which are coming up on the next few pages.

Q1 Video Solution

Q1 An engineer wants to make a circuit containing a bulb that will get brighter when the surroundings get darker. Draw a diagram of a circuit the engineer could make. You can assume the light from the bulb does not affect any of the circuit components. [3 marks]

Topic P2 — Electricity

Series Circuits

You need to be able to tell if components are connected in series or parallel *just by looking at circuit diagrams*. You also need to know the **rules** about what happens with both types. Read on to find out more.

Series Circuits — All or Nothing

1) In series circuits, the components are all connected in a line between the ends of the power supply.
2) Only voltmeters break this rule. They're always in parallel (see p.316).
3) If you remove one component, the circuit is broken. So all the components stop working.
4) You can use the rules on this page and the next to design series circuits to measure and test all sorts of things.
5) For example the test circuit on p.310 and the sensing circuits on p.313.

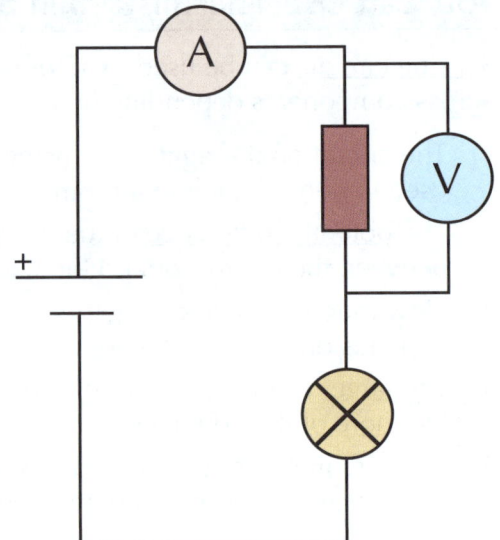

Cell Potential Differences Add Up

1) There is a bigger potential difference when more cells are in series, provided the cells are all connected the same way.
2) For example when two batteries of voltage 1.5 V are connected in series they supply a total of 3 V.

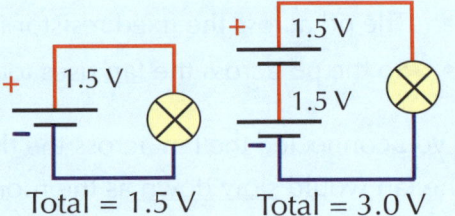

Potential Difference is Shared

1) In series circuits the total pd of the supply is shared between all of the components.
2) If you add up the pd across each component, you get the pd of the power supply.

$$V_{total} = V_1 + V_2 + ...$$

EXAMPLE

In the circuit diagram on the right $V_1 = 3.5$ V.
Calculate the potential difference across the filament lamp, V_2.

Substitute into the equation above, then rearrange for the unknown voltage.

$8 = 3.5 + V_2$
$V_2 = 8 - 3.5 = 4.5$ V

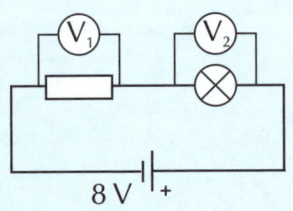

3) The bigger a component's resistance, the bigger its share of the total pd.

There are two main types of circuit — series and parallel...

Ammeters and voltmeters don't affect how you describe a circuit — you can have a parallel circuit (p.316) with ammeters connected in series, or a series circuit with voltmeters connected in parallel with components.

Topic P2 — Electricity

Series Circuits

We're not done with **series circuits** yet. Here's the low-down on **current** and **resistance**...

Current is the Same Everywhere

In series circuits the same current flows through all components.

$$I_1 = I_2 = \ldots$$

Resistance Adds Up

1) In series circuits, the total resistance of two components is found by adding up their resistances.
2) R_{total} is the total resistance of the circuit. R_1 and R_2 are resistances of the components:

$$R_{total} = R_1 + R_2$$

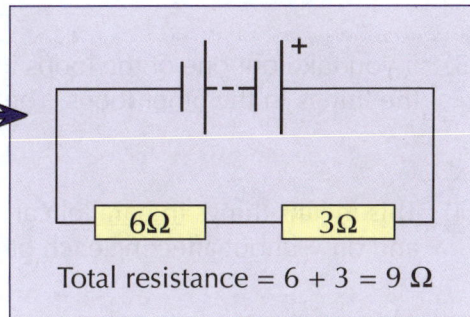

Total resistance = 6 + 3 = 9 Ω

EXAMPLE

For the circuit diagram on the right, calculate the current passing through the circuit.

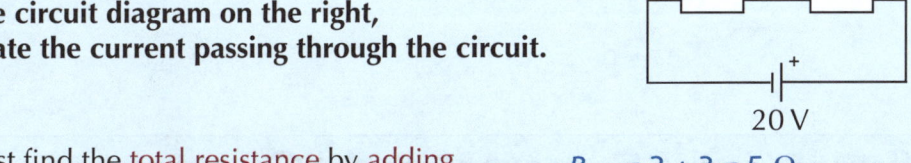

1) First find the total resistance by adding together the resistance of the two resistors. $R_{total} = 2 + 3 = 5\ \Omega$
2) Then rearrange $V = IR$ for I. $I = V \div R$
3) Substitute in the values you have and calculate the current. $= 20 \div 5$
   $= 4\ A$

3) You need to be able to explain why adding resistors in series increases the total resistance of the circuit:

- Adding a resistor in series means the resistors have to share the total pd.
- This means the pd across each resistor is lower, so the current through each resistor is lower ($V = IR$).
- The current is the same everywhere.
- So the total current in the circuit is reduced when a resistor is added.
- This means the total resistance of the circuit has gone up.

Series circuits aren't used very much in the real world...

Since series circuits put all components on the same loop of wire, if one component breaks, it'll break the circuit, and all other components will stop working too.

Q1 A battery is connected in series with a 4 Ω resistor, a 5 Ω resistor and a 6 Ω resistor. A current of 0.6 A flows through the circuit. Calculate the potential difference of the battery. [3 marks]

Q1 Video Solution

Topic P2 — Electricity

Parallel Circuits

Parallel circuits can be a little bit trickier to wrap your head around, but they're much more **useful** than series circuits. Most electronics use a combination of series and parallel circuitry.

Parallel Circuits — Every Component Connected Separately

1) In parallel circuits, each component is separately connected to the ends of the power supply.

2) Only ammeters break this rule, they're always in series (see p.314).

3) If you take out one of the loops in a parallel circuit, the things in the other loops won't be affected.

4) This means things in parallel can be switched on and off without affecting each other.

5) Everyday circuits often include a mixture of series and parallel parts.

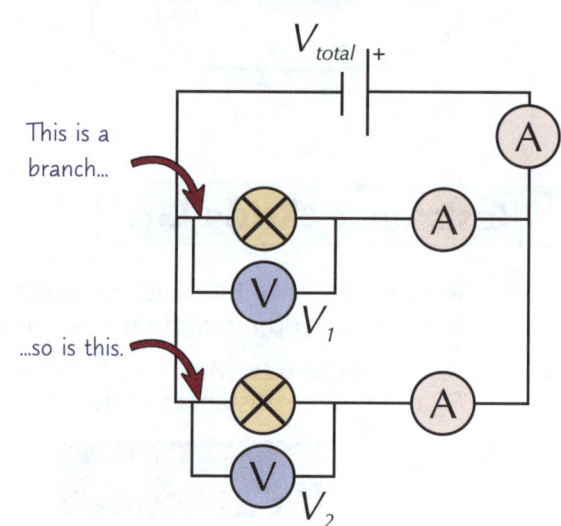

Potential Difference is the Same Across All Components

1) In parallel circuits all components get the full source pd.

2) So the potential difference is the same across all components.

$$V_1 = V_2 = V_3 = ...$$

3) This means that identical bulbs connected in parallel will all be at the same brightness.

Parallel circuits have a big advantage over series circuits...

If a component in a parallel circuit breaks, only that component's branch will be affected. Current will continue to flow through all of the circuit's other branches. A classic example of this is fairy lights. If a bulb breaks, only the broken bulb will go out, since they're wired in parallel.

Topic P2 — Electricity

Parallel Circuits

*In some ways current in a circuit is a little like water flowing in a river. And, just as like flowing water splits when it reaches a fork in the river, **current** will split **when** it reaches a **branch** in the circuit.*

Current is **Shared** Between Branches

1) In parallel circuits the <u>total current</u> in a circuit is equal to the <u>sum</u> of all the currents through the <u>separate components</u>.

2) At <u>junctions</u>, the current either <u>splits</u> or <u>rejoins</u>.

3) The total current going <u>into</u> a junction must equal the total current <u>leaving</u> it.

$$I_{total} = I_1 + I_2 + ...$$

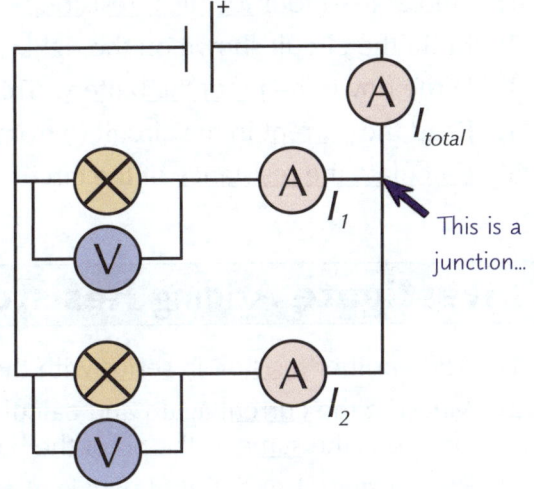

EXAMPLE

For the circuit diagram on the right, find the pd measured by V_1 and the current measured by A_3.

1) The resistors are in <u>parallel</u>, so the pd across each resistor is the <u>same</u> as the <u>cell pd</u>.
   Cell pd = 6 V
   V_1 = 6 V

2) The current <u>into</u> the first junction is the <u>same</u> as the current <u>out</u> of it.
   In: 2 A   Out: 1 A + A_3
   So A_3 = 2 − 1 = 1 A

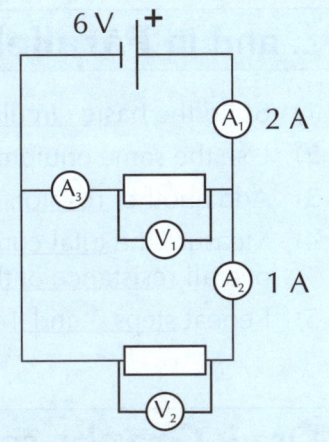

Adding a Resistor in Parallel **Reduces** the **Total Resistance**

The total resistance of <u>two resistors connected in parallel</u> is <u>less than</u> the resistance of the <u>smallest</u> of the two resistors. Here's why:

1) If you add a resistor in parallel, both resistors still have the <u>same potential difference</u> across them as the power supply.
2) This means the 'pushing force' making the current flow is still the <u>same</u>.
3) But by adding another loop, the <u>current</u> has <u>more than one</u> direction to go in.
4) More <u>current</u> can flow around the circuit, so the total current <u>increases</u>.
5) This means the <u>total resistance</u> of the circuit is <u>lower</u> (as $R = V \div I$).

The current flowing into a junction equals the current out

Q1 A circuit contains three resistors, each connected in parallel with a cell. Explain what happens to the total current and resistance in the circuit when one resistor is removed. [4 marks]

Topic P2 — Electricity

 # Investigating Circuits

You saw on page 308 how the length of the wire used in a circuit affects its resistance. Now it's time to do an **experiment** to see how placing **resistors** in series or in parallel can affect the resistance of the **whole circuit**.

First Set Up the Basic Circuit

1) Find at least four identical resistors.
2) Build the circuit shown on the right.
3) Write down the pd of the battery. This is the pd of the circuit (V).
4) Read the current in the circuit (I) from the ammeter.
5) Calculate the resistance of the circuit using R = V ÷ I.

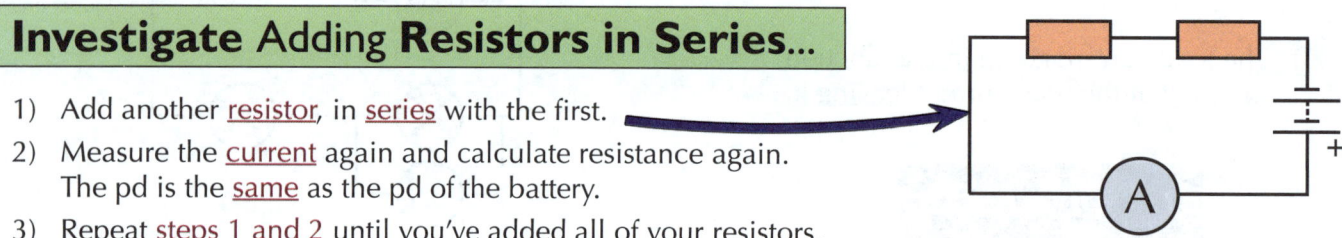

Investigate Adding Resistors in Series...

1) Add another resistor, in series with the first.
2) Measure the current again and calculate resistance again. The pd is the same as the pd of the battery.
3) Repeat steps 1 and 2 until you've added all of your resistors.

... and in Parallel

1) Build the basic circuit again. You already know its resistance.
2) Use the same equipment so it's a fair test.
3) Add another resistor, in parallel with the first.
4) Measure the total current through the circuit and calculate the overall resistance of the circuit. The pd is still the same as before.
5) Repeat steps 3 and 4 until you've added all of your resistors.

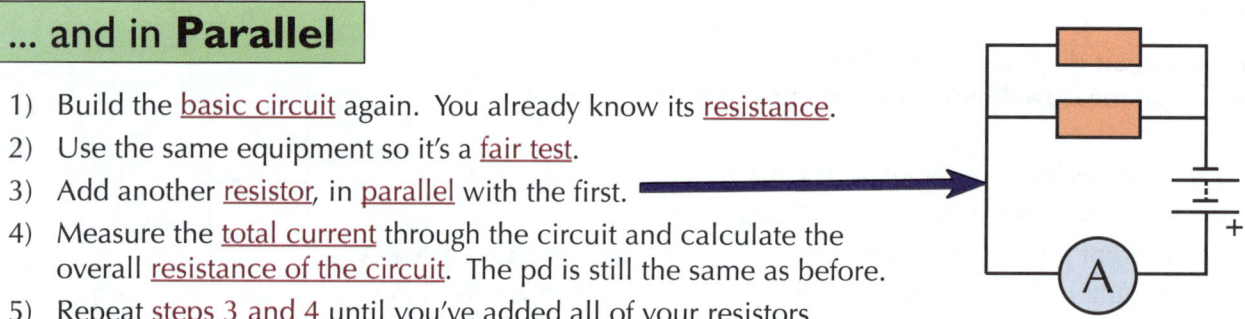

Draw Graphs so you can Compare your Results

1) Plot a graph of the number of resistors in the circuit against the total resistance.
2) You should get graphs that look like this:

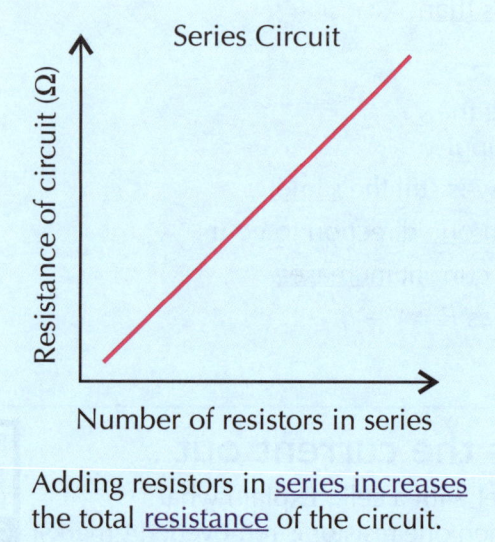 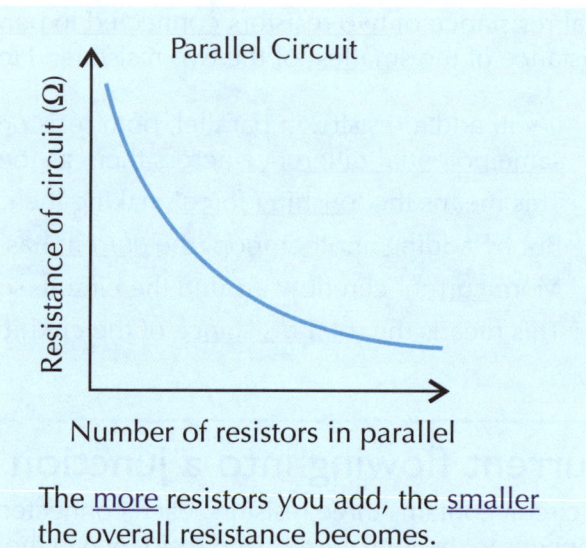

Adding resistors in series increases the total resistance of the circuit.

The more resistors you add, the smaller the overall resistance becomes.

Topic P2 — Electricity

Warm-Up & Exam Questions

Time to check and see what you can remember about those circuit devices, plus parallel and series circuits.

Warm-Up Questions

1) Give one use of a light-dependent resistor (LDR).
2) What happens to the resistance of a thermistor as its temperature increases?
3) Describe what a series circuit is.
4) True or false? In a series circuit, if you add up the pd across each component, you get the pd of the power supply.
5) How do you work out the total resistance in a series circuit?
6) Which has the higher total resistance: two resistors in series, or the same two resistors in parallel?

Exam Questions

1 Which of the following statements does **not** apply to parallel circuits? Tick **one** box. (Grade 3-4)

If one branch is disconnected, the other branches will not be affected. ☐

Total resistance is equal to the sum of the individual resistances. ☐

Current splits when it reaches a junction. ☐

Each branch experiences the whole supply pd. ☐

[1 mark]

2 **Figure 1** shows a series circuit. (Grade 3-4)

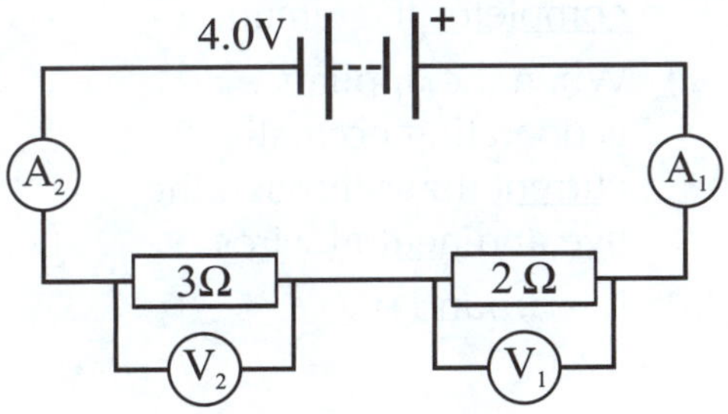

2.1 Calculate the total resistance in the circuit.
[2 marks]

2.2 The current through A_1 is 0.4 A.
What is the current through A_2? Explain your answer.
[2 marks]

2.3 V_1 reads 1.6 V.
Calculate the reading on V_2.
[2 marks]

Figure 1

3 **Figure 2** shows a circuit that contains an LED, a light-dependent resistor and a cell. (Grade 4-5)

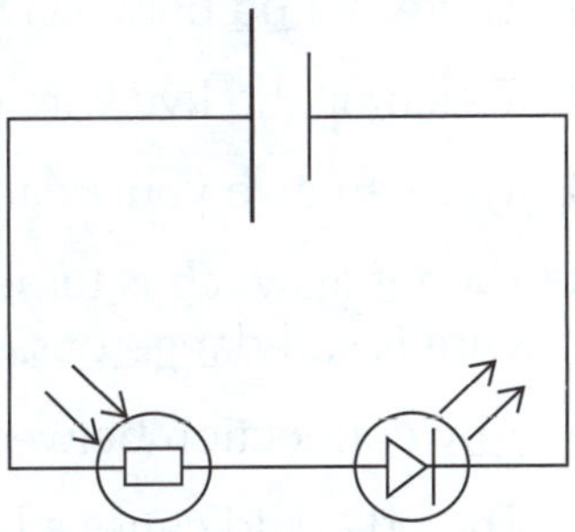

3.1 Describe how you could tell that a current is flowing in the circuit.
[1 mark]

3.2 The circuit is placed in a well lit room. At the end of the day, the lights in the room are turned off. Describe and explain how the resistance of the circuit changes when the room lights are switched off.
[2 marks]

Figure 2

Topic P2 — Electricity

Electricity in the Home

*Now you've learnt the basics of **electrical circuits**, it's time to see how **electricity** is used in **everyday life**.*

Mains Supply is ac, Battery Supply is dc

1) An alternating potential difference is a potential difference that is constantly changing direction. It produces an alternating current (ac).
2) In an alternating current, the current (flow of charge) is also constantly changing direction.
3) The UK mains supply (the electricity in your home) is an ac supply at around 230 V.
4) The frequency (p.371) of the ac mains supply is 50 Hz (hertz).
5) Direct current (dc) is a current that is always flowing in the same direction.
6) It's created by a direct potential difference. The direction of a direct potential difference is always the same.

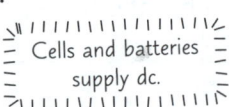
Cells and batteries supply dc.

Most Cables Have Three Separate Wires

1) Most electrical appliances are connected to the mains supply by three-core cables.
2) They have three wires covered with plastic insulation inside them.
3) They are coloured so that it is easy to tell the different wires apart.

LIVE WIRE — brown.
1) The live wire provides the alternating potential difference from the mains supply.
2) It is at about 230 V.

NEUTRAL WIRE — blue.
1) The neutral wire completes the circuit.
2) When the appliance is operating normally, current flows through the live and neutral wires.
3) It is around 0 V.

EARTH WIRE — green and yellow.
1) The earth wire is a safety wire.
2) It stops the appliance becoming live:
   - It is connected to the metal casing of an appliance.
   - If a fault causes the live wire to touch the casing, the current flows away through the earth wire.
3) It's also at 0 V.

The Live Wire Can Give You an Electric Shock

1) There is a pd between the live wire and your body (which is at 0 V).
2) Touching the live wire can cause a current to flow through your body.
3) This can give you a dangerous electric shock.
4) Even if a switch is turned off (the switch is open), touching the live wire is still dangerous. This is because it still has a pd of 230 V.
5) Any connection between live and earth can be dangerous.
6) The pd could cause a huge current to flow, which could result in a fire.

Power of Electrical Appliances

Energy is transferred between stores *electrically* (like you saw on page 284) by *electrical appliances*.

Energy is Transferred from Cells and Other Sources

1) When a charge moves around a circuit, work is done against the resistance of the circuit.
2) Whenever work is done, energy is transferred.
3) When the work is done by a charge, the energy is transferred electrically.
4) Electrical appliances transfer energy to components in the circuit when a current flows.

Kettles transfer energy electrically from the mains supply to the thermal energy store of the heating element inside the kettle.

Energy is transferred electrically from the battery of a handheld fan to the kinetic energy store of the fan's motor.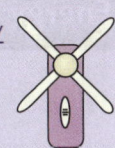

Energy Transferred Depends on the Power

1) The total energy transferred by an appliance depends on how long the appliance is on for and its power.
2) The power of an appliance is the energy that it transfers per second.
3) So the more energy it transfers in a given time, the higher its power.
4) The amount of energy transferred by electrical work is given by:

This equation should be familiar from page 293.

Energy transferred (J) = Power (W) × Time (s) $E = Pt$

EXAMPLE

A 600 W microwave is used for 5 minutes. How much energy does it transfer?

1) Convert the time into seconds. $t = 5 \times 60 = 300$ s
2) Substitute the numbers into $E = Pt$ $E = Pt = 600 \times 300$
   to find the energy transferred. $= 180\ 000$ J

5) Appliances are often given a power rating. This is the power that they work at.
6) The power rating tells you how much energy is transferred between stores when the appliance is used.
7) An appliance with a higher power will cost more to run for a given time, as it uses more energy.

A 850 W microwave will transfer more energy between stores during 5 minutes than the 600 W microwave in the example above. This means it will cost more to use it for 5 minutes.

Power is the rate of energy transfer...

Remember, the power rating of an electrical appliance is the amount of energy transferred to the appliance per second, not the amount that it transfers to useful energy stores.

Q1 An appliance transfers 6000 J of energy in 30 seconds. Calculate its power. [3 marks]

Q2 Calculate the difference in the amount of energy transferred by a 250 W TV and a 375 W TV when they are both used for two hours. [4 marks]

Q2 Video Solution

Topic P2 — Electricity

More on Power

*And we're not done yet. There are even more **power equations** for you to get your head around. How fun.*

Potential Difference is Energy Transferred per Charge Passed

1) As a charge moves around a circuit, energy is transferred to or from it.
2) The energy transferred by a component depends on the potential difference across it and the charge flowing through it.
3) The formula is really simple:

Energy transferred (J) = Charge Flow (C) × Potential Difference (V) $E = QV$

EXAMPLE

An electric toothbrush contains a 3.0 V battery. 140 C of charge passes through the toothbrush as it is used. Calculate the energy transferred.

Substitute the numbers into $E = QV$ to find the energy transferred. $E = QV = 140 × 3.0 = 420$ J

Power Also Depends on Current and Potential Difference

1) You saw on the previous page that power is energy transferred in a given time.
2) The power of an appliance can also be found using:

Power (W) = Potential difference (V) × Current (A) $P = VI$

3) You can also find the power if you don't know the potential difference, using: $P = I^2R$ — Resistance (Ω)

EXAMPLE

A motor with a power of 1250 W has a resistance of 50 Ω. Calculate the current flowing through the motor.

1) First rearrange the formula $P = I^2R$ to make I the subject.
   - Divide both sides by R. $P ÷ R = I^2$ so $I^2 = P ÷ R$
   - Find the square root of both sides. $I = \sqrt{P ÷ R}$
2) Now just plug in the numbers. $I = \sqrt{1250 ÷ 50} = \sqrt{25}$
                                              $= 5$ A

Your calculator should have a '√' (square root) button to help with these calculations.

Power is measured in watts, W — one W is equal to one J/s...

The best way to learn all of this is to just practise using these equations again and again.

Q1 Calculate the energy transferred in a circuit with a pd of 200 V when 10 000 C of charge flows. [2 marks]

Q2 An appliance is connected to a 12 V source. A current of 4.0 A flows through it. Calculate the power of the appliance. [2 marks]

Q3 An appliance has a power of 2300 W and has a current of 10.0 A flowing through it. Calculate the resistance of the appliance. [3 marks]

Q3 Video Solution

Topic P2 — Electricity

The National Grid

*The **national grid** is a giant web of wires that covers **the whole of Britain**, getting electricity from power stations to homes everywhere. Whoever you pay for your electricity, it's the national grid that gets it to you.*

Electricity is Distributed via the National Grid

1) The national grid is a giant system of cables and transformers that covers the UK.

2) It transfers electrical power from power stations to consumers (anyone who is using electricity) across the UK.

Electricity Production has to Meet Demand

1) Throughout the day, the amount of electricity used (the demand) changes.

2) Power stations have to produce enough electricity for everyone to have it when they need it.

3) More electricity is used when people get up in the morning, come home from school or work and when it starts to get dark or cold outside.

4) Power stations often run at well below their maximum power output, so that they can increase their power if needed.

5) This means that the national grid can cope with a high demand, even if another station shuts down without warning.

Energy demands are ever increasing...

The national grid has been working since the 1930s and has gone through many changes and updates since then to meet increasing energy demands. Using energy-efficient appliances and switching unneeded lights off are some ways we might ensure that supply and demand stay in balance. It'll do wonders for the electricity bill too, so it's worth bearing in mind.

Topic P2 — Electricity

The National Grid

*To transfer electricity **efficiently**, the national grid makes use of some clever tech called **transformers**.*

The National Grid Uses a High Pd and a Low Current

1) The national grid transfers loads of energy, so the power has to be very high.

2) To transmit this huge amount of power you need either a high potential difference or a high current.

3) This is because $P = VI$ (from page 322).

4) A high current means loads of energy is lost to thermal energy stores as the wires heat up.

5) So the national grid transmits electricity at a very high pd. For a given power, the higher the pd the lower the current.

6) This reduces the energy lost, making the national grid an efficient way of transferring energy.

Potential Difference is Changed by Transformers

1) Step-up transformers are used to increase the pd from power stations to electric cables.
2) Step-down transformers bring the pd back down to safe levels before the electricity gets to homes.

3) Transformers all have two coils, a primary coil and a secondary coil, joined with an iron core.
4) The power of a primary coil is given by power = pd × current.
5) The power in primary coil = power in secondary coil.

The national grid — it's a powerful thing...

The key to the efficiency of the national grid is the power equation, $P = VI$ (see page 322). If you have a constant power, but increase the potential difference using a transformer, the current must decrease. Having as low a current as possible makes sure that power is transferred as efficiently as possible.

Warm-Up & Exam Questions

Who knew there was so much to learn about electricity in the home and across the country? See if it's switched on a light bulb in your brain by trying out these questions.

Warm-Up Questions

1) Name the three wires in a three-core cable that connect electrical appliances to the mains supply.
2) What is the main energy transfer when electric current flows through an electric kettle?
3) What is the equation linking power, current and resistance?
4) What is the national grid?

Exam Questions

1 This question is about the national grid. *Grade 1-3*

1.1 Use words from the box to complete the sentence.

| current | potential difference | step-up | step-down |

_____ transformers are used between the power station and the transmission cables. This increases the _____ , so that power may be transferred more efficiently.

[2 marks]

1.2 Which describes the UK mains electricity supply? Tick **one** box.

☐ 230 V ac ☐ 170 V ac ☐ 230 V dc ☐ 170 V dc

[1 mark]

2 Table 1 shows the power and potential difference ratings for two kettles. *Grade 4-5*

Table 1

	Power (W)	Potential Difference (V)
Kettle A	2760	230
Kettle B	3000	230

2.1 Write down the equation linking power, potential difference and current.

[1 mark]

2.2 Calculate the current, in amps, drawn from the mains supply by kettle A.

[2 marks]

2.3 A student wants to test which kettle heats water most quickly. He measures the increase in water temperature in each kettle after 1 minute. Identify one variable the student must control.

[1 mark]

2.4 The kettles are equally efficient.
A second student states: 'Kettle B has the greatest power so it will heat water most quickly.'
Justify the student's statement.

[2 marks]

Topic P2 — Electricity

Topic P3 — Particle Model of Matter

Particle Model

Everything is made up of *small particles*. The particle model *describes* how these particles behave.

There are **Three States of Matter**

1) The three states of matter are solid (e.g. ice), liquid (e.g. water) and gas (e.g. water vapour).
2) The particle model explains the differences between the states of matter:
   - The particles of a certain material are always the same, no matter what state it is in.
   - But the particles have different amounts of energy in different states.
   - And the forces between particles are different in each state.
   - This means that the particles are arranged (laid out) differently in different states.

Solids

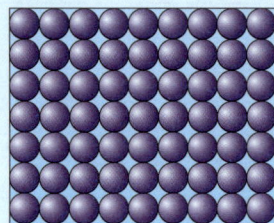

1) Particles are held close together by strong forces in a regular, fixed pattern.
2) The particles don't have much energy.
3) So they can only vibrate (jiggle about) around a fixed position.

Liquids

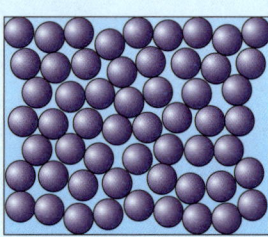

1) The particles are held close together in an irregular pattern.
2) The particles have more energy than the particles in a solid.
3) They can move past each other in random directions at low speeds.

Gases

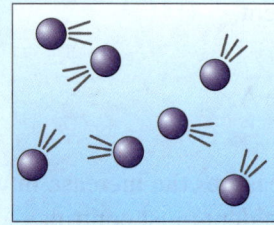

1) The particles aren't held close together. There are no forces between them.
2) The particles have more energy than in liquids and solids.
3) The particles constantly move around in random directions at a range of speeds.

The more energy in their kinetic stores, the faster the particles move...

Learn those diagrams above and make sure that you can describe the arrangement and movement of particles in solids, liquids and gases. You should also be able to talk about how much energy they have.

Particle Motion in Gases

*The **particle model** explains how **temperature**, **pressure** and **energy in kinetic stores** are all related. And this page is here to explain it all to you. I bet you're just itching to find out more...*

Gas Particles Bump into Things and Create Pressure

1) Particles in a gas are <u>free to move</u> around.

2) They <u>collide with</u> (bump into) each other and the sides of the <u>container</u> they're in.

3) When they hit something, they <u>apply a force</u> to it. <u>Pressure</u> is the <u>force</u> applied over a <u>given area</u>.

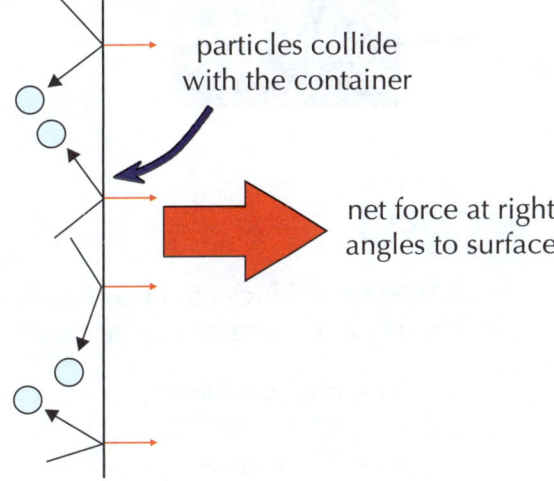

Increasing the Temperature of a Gas Increases its Pressure

1) The <u>temperature</u> of a gas depends on the <u>average energy</u> in the <u>kinetic energy stores</u> of the gas particles.

- The <u>hotter</u> the gas, the <u>higher</u> the average energy.
- If particles have <u>more energy</u> in their kinetic stores, they <u>move faster</u>.
- So the <u>hotter</u> the gas, the <u>faster</u> the particles move on average.
- <u>Faster particles</u> hit the sides of the container <u>more often</u>. This <u>increases</u> the <u>force</u> on the container.
- So increasing the <u>temperature</u> of a gas increases its <u>pressure</u>.

2) This <u>only</u> works if the <u>space</u> the gas takes up (the <u>volume</u>) <u>doesn't change</u>.

Higher temperatures mean higher average energies in kinetic stores...

The <u>particle model</u> can be used to explain what happens when you <u>change</u> the <u>temperature</u> of a <u>gas</u> which is kept at a <u>constant volume</u>. Have a look back on page 326 for more about the particle model.

Topic P3 — Particle Model of Matter

Density of Materials

*The **density** of an object tells you how many of its **particles** are squished into a **given space**.*

The Particle Model can also Explain Density

1) <u>Density</u> is a measure of <u>how much mass</u> there is in a <u>certain space</u>.
2) You can work out density using:

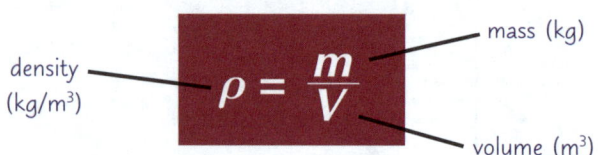

EXAMPLE

A 0.0020 m³ block of aluminium has a mass of 5.4 kg. Calculate the density of aluminium.

density = mass ÷ volume
= 5.4 ÷ 0.0020
= 2700 kg/m³

3) The density of an object depends on <u>what it's made of</u> and how its <u>particles</u> are <u>arranged</u>.
4) A <u>dense</u> material has its particles <u>packed tightly</u> together.
5) So, <u>solids</u> are generally <u>denser</u> than <u>liquids</u>.
6) And <u>liquids</u> are generally <u>denser</u> than <u>gases</u>.

You Can Find the Density of a Regularly Shaped Object

1) Use a <u>balance</u> to measure its <u>mass</u> (see p.392).
2) Measure its <u>length</u>, <u>width</u> and <u>height</u> with a <u>ruler</u>.
3) Then calculate its <u>volume</u> using the <u>formula</u> for that shape.
4) Use <u>density = mass ÷ volume</u> to find the density.

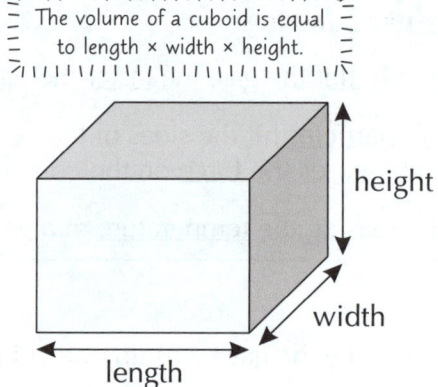

The volume of a cuboid is equal to length × width × height.

The denser the substance, the closer together its particles are...

When you're measuring the <u>dimensions</u> of an object, think about which apparatus is best to use. If what you're measuring is pretty <u>small</u>, a <u>ruler</u> won't give you very <u>accurate</u> measurements. For small measurements, a <u>micrometer</u> or <u>Vernier callipers</u> would be more appropriate.

Topic P3 — Particle Model of Matter

Measuring Density

PRACTICAL

You need to be able to carry out practicals to work out the **densities** of different **solids** and **liquids**.

Find the Density of an **Irregularly Shaped Object** Using a **Eureka Can**

1) Use a balance to measure the object's mass.
2) Fill a eureka can (a can with a spout in its side) with water.
3) The water level should end up just below the start of the spout.
4) Place a measuring cylinder (p.392) under the end of the spout.
5) Place your object into the water.
   This will push some of the water out through the spout.

There's more about eureka cans on page 393.

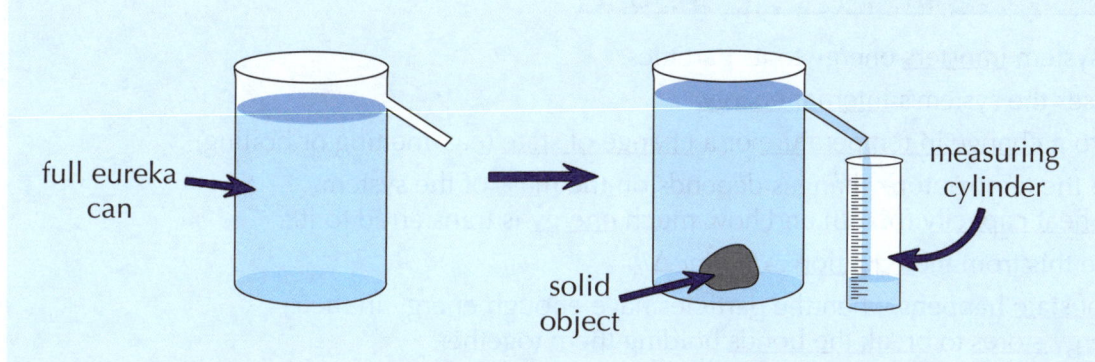

6) Measure the volume of water that has collected in the measuring cylinder.
7) This is equal to the volume of the object.
8) Use the formula on the previous page to find the object's density.

For both of these experiments, you'll need to know that 1 ml = 1 cm³ and that 1 cm³ = 0.000001 m³.

Find the Density of a **Liquid** Using a **Measuring Cylinder**

1) Place a measuring cylinder on a balance and zero the balance (p.392).
2) Pour 50 ml of the liquid into the measuring cylinder.
3) Record the liquid's mass shown on the mass balance.
4) Use the formula on the previous page to find the density.
   The volume is 50 cm³, or 0.00005 m³.

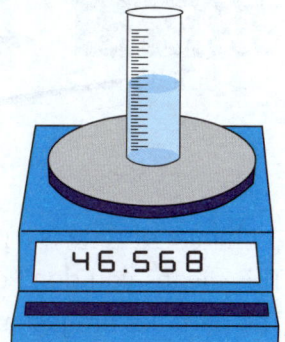

A eureka can makes it pretty easy to find an object's volume...

Try not to put your fingers in the water as you're dropping your object into the eureka can. This will displace more water, and make your volume measurement too big.

Q1 A 0.019 kg gemstone is placed into a full eureka can, causing 7.0 cm³ of water to be pushed out the spout into a measuring cylinder. Calculate the density of the gemstone in g/cm³.

[3 marks]

Q1 Video Solution

Topic P3 — Particle Model of Matter

Internal Energy and Changes of State

*This page is all about heating things. Take a look at your **specific heat capacity** notes (p.288) before you start. You need to understand it and be able to use $\Delta E = mc\Delta\theta$ for this topic too I'm afraid.*

Internal Energy is the Total Energy Stored by Particles in a System

1) The energy stored in a system (p.284) is stored by its particles (atoms and molecules).
2) The particles have energy in their kinetic energy stores.
3) They also have energy in their potential energy stores because of their positions.
4) The internal energy of a system is the total energy that all its particles have in their kinetic and potential energy stores.

Heating Increases Internal Energy

1) Heating a system transfers energy to its particles.
2) This increases the system's internal energy.
3) This leads to a change in temperature or a change of state (e.g. melting or boiling).
4) How much the temperature changes depends on the mass of the system, its specific heat capacity (p.288) and how much energy is transferred to it.
5) You can see this from the equation $\Delta E = mc\Delta\theta$.
6) A change of state happens when the particles have enough energy in their kinetic energy stores to break the bonds holding them together.

Mass Doesn't Change in a Change of State

1) A change of state can happen because of cooling, as well as heating.
2) The changes of state are:

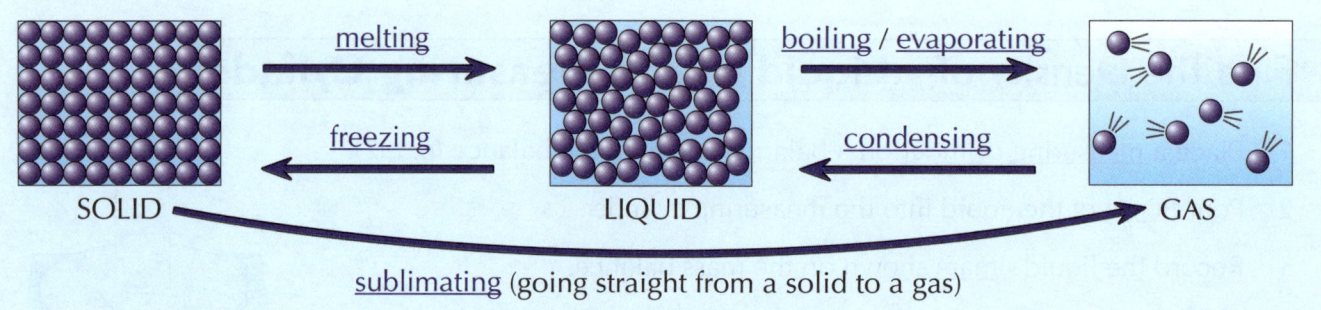

3) A change of state is a physical change (not a chemical change).
4) This means you don't end up with a new material.
5) The particles are just arranged in a different way (p.326).
6) The number of particles stays the same when the state changes.
7) This means the mass is conserved (it doesn't change).
8) If you reverse a change of state, the material will get back all the properties it had before the change.

Physical changes are different from chemical changes...

You should make sure you know all the changes of state, and why they are physical changes.

Topic P3 — Particle Model of Matter

Specific Latent Heat

*The **energy needed** to change the state of a substance is called **latent heat**. This is exciting stuff I tell you...*

Temperature Doesn't Change During a Change of State

1) Heating a material transfers energy to the material.
2) This either increases the temperature of the material or changes its state (p.330).
3) During a change of state, the temperature doesn't change. But the internal energy does.
4) The energy transferred is used to break bonds between particles. It is not used to raise the temperature.

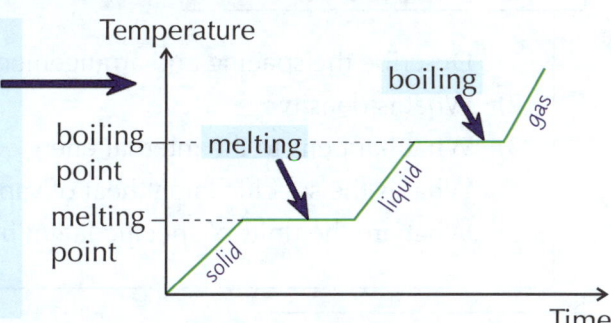

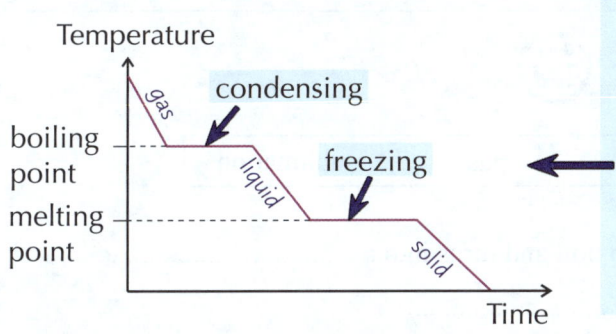

5) When a material cools, energy is transferred away from it.
6) As a material condenses or freezes, bonds form between particles. This causes energy to be released.
7) So its internal energy decreases, but its temperature stays the same during the change of state.
8) The flat spots on these graphs show that the temperature doesn't change during a change of state.

Specific Latent Heat is the Energy Needed to Change State

1) The energy transferred during a change of state is called latent heat.
2) For heating, latent heat is the energy gained to cause a change of state.
3) For cooling, it is the energy released by a change in state.
4) The specific latent heat of a material is the amount of energy needed to change the state of 1 kg of the material without changing its temperature.
5) You can work out the energy needed (or released) during a change of state using this formula:

| Energy (J) = Mass (kg) × Specific Latent Heat (J/kg) | or: | $E = mL$ |

6) Specific latent heat has different names for different changes of state:
7) For changing between a solid and a liquid it is called the specific latent heat of fusion.
8) For changing between a liquid and a gas it is called the specific latent heat of vaporisation.

EXAMPLE The specific latent heat of vaporisation for water is 2 260 000 J/kg. How much energy is needed to completely boil 1.50 kg of water once it has reached its boiling point?

1) The mass and specific latent heat are in the right units, so just put them into the formula.
2) The units for the answer are joules because it's energy.

$E = mL$
$= 1.50 \times 2\,260\,000$
$= 3\,390\,000$ J

Temperature doesn't change during a change of state

Q1 The SLH of fusion for a particular substance is 120 000 J/kg. How much energy is needed to melt 250 g of the substance when it is already at its melting temperature? [2 marks]

Topic P3 — Particle Model of Matter

Warm-Up & Exam Questions

So that's it, you've covered everything from the particle model to changes of state. Once you think you've got to grips with the stuff in this topic, it's time to test yourself with these questions.

Warm-Up Questions

1) Describe the spacing and arrangement of particles in a solid.
2) What is density?
3) What happens to the internal energy of a system when it is heated?
4) What is the specific latent heat of vaporisation?
5) What are the units of specific latent heat?

Exam Questions

1 Use words in the box to complete the sentence below. *(Grade 1-3)*

| solid | condensation | evaporation | gas | sublimation |

If a liquid is heated to a certain temperature it starts to boil and turns into a ....

Another process that causes this change of state is ....

[2 marks]

2 A student has a collection of metal toy soldiers of different sizes made from the same metal. *(Grade 3-4)*

2.1 Which of the following statements is true? Tick **one** box.

☐ The masses and densities of each of the toy soldiers are the same.

☐ The masses of each of the toy soldiers are the same, but their densities may vary.

☐ The densities of each of the toy soldiers are the same, but their masses may vary.

[1 mark]

2.2 One of the soldiers has a mass of 200 g and a volume of 2.5×10^{-5} m^3.

The equation below can be used to calculate density.

density = mass ÷ volume

Calculate the density of the soldier.

[2 marks]

3 0.40 kg of liquid methanol is at its boiling point. *(Grade 3-4)*
The specific latent heat of vaporisation of methanol is 1200 J/kg.

Calculate the amount of energy needed to convert the liquid methanol to gaseous methanol.
Use the correct equation from the Physics Equation Sheet on page 538.

[2 marks]

Topic P3 — Particle Model of Matter

Exam Questions

4 A student wants to measure the density of a pendant. He can use the equipment shown in **Figure 1**. **PRACTICAL**

Figure 1

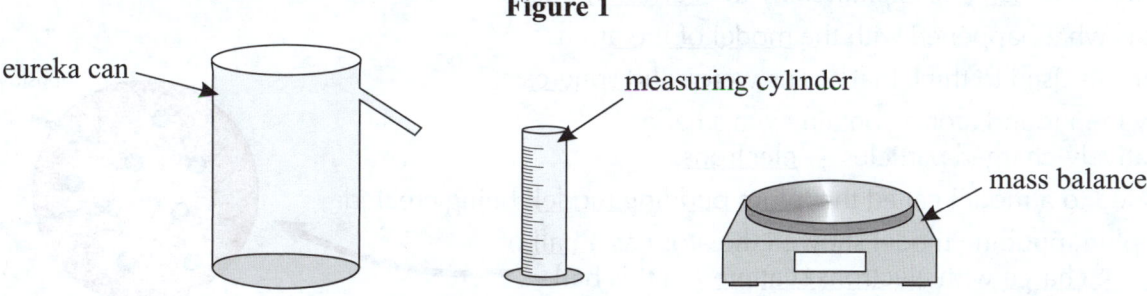

4.1 Name the **two** quantities the student should measure.

[2 marks]

4.2 Describe the steps the student could take to find the density of the pendant with the equipment shown.

[5 marks]

5 The graph in **Figure 2** shows the temperature of a substance against time as it is heated.

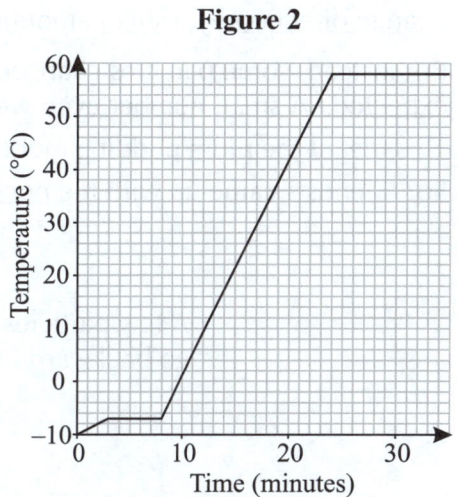

5.1 Describe what is happening during the period between 3 and 8 minutes from the beginning of heating.

[1 mark]

5.2 Give the melting and boiling points of the substance.

[2 marks]

6* A gas is sealed in a rigid container. A scientist heats the container.

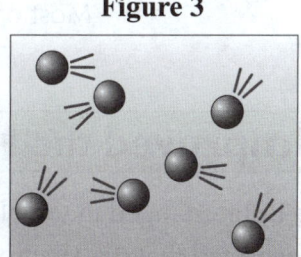

Figure 3 shows gas particles as represented in the particle model. Use the particle model to explain why the pressure of the gas increases as the container is heated. In your answer, you should include a description of how gas particles create pressure in the container.

[6 marks]

Topic P3 — Particle Model of Matter

Topic P4 — Atomic Structure

Developing the Model of the Atom

You might have thought **atomic structure** was all Chemistry. But you need to know it for Physics too.

Models of the Atom Have Changed Over Time

1) Scientific models (p.3) change over time. This happens when new evidence is found that can't be explained by the current model.
2) This is what happened with the model of the atom.
3) Scientists used to think that atoms were solid spheres.
4) They then found atoms contain even smaller, negatively charged particles — electrons.
5) This led to a model called the 'plum pudding model' being created.
6) The plum pudding model showed the atom as a ball of positive charge with electrons scattered in this ball.

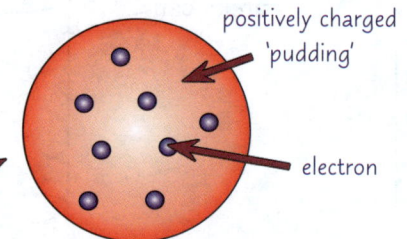

Experiments Showed that the Plum Pudding Model Was Wrong

1) Later, scientists carried out alpha particle scattering experiments. They fired positively charged alpha particles at a very thin sheet of gold.
2) From the plum pudding model, they expected most of the particles to go straight through the sheet. They predicted that a few particles would change direction by a small amount.
3) But instead, some particles changed direction more than expected. A small number even went backwards.
4) This meant the plum pudding model couldn't be right.
5) So, scientists came up with the nuclear model of the atom:

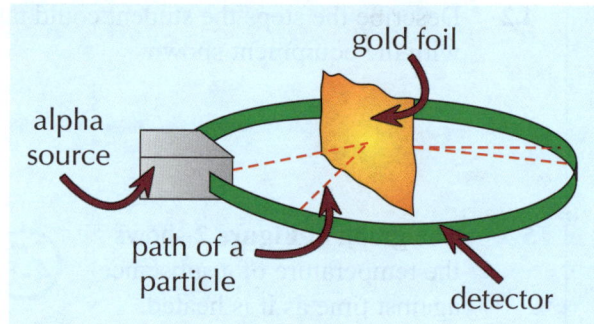

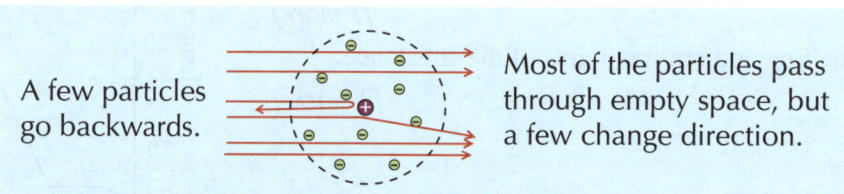

Nuclear Model of Atom
- There's a tiny, positively charged nucleus at the centre of the atom.
- Most of the mass is in the nucleus.
- The nucleus is surrounded by a 'cloud' of negative electrons.
- Most of the atom is empty space.

Bohr Improved the Nuclear Model

1) Niels Bohr changed the nuclear model of the atom.
2) He suggested that the electrons orbit (go around) the nucleus in energy levels:
3) Each energy level (or shell) is a fixed distance from the nucleus.
4) Bohr's theory was supported by many experiments. Experiments later showed that Bohr's theory was correct.

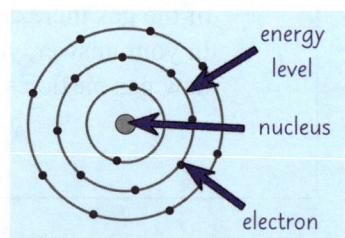

Developing the Model of the Atom

*Bohr's model was pretty good. But the discovery of **protons** and **neutrons** made it even better.*

The Nucleus was Found to Contain Protons and Neutrons

1) Eventually experiments by scientists showed that the nucleus can be divided into smaller particles. Each particle has the same positive charge. These particles were named protons.
2) Experiments by James Chadwick showed that the nucleus also contained neutral particles — neutrons. This happened about 20 years after scientists agreed that atoms have nuclei.
3) This led to a model of the atom which was pretty close to the one we have today.

You Need to Know the Current Model of the Atom

The current model of the atom is a nuclear model.
This means there is a nucleus in the centre surrounded by electrons.

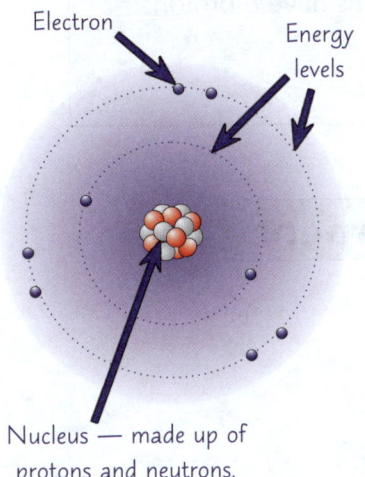

Electron
Energy levels
Nucleus — made up of protons and neutrons.

1) Atoms are very small. The radius of an atom is about 1×10^{-10} m (see p.19 for more on standard form).
2) The nucleus is tiny, but it makes up most of the mass of the atom.
3) The radius of the nucleus is about 10 000 times smaller than the radius of the atom.
4) The nucleus is made up of protons and neutrons.
5) Protons are positively charged and neutrons have no charge. So the nucleus is positively charged.
6) Electrons have a negative charge. They move around (orbit) the nucleus at different distances.
7) These distances are known as energy levels.
8) Atoms have no overall charge. The number of protons = the number of electrons.

Electrons can Move Between Energy Levels

1) The further an energy level is from the nucleus, the more energy an electron in that energy level has.
2) Electrons can move between energy levels by absorbing (taking in) or releasing electromagnetic radiation (p.378).

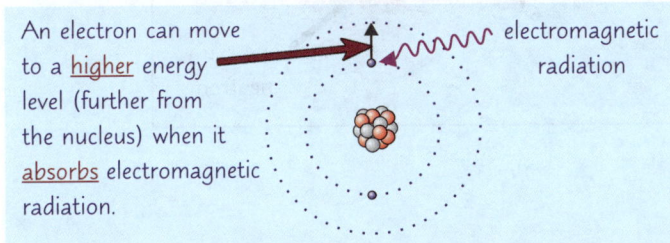

An electron can move to a higher energy level (further from the nucleus) when it absorbs electromagnetic radiation.

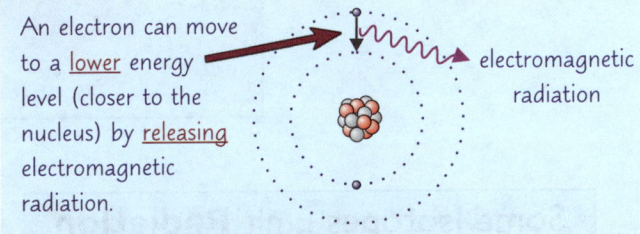

An electron can move to a lower energy level (closer to the nucleus) by releasing electromagnetic radiation.

3) If an electron in an outer energy level absorbs electromagnetic radiation, it can leave the atom.
4) If an atom loses one or more electrons it turns into a positively charged ion.

The model of the atom has developed over time...

Due to lots of scientists doing lots of experiments, we now have a better idea of what the atom's really like. We now know about the particles in atoms — protons, neutrons and electrons.

Topic P4 — Atomic Structure

Isotopes

Isotopes of an element look pretty similar, but watch out — they have **different numbers of neutrons**.

Atoms of the Same Element have the Same Number of Protons

1) The number of protons in an atom is called its atomic number.
2) The protons in a nucleus give the nucleus its positive charge.
   So the atomic number of an atom tells you the charge on the nucleus.
3) The mass number of an atom is the sum of the number of protons and the number of neutrons.
4) An element is a substance only containing atoms with the same number of protons.
5) You can show information about an atom of an element like this:

- Oxygen has an atomic number of 8, this means all oxygen atoms have 8 protons.
- This atom of oxygen has a mass number of 16.
  Since it has 8 protons, it must have 16 − 8 = 8 neutrons.

Isotopes are Different Forms of the Same Element

1) Atoms of an element with the same number of protons but a different number of neutrons are called isotopes.

2) Isotopes of an element have the same atomic number, but a different mass number.

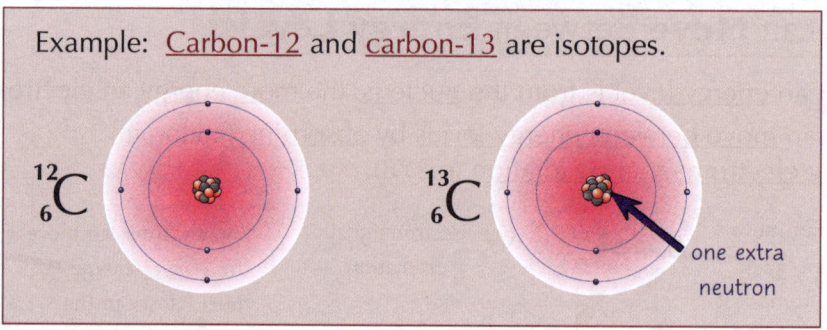

Example: Carbon-12 and carbon-13 are isotopes.

Some Isotopes Emit Radiation

1) Some isotopes are unstable.
2) They emit (give out) radiation from their nuclei to try and become more stable.
3) This process is called radioactive decay.
4) The radiation emitted is called nuclear radiation.
5) There are four different types of nuclear radiation — see the next page.
6) Isotopes that give out nuclear radiation are called radioactive isotopes.

Topic P4 — Atomic Structure

Types of Nuclear Radiation

*The four types of **nuclear radiation** all come from the nucleus, but they have some **key differences**...*

There are Four Types of Nuclear Radiation

Alpha Particles, (α)

An alpha particle is two neutrons and two protons (like a helium nucleus).

Beta Particles, (β)

A beta particle is a fast-moving electron.

Gamma Rays, (γ)

Gamma rays are waves of electromagnetic radiation (p.378).

Neutrons, (n)

Isotopes can also give out neutrons when they decay.

You Need to Know the Properties of Ionising Nuclear Radiation

1) Ionising radiation is radiation that can knock electrons off atoms and turn them into ions.
2) The ionising power of radiation is how easily it can do this.
3) Alpha particles, beta particles and gamma rays are all types of ionising radiation.
4) They all have different properties:

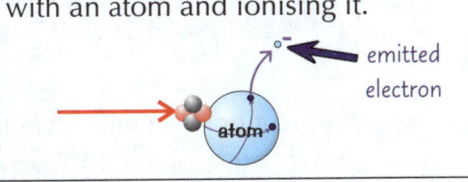

Example: an alpha particle colliding with an atom and ionising it.

Type of radiation	Ionising power	Range in air	Stopped by
alpha particles	strong	a few centimetres	a sheet of paper
beta particles	moderate	a few metres	a sheet of aluminium
gamma rays	weak	a long distance	thick sheets of lead or metres of concrete

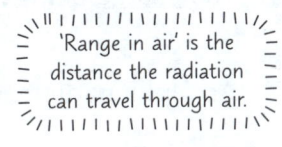

'Range in air' is the distance the radiation can travel through air.

5) Their different ionising powers, ranges and abilities to penetrate (get through) materials make them suitable for different uses.

> E.g. A medical tracer is a radioactive isotope that is injected into a patient. The radiation it emits needs to be detected outside the patient's body.
> You couldn't use a source of alpha radiation as a medical tracer.
> Alpha particles wouldn't be able to pass through the body and be detected outside the body. They are also very ionising and so could do a lot of damage (see p.342).
> Medical tracers usually emit gamma rays. Gamma rays are only weakly ionising and easily pass through the body. This means they can be easily detected and do less harm than alpha particles.

Alpha particles are more ionising than beta particles...

...and beta particles are more ionising than gamma rays. Make sure you've got that memorised, as well as what makes up each type of radiation, as this isn't the last you'll see of this stuff. No siree.

Q1 In order to sterilise medical equipment, radiation is directed at the equipment while it is sealed in packaging. Explain whether alpha radiation would be suitable for this use.

[2 marks]

Topic P4 — Atomic Structure

Nuclear Equations

Nuclear equations show *radioactive decay*. The next few pages will help you get the hang of them.

Mass and Atomic Numbers Have to Balance

1) Nuclear equations are a way of showing radioactive decay (p.336).
2) They're normally written like this:

> nucleus before decay → nucleus after decay + radiation emitted

3) There is one golden rule to remember:

> The total mass and atomic numbers must be equal on both sides of the arrow.

Alpha Decay Decreases the Charge and Mass of the Nucleus

1) Alpha decay is when an alpha particle is emitted from a radioactive nucleus.

2) An alpha particle is made up of two protons and two neutrons. It is the same as a helium nucleus.

3) When a nucleus emits an alpha particle, its atomic number goes down by 2 and its mass number goes down by 4.

Gamma rays are sometimes also released when a nucleus decays by alpha or beta decay.

4) The charge of the nucleus decreases when it gives out an alpha particle.

5) An alpha particle is usually written as a helium nucleus in a nuclear equation: $^{4}_{2}\text{He}$.

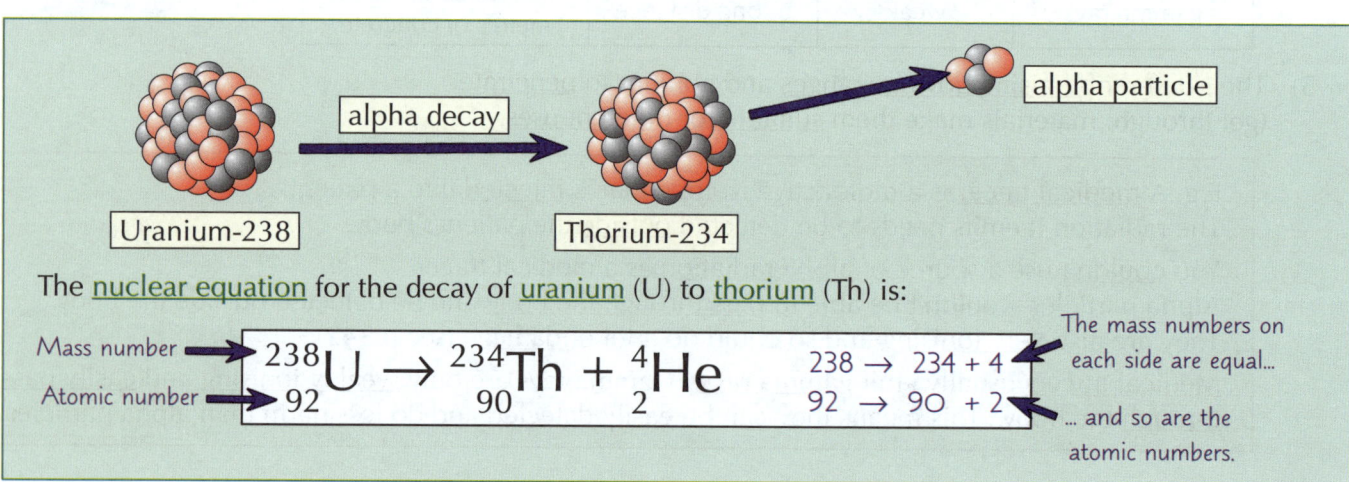

The nuclear equation for the decay of uranium (U) to thorium (Th) is:

$$^{238}_{92}\text{U} \rightarrow \,^{234}_{90}\text{Th} + \,^{4}_{2}\text{He}$$

238 → 234 + 4   The mass numbers on each side are equal...
92 → 90 + 2   ... and so are the atomic numbers.

Mass and atomic numbers must balance in nuclear equations

Q1 After undergoing alpha decay, a nucleus has a mass number of 14. What was the mass number of the nucleus before the decay? [1 mark]

Q2 Write the nuclear equation for $^{219}_{86}\text{Rn}$ decaying to polonium (Po) by emitting an alpha particle. [3 marks]

Nuclear Equations

*On the last page you saw how to write **nuclear equations** for **alpha decay**. Well, now it's time for **beta**.*

Beta Decay Increases the Charge of the Nucleus

1) Beta decay is when a beta particle is emitted from a radioactive nucleus.

2) During beta decay, a neutron in the nucleus turns into a proton.

3) This means the nucleus has one more proton, so its atomic number goes up by 1.

4) It also means the positive charge of the nucleus increases.

5) A beta particle has an atomic number of –1 so the atomic numbers balance on each side of the equation.

6) Protons and neutrons have the same mass, so the mass of the nucleus doesn't change.

7) A beta particle is an electron. It is written as $_{-1}^{0}e$ in nuclear equations.

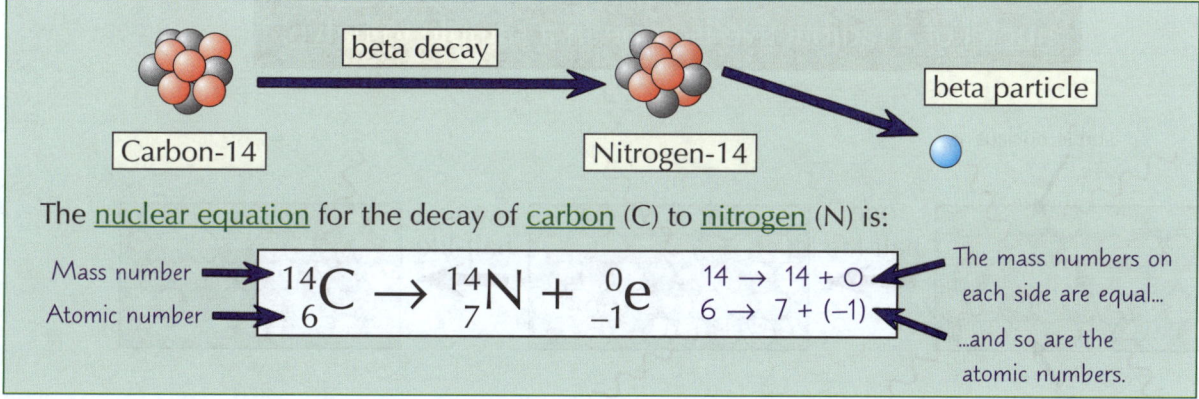

The nuclear equation for the decay of carbon (C) to nitrogen (N) is:

$$_{6}^{14}C \rightarrow\ _{7}^{14}N +\ _{-1}^{0}e$$

Mass number ↑
Atomic number ↑

$14 \rightarrow 14 + 0$ — The mass numbers on each side are equal...
$6 \rightarrow 7 + (-1)$ — ...and so are the atomic numbers.

Gamma Rays Don't Change the Charge or Mass of the Nucleus

1) Gamma rays are a way of getting rid of extra energy from a nucleus.
2) When they are emitted, they don't change the mass or charge of the atom and nucleus.

Beta particles are electrons but they come from the nucleus...
It seems a bit odd that beta decay causes the atomic number to increase. It makes sense if you understand what's happening to the particles — beta decay happens when a neutron turns into a proton and an electron. So there's one more proton, and the atomic number = number of protons.

Topic P4 — Atomic Structure

Half-Life

*How quickly **unstable nuclei** decay is measured using **activity** and **half-life** — two very important terms.*

The Activity of a Source is the Number of Decays per Second

1) The radiation given out by a radioactive decay can be measured with a Geiger-Muller tube and counter detector.

2) The number of decays the counter measures every second is called the count-rate.

3) The activity of a radioactive source is the rate at which it decays. This means how many unstable nuclei decay every second.

4) Activity is measured in becquerels, Bq. 1 Bq is 1 decay per second.

Radioactivity is a Totally Random Process

1) Radioactive decay is entirely random. So you can't predict exactly which nucleus in a sample will decay next, or when any one of them will decay.

2) But you can predict how long it will take for half of the nuclei to decay. This is known as a half-life.

> The half-life is the time taken for the number of nuclei of a radioactive isotope in a sample to halve.

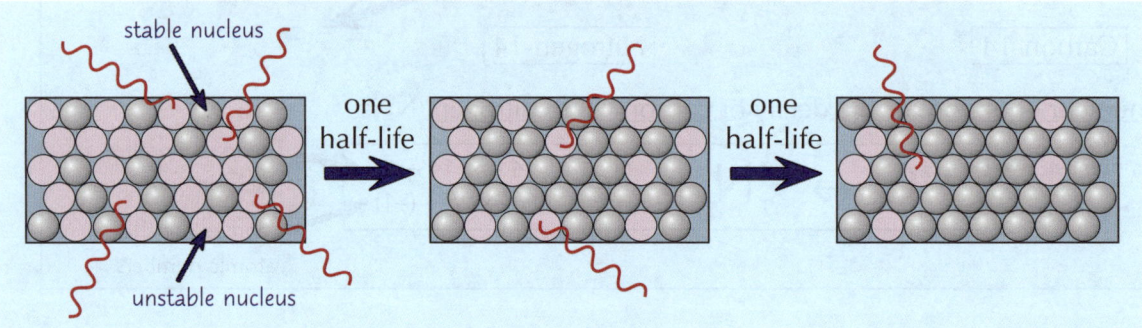

3) Half-life is also the time taken for the count-rate or activity of a sample to fall to half of its initial (starting) value.

4) The half-life of a radioactive sample will always be the same. This means it doesn't matter what activity you start with when doing half-life calculations (see the next page).

Different substances have different half-lives...

Some substances take a long time to decay, giving them a long half-life, while others decay in the blink of an eye. For example, neodymium-144 has a half-life of 2 million billion years, while helium-5 has a half-life of 7.6×10^{-22} seconds. That's 0.00000000000000000000076 seconds. Pretty speedy eh?

Topic P4 — Atomic Structure

Half-Life

*You learnt all about what **half-life** is on the last page, but now it's time to find out how to **calculate** it.*

You Need to be able to Calculate a Half-Life

1) You may be asked to calculate the half-life of a source.
2) You just need to find out how long it takes for the activity or count-rate of the source to halve.

 **The activity of a radioactive isotope was measured. Initially it was 64 Bq. 12 seconds later it had fallen to 16 Bq. Calculate the half-life of the sample.**

1) First, find how many half-lives it takes for the activity to fall to from 64 Bq to 16 Bq.

   After one half life, the activity will be 64 ÷ 2 = 32 Bq
   After two half lives, the activity will be 32 ÷ 2 = 16 Bq

2) So you know 12 s is equal to two half-lives. Divide 12 by 2 to find one half-life.

   Time for one half life = 12 ÷ 2 = 6 s

You Can Find the Half-Life of a Sample from a Graph of its Activity

1) A graph of activity against time is always shaped like the one below.
2) The half-life is found from the graph by finding the time interval on the bottom axis corresponding to a halving of the activity on the vertical axis. Easy.

 **The activity of a sample of a radioactive material, X, is shown on the graph below. Calculate the half-life of material X.**

1) Read the initial activity off the graph. This is the activity when time = 0.
2) Divide the initial activity by 2 to find the value of half the initial activity.
   80 ÷ 2 = 40
3) Find this value on the y-axis and read along horizontally to the curve.
4) Then read down from the curve at this point to find the half-life.

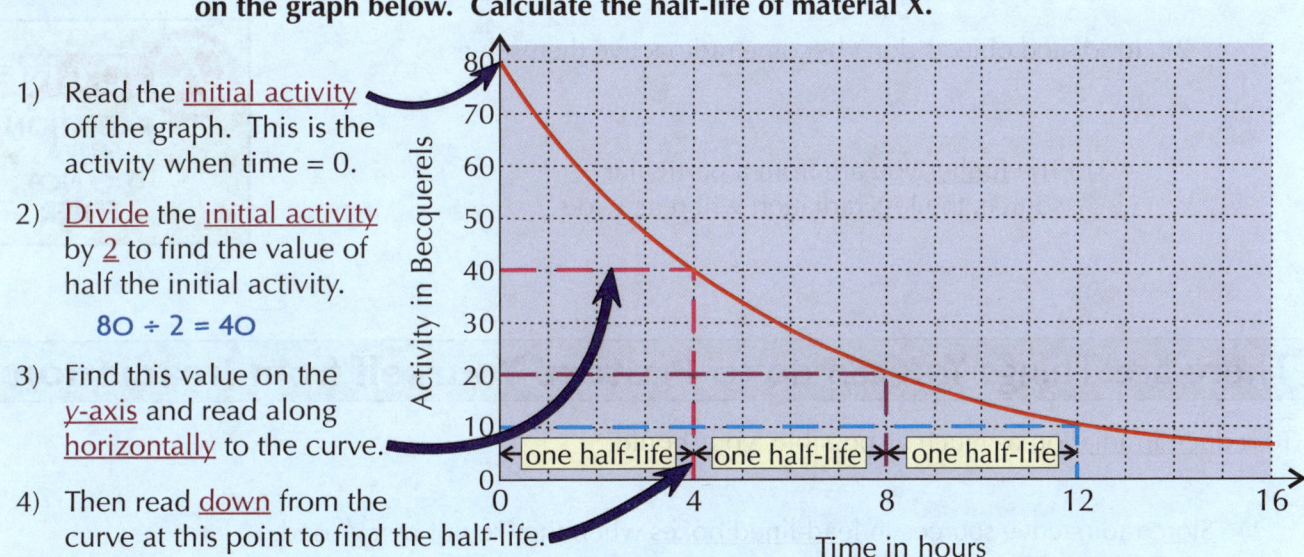

So the half-life of the sample is **4 hours**.

You can get useful information from activity-time graphs...

Make sure you can use graphs like the one above to work out half-lives. All you've got to do is read off the initial activity from the y-axis, then work out what half this activity would be by dividing by two. Then, just read off the time from the x-axis for this value, which is one half-life.

Q1 The activity of a radioactive isotope fell from 82 Bq to 41 Bq in 16 seconds. What is the half-life of the sample? [1 mark]

Q2 The initial count-rate of a sample is 168 counts per second (cps). After 60 minutes, the count-rate of the sample is 21 cps. Calculate the half-life of the sample. [3 marks]

Topic P4 — Atomic Structure

Irradiation and Contamination

*You need to be **very careful** when working with radiation. You don't even need to touch a source to be **irradiated** by it — just being near it is enough. Make sure you know how to **protect yourself**.*

Ionising Radiation can Damage Cells

1) There are hazards (dangers) you need to protect yourself from when working with radioactive sources.
2) Ionising radiation can enter living cells and ionise (p.337) atoms in them.
3) This can damage the cells. This may cause cancer or kill cells off completely.

> We need to know how radiation affects the human body, so we can protect ourselves from harm.
> So it's really important for research into the effects of radiation to be published and checked by peer review — see page 2.

Exposure to Radiation is called Irradiation

1) Objects near a radioactive source can be irradiated by it. This means radiation from the radioactive source will reach the object.

2) Irradiated objects don't become radioactive themselves.

3) The further you are from a particular source, the less radiation will reach you.

There are Things You can do to Protect Yourself from Irradiation

To reduce irradiation as much as possible, you should:

1) Store radioactive sources in lead-lined boxes when they're not being used.

2) Stand behind barriers that will absorb radiation when using sources.

3) Keep the source as far away from you as possible, e.g. hold it at arm's length.

Safety precautions can help protect against hazards from radiation...

Radiation can be pretty dangerous stuff, so it's important to protect yourself when you're working with radioactive substances. Lead is often used to line storage boxes and in protective screens because it is very good at absorbing radiation, so it stops a lot of the radiation from reaching what's on the other side.

Topic P4 — Atomic Structure

Irradiation and Contamination

*On the previous page you saw how **irradiation** can result from even just being near a source. Now for **radioactive contamination**. This only happens if you **touch** or **breathe in** a radioactive substance.*

Contamination is Radioactive Particles Getting Onto Objects

1) If unwanted radioactive atoms get onto or into an object, the object is contaminated.

2) These contaminating atoms might then decay and release radiation which could harm you.

3) Being contaminated by a source may cause more damage than if you are irradiated by the same source, as you may carry it for a long time.

4) To help stop contamination, you should wear gloves and use tongs when handling radioactive sources.

People whose jobs involve radioactive materials often wear protective suits and face masks to help stop them being contaminated.

Irradiation and Contamination Dangers Depend on the Source

The amount of harm contamination or irradiation by a source can cause depends on the radiation type.

Irradiation

1) Beta and gamma sources are the most dangerous to be irradiated by. These types of radiation have long ranges (p.337). That means more radiation will reach you from a beta or gamma source than from an alpha source at the same distance.

2) They can also penetrate (travel through) your body and may damage your organs. Alpha is less dangerous because it can't get through the skin and is easily blocked, e.g. by a small air gap (p.337).

Contamination

1) INSIDE the body, alpha sources are the most dangerous to be contaminated by. This is because alpha particles are the most ionising type of radiation.

2) Beta particles and gamma rays are less damaging because they are less ionising.

3) Gamma sources are the least dangerous inside the body. This is because gamma rays are the least ionising type of radiation and they mostly pass straight out without doing any damage.

4) OUTSIDE of the body, an alpha source is the least dangerous to be contaminated by. This is because alpha particles can't get through the skin and damage your organs.

Topic P4 — Atomic Structure

Warm-Up & Exam Questions

Atoms may be tiny, but you could bag some big marks in your exams if you know them inside-out. Here are some questions to check just how great your understanding of atoms and radiation really is...

Warm-Up Questions

1) Why was the plum pudding model of an atom replaced with a nuclear model?
2) Describe our current nuclear model of the atom.
3) What is emitted when an electron moves to a lower energy level?
4) What does an atom become if it loses an electron?
5) What does the mass number tell you about an atom?
6) Describe what happens to the mass number and atomic number of an atom if it undergoes alpha decay. What happens to them if the atom undergoes gamma decay?
7) What is the difference between contamination and irradiation by a radioactive source?
8) Why are gamma sources the least dangerous type of radioactive source to have inside you?

Exam Questions

1 Two different atoms can be isotopes of one another.

1.1 Name the particles found in the nucleus of an atom.

[2 marks]

1.2 What is meant by the term **isotopes**?

☐ Atoms with the same atomic number but a different mass number

☐ Atoms with the same mass number but a different atomic number

☐ Atoms with the same number of protons but a different atomic number

☐ Atoms with the same number of neutrons but a different number of electrons

[1 mark]

2 Table 1 shows some information about iodine-131 ($^{131}_{53}$I), an isotope of iodine.

Table 1

Particle	Type of Charge	Number present in an atom of iodine-131
Proton	positive	
Neutron	zero	
Electron		53

Complete **Table 1**.

[3 marks]

Topic P4 — Atomic Structure

Exam Questions

3 This question is about the effects on the nucleus of different types of radioactive decay. *Grade 4-5*
Draw **one** line from each description to the correct type of decay.

Both the mass and charge of the nucleus change.	alpha
Neither the mass or charge of the nucleus changes.	beta
	gamma
The charge of the nucleus changes but the mass stays the same.	neutron

[3 marks]

4 Sources A, B and C each emit a single type of radiation. Radiation from each source was directed at thin sheets of paper and aluminium. A detector was used to measure where radiation had passed through the sheets. The results are shown in **Figure 1**. *Grade 4-5*

Figure 1

- Source A: passes through paper and aluminium — radiation detected
- Source B: passes through paper, stopped by aluminium — no radiation detected after aluminium; radiation detected after paper
- Source C: stopped by paper — no radiation detected

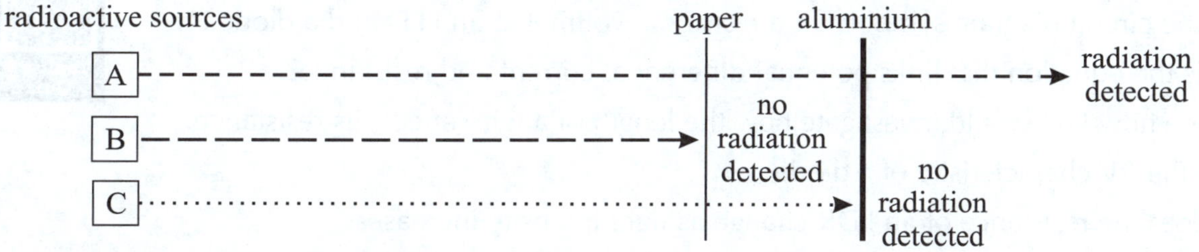

4.1 Name the type of radiation that source C emits. Explain your answer.

[2 marks]

4.2 Give **one** example of a detector that could have been used to detect the radiation.

[1 mark]

5 A student measured the activity of a radioactive source every 10 minutes. Her results are shown in **Table 2**. *Grade 4-5*

Table 2

Time (mins)	0	10	20	30	40	50
Activity (Bq)	740	575	450	350	270	210

5.1 Which value below is the half-life of the source?

☐ 20 minutes ☐ 28 minutes ☐ 34 minutes ☐ 40 minutes

[1 mark]

5.2 The source emits only alpha radiation. It must be removed from a lead-lined box before the activity can be measured. The student believes that they should be more concerned about avoiding contamination by the source rather than irradiation by it. Suggest why.

[2 marks]

Topic P4 — Atomic Structure

Revision Summary for Topics P1-4

Well, that's Topics P1-4 — hopefully they weren't too painful.
Time to see how much you've absorbed.
- Try these questions and tick off each one when you get it right.
- When you're completely happy with a topic, tick it off.

For even more practice, try the Retrieval Quizzes for Topics P1-4 — just scan the QR codes!

Topic P1 — Energy (p.283-303) ☑
1) Write down four energy stores.
2) What kind of energy store is energy transferred to when you heat an object?
3) Give the equation for finding the energy in an object's gravitational potential energy store.
4) What units must extension, e, be in before it can be used in the equation $E_e = \frac{1}{2}ke^2$?
5) Describe an experiment to find the specific heat capacity of a material.
6) Give two equations you could use to calculate power.
7) How can you reduce unwanted energy transfers in a machine with moving parts?
8) Name four renewable energy resources and four non-renewable energy resources.
9) Give one environmental issue associated with using biofuels to generate electricity.
10) Explain why the UK is trying to use more renewable energy resources in the future.

Topic P2 — Electricity (p.305-324) ☑
11) Draw the circuit diagram symbols for a resistor, a voltmeter, an LED and a diode.
12) What is the equation that links potential difference, current and resistance?
13) Describe how you would investigate how the length of a wire affects its resistance.
14) Sketch the I-V characteristic of a diode.
15) How does the resistance of an LDR change as light intensity increases?
16) A circuit consists of a battery and a resistor. How does the overall resistance of the circuit change when a second resistor is added in parallel?
17) What is the potential difference and the frequency of the UK mains supply?
18) Give the potential differences for the three wires in a three-core mains cable.
19) Describe the useful energy transfer that occurs for a battery-powered fan.
20) Explain why electricity is transferred by the national grid at a high pd but low current.

Topic P3 — Particle Model of Matter (p.326-331) ☑
21) Describe how particles are arranged in a liquid.
22) Briefly describe an experiment to find the density of a liquid.
23) What happens to the particles in a substance when that substance is heated?
24) Is a change of state a physical change or a chemical change?

Topic P4 — Atomic Structure (p.334-343) ☑
25) What happens when an electron in an atom absorbs electromagnetic radiation?
26) How do atoms of two isotopes of an element differ from each other?
27) How does the emission of a beta particle change the atomic number of an atom?
28) Describe how to find a radioactive source's half-life, given a graph of its activity over time.
29) Give two examples of how to stop: a) irradiation, b) contamination.
30) Compare the hazards of being irradiated and contaminated by a gamma source.

Topic P5 — Forces

Contact and Non-Contact Forces

*When you're talking about the **forces** acting on an object, it's not enough to just talk about the **size** of each force. You need to know their **direction** too — force is a **vector**, with a size and a direction.*

Force is a Vector Quantity

1) Vector quantities have a magnitude (size) and a direction. For example:

> Vector quantities: force, velocity, displacement, acceleration

2) Some quantities have a magnitude but no direction. These are called scalar quantities. Here are some examples of scalar quantities:

> Scalar quantities: speed, distance, mass, temperature, time

3) Vectors are usually represented by an arrow.
4) The length of the arrow shows the magnitude.
5) The direction of the arrow shows the direction of the quantity.

Forces Can be Contact or Non-Contact

1) A force is a push or a pull that acts on an object.
2) Forces are caused by objects interacting with each other.
3) All forces are either contact or non-contact forces.
4) When two objects have to be touching for a force to act, the force is a contact force.

> Contact force examples: friction, air resistance, tension, normal contact force

When an object exerts a force on a second object, the second object pushes back. This is the normal contact force.

5) If the objects do not need to be touching for the force to act, the force is a non-contact force.

> Non-contact force examples: magnetic force, gravitational force, electrostatic force

6) When two objects interact, a force is produced on both objects. The forces on the two objects are equal in size but act in opposite directions.
7) These two forces are called an interaction pair.

The gravitational attraction between the Earth and the Sun is an example of an interaction pair.

A gravitational force acts on the Earth attracting it to the Sun.

At the same time a force acts on the Sun attracting it towards the Earth.

These forces are the same size but act in opposite directions.

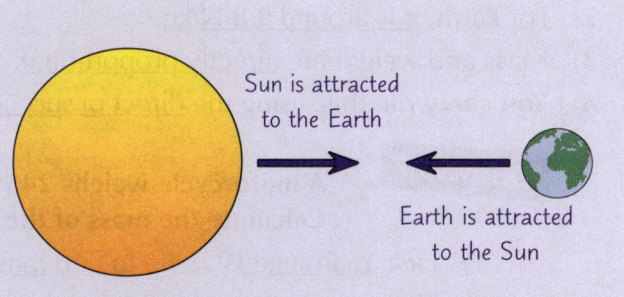

Sun is attracted to the Earth

Earth is attracted to the Sun

An interaction between two objects produces a force on each object...

Whether it's a contact or non-contact force, it's always acting as part of a pair. Remember it's the two forces, not the objects, which are called the interaction pair — and they are always equal and opposite.

Weight, Mass and Gravity

*Mass and weight are **NOT** the same... Read on to find out **why**. You know you want to.*

Mass is Measured in kg

1) Mass is just the amount of matter (stuff) in an object.

2) It's measured in kilograms, kg.

3) Scientists sometimes think of all the mass in an object as being at one single point in the object.

4) This point is called the centre of mass.

Weight is Measured in Newtons

1) Weight is the force acting on an object due to gravity.
2) You can think of this force as acting from an object's centre of mass.
3) Close to Earth, this force is caused by the gravitational field around the Earth.
4) The weight of an object depends on its mass and the strength of the gravitational field it's in.
5) The gravitational field strength of Earth changes slightly depending on where you are.
6) So the weight of an object depends on its location.
7) Unlike the mass of an object which is always the same.
8) Weight is measured in newtons, N.
9) It can be measured with a calibrated spring balance (or newtonmeter).

Mass and Weight are Directly Proportional

1) You can calculate the weight of an object if you know its mass (*m*) and the strength of the gravitational field that it is in (*g*):

Weight (N) = Mass (kg) × Gravitational Field Strength (N/kg)

2) For Earth, *g* is around 9.8 N/kg.
3) Mass and weight are directly proportional.
4) You can write this, using the direct proportionality symbol, as $W \propto m$.

EXAMPLE A motorcycle weighs 2450 N on Earth. Calculate the mass of the motorcycle. (*g* = 9.8 N/kg)

1) First, rearrange $W = mg$ to find mass.   mass = weight ÷ gravitational field strength
2) Then, put in the numbers to calculate the mass.   mass = 2450 ÷ 9.8 = 250 kg

---

An object's weight depends on the gravitational field it is in

Q1 Calculate the weight in newtons of a 5 kg mass:
   a) on Earth (*g* ≈ 9.8 N/kg)   b) on the Moon (*g* ≈ 1.6 N/kg)   [4 marks]

Resultant Forces and Work Done

*I'm sure you're no stranger to **doing work**, but in physics it's all to do with **transferring energy**.*

A Resultant Force is the Overall Force on a Point or Object

1) If a number of forces act at a single point, you can replace them with a single force.
2) This single force is called the resultant force.
3) It has the same effect as all the original forces added together.
4) You can find the resultant force when forces are acting in a straight line. Add together forces acting in the same direction and take away any going in the opposite direction.

A trolley is pulled by two children. One child pulls the trolley with a force of 5 N to the left. The other pulls the trolley with a force of 10 N to the right.

So the resultant force, F, is:
$F = 10\ N - 5\ N$
$= \underline{5\ N\ \text{to the right}}$.

If a Force Moves an Object, Work is Done

> When a force moves an object through a distance, energy is transferred and work is done on the object.

1) To make an object move, a force must act on it.
2) The force does 'work' to move the object.
3) This causes energy to be transferred to the object.
4) The force usually does work against frictional forces too.
5) Doing work against frictional forces causes energy to be transferred to the thermal energy store of the object.
6) This causes the temperature of the object to increase.

'Work done' and 'energy transferred' are the same thing. You need to be able to describe how energy is transferred when work is done. Look back at p.285 for more on this.

- When you push something along a rough surface (like a carpet) you are doing work against frictional forces.
- Some energy is transferred to the kinetic energy store of the object because it starts moving.
- Some is also transferred to thermal energy stores due to the work done against friction.
- This causes the overall temperature of the object to increase.

7) You can find out how much work has been done using:

Work done (J) — $W = Fs$ — Force (N), Distance (moved along the line of action of the force) (m)

The line of action of the force is the direction of the force.

8) One joule of work is done when a force of one newton causes an object to move a distance of one metre in the direction of the force.
9) You need to be able to convert joules to newton metres: $1\ J = 1\ Nm$.

Work done is measured in joules, J — one J equals one Nm...

Although "work done" may sound like an odd phrase, all it means is "energy transferred".

Q1 A force of 20 N pushes an object 20 cm. Calculate the work done on the object. [3 marks]

Q1 Video Solution

Warm-Up & Exam Questions

Now you've learnt the basics of forces, it's time to act on your new knowledge.
Give these questions a go and test how well you've forced those facts into your brain.

Warm-Up Questions

1) Which of the following is a contact force: magnetic force, gravitational force or friction?
2) Vectors can be drawn as arrows. What is represented by the length of the arrow?
3) State the units of: a) mass, b) weight.
4) If two forces are acting in the same direction, how can you find the resultant force?

Exam Questions

1 Most quantities can be divided into two groups: scalars and vectors. *(Grade 1-3)*

1.1 Which of the following is a vector? Tick **one** box.

☐ speed   ☐ time   ☐ mass   ☐ force

[1 mark]

1.2 Which of the following is a scalar? Tick **one** box.

☐ 14 kg   ☐ 300 kN down   ☐ 24 m/s west   ☐ 1 m/s^2 up

[1 mark]

2 **Figure 1** shows two hot air balloons, labelled with the forces acting on them. *(Grade 3-4)*

Figure 1

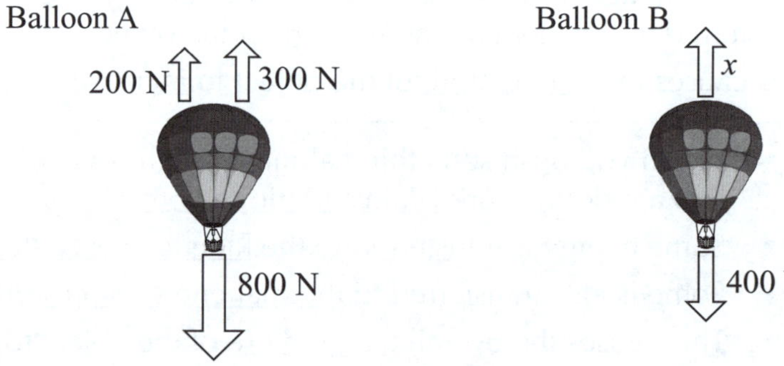

Balloon A: 200 N ↑, 300 N ↑, 800 N ↓

Balloon B: x ↑, 400 N ↓

2.1 Calculate the size of the resultant force acting on Balloon A and give its direction.

[2 marks]

2.2 The resultant force acting on Balloon B is zero. Calculate the size of force x.

[1 mark]

3 A train moves 750 m in a straight line along a flat track.
The resultant force acting on the train is 44 000 N forwards along the track. *(Grade 4-5)*

3.1 Write down the equation that links work done, force and distance.

[1 mark]

3.2 Calculate the work done by the resultant force as the train moves. Give your answer in kilojoules.

[3 marks]

Topic P5 — Forces

Forces and Elasticity

Forces don't just make objects **move**, they can also make them **deform** (change shape).

Stretching, Compressing or Bending Transfers Energy

1) When you apply a force to an object you may cause it to <u>deform</u> (<u>stretch</u>, <u>compress</u> or <u>bend</u>).

2) To do this, you need <u>more than one</u> force acting on the object. <u>One</u> force would just make the object <u>move</u>, not change its shape.

3) An object has been <u>elastically deformed</u> if it can <u>go back</u> to its <u>original shape</u> and <u>length</u> after the force has been removed.

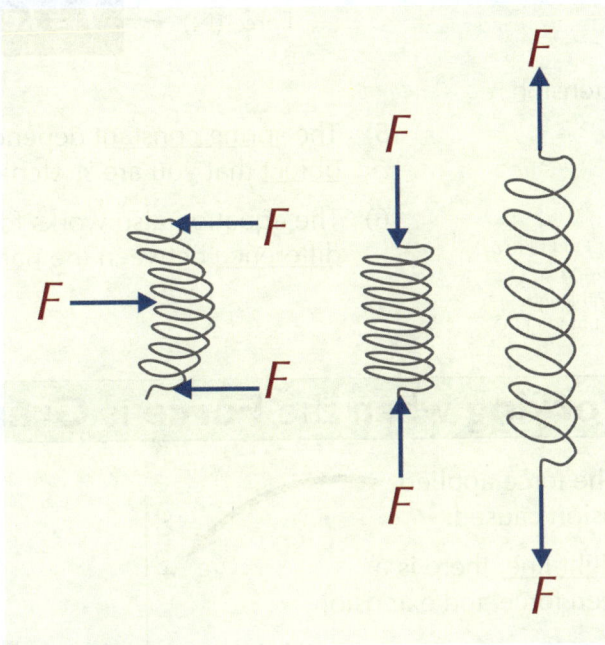

4) If the object <u>doesn't</u> go back to how it was, it has been <u>inelastically deformed</u>.

5) Objects that can be elastically deformed are called <u>elastic objects</u> (e.g. a spring).

6) <u>Work is done</u> when a force stretches or compresses an object. This causes energy to be transferred to the <u>elastic potential energy</u> store of the object.

Elastic objects are only elastic up to a certain point...
Remember the difference between <u>elastic deformation</u> and <u>inelastic deformation</u>. If an <u>object</u> has been <u>elastically deformed</u>, it will <u>return</u> to its <u>original shape</u> when you <u>remove the force</u>. If it's been <u>inelastically deformed</u>, its shape will have been <u>changed permanently</u> — for example, an over-stretched spring will stay stretched even after you remove the force.

Topic P5 — Forces

Forces and Elasticity

Springs obey a really handy little *equation* that relates the *force* on them to their *extension* — for a while at least. Thankfully, you can *plot a graph* to see where this equation is *true*.

Extension is Directly Proportional to Force...

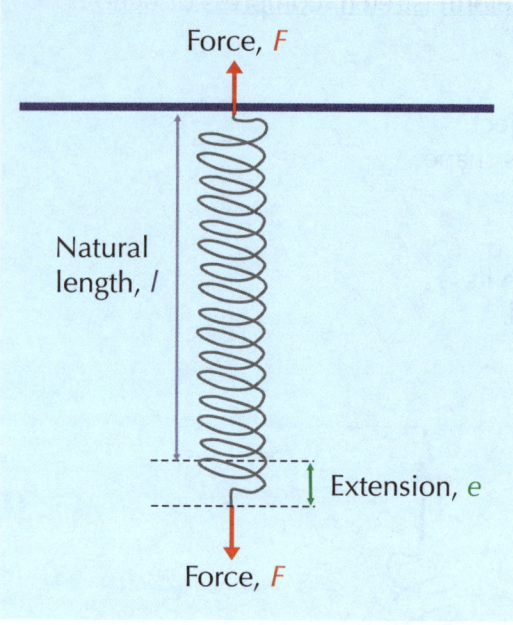

The length of the unstretched spring is sometimes called the spring's natural length.

1) When a force stretches a spring, it causes it to extend.
2) This extension is the difference in length between the stretched and unstretched spring.
3) Up to a given force, the extension is directly proportional to force.
4) So long as a spring hasn't been stretched past its limit of proportionality (see below), you can use:

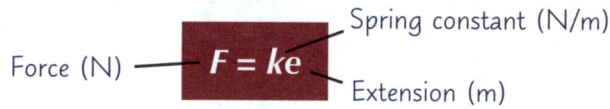

$$F = ke$$

Force (N), Spring constant (N/m), Extension (m)

5) The spring constant depends on the object that you are stretching.
6) The equation also works for compression (where e is the difference between the natural and compressed lengths).

...but this Stops Working when the Force is Great Enough

1) You can plot a graph of the force applied to a spring and the extension caused.
2) When the graph is a straight line, there is a linear relationship between force and extension.
3) This shows force and extension are directly proportional.
4) The gradient of the straight line is equal to k, the spring constant.
5) When the line begins to bend, the relationship is now non-linear. Force and extension are no longer directly proportional.
6) Point P on the graph (when the line starts to bend) is the limit of proportionality. Past this point, the equation $F = ke$ is no longer true.

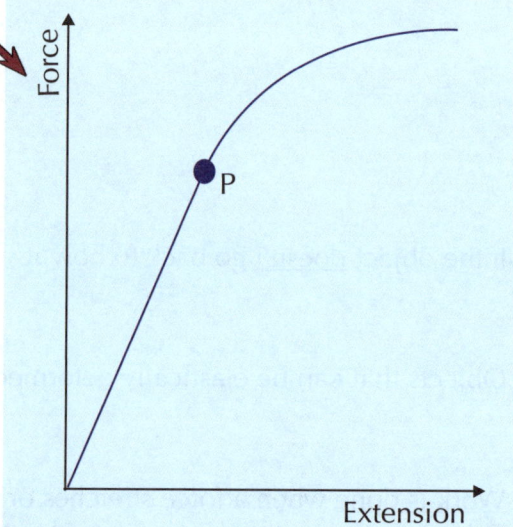

The spring constant is measured in N/m...

Be careful with units when doing calculations with springs. Your values for extension will usually be in centimetres or millimetres, but the spring constant is measured in newtons per metre. So convert the extension into metres before you do any calculations, or you'll get the wrong answer.

Q1 A spring is fixed at one end and a force of 1 N is applied to the other end, causing it to stretch. The spring extends by 2 cm. Calculate the spring constant of the spring. [4 marks]

Topic P5 — Forces

Investigating Springs

Oh look, here's one of those **Required Practicals**... There might be a few ways this experiment can *stretch* you. **Read on** and take your time with **each step**, so you won't be past your limits in the exam.

You Can **Investigate** the Link Between **Force** and **Extension**

1) Set up the apparatus as shown in the diagram.

2) Measure the mass of each mass.

3) Calculate its weight (the force applied) using $W = mg$ (p.348).

4) Measure the original (natural) length of the spring.

5) Add a mass to the spring and allow it to come to rest.

6) Record the force and measure the new length of the spring.

7) Find the extension.

extension = new length − original length

8) Repeat steps 5 to 7 until you've added all the masses.

9) Plot a force-extension graph of your results.

10) You should make sure you have at least 5 measurements before the limit of proportionality (where the line starts to curve).

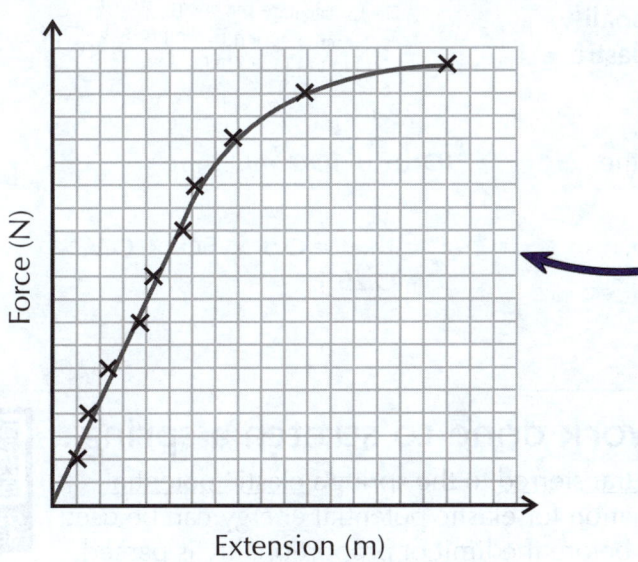

Topic P5 — Forces

Investigating Springs

*Believe it or not there's another **equation** involving the **extension** of a spring — lucky you.*

You Can **Work Out Energy Stored** for **Linear** Relationships

1) If a spring is not stretched <u>past</u> its <u>limit of proportionality</u>, the <u>work done</u> in stretching the spring can be found using:

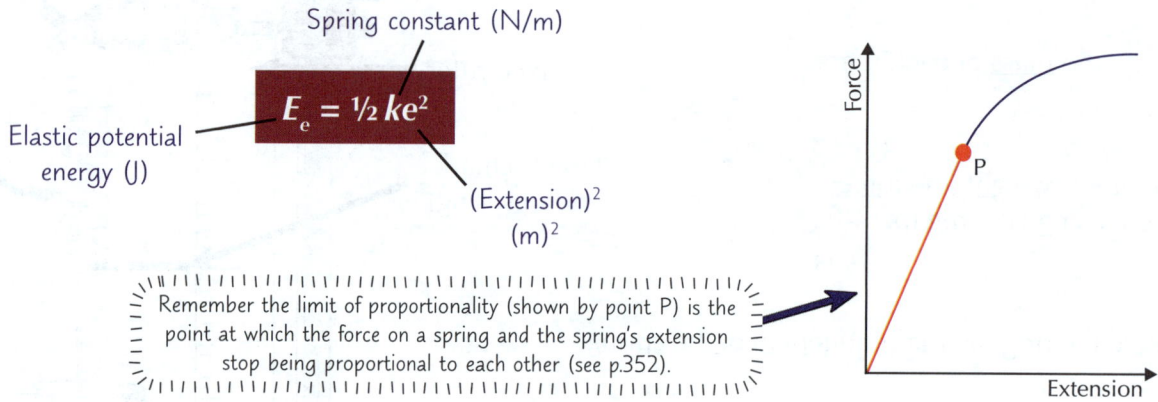

$$E_e = \tfrac{1}{2} k e^2$$

- Elastic potential energy (J)
- Spring constant (N/m)
- (Extension)² (m)²

Remember the limit of proportionality (shown by point P) is the point at which the force on a spring and the spring's extension stop being proportional to each other (see p.352).

2) For an <u>elastic deformation</u>, this formula can be used to calculate the energy stored in a spring's <u>elastic potential energy store</u>.

3) It's also the energy <u>transferred to</u> the spring as it's <u>deformed</u>, or <u>transferred by</u> the spring as it returns to its <u>original shape</u>.

EXAMPLE

A spring with a spring constant of 50 N/m extends elastically by 10 cm. It doesn't pass its limit of proportionality. Calculate the amount of energy stored in its elastic potential energy store.

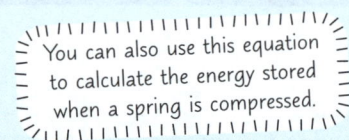
You can also use this equation to calculate the energy stored when a spring is compressed.

1) First, you need to <u>convert</u> the extension of the spring into <u>metres</u>.

   10 cm ÷ 100 = 0.1 m

2) Then put in the <u>numbers</u> you've been given.

   $E_e = \tfrac{1}{2} k e^2 = 0.5 \times 50 \times 0.1^2$
   $= 0.25$ J

Elastic potential energy equals the work done to stretch a spring...

As a spring is stretched <u>work is done</u> and the <u>energy is transferred</u> to the spring's <u>elastic potential energy store</u>, so the two are <u>equal</u>. This means the equation for elastic potential energy can be used to calculate work done — but remember it only works before the <u>limit of proportionality</u> is passed.

Q1 Video Solution

Q1   A spring with a spring constant of 40 N/m extends elastically by 2.5 cm. Calculate the amount of energy stored in its elastic potential energy store.   [3 marks]

Topic P5 — Forces

Warm-Up & Exam Questions

It's time to stretch those thinking muscles with another round of questions. Give these a go to test the limits of your newly extended knowledge of springs and elasticity.

Warm-Up Questions

1) True or false? An object which is inelastically deformed will not return to its original shape when the force is removed.
2) True or false? The formula $F = ke$ cannot be used if a spring is compressed.
3) What units must extension be in before you use the formula for work done on a spring?

Exam Questions

1 A force of 4 N is applied to a spring which causes the spring to extend elastically by 0.05 m.

1.1 Write down the equation which links force, extension and the spring constant.

[1 mark]

1.2 Calculate the spring constant of the spring.

[3 marks]

PRACTICAL

2 A teacher does an experiment to show how a spring extends when masses are hung from it, using the setup shown in **Figure 1**.

Figure 1

He hangs a number of 90 g masses from a 50 g hook attached to the base of the spring. He records the extension of the spring and the total weight of the masses and hook each time he adds a mass to the bottom of the spring.

2.1 State **one** control variable in this experiment.

[1 mark]

2.2 State the independent variable in this experiment.

[1 mark]

Figure 2 shows the force-extension graph of the results of the experiment.

2.3 Calculate the work done when the spring is stretched with a force of 7.0 N. The spring constant is 175 N/m. Use the correct equation from the Physics Equation Sheet on page 538.

Choose the correct unit from the box.

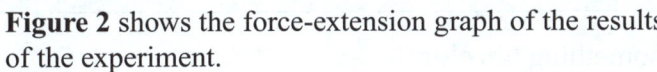

[3 marks]

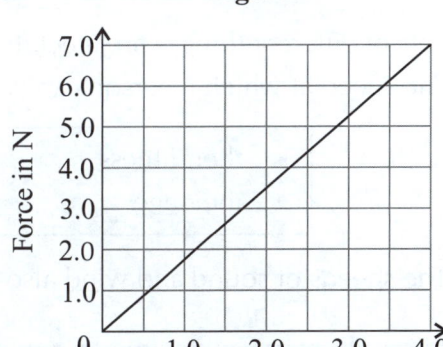

Figure 2

Topic P5 — Forces

Distance, Displacement, Speed, Velocity

*This page is about a bunch of **scalars** and **vectors**. Still a little unsure on the difference between **scalar** and **vector quantities**? Take a look back at p.347 before starting this page.*

Distance is a Scalar, Displacement is a Vector

1) Distance is just how far an object has moved.
2) Distance is a scalar quantity (p.347), so it doesn't involve direction.
3) Displacement is a vector quantity.
4) It measures the distance and direction in a straight line from an object's starting point to its finishing point.
5) The direction could be in relation to a point, e.g. towards the school.
6) If you walk 5 m north, then 5 m south, your displacement is 0 m but the distance travelled is 10 m.

Speed and Velocity are Both How Fast You're Going

1) Speed is a scalar and velocity is a vector:

> Speed is just how fast you're going (e.g. 30 mph or 20 m/s) with no regard to the direction.
> Velocity is speed in a given direction, e.g. 30 mph north or 20 m/s to the right.

2) To measure the speed of an object that's moving with a constant speed, time how long it takes the object to travel a certain distance. Make sure you use the correct equipment (see p.399).
3) You can then calculate the object's speed using this formula:

$$s = vt \qquad \text{distance travelled (m) = speed (m/s) × time (s)}$$

4) Objects rarely travel at a constant speed.
5) When you walk, run or travel in a car, your speed is always changing.
6) In these cases, the formula above gives the average (mean) speed during that time.

You Need to Know Some Typical Everyday Speeds

1) You need to know the typical (usual) speeds of objects:

> A person walking — 1.5 m/s A person cycling — 6 m/s   A train — 30 m/s
> A person running — 3 m/s   A car — 25 m/s   A passenger plane — 250 m/s

2) Lots of different things can affect the speed something travels at.
3) The speed at which a person can walk, run or cycle depends on, among other things:

> • their fitness   • the distance they've travelled
> • their age   • the terrain (what type of ground they are on)

4) The speeds of sound and wind also vary. A typical speed for sound in air is 330 m/s.

You can also think of that equation as speed = distance ÷ time...

Remember those typical speeds of objects — you might need to use them to make estimates.
Q1   A sprinter runs 200 m in 25 s. Calculate his speed. [3 marks]

Topic P5 — Forces

Acceleration

*Acceleration is the **rate of change** of velocity — if it stays **constant**, there's a handy **equation** you can use.*

Acceleration is How Quickly You're Speeding Up

1) Acceleration is the change in velocity in a certain amount of time.
2) You can find the average acceleration of an object using:

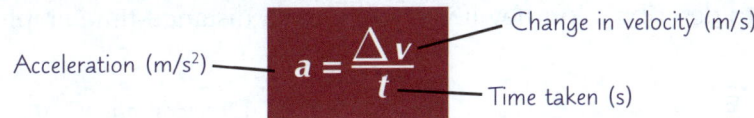

Acceleration (m/s²) — $a = \dfrac{\Delta v}{t}$ — Change in velocity (m/s), Time taken (s)

3) Deceleration (when something slows down) is just negative acceleration.

You Need to be Able to Estimate Accelerations

You might have to estimate the acceleration of an object. To do this, you need the typical speeds from the previous page:

An estimate is just a guess using rough numbers for things.

EXAMPLE A woman gets onto a bike and accelerates to a typical speed from stationary in 10 seconds. Estimate the acceleration of the bicycle.

1) First, give a sensible speed for the bicycle to be travelling at.
2) Put these numbers into the acceleration equation.
3) The ~ symbol just means it's an approximate answer.

The typical speed of a bike is about 6 m/s.
The bicycle accelerates in 10 s.
$a = \Delta v \div t$
$= 6 \div 10 = 0.6$ m/s²
So the acceleration is ~0.6 m/s²

Uniform Acceleration Means a Constant Acceleration

1) Constant acceleration is sometimes called uniform acceleration.
2) Acceleration due to gravity (*g*) is uniform for objects falling freely.
3) It's roughly equal to 9.8 m/s² near the Earth's surface.
4) You can use this equation for uniform acceleration:

(Final velocity)² (m/s)² — $v^2 - u^2 = 2as$ — Acceleration (m/s²), Distance (m)
(Initial velocity)² (m/s)²

Initial velocity is just the starting velocity of the object.

EXAMPLE A van travelling at 23 m/s starts decelerating uniformly at 2.0 m/s² as it heads towards a built-up area 112 m away. What will its speed be when it reaches the built-up area?

1) First, rearrange the equation so v^2 is on one side.
2) Now put the numbers in — remember *a* is negative because it's a deceleration.
3) Finally, square root the whole thing.

$v^2 = u^2 + 2as$
$v^2 = 23^2 + (2 \times -2.0 \times 112)$
$= 81$
$v = \sqrt{81} = 9$ m/s

An object is accelerating if its velocity is changing...

Don't forget — you can only use that second equation when the acceleration is uniform.

Q1 A ball is dropped from a height, *h*, above the ground. The speed of the ball just before it hits the ground is 7 m/s. Calculate the height the ball is dropped from. (acceleration due to gravity ≈ 9.8 m/s²)

[3 marks]

Q1 Video Solution

Topic P5 — Forces

Distance-Time Graphs

*You need to be able to **draw** and **understand** distance-time graphs.*

You Can Show Journeys on Distance-Time Graphs

1) If an object moves in a straight line, the distance it travels can be plotted on a distance-time graph.
2) You may be asked to draw a distance-time graph for a journey.
3) Or you might have to describe a journey if you're shown a distance-time graph.

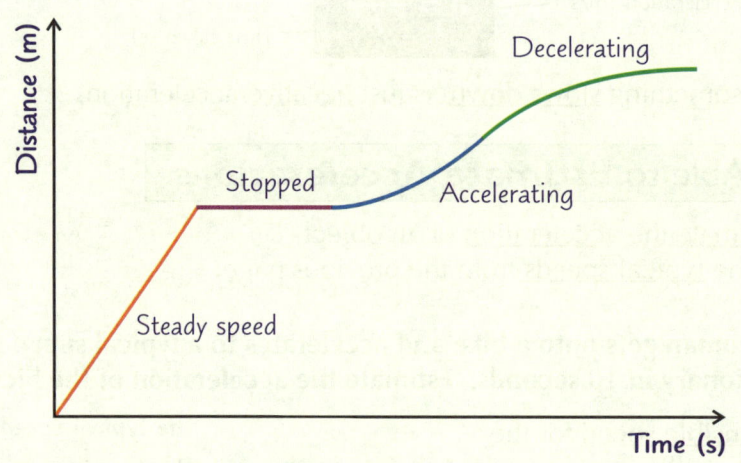

- Gradient = speed. The steeper the graph, the faster it's going.
- Flat sections are where it's stationary — it's stopped.
- Straight uphill (/) sections mean it is travelling at a steady speed.
- Curves represent acceleration or deceleration (p.357).
- A curve that is getting steeper means it's speeding up (accelerating).
- A levelling off curve means it's slowing down (decelerating).

4) You might also have to calculate an object's speed from the graph:

EXAMPLE

Using the distance-time graph on the right, calculate the speed of the car.

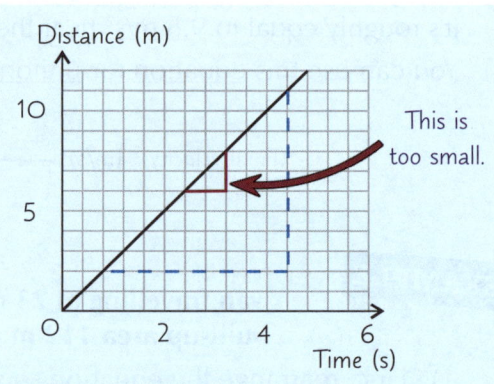

1) The gradient of the graph is the speed of the car.
2) Gradient = $\dfrac{\text{change in vertical axis}}{\text{change in horizontal axis}}$.
3) Draw a large triangle, that takes up most of the straight line.
4) Use the horizontal side of the triangle to find the change in time.
5) Use the vertical side of the triangle to find the change in distance.
6) Put the values for vertical and horizontal into the equation.

Change in time = 4.4 − 0.8 = 3.6 s
Change in distance = 11 − 2 = 9 m
Gradient = 9 ÷ 3.6 = 2.5
So speed = 2.5 m/s

A steeper gradient means a faster speed...

Make sure you know how to use a distance-time graph to find an object's speed.

Q1 Sketch the distance-time graph for an object that accelerates, then travels at a steady speed, and then comes to a stop. [3 marks]

Q1 Video Solution

Velocity-Time Graphs

You also need to know about **velocity-time graphs** — they can be used to calculate **acceleration**.

Journeys Can be Shown on a **Velocity-Time Graph**

1) How an object's velocity changes as it travels can be plotted on a velocity-time graph.
2) You might have to draw a velocity-time graph for a journey.
3) Or you might have to describe a journey from a velocity-time graph.

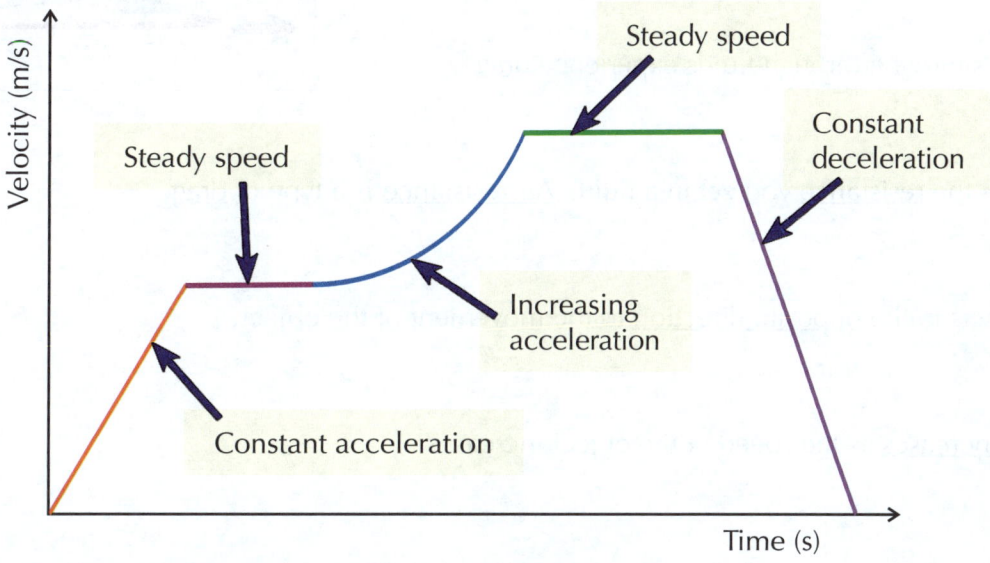

1) Gradient = acceleration. You can calculate this with a similar method to the example on the last page.

2) Flat sections represent travelling at a steady speed.

3) The steeper the graph, the greater the acceleration or deceleration.

4) Uphill sections (/) are acceleration.

5) Downhill sections (\) are deceleration.

6) A curve means changing acceleration.

Read the axes of any graph you get given carefully...

Don't get confused between distance-time graphs and velocity-time graphs. They can look similar, but different parts of the graphs tell you very different things about the motion of an object.

Q1 Sketch a velocity-time graph for an object that travels at a constant speed, then accelerates at a constant rate, then moves at a constant speed (that is different to the initial speed). [3 marks]

Q2 A stationary car starts accelerating increasingly for 10 s until it reaches a speed of 20 m/s. It travels at this speed for 20 s until the driver sees a hazard and brakes. He decelerates uniformly, coming to a stop 4 s after braking. Draw the velocity-time graph for this journey. [3 marks]

Topic P5 — Forces

Drag and Terminal Velocity

If an object **falls** for long enough, it will reach its **terminal velocity**. It's all about **balance** between **weight** and **air resistance**. Parachutes work by **decreasing** your terminal velocity.

Drag Always Slows Things Down

1) Gases and liquids are both fluids.

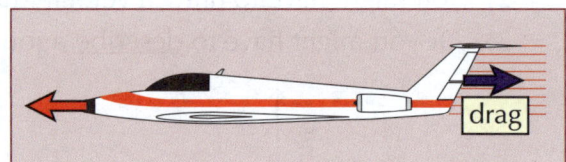

2) Objects moving through fluids experience drag.

3) Drag is the resistance you get in a fluid. Air resistance is a type of drag.

4) Drag acts in the opposite direction to the movement of the object.

5) Drag increases as the speed of the object increases.

Falling Objects Reach a Terminal Velocity

1) When an object first starts falling, the force of gravity is much larger than the drag slowing it down.

2) This means the object accelerates (the object speeds up).

3) As the speed increases, so does the drag.

4) This reduces the acceleration until the drag is equal to the gravitational force. The resultant force (p.349) on the object is then zero.

5) The object will fall at a constant speed. This speed is called its terminal velocity.

All objects falling in air have a terminal velocity...

...but some hit the ground before they reach it. When a falling object is at its terminal velocity the resultant force acting on the object is zero. This doesn't mean the object stops moving, it just means that it stops accelerating — so it continues to fall at a constant speed (its terminal velocity).

Warm-Up & Exam Questions

Slow down, it's not quite time to move on to Newton's Laws just yet. First it's time to check that all the stuff you've just read is still running around your brain. Dive into these questions.

Warm-Up Questions

1) What is the difference between speed and velocity?
2) Suggest the typical speeds of: a) a person running,   b) a train,   c) a plane.
3) True or false? Acceleration is shown as a curve on a distance-time graph.
4) An object travelling at a steady speed will be represented by which of the following on a velocity-time graph?   a) an uphill section,   b) a downhill section,   c) a flat section.
5) What is the resultant force acting on object falling at its terminal velocity?

Exam Questions

1   A cyclist travels from his house to his local shops.

It takes the cyclist 20 seconds to accelerate from 2.0 m/s to 10 m/s with a steady acceleration. Using the following formula, calculate the cyclist's acceleration:
acceleration = change in velocity ÷ time

[2 marks]

2   **Figure 1** shows the velocity-time graph of a cyclist.

2.1   Describe the motion of the cyclist between 5 and 10 seconds.

[2 marks]

2.2   Calculate the acceleration of the cyclist between 2 and 5 seconds.

[2 marks]

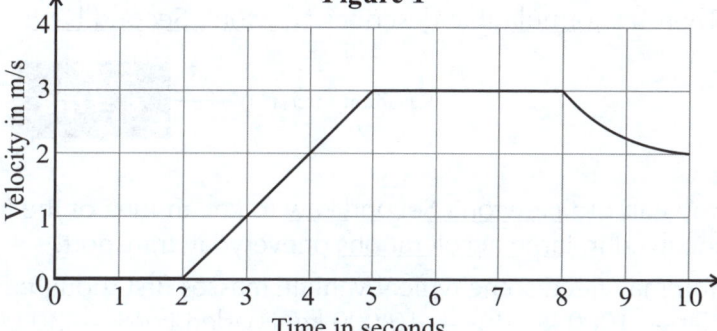

Figure 1

3   A coin is rolled in a straight line along a balcony edge at a steady speed of 0.46 m/s.

3.1   Write the equation which links speed, distance and time.

[1 mark]

3.2   Calculate how far the coin rolls in 2.4 s. Give your answer to two significant figures.

[3 marks]

3.3   Another coin is dropped from a height of 8 m.
It accelerates from rest at a constant rate and hits the ground at a speed of 12 m/s.
Calculate the acceleration of the coin during its fall.
Use the correct equation from the Physics Equation Sheet on page 538.

[3 marks]

Topic P5 — Forces

Newton's First and Second Law

In the 1660s, Isaac Newton worked out some really useful Laws of Motion. Here are the first two.

A Force is Needed to Change Motion

1) Newton's First Law says that a resultant force (p.349) is needed to make something start moving, speed up or slow down:

> If the resultant force on a stationary object is zero, the object will remain stationary. If the resultant force on a moving object is zero, it'll just carry on moving at the same velocity (the same speed and direction).

2) So, when a train or car or bus or anything else is moving at a constant velocity, the driving and resistive forces on it must be balanced.
3) Its velocity will only change if there's a non-zero resultant force acting on it.
4) A non-zero resultant force will always produce acceleration (or deceleration) in the direction of the force.

Acceleration is Proportional to the Resultant Force

1) The larger the resultant force acting on an object, the more the object accelerates.
2) Newton's Second Law says that the force acting on an object and the acceleration of the object are directly proportional. This can be shown as $F \propto a$.
3) Newton's Second Law also says that acceleration is inversely proportional to the mass of the object.
4) So an object with a larger mass will accelerate less than one with a smaller mass, for a given force.
5) There's a formula that describes Newton's Second Law:

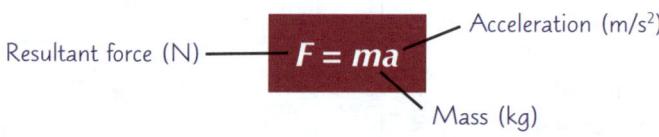

Resultant force (N) — $F = ma$ — Acceleration (m/s²), Mass (kg)

6) You can use Newton's Second Law to get an idea of the forces involved in large accelerations of everyday transport.
7) You may need some typical vehicle masses first though:
Car — 1000 kg, Bus — 10 000 kg, Loaded Lorry — 30 000 kg

EXAMPLE Estimate the resultant force on a car as it accelerates from rest to a typical speed.

1) Estimate the speed of the car and the time taken to reach that speed.
2) Use the speed and time taken to estimate the acceleration of the car.
3) Put the acceleration and the mass of the car into $F = ma$.

A typical speed of a car is ~25 m/s. It takes ~10 s to reach this.

So $a = \Delta v \div t = 25 \div 10 = 2.5$ m/s²

Mass of a car is ~1000 kg.
$F = ma = 1000 \times 2.5 = 2500$ N
So the resultant force is ~2500 N.

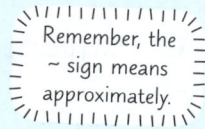
Remember, the ~ sign means approximately.

Newton's second law says F = ma...

Make sure you've got your head around both of these laws, before moving on to Newton's third law.
Q1 Find the force needed for an 80 kg man on a 10 kg bike to accelerate at 0.25 m/s². [2 marks]

Newton's Third Law

*This page is on **Newton's Third Law**. Make sure you really understand what's going on with it.*

Newton's Third Law — Interaction Pairs are Equal and Opposite

Newton's Third Law says:

> When two objects interact, the forces they exert on each other are equal and opposite.

1) This means if you push something, it will push back against you, just as hard.
2) And as soon as you stop pushing, so does the object.
3) You may be thinking "if the forces are always equal, how does anything ever go anywhere?".
4) The important thing to remember is that the two forces are acting on different objects.

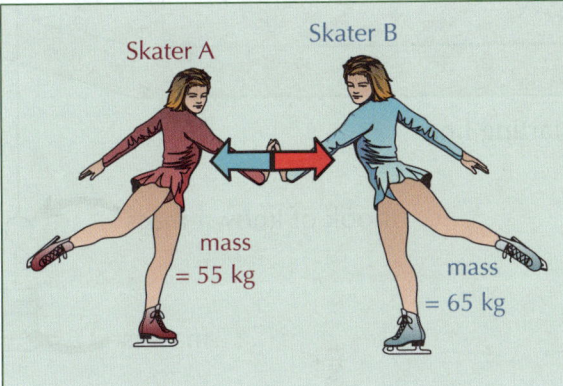

- Skater A pushes on skater B.
- When she does, she feels an equal and opposite force from skater B's hand.
- Both skaters feel the same sized force, in opposite directions.
- This causes them to accelerate away from each other.
- Skater A will be accelerated more than skater B, because she has a smaller mass.
- Remember $a = F \div m$.

It's More Complicated for an Object in Equilibrium

Imagine a book sat on a table in equilibrium (the resultant force on the book is zero):

1) The weight of the book pulls it down, and the normal contact force from the table pushes it up.
2) This is NOT Newton's Third Law.
3) These forces are different types and they're both acting on the book.

The pairs of forces due to Newton's Third Law in this case are:
- The book being pulled down by gravity towards the Earth (W_B) and the Earth being pulled up by the book (W_E).
- The normal contact force from the table pushing up on the book (R_B) and the normal contact force from the book pushing down on the table (R_T).

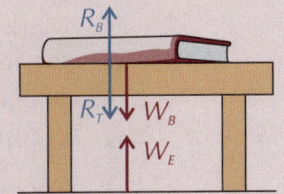

Newton's Third Law is only true for interaction pairs...

Make sure you get your head around Newton's laws of motion. His third law can be tricky — take your time, look at each object one by one, and work out all the forces acting on it.

Q1 A car moves at a constant velocity along a road, so that it is in equilibrium. Give an example of a pair of forces that demonstrate Newton's Third Law in this situation. [1 mark]

Q1 Video Solution

Investigating Motion

Here comes another **Required Practical**. This one's all about testing **Newton's Second Law**. It uses some nifty bits of kit that you may not have seen before, so make sure you follow the instructions closely.

You can Investigate how Mass and Force Affect Acceleration

This page shows you how to set up the experiment that tests Newton's Second Law, $F = ma$ (p.362). On the next page you can find out how to investigate the variables that affect Newton's Second Law.

1) Set up the apparatus as shown on the right.

2) The mass, m, that you'll be accelerating is the total mass of the trolley, hook and the added masses.

3) You can measure m using a mass balance.

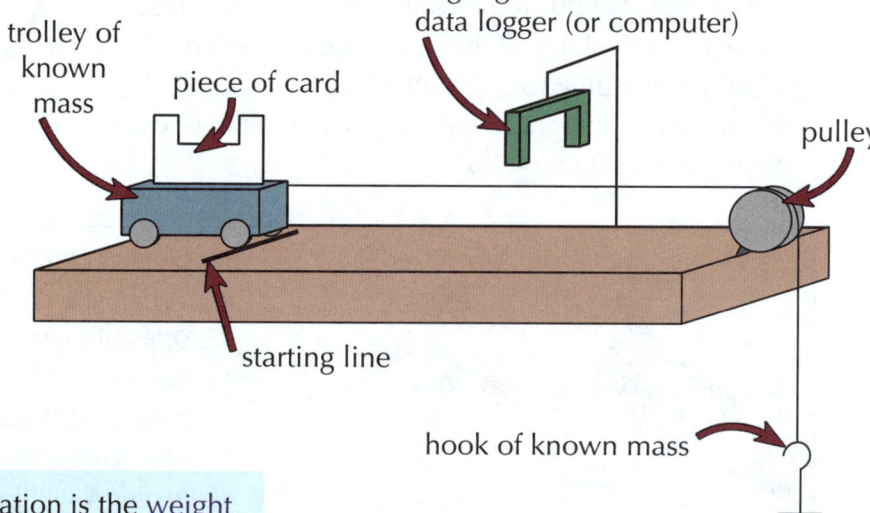

4) The force, F, causing the acceleration is the weight of the hook and the masses on the hook.

5) To find F, first measure the mass of the hook and any masses on the hook. Then multiply this by g (as $W = mg$, see page 348).

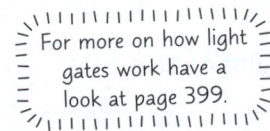

For more on how light gates work have a look at page 399.

6) The acceleration, a, is found by following this method:

- Mark a starting line on the table the trolley is on. This is so that the trolley always travels the same distance to the light gate.
- Place the trolley on the starting line.
- Hold the trolley so the string is tight and not touching the table. Then release it.
- Record the acceleration measured by the light gate as the trolley passes through it.

This experiment has a lot of steps, so don't speed through it...

Make sure the string is the right length and there's enough space for the hanging masses to fall. There needs to be enough space so that the masses don't hit the floor before the trolley has passed through the light gate fully — if they hit the floor, the force won't be applied the whole way through the trolley's journey, so you won't get an accurate measurement for the acceleration.

Topic P5 — Forces

Investigating Motion

*Now you know how to set up the **equipment** (p.364) it's time to start looking at **adjusting** your **variables**. Take care with the **method** here — there are some important points you don't want to miss.*

Investigating How Mass Affects Acceleration

To investigate the effect of mass, you need to change the mass but keep the force the same. Remember, the mass (m) is the mass of the trolley, the hook and any extra masses added together.

1) The force is the weight of the hook and any masses on the hook.

2) So, keep the mass on the hook the same.

3) Add masses to the trolley one at a time to increase the total mass being accelerated.

4) Record the acceleration, a, for each total mass, m.

5) You should find that as the mass goes up, the acceleration goes down.

6) This agrees with Newton's Second Law — mass and acceleration are inversely proportional.

Investigating How Force Affects Acceleration

This time, you need to change the force without changing the total mass of the trolley, hook and masses.

1) Start with all the extra masses loaded onto the trolley.

2) Moving the masses from the trolley to the hook will keep the total mass, m, the same.

3) But it will increase the force, F (the weight of the hook and the masses on the hook).

4) Each time you move a mass, record the new force, and measure the acceleration.

5) You should find that as the force goes up, the acceleration goes up.

6) This agrees with Newton's Second Law — force and acceleration are directly proportional.

Topic P5 — Forces

Warm-Up & Exam Questions

Now you've gotten yourself on the right side of the law(s) of motion), it's time to put your knowledge on trial. Have a go at the warm-up questions first and then knuckle down to those exam questions.

Warm-Up Questions

1) True or false? The resultant force acting on an accelerating object must have a magnitude greater than zero.
2) Boulders A and B are accelerated at 2 m/s². Boulder A required a force of 70 N, and Boulder B required a force of 95 N. Which boulder has a greater mass?
3) True or false? Two interacting objects exert equal and opposite forces on each other.

Exam Questions

1   A ball hits a stationary cricket bat. The bat exerts a force of 520 N on the ball when they are in contact. The cricket bat has a mass of 1.6 kg and the ball has a mass of 160 g.   *Grade 3-4*

1.1   State the force that the ball exerts on the bat.

[1 mark]

1.2   Complete the sentence below using words from the box.

| less than | the same as | greater than |

The acceleration of the ball will be .... that of the bat because it has a smaller mass.

[1 mark]

1.3   Use the following formula to calculate the size of the acceleration of the bat when the ball hits it.
Force = mass × acceleration

[3 marks]

2   The camper van shown in **Figure 1** has a mass of 2500 kg. It is driven along a straight, level road at a constant speed of 90.0 kilometres per hour.   *Grade 4-5*

Figure 1
90.0 km/h
2500 kg

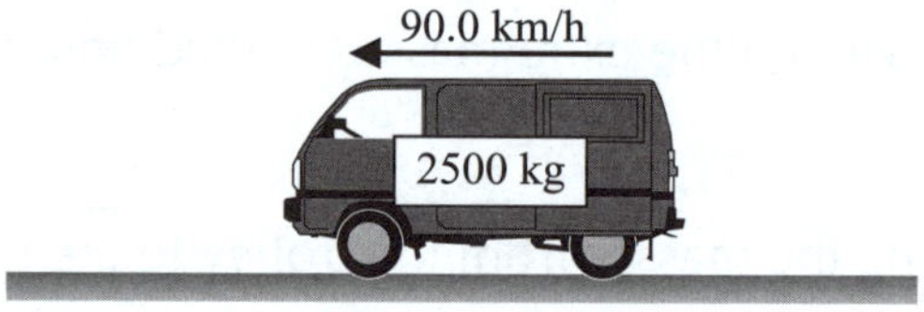

2.1   State the resultant force acting on the camper van.

[1 mark]

2.2   The camper van slows down. What changes from those below could have caused the van to slow down? Tick **two** boxes.

☐ The driving force decreases
☐ The wind blowing in the opposite direction to the van's movement increases
☐ The drag acting on the van decreases
☐ The van becomes lighter

[2 marks]

2.3   The camper van slows down with a deceleration of 1.4 m/s². What is the resultant force acting on the van during its deceleration? Give the direction of the resultant force.

[3 marks]

Topic P5 — Forces

Stopping Distance and Thinking Distance

*This page is all about cars, but unfortunately it's not as fun as it sounds... It's even better — it's about **safety**...*

Stopping Distance = Thinking Distance + Braking Distance

1) In an emergency, a driver may perform an emergency stop.
2) During an emergency stop, the maximum force is applied by the brakes. This is so the vehicle stops in the shortest possible distance.
3) The distance it takes to stop a vehicle in an emergency is its stopping distance. It is found by:

> Stopping Distance = Thinking Distance + Braking Distance

4) Thinking distance is how far the vehicle travels during the driver's reaction time.
5) The reaction time is the time between the driver seeing a hazard and applying the brakes.
6) Braking distance is the distance taken to stop under the braking force (once the brakes are applied).

Stopping Distances for Different Speeds can be Estimated

1) You need to be able to estimate the stopping distance of vehicles.
2) The heavier a vehicle is, or the faster it's travelling, the longer its stopping distance will be.
3) As a guide, typical car stopping distances are: 23 m at 30 mph, 73 m at 60 mph, 96 m at 70 mph.

Stopping Distances Affect Safety

1) The longer it takes to perform an emergency stop, the higher the risk of crashing into whatever's in front.
2) So the shorter a vehicle's stopping distance, the safer it is.
3) You need to be able to describe how different factors can affect the safety of a journey.
4) For example, how driving if you're tired is unsafe. There's more on this below.

Thinking Distance is Determined by the Driver's Reactions

1) Thinking distance is affected by:
   - Your speed — the faster you're going, the further you'll travel during the time you take to react.
   - Your reaction time — the longer your reaction time (p.369), the longer your thinking distance.

2) A driver's reaction times can be affected by tiredness, drugs or alcohol.
3) Distractions can also affect your ability to react.

> - Driving while tired is unsafe as it makes you slower to react.
> - This increases your reaction time, which increases your thinking distance.
> - This means your stopping distance is longer, so you're more likely to crash.

> - Driving above the speed limit is unsafe.
> - You travel further in your reaction time than you would at a lower speed.
> - This increases your thinking (and so stopping) distance.

Topic P5 — Forces

Braking Distance

*So you know the basics of **stopping distances** now, but how do the brakes actually work to **slow down a car**? Well, it's all down to **friction** and **transferring energy** away from the wheels to the brakes.*

Braking Distance Depends on a Few Different Factors

Braking distance is affected by:

1) Your speed: for a given braking force, the faster a vehicle travels, the longer it takes to stop.
2) Weather or road surface:
   - If there is less grip between a vehicle's tyres and the road, it can cause the vehicle to skid.
   - Skidding increases the braking distance of a car.
   - Water, ice, oil or leaves on the road all reduce grip.

   > Icy conditions increase the chance of skidding. This increases the braking distance, which increases the stopping distance. So more room should be left between cars to be safe.

3) The condition of your tyres:
   - Bald tyres (ones that don't have any tread left) cannot get rid of water in wet conditions.
   - This leads to them skidding on top of the water.
4) How good your brakes are:
   - If brakes are worn, they won't be able to apply as much force.
   - So it takes longer to stop a vehicle travelling at a given speed (see below).

Braking Relies on Friction Between the Brakes and Wheels

1) When the brake pedal is pushed, brake pads are pressed onto the wheels.
2) The brake pads cause friction, which causes work to be done (p.349).
3) Remember, when work is done, energy is transferred (p.285).
4) Energy is transferred from the kinetic energy store of the vehicle to the thermal energy stores of the brakes.
5) The brakes increase in temperature.
6) To stop a vehicle, the brakes must transfer all of the energy from the kinetic store, so:

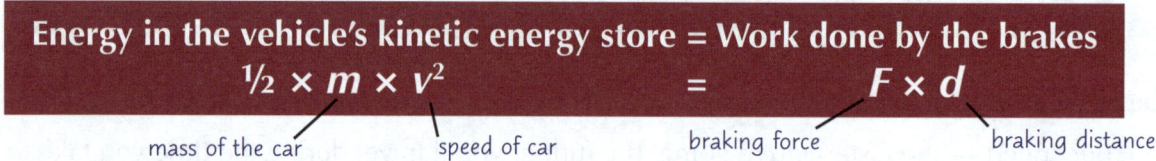

Energy in the vehicle's kinetic energy store = Work done by the brakes
$$\tfrac{1}{2} \times m \times v^2 = F \times d$$
mass of the car — speed of car — braking force — braking distance

7) The faster a vehicle is going, the more energy it has in its kinetic energy store.
8) So more work needs to be done to stop it.
9) This means that as the speed of a vehicle increases, the force needed to make it stop within a certain distance also increases.
10) A larger braking force means a larger deceleration.
11) Very large decelerations can be dangerous because they may cause brakes to overheat. This means the brakes won't work as well.
12) Very large decelerations may also cause the vehicle to skid.

The faster a car is travelling, the more work is needed to stop it

Q1 A car with a mass of 1000 kg is travelling in a straight line at a speed of 10 m/s. The driver spots a hazard and applies the brakes. The car is 50 m from the hazard when the brakes are applied. Calculate the minimum braking force required for the car to stop before the hazard.
[5 marks]

Q1 Video Solution

Topic P5 — Forces

Reaction Times

Reaction times are an *important factor* in *thinking distances*. They're also super easy to *test* for yourself.

A Typical Reaction Time is 0.2 s – 0.9 s

1) Everyone's reaction time is different.
2) A typical reaction time is between 0.2 and 0.9 s.
3) You can do simple experiments to investigate your reaction time — more on these below.

You can Measure Reaction Times with the Ruler Drop Test

1) As reaction times are so short, you haven't got a chance of measuring one with a stopwatch.
2) One way of measuring reaction times is to use a computer-based test. For example, clicking a mouse when the screen changes colour.

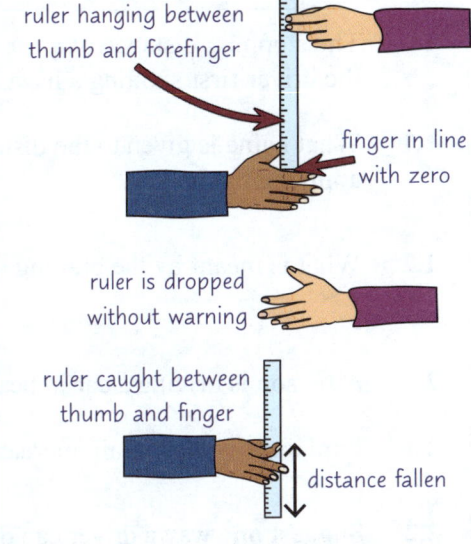

ruler hanging between thumb and forefinger

finger in line with zero

ruler is dropped without warning

ruler caught between thumb and finger

distance fallen

Another method to measure reaction times is the ruler drop test:
1) Sit with your arm resting on the edge of a table.
2) Get someone else to hold a ruler so it hangs between your thumb and forefinger, lined up with zero.
3) You may need a third person to be at eye level with the ruler to check it's lined up.
4) Without giving any warning, the person holding the ruler should drop it.
5) Close your thumb and finger to try to catch the ruler as quickly as possible.
6) The measurement on the ruler at the point where it is caught is how far the ruler dropped in the time it took you to react.
7) The longer the distance, the longer the reaction time.
8) You can calculate how long the ruler falls for (the reaction time) because acceleration due to gravity is constant.
9) To find the reaction time, you can use the equation:

$$t = \frac{\sqrt{2as}}{a}$$

- t is the reaction time in seconds, s.
- a is the acceleration due to gravity. $a = 9.8$ m/s².
- s is how far the ruler fell before it was caught, in metres, m.

You *don't* need to learn this equation. It comes from squishing $v^2 - u^2 = 2as$ and $a = \Delta v \div t$ from p.357 together.

It's hard to do this experiment accurately, but you can do a few things to improve your results.
- Do a lot of repeats and calculate an average reaction time.
- Add a blob of modelling clay to the bottom to help the ruler to fall straight down.
- Make it a fair test — use the same ruler for each repeat, and have the same person dropping it.

The further the ruler falls, the longer the reaction time

Q1 Mark's reaction time is tested using the ruler drop test. He is tested in the early afternoon and at night. In the afternoon, he catches the ruler after it has fallen a distance of 16.2 cm. At night, he catches the ruler after it has fallen 18.5 cm.
  a) Calculate Mark's reaction time in the afternoon.
     Give your answer to 2 significant figures. [5 marks]
  b) Explain why Mark's thinking distance might be longer when driving in the evening. [2 marks]

Topic P5 — Forces

Warm-Up & Exam Questions

Time to apply the brakes for a second and put your brain through its paces. Try out these questions. If you can handle these, your exam should be clear of hazards.

Warm-Up Questions

1) Give one factor that can affect thinking distance.
2) What must be added to the thinking distance to find the total stopping distance of a car?
3) Give one reason that the large deceleration of a car can be dangerous.
4) True or false? The larger the braking force, the greater the deceleration of a car.

Exam Questions

1 The stopping distance of a car is the distance covered in the time between the driver first spotting a hazard and the car coming to a complete stop. *(Grade 1-3)*

1.1 What name is given to the distance travelled between the driver first spotting a hazard and then applying the brakes?

[1 mark]

1.2 What is meant by the braking distance of a car?

[1 mark]

2 A person is driving a car in heavy rain. *(Grade 3-4)*

2.1 Explain why heavy rain increases a car's stopping distance.

[1 mark]

2.2 Suggest **one** way a driver can decrease their stopping distance when driving in heavy rain.

[1 mark]

2.3 When a car brakes, work is done. What energy transfer occurs?
Tick **one** box in each row.

	The car's kinetic energy stores	The car's chemical energy stores	The brakes' thermal energy stores	The brakes' elastic energy stores
Energy transferred from	☐	☐	☐	☐
Energy transferred to	☐	☐	☐	☐

[2 marks]

3 A student tries to catch a falling ruler as quickly as possible. The distance the ruler falls can be used to calculate the student's reaction time. The student repeats the test three times. Their results are shown below. *(Grade 4-5)*

0.24 s   0.19 s   0.23 s

3.1 Calculate the student's mean reaction time.

[2 marks]

3.2 State **one** way the student could improve their results.

[1 mark]

Topic P5 — Forces

Topic P6 — Waves

Wave Basics

*Waves **transfer energy** from one place to another **without** transferring any **matter** (stuff).*

Waves Transfer Energy but not Matter

1) When a wave travels through a medium, the particles of the medium vibrate.
2) A medium is just a fancy word for whatever the wave is travelling through (e.g. water, air).
3) The particles transfer energy between each other as they vibrate (see p.284).
4) BUT overall, the particles stay in the same place — only energy is transferred.

- For example, if you drop a twig into calm water, ripples spread out. The ripples don't carry the water (or the twig) away with them though.
- And if you strum a guitar string, the sound waves don't carry the air away from the guitar. If they did, you'd feel a wind whenever there was a sound.

Waves have an Amplitude, Wavelength and Frequency

1) The amplitude of a wave is the maximum displacement of a point on the wave from its undisturbed (rest) position.

2) The wavelength is the distance between one point on a wave and the same point on the next wave. For example, the distance between the trough of one wave and the trough of the wave next to it.

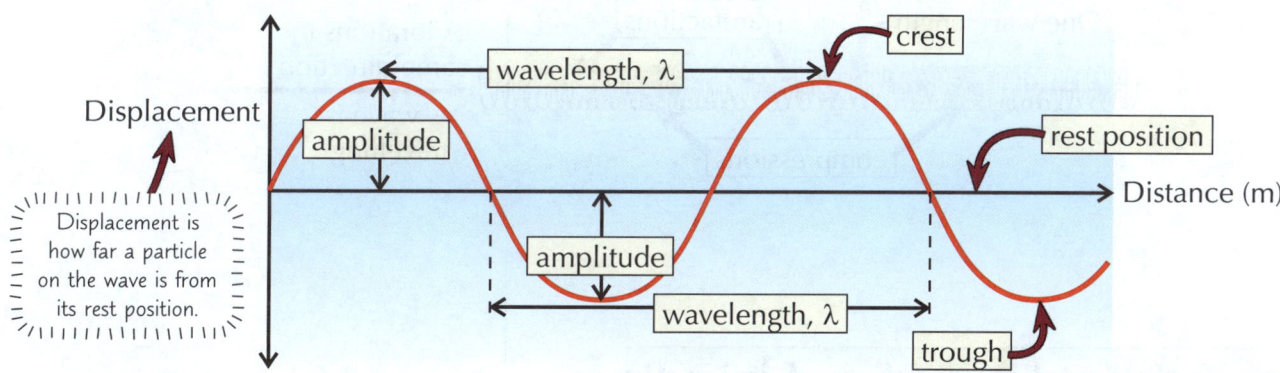

Displacement is how far a particle on the wave is from its rest position.

3) Frequency is the number of complete waves passing a certain point each second. Frequency is measured in hertz (Hz). 1 Hz is 1 wave per second.

4) Period is the amount of time it takes for one complete wave to pass a certain point.

 **The only thing a wave transfers is energy...**
It's really important that you understand this stuff really well, or the rest of this topic will be a blur. Make sure you can sketch the wave diagram above and can label all the features from memory.

Transverse and Longitudinal Waves

*All waves are either **transverse** or **longitudinal**. Read on to find out more...*

Transverse Waves Have Perpendicular Vibrations

1) In transverse waves, the vibrations are perpendicular (at right angles) to the direction of energy transfer.
2) Examples of transverse waves are light (p.378), ripples on water (see p.374) and waves on a string (p.375).
3) A spring wiggled up and down gives a transverse wave:

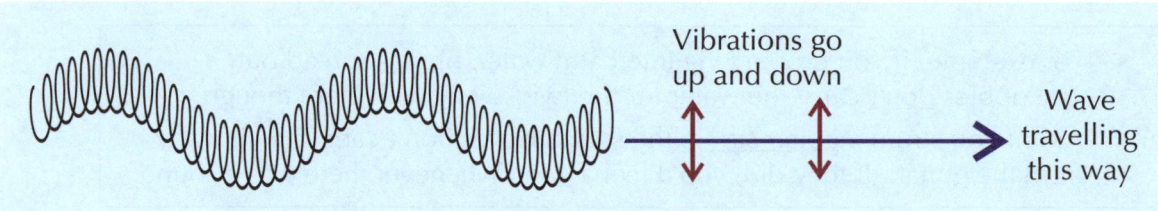

Longitudinal Waves Have Parallel Vibrations

1) In longitudinal waves, the vibrations are in the same direction as the energy transfer.
2) They have compressions (where the particles squish together), and rarefactions (where they spread out).
3) A sound wave is an example of a longitudinal wave.
4) If you push the end of a spring, you get a longitudinal wave:

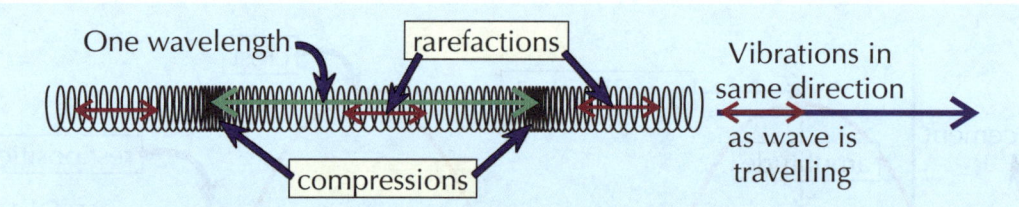

Frequency and Period are Linked

You can find the period of a wave from its frequency:

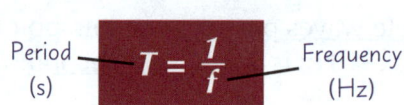

Period (s) — $T = \dfrac{1}{f}$ — Frequency (Hz)

EXAMPLE

Calculate the period of a wave with a frequency of 2 Hz.

$T = 1 \div f = 1 \div 2 = $ **0.5 s**

Light waves are transverse and sound waves are longitudinal...

The equation that links period and frequency together is pretty useful — it means that you can always work one out from the other. Make sure you practise using the equation until you're comfortable with it.

Topic P6 — Waves

Wave Speed

*Measuring the **speed of waves** isn't that simple. It calls for crafty methods...*

Wave Speed = Frequency × Wavelength

1) The wave speed is how fast the wave is moving.
2) It is the speed at which energy is being transferred through the medium.
3) The wave equation applies to all waves:

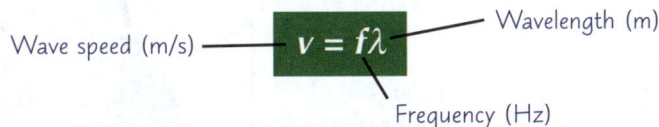

Wave speed (m/s) — $v = f\lambda$ — Wavelength (m)
Frequency (Hz)

EXAMPLE A radio wave has a frequency of 12 000 000 Hz. Find its wavelength. (The speed of radio waves in air is 3×10^8 m/s.)

1) Rearrange the wave speed equation for wavelength. $\lambda = v \div f$
2) Put in the values you've been given. $= (3 \times 10^8) \div (12\,000\,000)$
   Watch out — the speed is in standard form (p.19). $= 25$ m

Use an Oscilloscope to Measure the Speed of Sound

1) Connect two microphones to an oscilloscope (a device which shows waves on a screen).
2) Connect a signal generator to a speaker. This will let you generate sound waves at a set frequency.

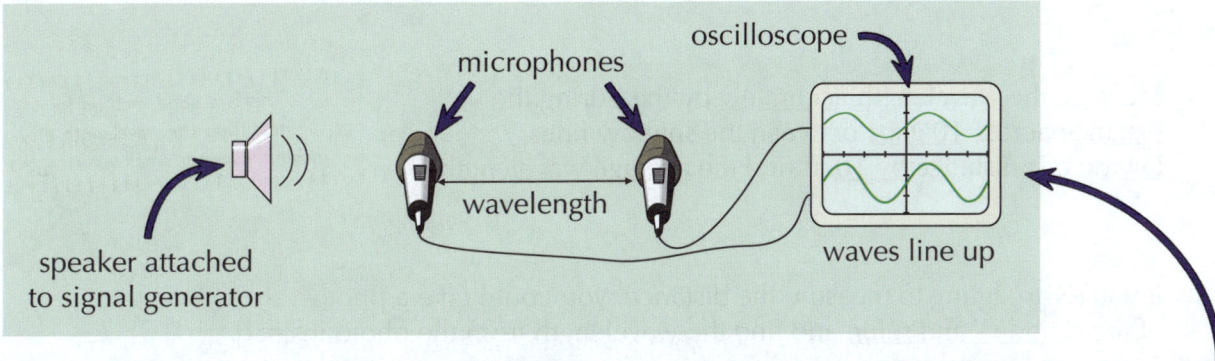

3) Set up the oscilloscope so the waves reaching each microphone are shown separately.
4) Start with both microphones next to the speaker. The waves on the oscilloscope should line up.
5) Slowly move one microphone away. Stop when the two waves line up again on the display.
6) This means the microphones are now exactly one wavelength apart.
7) Measure the distance between the microphones to find the wavelength (λ).
8) Use the formula $v = f\lambda$ to find the speed (v) of the sound waves passing through the air.
9) The frequency (f) is whatever you set the signal generator to.
10) The speed of sound in air is around 330 m/s, so check your results roughly agree with this.

That was the first of three wave speed practicals to learn...

Make sure you understand each step of that method above — you could be tested on it in the exams.

Q1 A wave has a speed of 0.15 m/s and a wavelength of 7.5 cm.
Calculate its frequency. [4 marks]

PRACTICAL: Investigating Waves

*Choosing **suitable equipment** means making sure it's **right** for the job. It's important here.*

Measure the Speed of Water Ripples Using a Ripple Tank

1) Attach a signal generator to the ripple tank dipper. Turn on the signal generator to create waves.

   Make sure you do this experiment in a darkened room.

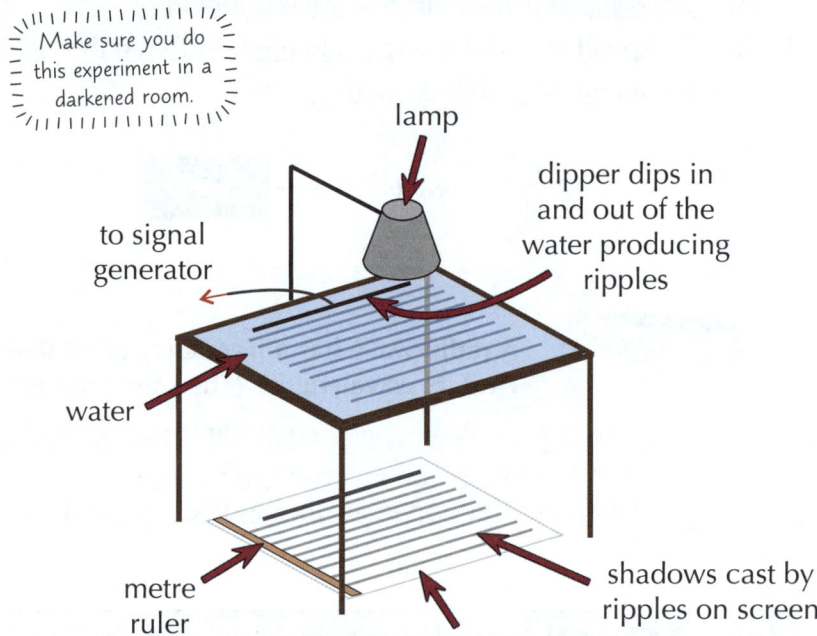

2) Find the frequency of the waves by counting the number of ripples that pass a point in 10 seconds and dividing by 10.

3) Use a lamp to create shadows of the ripples on a screen below the tank. Place a metre ruler beside the shadows.

4) The distance between each shadow line is equal to one wavelength.

5) Measure the wavelength accurately by measuring the distance across 10 gaps between the shadow lines. Divide this distance by 10 to find the average wavelength.

   This is a good method for measuring the wavelength of moving waves or small wavelengths.

6) If you're struggling to measure the distance, you could take a photo of the shadows and ruler, and find the wavelength from the photo instead.

7) Use $v = f\lambda$ to calculate the wave speed of the waves.

8) This set-up is suitable for investigating waves, because it allows you to measure the wavelength without disturbing the waves.

Make sure your measurements are accurate...

You should be careful that the screen beneath the tank is flat, or the shadows will become distorted and difficult to measure properly. You should also make sure that the metre ruler is at a right angle to the shadow wavefronts — if the ruler is skewed then your measured wavelength will be too long.

Topic P6 — Waves

Investigating Waves

One more wave experiment coming up. This time, it's to do with waves on strings.

You can Use the Wave Equation for Waves on Strings

In this practical, you create a wave on a string. Again, you use a signal generator, but this time you attach it to a vibration generator which converts the signals to vibrations.

1) Set up the equipment shown below.

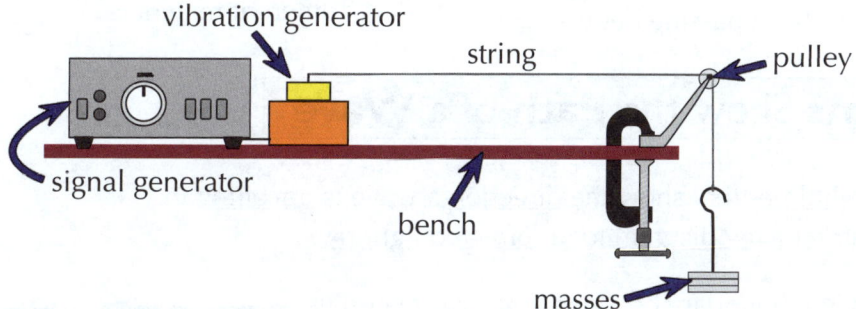

2) The vibration generator vibrates at a fixed frequency, set by the signal generator.

3) Turn on the signal generator and the string will start to vibrate.

4) Adjust the frequency of the signal generator until there's a clear wave on the string.

5) To measure the wavelength of these waves accurately:
   - Count how many wavelengths are on the string. Each vibrating loop is half a wavelength.
   - Measure the length of the whole vibrating string.
   - Divide by the number of wavelengths to give the length of one wavelength.

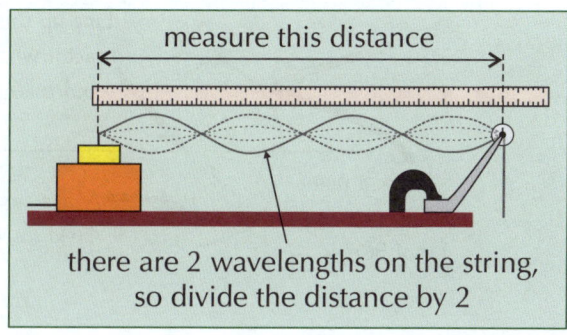

there are 2 wavelengths on the string, so divide the distance by 2

6) The frequency of the wave is whatever the signal generator is set to.

7) You can find the speed of the wave using $v = f\lambda$.

It's ok if you don't have a whole number of wavelengths on the string. If there are 3 loops, there are one-and-a-half wavelengths on the string. Divide the length of the string by 1.5.

8) This set-up is suitable for investigating waves on a string because it's easy to see and measure the wavelength (and frequency).

Learn the methods for all the wave speed practicals...

The experiments on pages 373-375 seem complicated, but they all have a few things in common. First, you set the frequency on the signal generator, then find the length of the resulting wave (this tends to be the fiddly bit). You can then use the equation $v = f\lambda$ to find the wave speed.

Topic P6 — Waves

Refraction

*Grab a glass of water and put a straw in it. The straw looks like it's **bending**. But it's not magic, it's refraction.*

Refraction — Waves **Changing Direction**

1) When a wave crosses a boundary between two materials it can change direction.
2) This is known as refraction.
3) Waves are only refracted if they meet the boundary at an angle.
4) How much a wave is refracted by depends on the two materials it's passing between.

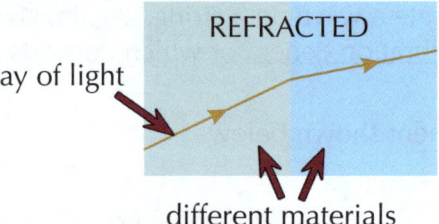
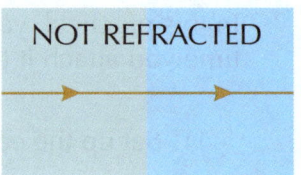

Ray Diagrams Show the **Path** of a **Wave**

- Rays are straight lines that show the direction a wave is travelling in.
- You can construct a ray diagram for a refracted light ray.

1) Start by drawing the boundary between your two materials.
2) Then draw a dotted line at right angles to the boundary.
3) This line is known as the 'normal' to the boundary. Normal just means 'at right angles'.
4) Next draw the incident ray. This is the ray that meets the boundary at the normal.
5) The angle between the incident ray and the normal is called the angle of incidence.
6) You need to use a protractor to draw or measure it.
7) So to draw an incident ray with an angle of incidence of 50°:

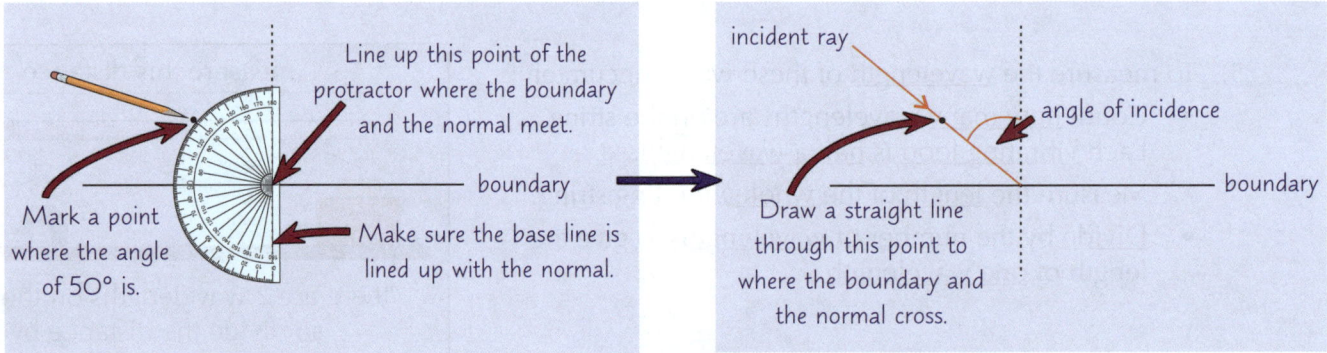

8) Now draw the refracted ray on the other side of the boundary.
9) The angle of refraction is the angle between the refracted ray and the normal.

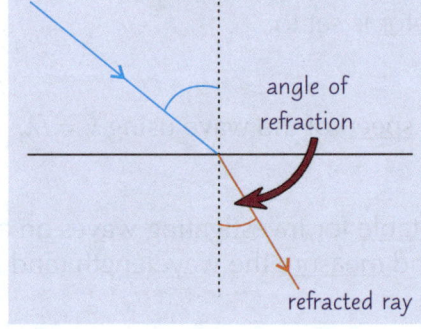

Hitting a boundary at an angle leads to refraction...

Refraction is a common behaviour of waves, so make sure you really understand it before moving on.

Q1 Draw a ray diagram for a ray of light meeting a boundary at an angle of incidence of 55°, and crossing into a second material with an angle of refraction of 35°. **[3 marks]**

Topic P6 — Waves

Warm-Up & Exam Questions

Now to check what's actually stuck in your mind over the last few pages...

Warm-Up Questions

1) Waves travel through a medium. Describe one piece of evidence that shows the medium's particles do not travel with the wave.
2) Describe the direction of vibrations in a longitudinal wave.
3) What is the name for the stretched out areas of a longitudinal wave?
4) True or false? A wave entering a new medium along the path of the normal won't be refracted.

Exam Questions

1 **Figure 1** shows a graph of a water wave.

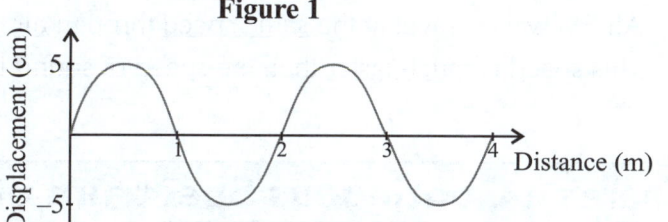

1.1 Give **one** reason why water waves are classified as transverse waves rather than longitudinal waves.

[1 mark]

1.2 Give the amplitude of this wave.

[1 mark]

1.3 Give the wavelength of this wave.

[1 mark]

2 **Figure 2** shows how an oscilloscope can be used to display sound waves by connecting microphones to it. Trace 1 shows the sound waves detected by microphone 1 and trace 2 shows the sound waves detected by microphone 2.

Figure 2

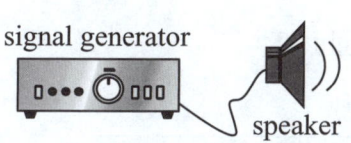

A student begins with both microphones at equal distances from the speaker and the signal generator set at a fixed frequency. He gradually moves microphone 2 away from the speaker, which causes trace 2 to move. He stops moving microphone 2 when both traces line up again as shown in **Figure 2**. He then measures the distance between the microphones.

2.1 What does this measurement tell the student about the sound waves detected?

[1 mark]

2.2 Give the equation that links wavelength, frequency and wave speed.

[1 mark]

2.3 With the signal generator set to 50.0 Hz, the distance between the microphones was measured as 6.8 m. Calculate the speed of sound in air. Give the correct unit.

[3 marks]

Electromagnetic Waves

*The **light waves** that we see are just one small part of a big group of **electromagnetic waves**...*

Electromagnetic Waves Transfer Energy

1) Electromagnetic (EM) waves are transverse waves (p.372).
2) They transfer energy from a source to an absorber.

- A camp fire is a source.
- It transfers energy to its surroundings by giving out infrared radiation.
- Infrared radiation is a type of EM wave.
- These infrared waves are absorbed by objects.
- Energy is transferred to the objects' thermal energy stores.
- This causes the objects to warm up.

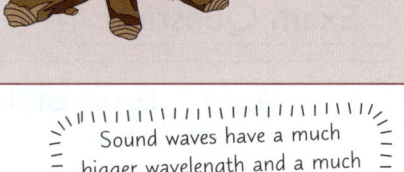

3) All EM waves travel at the same speed through air or a vacuum (space).
4) This speed is much faster than the speed of sound in air.

Sound waves have a much bigger wavelength and a much lower frequency than light.

There's a Continuous Spectrum of EM Waves

1) EM waves vary in wavelength and frequency.
2) There are EM waves of every wavelength within a certain range.
3) This is known as a continuous spectrum.
4) The spectrum is split up into seven groups based on wavelength and frequency.

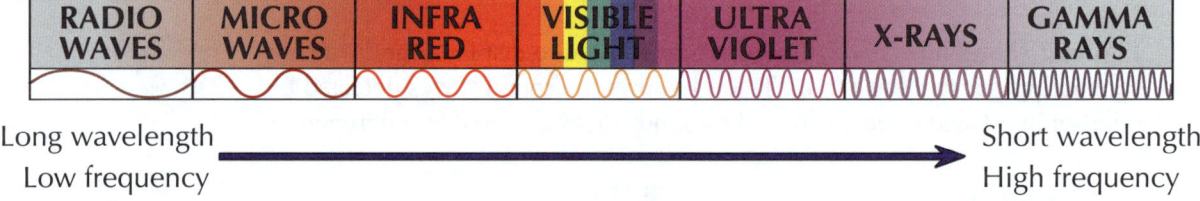

Long wavelength / Low frequency → Short wavelength / High frequency

RADIO WAVES | MICRO WAVES | INFRA RED | VISIBLE LIGHT | ULTRA VIOLET | X-RAYS | GAMMA RAYS

5) Our eyes can only detect a small part of this spectrum — visible light.

Changes In Atoms Produce the Spectrum of EM Waves

1) EM radiation can be absorbed or produced by changes in atoms and their nuclei.
2) There are lots of different changes that can happen in atoms. For example:
   - Electrons can move between energy levels in atoms (see p.335).
   - Changes in the nucleus of an atom can create gamma rays (see p.339).
3) Each different change produces or absorbs a different frequency of EM wave.
4) This is why atoms can generate (create) EM waves over a large range of frequencies.
5) It is also why atoms can absorb a range of frequencies.

 **Study the EM spectrum and wave goodbye to exam stress...**
There are a lot of facts to remember here, and you need to know them all. Here's a handy way to remember the order of EM waves: 'Rock Music Is Very Useful for eXperience with Guitars'.

Uses of EM Waves

EM waves are used for all sorts of stuff — and **radio waves** are definitely the most fun. They make your car **radio** and your **TV** work. Life would be pretty quiet without them.

Radio Waves are Used Mainly for Communication

1) Radio and TV signals can be sent by radio waves.
2) Very short wavelength signals are used for FM radio and TV.
3) They have to be in direct sight of the receiver when they're sent, with nothing in the way, so they can't travel very far.
4) Longer wavelength radio waves can travel further.
5) They can be used to send radio signals around the world.

Microwaves are Used for Satellites and Cooking

1) Communication with satellites uses microwaves, e.g. for satellite TV and satellite phones.
   - A signal is sent into space to a satellite dish high above the Earth.
   - The satellite sends the signal back to Earth in a different direction.
   - A satellite dish on the ground receives the signal.

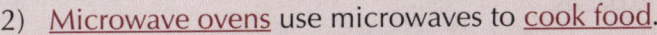

2) Microwave ovens use microwaves to cook food.
   - The oven gives out microwaves, which are absorbed by water in the food.
   - Energy carried by the microwaves is transferred to the water molecules, causing them to heat up.
   - This causes the rest of the food to heat up and quickly cooks it.

Infrared Radiation Can be Used to Cook and Heat Things

1) Infrared (IR) radiation is given out by all objects.
2) The hotter the object, the more infrared radiation it gives out.
3) When an object absorbs infrared radiation, energy is transferred to the object's thermal energy store. This makes it warm up.
4) Infrared radiation can be used in many ways:

   - Infrared cameras detect IR radiation and create a picture.
   - This is useful for seeing where a house is losing energy.
   - It can also allow you to see hot objects in the dark.

The different colours mean different amounts of IR radiation are being detected from those areas. Here, the redder the colour, the more infrared radiation is being detected.

   - Infrared radiation can also be used to warm things.
   - Electric heaters release lots of IR radiation to warm a room.
   - And food can be cooked using infrared radiation.

The uses of EM waves depend on their properties...

Differences in wavelength, frequency and energy between types of EM wave give them different properties. For example, some types of EM wave are very harmful (see page 383). Luckily, radio waves are considered safe to beam round the world. IR radiation is generally fairly safe, although too much of it will burn you.

More Uses of EM Waves

And we're still not finished with **uses** of **EM waves** — there's just no end to their talents...

Fibre Optic Cables Use Visible Light to Send Data

1) Optical fibres are thin glass or plastic tubes that can carry data over long distances.
2) They're often used to send information to telephones or computers.
3) Information is sent as light rays that bounce back and forth along the fibre.

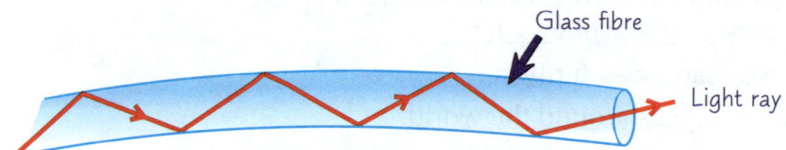

Ultraviolet Radiation Gives You a Suntan

1) When some materials absorb UV light, they give off visible light.
2) This can be pretty useful:

- Energy-efficient lights use UV radiation to produce visible light.
- Security pens can be used to mark property with your name (e.g. laptops).
- Under UV light the ink will glow, but it's invisible otherwise.
- This can help the police find out who stolen property belongs to.

3) Ultraviolet radiation (UV) is also produced by the Sun. It's what gives you a suntan.
4) UV lamps can be used to give people a suntan without the Sun (but this can be dangerous).

X-rays and Gamma Rays are Used in Medicine

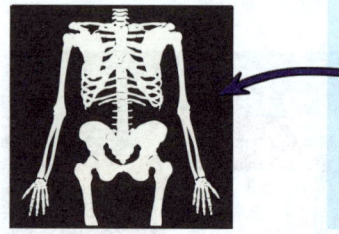

1) X-rays pass easily through flesh but not through bones or metal.
2) This can be used to create an X-ray image to check for broken bones.
3) X-rays can also treat people with cancer.
4) This is because they can kill cells. They are aimed at the cancer cells to kill them.

1) Gamma rays (p.339) can also kill cells.
2) They can be used to treat cancer in the same way as X-rays.
3) They can also be used to sterilise (remove germs from) medical equipment. The equipment is blasted with gamma rays which kills any living things on it.
4) Gamma rays are also really good at passing through your body.
5) This is why small amounts of them are used in 'medical tracers'. How they move around the body can be tracked. This can tell doctors if organs are working as they should.

Communications, security and imaging — pretty important stuff...

I hate to say it, but go back to page 379 and read all of the uses for EM waves again to really learn them.

Topic P6 — Waves

Investigating IR Radiation

*Time for another **Required Practical**. In this one, you'll meet a fun, new piece of kit called a **Leslie cube**. Read on to find out more about how you can use this equipment to investigate **infrared radiation emissions**.*

Different Surfaces Emit Different Amounts of IR Radiation

1) The amount of infrared radiation an object gives out depends on its temperature — see p.379.
2) It also depends on its surface.
3) This includes how rough or shiny it is, and its colour.
4) You can investigate how much IR radiation different surfaces emit using a Leslie cube.
5) A Leslie cube is a hollow, metal cube.
6) The four side faces have different surfaces.
7) For example, matt (dull) black paint, matt white paint, shiny metal and dull metal.

You Can Investigate Emission With a Leslie Cube

1) Place an empty Leslie cube on a heat-proof mat. Draw a square around the cube that is 10 cm from all faces of the cube.

2) Fill the Leslie cube with boiling water and wait for the cube to warm up.

3) Use the square you've drawn to hold an infrared detector 10 cm away from one of the cube's vertical faces. Record the amount of IR radiation it detects.

4) Repeat step 3) for each of the four faces.

5) The face that had the highest reading is giving off the most IR radiation.

6) You should find that the black surface is radiating more IR radiation than the white one. Matt surfaces should give off more than shiny ones.

7) As always, you should repeat the experiment to check your results.

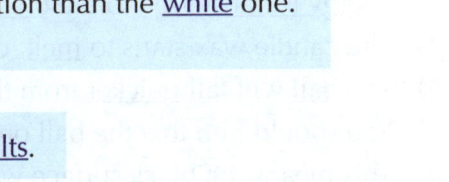

You can feel which face is giving off the most heat...

When doing this experiment, you could also place your hand near each surface of the cube (but not touching, it'll be super hot). You'll be able to feel which surface is giving off more infrared radiation.

Topic P6 — Waves

 **Investigating IR Absorption**

*Have you ever noticed that wearing **black clothes** on a **hot day** makes you feel really warm? Turns out there's some **science** behind it...*

You Can Investigate **Absorption** with the **Melting Wax Trick**

The amount of infrared radiation absorbed by different materials also depends on the surface. You can do an experiment to show this.

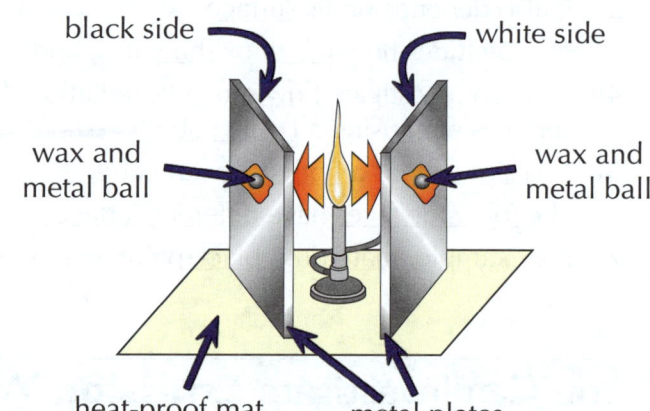

1) You'll need a Bunsen burner, candle wax, metal plates and metal balls.

2) The metal plates should be identical except their back surface, e.g. one plate will have a black back, and the other will have a white back.

3) Place the Bunsen burner on a heat-proof mat.

4) Stick a metal ball to each identical side of the metal plates with hot candle wax.

5) Leave the candle wax to cool. The wax will harden and hold the ball in place.

6) Then face the back of these plates towards the flame. They should both be the same distance away from the flame.

7) Record which ball falls first.

Explaining the **Results**

1) The plates absorb infrared radiation given out by the Bunsen burner.
2) Energy is transferred to the thermal energy stores of the candle wax.
3) The candle wax starts to melt, causing the balls to fall.
4) The ball will fall quicker from the plate that is better at absorbing infrared radiation.
5) You should find that the ball on the plate with the black back falls first.
6) This means the black surface was better at absorbing infrared radiation than the white surface.

 **Make sure you work safely during this practical...**
You need to be really careful when handling the hot objects in this practical — make sure you use protective gloves. You should also turn off the Bunsen burner when it's not in use.

Topic P6 — Waves

Dangers of Electromagnetic Waves

*Okay, so you know how **useful** electromagnetic radiation can be — well, it can also be pretty **dangerous**.*

Some EM Radiation Can be Harmful to People

1) When EM radiation enters living tissue — like you — it can be dangerous.
2) High frequency waves like UV, X-rays and gamma rays can all cause lots of damage.
3) UV radiation damages surface cells.
4) This can lead to sunburn and can cause skin to age faster than it should.
5) Some more serious effects are blindness and a higher risk of skin cancer.
6) X-rays and gamma rays are types of ionising radiation. This means they can knock electrons off atoms, p.337.
7) This can destroy cells or mutate (change) genes. This can cause cancer.

You Can Measure Risk Using the Radiation Dose in Sieverts

1) UV radiation, X-rays and gamma rays can all be useful as well as harmful (see page 380).
2) Radiation dose (measured in sieverts) is a measure of the risk of harm from the body being exposed to radiation.
3) The risk depends on the total amount of radiation absorbed and how harmful the type of radiation is.
4) A sievert is pretty big, so millisieverts (mSv) are often used. 1000 mSv = 1 Sv.

Risk can be Different for Different Parts of the Body

1) A CT scan uses X-rays to create a detailed picture of inside a patient's body.
2) The table shows the radiation dose received by two different parts of a patient's body when having CT scans.

	Radiation dose (mSv)
Head	2.0
Chest	8.0

3) You can see that the radiation dose from a chest scan is 4 times larger than from a head scan — (2.0 mSv × 4 = 8.0 mSv).
4) Remember, radiation dose measures the risk of harm.
5) This means that if a patient has a CT scan on their chest, they are four times more likely to be harmed than if they have a head scan.

The risks and benefits must be weighed up...

Ionising radiation can be dangerous, but the risk can be worth taking. X-ray machines used to be installed in shoe shops for use in shoe fittings. They were removed when people realised X-rays were harmful and the risks far outweighed the benefits of using X-rays rather than tape measures...

Topic P6 — Waves

Warm-Up & Exam Questions

There are lots of electromagnetic waves — you never know which ones might come up in the exams so make sure you know about all of them. See how much you can remember by trying these questions.

Warm-Up Questions

1) Which type of EM wave has the highest frequency?
2) How many types of EM wave are there?
3) How are gamma rays created?
4) Give one use of infrared radiation.
5) Describe two effects that ultraviolet exposure has on the skin.

Exam Questions

1 Optical fibres have many practical uses. *(Grade 1-3)*

Which type of electromagnetic wave is typically transmitted in optical fibres?

☐ radio waves ☐ microwaves

☐ visible light ☐ X-rays

[1 mark]

2 **Figure 1** shows an image of the bones in a patient's foot. *(Grade 1-3)*

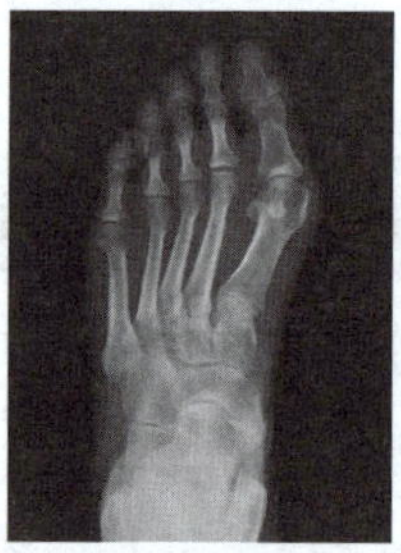

2.1 Which type of EM radiation could have been used to produce this image?

☐ radio waves ☐ visible light

☐ microwaves ☐ X-rays

[1 mark]

Figure 1

2.2 Give **one** risk to the patient from being exposed to this type of radiation.

[1 mark]

3 Some satellite signals are microwaves. *(Grade 4-5)*

3.1 Which of the following is high-frequency microwave radiation closest in frequency to?

☐ high-frequency ultraviolet ☐ low-frequency visible light

☐ high-frequency radio waves ☐ low-frequency infrared

[1 mark]

3.2 Do microwave signals travel faster in a vacuum than radio signals? Explain your answer.

[1 mark]

Topic P6 — Waves

Exam Questions

PRACTICAL

4 A student uses four identical beakers to investigate the infrared radiation emitted by different surfaces.

He wraps a piece of card around each beaker. Each piece of card has a different surface.

The student fills the beakers with boiling water and places a thermometer next to each one, as shown in **Figure 2**. He then records the temperature increase measured by each thermometer after two minutes.

Figure 2

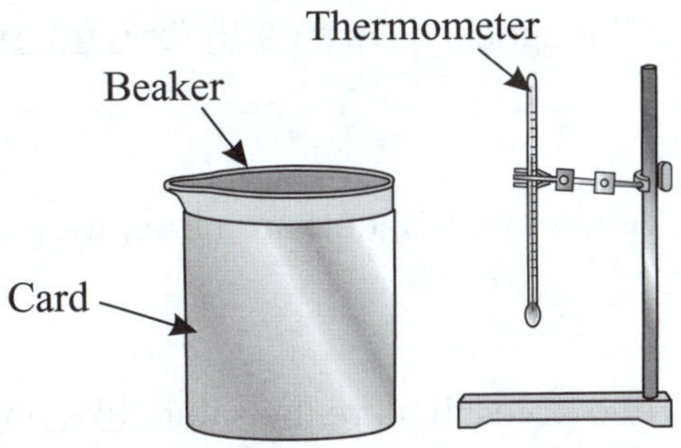

4.1 Name **one** control variable for this experiment.

[1 mark]

4.2 Name **one** safety precaution that needs to be taken in this experiment.

[1 mark]

Figure 3 shows the reading on a thermometer after it has been placed next to one of the beakers. The scale on the thermometer is in °C.

4.3 Write down the temperature shown on the thermometer. Give your answer to the nearest °C.

[1 mark]

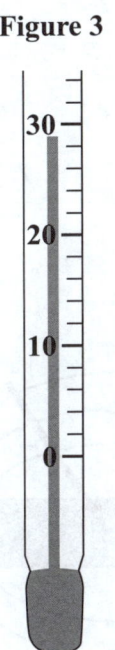

Figure 3

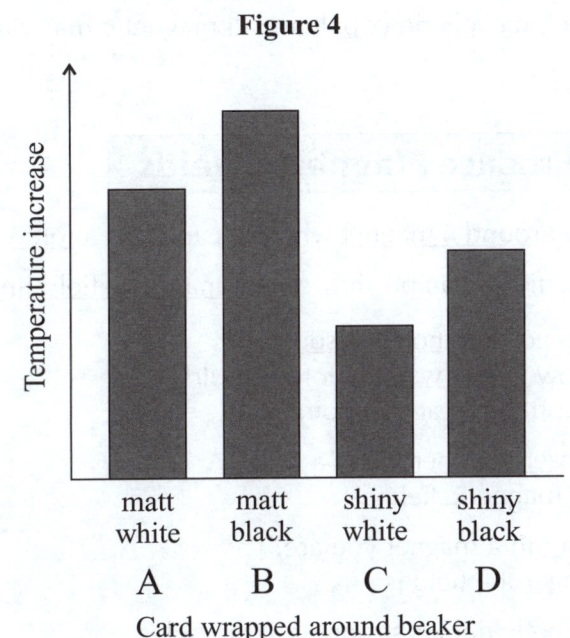

Figure 4

The student's results are displayed in **Figure 4**.

4.4 Some of the surfaces used in the experiment were shiny and others were matt (not shiny). Draw a conclusion from the data in **Figure 4** about how this characteristic affects infrared radiation emission.

[1 mark]

4.5 Suggest **one** way that the experiment could be made more accurate.

[1 mark]

Topic P7 — Magnetism and Electromagnetism

Magnetism

*I think magnetism is an **attractive** subject, but don't get **repelled** by the exam — **revise**.*

Magnets Exert **Forces** on Each Other

1) All magnets have a north (or north seeking) pole and a south (or south seeking) pole.

2) When two magnets are close, they exert a non-contact force on each other.

3) Two poles that are the same (like poles) repel each other.

Objects don't need to be touching for a non-contact force to act between them — see p.347.

4) Two different (unlike) poles attract each other.

5) There are also forces between magnets and magnetic materials. These forces are always attractive.

6) Iron, steel, nickel and cobalt are all magnetic materials.

Magnets Produce **Magnetic Fields**

1) The area around a magnet where it can exert a force is its magnetic field.
2) Magnetic fields can be shown with magnetic field lines.
3) The lines go from north to south. They show which way the force would push a north pole at that point.
4) Lines closer together mean a stronger magnetic field.
5) The closer to a magnet you are, the stronger the field is.
6) The magnetic field is strongest at the poles.
7) This means the magnetic force is strongest here too.

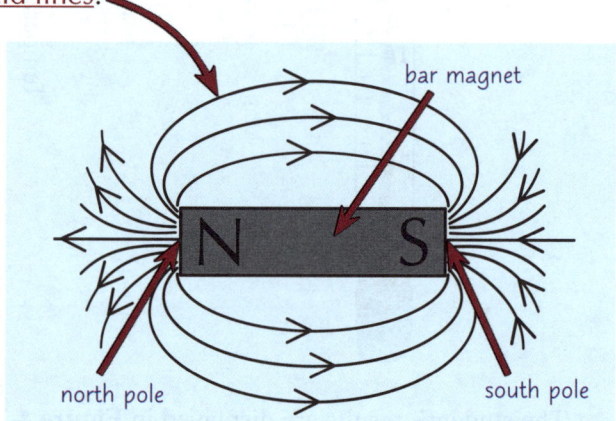

Remember magnetic field lines go from north to south...

You can see the shape of a magnetic field using iron filings. They give you a pretty pattern but they don't show you the direction of the magnetic field. That's where compasses really shine — there's more on using compasses to investigate magnetic fields on the next page.

Magnetism

Compasses aren't just useful when you're lost up a hill. They're also great tools to use to **plot** a **magnetic field**.

A Compass Shows the Direction of a Magnetic Field

1) The needle of a compass is a tiny bar magnet. It points in the direction of any magnetic field that it's in.

2) So you can use a compass to plot magnetic field patterns:

- Draw around a magnet on a piece of paper.
- Put a compass by the magnet.
- Mark the direction the compass needle points in by drawing a dot at each end of the needle.
- Move the compass so that the tail end of the needle is where the tip of the needle was before.
- Repeat this lots of times. Join up all the marks. You will end up with a drawing of one field line.

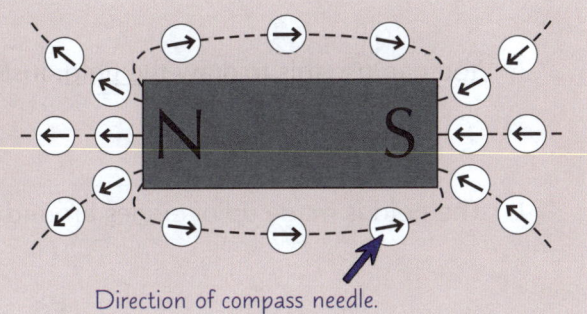

Direction of compass needle.

3) When they're not near a magnet, compasses always point north.

4) This is because they point in the direction of the Earth's magnetic field.

5) So the inside (core) of the Earth must be magnetic.

Magnets Can be Permanent or Induced

1) Permanent magnets create their own magnetic field.

2) An induced magnet turns into a magnet when it's put into another magnetic field.

3) When you take away the magnetic field, induced magnets quickly stop being magnets. A fancy way to say this is to say that they lose their magnetism (or most of it).

4) Permanent magnets and induced magnets always attract each other.

Topic P7 — Magnetism and Electromagnetism

Electromagnetism

A *magnetic field* is also found around a *wire* that has a *current* passing through it.

A Current Creates a Magnetic Field

1) A current flowing through a wire creates a magnetic field.

2) You can see this by placing a compass near to the wire. The compass will move to point in the direction of the field.

3) You can use this to draw the field, just like on the previous page.

4) The field is made up of circles around the wire (see below).

5) You can also use the right-hand thumb rule to quickly work out which way the field goes:

The Right-Hand Thumb Rule
- Point your right thumb in the direction of current.
- Curl your fingers.
- The direction of your fingers is the direction of the field.

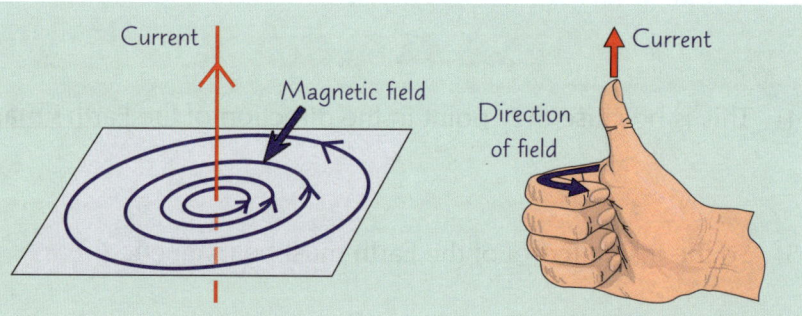

6) Reversing (swapping) the direction of the current reverses the direction of the magnetic field.

7) The closer to the wire you are, the stronger the magnetic field gets.

8) And the larger the current through the wire is, the stronger the field is.

Just point your thumb in the direction of the current...

...and your fingers show the direction of the field. Remember, it's always your right thumb.

Q1 Draw the magnetic field for a current-carrying wire. [2 marks]

Topic P7 — Magnetism and Electromagnetism

Solenoids

Electric currents can create **magnetic fields** (see previous page). We can use this effect to make magnets that can be switched on and off — these are **electromagnets** (which are made using **solenoids**).

A Solenoid is a Coil of Wire

1) If you <u>wrap</u> a wire into a <u>coil</u> it's called a <u>solenoid</u>.

2) The magnetic field <u>inside</u> a solenoid is <u>strong</u> and <u>uniform</u>.

3) <u>Uniform</u> means the field has the <u>same strength</u> and <u>direction</u> everywhere.

4) The magnetic field <u>outside</u> a coil is just like the one around a <u>bar magnet</u>.

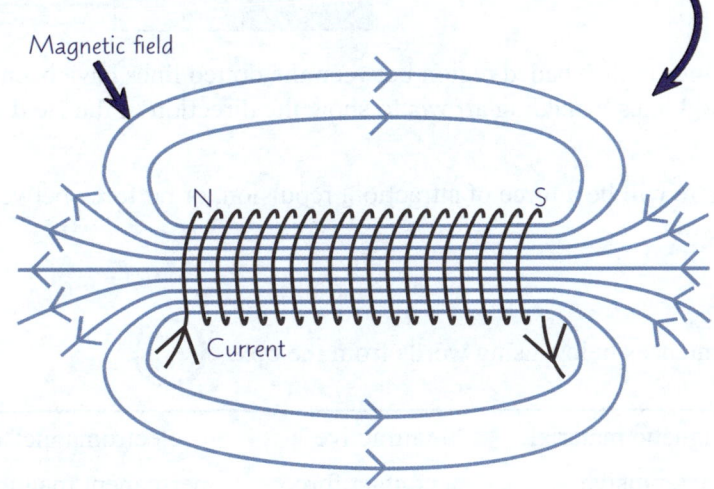

5) Wrapping a wire into a solenoid <u>increases the strength</u> of the magnetic field produced by the current in the wire.

- This is because the <u>field lines</u> around <u>each loop</u> of wire <u>line up</u> with each other.
- So <u>lots</u> of <u>field lines</u> end up <u>close</u> to each other and pointing in the <u>same direction</u>.
- The <u>closer together</u> field lines are, the <u>stronger the field</u> is.

6) You can <u>increase</u> the field strength <u>even more</u> by putting a block of <u>iron</u> in the coil.

7) A <u>solenoid with an iron core</u> is called an <u>electromagnet</u>.

Fields around solenoids and bar magnets are the same shape...

Electromagnets pop up in lots of different places — they're used in <u>electric bells</u>, <u>car ignition circuits</u> and some <u>security doors</u>. Electromagnets aren't all the same strength though — how <u>strong</u> they are depends on stuff like the number of <u>turns</u> of wire there are and the <u>size</u> of the <u>current</u> going through the wire.

Topic P7 — Magnetism and Electromagnetism

Warm-Up & Exam Questions

Time to test your knowledge — as usual, make sure you're switched on by doing the warm-up questions, then dive into the exam questions. Don't forget to go back and check up on any bits you can't do.

Warm-Up Questions

1) Draw a diagram to show the magnetic field around a single bar magnet. Indicate its direction with arrows.
2) When using the right-hand thumb rule, what does the direction of your fingers represent?
3) How can you form a solenoid from a current-carrying wire?

Exam Questions

1 A student arranges two magnets as shown in **Figure 1**.

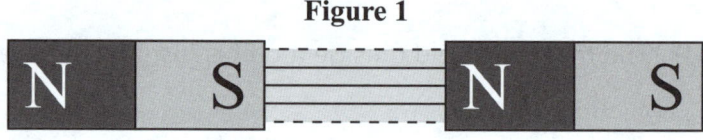

Figure 1

1.1 Magnetic field lines in the shaded region between the dotted lines have been drawn on **Figure 1**. Complete the field lines by adding arrows to show the direction of the field.

[1 mark]

1.2 State whether there will be a force of attraction, repulsion, or no force between the two magnets. Explain your answer.

[2 marks]

2 Complete the sentences below using words from the box.

| magnetic material | attractive | electromagnet |
| repulsive | a contact force | permanent magnet |

Cobalt is an example of a .... The force between

a piece of cobalt and a bar magnet is always ....

[2 marks]

3 A student is investigating magnetic fields. She passes a wire through a horizontal piece of card and connects it in an electrical circuit, as shown in **Figure 2**.

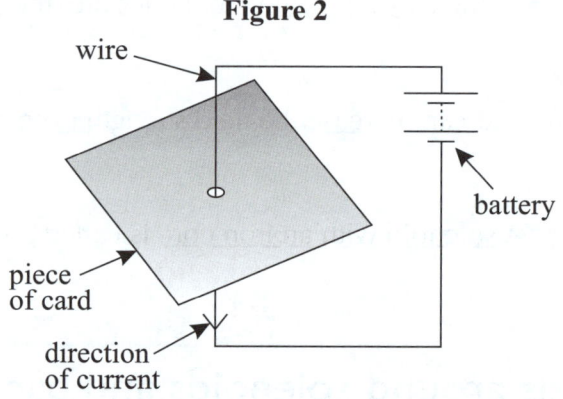

Figure 2

3.1 The student uses a compass to plot the magnetic field lines on the piece of card. Describe the shape and direction of the magnetic field the student will plot.

[2 marks]

3.2 The student forms the wire into a coil. Explain why this leads to an increase in the magnetic field strength.

[2 marks]

Topic P7 — Magnetism and Electromagnetism

Revision Summary for Topics P5-7

That wraps up Topics P5-7 — time to put yourself to the test and find out how much you really know.
- Try these questions and tick off each one when you get it right.
- When you're completely happy with a topic, tick it off.

For even more practice, try the Retrieval Quizzes for Topics P5-7 — just scan the QR codes!

Topic P5 — Forces (p.347-369) ☑

1) Explain the difference between scalar and vector quantities.
2) What is the difference between contact and non-contact forces?
3) What is the formula for calculating the weight of an object?
4) What is a resultant force?
5) Why does the temperature of an object increase when it is pushed along a rough surface?
6) Sketch a typical force-extension graph for a spring where its limit of proportionality is exceeded.
7) Describe an experiment you could do to investigate the relationship between force and extension.
8) Estimate the speed of a person walking.
9) Write down the equation that links acceleration, velocity and time.
10) What does the gradient represent for: a) a distance-time graph? b) a velocity-time graph?
11) State Newton's three laws of motion.
12) Describe an experiment that you could do to investigate Newton's Second Law.
13) What is a typical stopping distance of a car travelling at 60 mph?
14) Give two things that affect a person's reaction time.
15) Briefly describe an experiment you could do to compare people's reaction times.

Topic P6 — Waves (p.371-383) ☑

16) What is: a) the amplitude of a wave? b) the wavelength of a wave?
17) What is the difference between transverse and longitudinal waves?
18) Give the units that each quantity in the following equations should be given in:
    a) period = $\frac{1}{\text{frequency}}$ b) $v = f\lambda$
19) Describe an experiment you could do to measure the speed of waves on a string.
20) Sketch a ray diagram showing the refraction of a light ray between two materials.
21) True or false? All electromagnetic waves are transverse.
22) Give a common use of radio waves.
23) Which type of electromagnetic wave is used in suntanning?
24) How could you use a Leslie cube to investigate infrared emission by different surfaces?
25) Name two types of ionising electromagnetic radiation.

Topic P7 — Magnetism and Electromagnetism (p.386-389) ☑

26) Give three magnetic materials.
27) In what direction do magnetic field lines point?
28) Where on a bar magnet is the magnetic force strongest?
29) What is meant by an 'induced magnet'?
30) How can the direction of the magnetic field around a current-carrying wire be reversed?
31) Sketch the shape of the magnetic field around a solenoid.

Practical Skills
Measuring Techniques

Get your lab coat on, it's time to find out about the skills you'll need in **experiments**...

Mass Should Be Measured Using a Balance

1) To measure mass, put the container you're measuring the substance into on the balance.
2) Set the balance to exactly zero. Then add your substance and read off the mass.
3) If you want to transfer the substance to a new container, you need to make sure that the mass you transfer is the same as the mass you measured. There are different ways you can do this. For example:

- If you're dissolving a mass of a solid in a solvent to make a solution, you could wash any remaining solid into the new container using the solvent.
- You could set the balance to zero before you put your weighing container on the balance. Then reweigh the weighing container after you've transferred the substance. Use the difference in mass to work out exactly how much substance you've transferred.

Different Ways to Measure Liquids

1) There are a few methods you might use to transfer a volume of liquid:

Dropping pipette — Use this if you only want a couple of drops of liquid. It's also used if you don't need an accurate volume of liquid.

Pipette — Use this if you want an accurate volume of liquid. The pipette filler lets you safely control the amount of liquid you're drawing up.

Measuring cylinders — These come in many different sizes. You need to use one that's the right size for the measurement you want to make (you don't want one that's too big).

2) To measure the volume of a liquid, read the volume from the bottom of the meniscus (the curved upper surface of the liquid) when it's at eye level.

You Can Measure Gas Volumes

There are a few ways you can measure the volume of a gas:

1) Gas syringe — this is the most accurate way to measure gas volume.

- Make sure the gas syringe is the right size for your measurements.
- Make sure the plunger moves smoothly.
- Read the volume from the scale on the syringe.

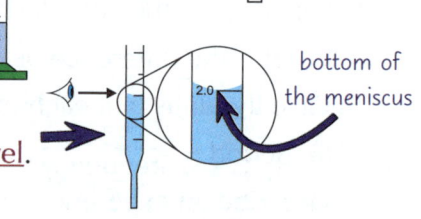

2) Upturned measuring cylinder filled with water — read more about this on page 237.
3) Counting the bubbles produced or measuring the length of a gas bubble drawn along a tube (see p.88).

- These methods are less accurate.
- But they will give you results that you can compare.

Always make sure your equipment is sealed so no gas can escape. This will make your results more accurate.

Measuring Techniques

Eureka Cans Measure the Volumes of Solids

1) A <u>eureka can</u> is a <u>beaker with a spout</u>.
2) It's used with a <u>measuring cylinder</u> to find the <u>volume</u> of an <u>irregularly shaped solid object</u> (p.329).
3) Here are a few things you need to do when you <u>use</u> one:

- Fill it with water so the water level is <u>above the spout</u>.
- Let the water <u>drain</u> from the spout, leaving the water level <u>just below</u> the start of the spout. (This means <u>all</u> the water moved by an object goes into the measuring cylinder.)
- After adding the object, wait until the spout has <u>stopped dripping</u> before measuring the volume. This will give you a more <u>accurate</u> result.

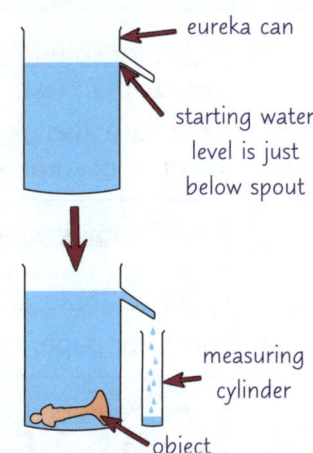

Measure Most Lengths with a Ruler

1) Make sure you <u>choose</u> the <u>right ruler</u> to measure length:
    - In most cases a <u>centimetre ruler</u> can be used.
    - <u>Metre rulers</u> are handy for <u>large</u> distances.
    - <u>Micrometers</u> are used for measuring <u>tiny</u> things (e.g. the <u>diameter of a wire</u>).
2) The ruler should always be <u>alongside</u> what you want to measure.
3) It may be <u>tricky</u> to measure just <u>one</u> of something (e.g. water ripples, p.374). Instead, you can measure the length of <u>ten</u> of them together. Then <u>divide by ten</u> to find the <u>length of one</u>.
4) You might need to take <u>lots of measurements</u> of the <u>same</u> object (e.g. a spring). If so, make sure you always measure from the <u>same point</u> on the object. Draw or stick small <u>markers</u> onto the object to line your ruler up against.
5) Make sure the ruler and the object are always at <u>eye level</u> when you take a reading.

Use a Protractor to Find Angles

1) Place the <u>middle</u> of the protractor on the <u>pointy bit</u> of the angle.
2) <u>Line up</u> the <u>base line</u> of the protractor with one line of the angle.
3) Use the <u>scale</u> on the protractor to measure the angle of the other line.
4) Use a <u>sharp pencil</u> to draw lines at an angle (e.g. in ray diagrams). This helps to <u>reduce errors</u> when measuring the angles.

Measure Temperature Using a Thermometer

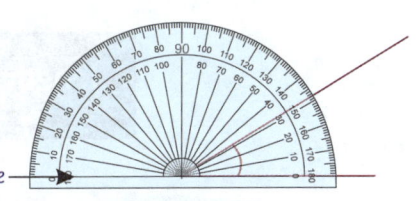

1) Make sure the <u>bulb</u> of your thermometer is <u>completely under the surface</u> of the substance.
2) If you're taking a <u>starting temperature</u>, you should wait for the temperature to <u>stop changing</u>.
3) Read your measurement off the <u>scale</u> at <u>eye level</u>.

You May Have to Measure the Time Taken for a Change

1) You should use a <u>stopwatch</u> to <u>time</u> experiments. These measure to the nearest <u>0.1 s</u>.
2) Always make sure you <u>start</u> and <u>stop</u> the stopwatch at exactly the right time.
3) You can set an <u>alarm</u> on the stopwatch so you know exactly when to stop an experiment or take a reading.

Practical Skills

Measuring Techniques

There are Different Methods for Measuring pH

1) <u>Indicator solutions</u> can be used to estimate pH. Add a <u>couple of drops</u> of the indicator to the solution you want to test. It will <u>change colour</u> depending on if it's in an <u>acid</u> or an <u>alkali</u> (see p.215).
2) There are also <u>paper indicators</u>. These are <u>strips of paper</u> that contain indicator. If you <u>spot</u> some solution onto indicator paper, the paper will <u>change colour</u> to show the pH.

- <u>Litmus paper</u> turns <u>red</u> in acidic conditions and <u>blue</u> in alkaline conditions.
- <u>Universal indicator paper</u> can be used to <u>estimate</u> the pH based on its colour.

litmus paper

3) Indicator paper is <u>useful</u> when:

- You <u>don't</u> want to change the colour of <u>all</u> of the substance.
- The substance is <u>already</u> coloured (so it might <u>hide</u> the colour of the indicator).
- You want to find the pH of a <u>gas</u> — hold a piece of <u>damp indicator paper</u> in a <u>gas sample</u>.

4) <u>pH probes</u> measure pH <u>electronically</u> (see page 217). They are more <u>accurate</u> than indicators.

You Can Measure the Size of a Single Cell

When viewing <u>cells</u> under a <u>microscope</u>, you might need to work out their <u>size</u>. To work out the size of a <u>single cell</u>:

1) Place a <u>clear, plastic ruler</u> on <u>top</u> of your microscope <u>slide</u>.
2) <u>Clip</u> the <u>ruler</u> and <u>slide</u> onto the <u>stage</u>.
3) Select the <u>objective lens</u> that gives an overall magnification of <u>× 100</u>.
4) Use the <u>coarse adjustment knob</u> and the <u>fine adjustment knob</u> to see a <u>clear image</u> of the cells.
5) <u>Move</u> the ruler so that the cells are <u>lined up</u> along <u>1 mm</u>.
6) <u>Count</u> the <u>number of cells</u> along this <u>1 mm sample</u>.
7) 1 mm = 1000 μm. So you can <u>calculate</u> the <u>size</u> of a <u>single cell</u> using this <u>formula</u>:

You can read all about using a microscope on pages 20-21.

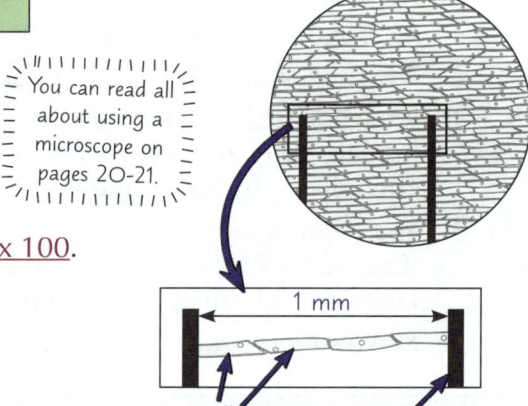

$$\text{length of cell (μm)} = \frac{1000 \text{ μm}}{\text{number of cells counted in sample}}$$

EXAMPLE

Under a microscope, 4 cells were counted in 1 mm. Calculate the size of one cell. Give your answer in μm.

1 mm is the <u>same</u> as 1000 μm, so you just need to put the <u>number of cells</u> into the formula.

$$\text{length of cell (μm)} = \frac{1000 \text{ μm}}{4} = 250 \text{ μm}$$

You might have to bend down a little...

Whether you're reading off a ruler, a pipette or a measuring cylinder, make sure you take all readings at <u>eye level</u>. And, if you're taking a reading for a volume, make sure you measure from the bottom of the <u>meniscus</u> (that's from the dip in the curved surface at the top of the liquid).

Safety and Ethics

Before you start any experiment, you need to know what **safety measures** you should be taking. They depend on your **method**, your **equipment** and the **chemicals** you're using.

To Make Sure You're Working Safely in the Lab You Need to...

1) Wear sensible clothing (e.g. shoes that will protect your feet from spillages). Also:
   - Wear a lab coat to protect your skin and clothing.
   - Wear safety goggles to protect your eyes, and gloves to protect your hands.

2) Be aware of general safety in the lab. E.g. don't touch any hot equipment.
3) Follow any instructions that your teacher gives you carefully.
4) Chemicals and equipment can be hazardous (dangerous). E.g. some chemicals are flammable (they catch fire easily) — this means you must be careful not to use a Bunsen burner near them.
5) Here are some tips for working with chemicals and equipment safely...

Working with chemicals

1) Make sure you're working in an area that's well ventilated (has a good flow of air).
2) If you're doing an experiment that produces nasty gases (such as chlorine), carry out the experiment in a fume hood. This means the gas can't escape out into the room you're working in.
3) Never touch any chemicals (even if you're wearing gloves):
   - Use a spatula to transfer solids between containers.
   - Carefully pour liquids between containers using a funnel. This will help prevent spillages.
4) Be careful when you're mixing chemicals, as a reaction might occur. E.g. if you're diluting a liquid, always add the concentrated substance to the water. This stops it getting hot.

Working with equipment

1) Use clamp stands to stop masses and equipment falling.
2) Make sure masses are not too heavy (so they don't break the equipment they're used with).
3) Use pulleys that are not too long (so hanging masses don't hit the floor during the experiment).
4) Let hot materials cool before moving them. Or wear insulated gloves while handling them.
5) If you're using an immersion heater, you should always let it dry out in air. This is just in case any liquid has leaked inside the heater.
6) When working with electronics, make sure you use a low voltage and current. This prevents the wires overheating. It also stops damage to components.

You Need to Think About Ethical Issues

Any organisms that you use in your experiments need to be treated safely and ethically. This means:
1) Animals should be handled carefully.
2) Any captured wild animals should be returned to their habitat after the experiment.
3) Any animals kept in the lab should be well cared for. E.g. they should have plenty of space.
4) Other students that take part in any experiment should be happy to do so.

BEWARE — hazardous experiments about...

Before you carry out an experiment, it's important to consider all of the hazards. They can be anything from chemicals to sharp objects to heating equipment. Whatever the hazard, make sure you know all the safety precautions you should follow to keep yourself, and others, safe.

Practical Skills

Setting Up Experiments

Setting up the equipment for an experiment in the right way is important. Learn these set-ups...

You Can Identify the Products of Electrolysis

There's more about electrolysis on p.223-226.

1) When you electrolyse a salt solution:

   - At the cathode, you'll get a pure metal coating the electrode OR bubbles of hydrogen gas.
   - At the anode, you'll get bubbles of oxygen gas OR a halogen.

2) You may have to do some tests to find out what's been made.

3) To do this, you need to set up the equipment correctly to collect any gas that's produced. The easiest way to collect the gas is in a test tube.

4) Here's how to set up the equipment...

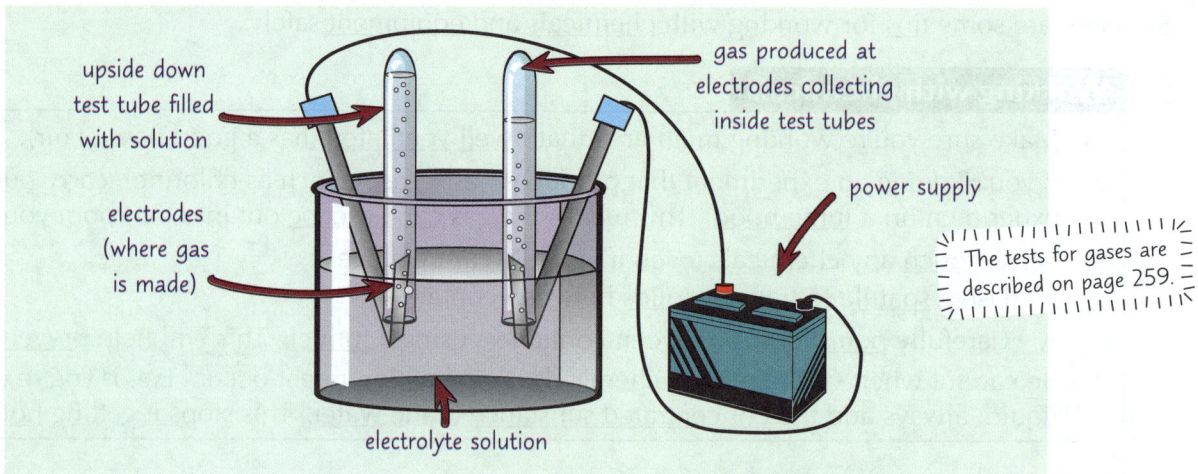

The tests for gases are described on page 259.

Set Up a Potometer to Measure Transpiration Rate

1) A potometer is a special piece of equipment.
2) You set it up as shown in the diagram.
3) You can use a potometer to estimate transpiration rate (see page 69). Here's what you do:

   1) Record the starting position of the air bubble.
   2) Start a stopwatch.
   3) As the plant takes up water, the air bubble gets sucked along the tube.
   4) Record how far the air bubble moves in a set time.
   5) Then you can estimate the transpiration rate.

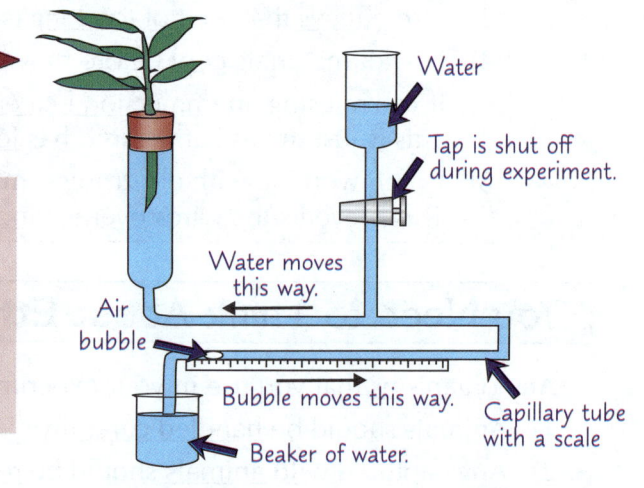

EXAMPLE

A potometer was used to estimate the transpiration rate of a plant cutting. The bubble moved 25 mm in 10 minutes. Estimate the transpiration rate.

To estimate the transpiration rate, divide the distance the bubble moved by the time taken.

$$\text{Transpiration rate} = \frac{\text{distance bubble moved}}{\text{time taken}} = \frac{25 \text{ mm}}{10 \text{ min}}$$

$$= 2.5 \text{ mm/min}$$

Setting Up Experiments

You Can Collect a Gas in a Measuring Cylinder

1) You can use a measuring cylinder turned upside down and filled with water to collect gas.
2) Then you can measure the gas volume. Here's how you do it:

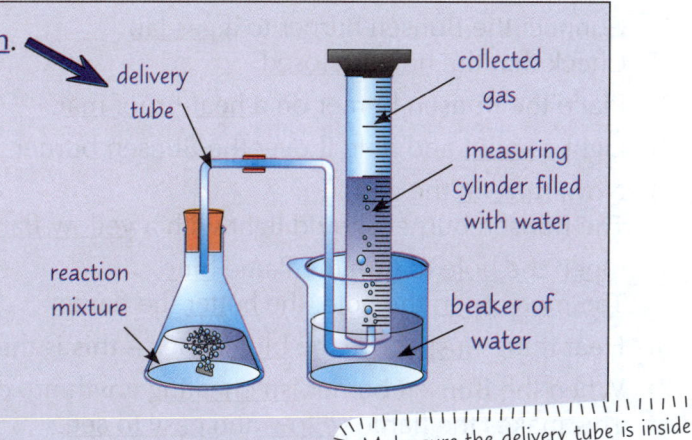

1) Set up the equipment like in the diagram.
2) Record the starting level of the water in the measuring cylinder.
3) Any gas from the reaction will pass through the delivery tube and into the measuring cylinder.
4) The gas will push the water out of the measuring cylinder.
5) Record the end level of water in the measuring cylinder.
6) Calculate the volume of gas produced — subtract the end level of water from the starting level of water.

Make sure the delivery tube is inside the measuring cylinder. This stops the gas escaping out into the air.

3) You can use the method above to collect a gas sample to test.
   - Use a test tube instead of a measuring cylinder.
   - When the test tube is full of gas, you can put a bung in it. This lets you store the gas for later.

Make Sure You Can Draw Diagrams of Your Equipment

1) Your method should include a labelled diagram of how your equipment will be set up.
2) Use scientific drawings — each piece of equipment is drawn as if you're looking at it from the side.
3) For example:

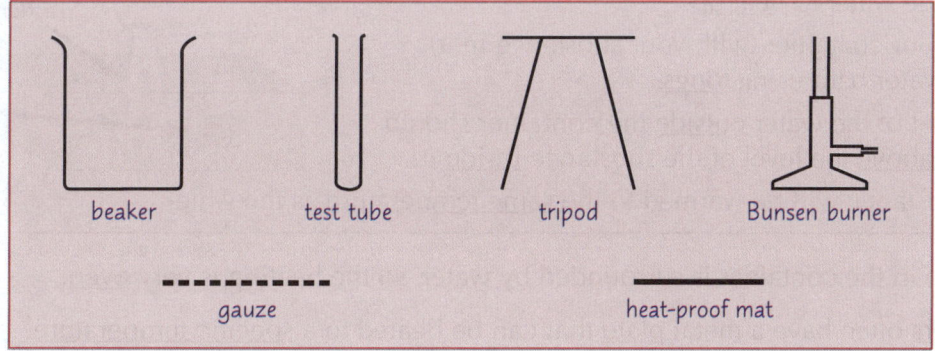

4) The beaker and test tube above aren't sealed. To show them sealed, draw a bung in the top.

These simple diagrams are clear and easy to draw...

Have a go at drawing diagrams of the experimental set-ups on these last few pages. It'll give you some practice at doing them and you can revise how to set up the experiments as well.

Practical Skills

Heating Substances

You need to be able to decide on the **best** and **safest** method for heating a substance...

Bunsen Burners Heat Things Quickly

Here's how to <u>use</u> a Bunsen burner...

1) Connect the Bunsen burner to a <u>gas tap</u>. Check that the <u>hole</u> is <u>closed</u>.
2) Place the Bunsen burner on a <u>heat-proof mat</u>.
3) Light a <u>splint</u> and hold it over the Bunsen burner.
4) Now, <u>turn on</u> the gas. The Bunsen burner should light with a <u>yellow flame</u>.
5) <u>Open</u> the <u>hole</u> to turn the flame <u>blue</u>. The <u>more open</u> the hole, the <u>hotter</u> the flame.
6) Heat things <u>just above</u> the <u>blue cone</u> — this is the <u>hottest</u> part of the flame.
7) When the Bunsen burner <u>isn't heating</u> anything, <u>close</u> the hole. This makes the flame <u>yellow</u> and <u>easy to see</u>.
8) If you're heating a container (with your substance in it) <u>in</u> the flame, hold it at the <u>top</u> with a pair of <u>tongs</u>.
9) If you're heating a container <u>over</u> the flame, put a <u>tripod and gauze</u> over the Bunsen burner before you light it. Then place the container on the gauze.

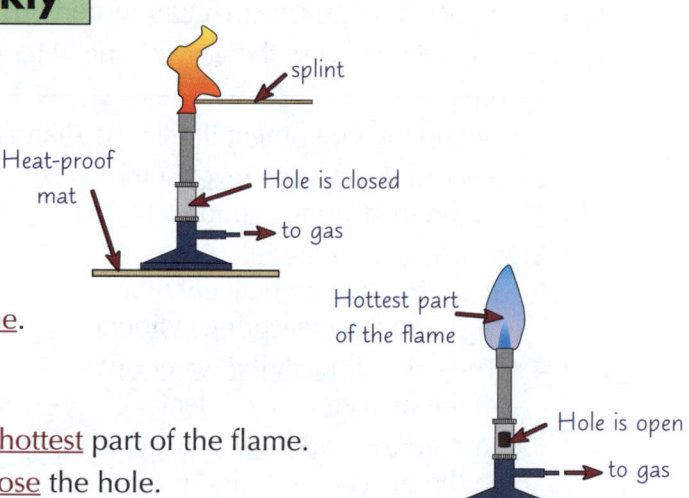

Water Baths & Electric Heaters Have Set Temperatures

1) A <u>water bath</u> is a <u>container</u> filled with <u>water</u>. It can be heated to a <u>specific temperature</u>.
2) A <u>simple</u> water bath can be made by heating a <u>beaker of water</u> over a <u>Bunsen burner</u>.
   - The temperature is checked with a <u>thermometer</u>.
   - However, it's <u>hard</u> to keep the temperature of the water <u>constant</u>.
3) An <u>electric water bath</u> will <u>check</u> and <u>change</u> the temperature for you. Here's how you use one:

> - <u>Set</u> the <u>temperature</u> on the water bath.
> - Allow the water to <u>heat up</u>.
> - Place your container (with your substance in it) in the water bath using <u>tongs</u>.
> - The level of the water <u>outside</u> the container should be <u>just above</u> the level of the substance <u>inside</u> it.
> - The substance will be warmed to the <u>same temperature</u> as the water.

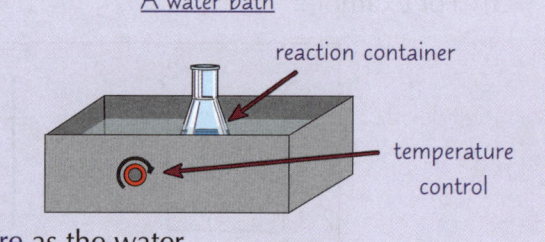

A water bath

The substance in the container is surrounded by water, so the heating is very <u>even</u>.

4) <u>Electric heaters</u> often have a metal <u>plate</u> that can be heated to a <u>specific temperature</u>.
   - Place your container on <u>top</u> of the <u>hot plate</u>.
   - You can heat substances to <u>higher temperatures</u> than you can in a water bath. (You <u>can't</u> use a water bath to heat something higher than <u>100 °C</u>.)
   - You have to <u>stir</u> the substance to make sure it's <u>heated evenly</u>.

Bunsen burners are useful, but need to be used in the right way...

When the Bunsen burner <u>isn't heating</u> anything, it's important to <u>close the hole</u>. This turns the flame <u>yellow</u> — that flame isn't as hot and is much <u>easier to spot</u>, so it helps to prevent any burns...

Practical Skills

Working with Electronics

Electrical devices are used in loads of *experiments*. Make sure you know how to use them.

There Are a Few Ways to Measure **Potential Difference** and **Current**

Voltmeters Measure Potential Difference

1) Connect the voltmeter in parallel (p.316) across the component you want to test.
2) The wires that come with a voltmeter are usually red (positive) and black (negative). These go into the red and black coloured ports on the voltmeter.
3) Then read the potential difference from the scale (or from the screen if the voltmeter is digital).

Ammeters Measure Current

1) Connect the ammeter in series (p.314) with the component you want to test.
2) Ammeters usually have red and black ports to show you where to connect your wires.
3) Read off the current shown on the scale (or screen).

Turn your circuit off between readings. This stops wires overheating and affecting your results (page 308).

Multimeters Measure Both

1) Multimeters measure a range of things — usually potential difference, current and resistance.
2) To find potential difference, plug the red wire into the port that has a 'V' (for volts).
3) To find the current, use the port labelled 'A' (for amps).
4) The dial on the multimeter should then be turned to the relevant section — for example, to measure the current in amps, turn the dial to 'A'.
5) The screen will display the value you're measuring.

Light Gates Measure Time, Speed and Acceleration

1) A light gate sends a beam of light from one side of the gate to a detector on the other side.
2) When something passes through the gate, the light beam is interrupted.
3) The gate measures when the beam was interrupted and how long it was interrupted for.
4) Light gates can be connected to a computer.
5) To find the speed of an object, type the length of the object into the computer. The computer will calculate the speed of the object as it passes through the beam.
6) To measure acceleration, use an object that interrupts the signal twice, e.g. a piece of card with a gap cut into the middle.
7) The light gate measures the speed for each section of the object. It uses this to calculate the object's acceleration. This can then be read from the computer screen.
8) Light gates can be used instead of a stop watch. This will reduce the errors in your experiment.

Have a look at page 364 for an example of a light gate being used.

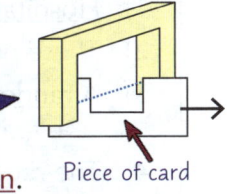

Light gate

Beam of light

Piece of card

Don't get your wires in a tangle when you're using circuits...

When you're dealing with voltmeters, ammeters and multimeters, you need to make sure that you wire them into your circuit correctly, otherwise you could mess up your readings. Just remember, the red wires should go into the red ports and the black wires should go into the black ports.

Practical Skills

Sampling

You need to be able to carry out **sampling** that'll give you **non-biased** results. First up why, then how...

Sampling Should be Random

1) When you're investigating a population, it's usually not possible to study every single organism in it.
2) This means that you need to take samples of the population.
3) The samples need to accurately represent the whole population.
   This is so you can use them to draw conclusions about the whole population.
4) To make sure a sample represents the population, it should be random.

Organisms Should Be Sampled At Random Sites in an Area

1) Quadrats can be used to take population samples of an organism in an area (see page 147).
2) If you're looking at plant species in a field...

- Divide the field into a grid.
- Label the grid along the bottom and up the side with numbers.
- Use a random number generator (e.g. on a computer or calculator) to select coordinates, e.g. (2,7).
- Place your quadrats at these coordinates to take your samples.

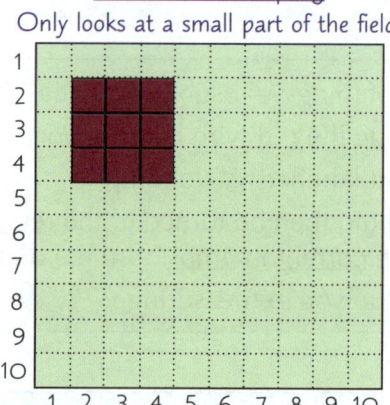

Non-random sampling
Only looks at a small part of the field.

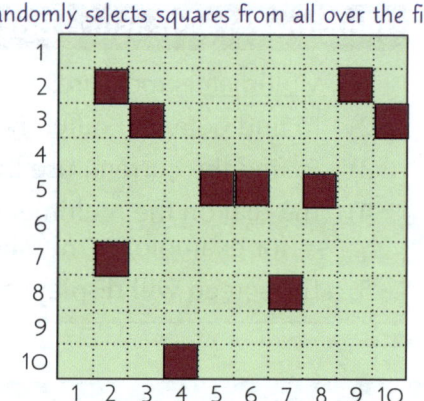

Random sampling
Randomly selects squares from all over the field.

Health Data Should be Taken from Randomly Selected People

You need to use random sampling to choose members of the population you're interested in. For example:

A scientist is looking at health data in country X. She wants to know how many people in the country have both Type 2 diabetes and heart disease:
1) Hospital records show that 270 196 people in the country have Type 2 diabetes.
2) These people are given a number between 1 and 270 196.
3) A random number generator is used to choose the sample group
   — e.g. it selects individuals with the numbers #72 063, #11 822, #193 123, etc.
4) The records of the sample group are used to find the number of people with heart disease in it.
5) The proportion of people in the sample group who have heart disease is worked out.
6) This can be used to estimate the total number of people with Type 2 diabetes who also have heart disease.

Sampling is an important part of an investigation...

Sampling is a really useful way to find out information about the whole population from a smaller group. It's important that the sample is selected randomly though, or the results won't be worth much.

Practical Skills

Comparing Results

*Being able to **compare** your results is really important. Here is one way you might do it.*

Percentage Change Allows you to Compare Results

1) When investigating the change in a variable, you may want to compare results that didn't have the same starting value.

> - For example, you may want to compare the change in mass of potato cylinders left in different concentrations of sugar solution (see page 32).
> - The cylinders probably all had different masses to start with.

2) To do this you can calculate the percentage change. You work it out like this:

$$\text{percentage (\%) change} = \frac{\text{final value} - \text{original value}}{\text{original value}} \times 100$$

3) A positive percentage change means that the value increased.
A negative percentage change means that the value decreased.

EXAMPLE

A student is investigating the effect of the concentration of sugar solution on potato cells. She records the mass of potato cylinders before and after placing them in sugar solutions of different concentrations. The table below shows some of her results.

Potato cylinder	Concentration (mol/dm^3)	Mass at start (g)	Mass at end (g)
1	0.0	7.5	8.7
2	1.0	8.0	6.8

Which potato cylinder had the largest percentage change?

Stick each set of results into the equation: $\quad \text{\% change} = \frac{\text{final value} - \text{original value}}{\text{original value}} \times 100$

The mass at the start is the original value.
The mass at the end is the final value.

potato cylinder 1: $\frac{8.7 - 7.5}{7.5} \times 100 = 16\%$

potato cylinder 2: $\frac{6.8 - 8.0}{8.0} \times 100 = -15\%$ — Here, the mass has decreased so the percentage change is negative.

Compare the results.

16% is greater than 15%. So potato cylinder 1 (in the 0.0 mol/dm^3 sugar solution) had the largest percentage change.

Good practical skills are needed when you're doing an investigation...

...but you also need to know about them for your exams. You're guaranteed to be tested on your practical knowledge, so if you've skipped through this section, you'd better go back and read through it again.

Practical Skills

Practice Exams

Once you've been through all the questions in this book, you should feel pretty confident about the exams. As final preparation, here is a set of **practice exams** to really get you set for the real thing. The time allowed for each paper is 1 hour 15 minutes. These papers are designed to give you the best possible preparation for your exams.

GCSE Combined Science
Biology Paper 1
Foundation Tier

In addition to this paper you should have:
- A ruler.
- A calculator.

Centre name				
Centre number				
Candidate number				

Time allowed:
- 1 hour 15 minutes

Surname	
Other names	
Candidate signature	

Instructions to candidates
- Write your name and other details in the spaces provided above.
- Answer **all** questions in the spaces provided.
- Do all rough work on the paper.
- Cross out any work you do not want to be marked.

Information for candidates
- The marks available are given in brackets at the end of each question.
- There are 70 marks available for this paper.
- You are allowed to use a calculator.
- You should use good English and present your answers in a clear and organised way.

Advice to candidates
- In calculations show clearly how you worked out your answers.

For examiner's use							
Q	Attempt Nº			Q	Attempt Nº		
	1	2	3		1	2	3
1				5			
2				6			
3				7			
4				8			
				Total			

1 Figure 1 shows the human respiratory system.

Figure 1

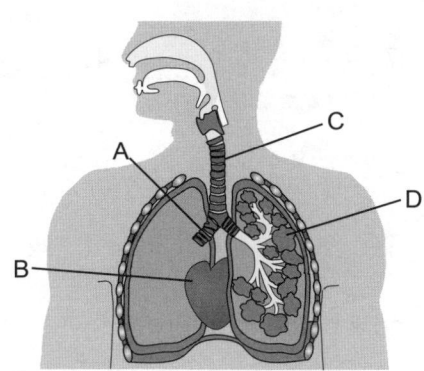

1.1 Which of the labels on **Figure 1** points to the trachea?
Tick **one** box.

☐ A ☐ B ☐ C ☐ D

[1 mark]

1.2 Which of the labels on **Figure 1** shows where gas exchange takes place?
Tick **one** box.

☐ A ☐ B ☐ C ☐ D

[1 mark]

1.3 Cells in the human body are organised into different levels.

Draw **one** line from each description of a level of cell organisation to the example of that level.

A group of similar cells that work together to carry out a function		Lungs
A group of organs working together to perform a function		Epithelial tissue
A group of different tissues that work together to perform a certain function		Respiratory system

[2 marks]

Question 1 continues on the next page

Turn over ▶

1.4 An athlete ran on a treadmill for 11 minutes. In that time she took 407 breaths.
What was the athlete's average breathing rate in breaths per minute?
Tick **one** box.

☐ 24   ☐ 37   ☐ 48   ☐ 61

[1 mark]

1.5 **Figure 2** shows an alveolus and a blood capillary.

Figure 2

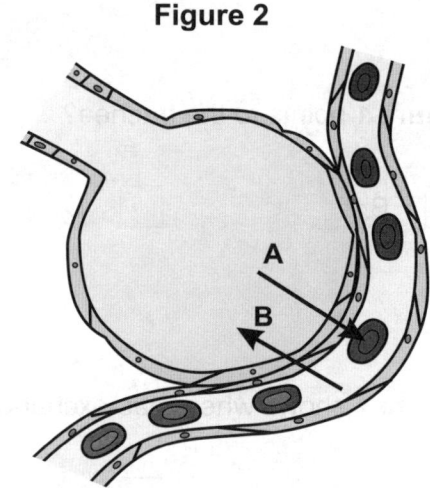

The arrows on the diagram show the net movement of two gases, **A** and **B**.
Name gases **A** and **B**.

Gas **A**: ....

Gas **B**: ....

[2 marks]

2 Plants produce glucose during photosynthesis.

2.1 Complete the word equation for photosynthesis.

carbon dioxide + water ⟶ glucose + ...

[1 mark]

2.2 Name the subcellular structure that absorbs light for photosynthesis.

....

[1 mark]

2.3 The glucose produced in photosynthesis can be used by plants to make cellulose.
What is cellulose used for in plant cells?

....

[1 mark]

2.4 Use a word from the box to complete the sentence below.

| damaged | reproduced | respired | differentiated |

Plants kept in a greenhouse that has become too hot may stop photosynthesising.

This is because the enzymes needed for photosynthesis are ....

[1 mark]

Question 2 continues on the next page

Turn over ▶

A student is investigating the effect of light intensity on the rate of photosynthesis.
Figure 3 shows the apparatus. **Table 1** shows the student's results.

Figure 3

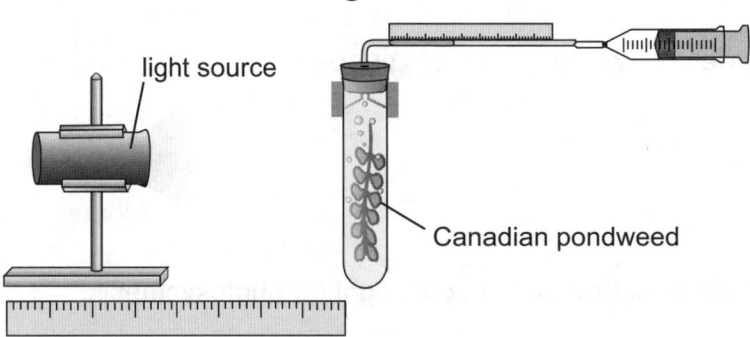

Table 1

Relative light intensity	1	2	3	4	5	6	7	8	9	10
Volume of gas produced in 10 minutes (cm³)	8	12	18	25	31	13	42	48	56	61

2.5 At which relative light intensity did the student record an anomalous result?

....

[1 mark]

2.6 The student plotted a graph of the rate of photosynthesis against relative light intensity.
Figure 4 shows the student's graph.

Figure 4

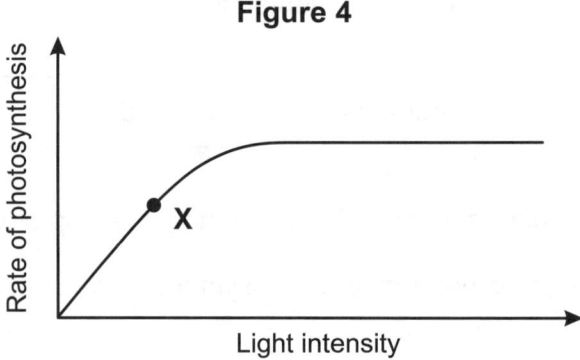

At point **X**, what is the limiting factor of photosynthesis?
Tick **one** box.

☐ oxygen concentration

☐ temperature

☐ light intensity

☐ carbon dioxide concentration

[1 mark]

3 There are a number of digestive enzymes found in the human body.

3.1 Draw **one** line from each digestive enzyme to its function.

Enzyme	Function
Protease	Converts lipids into fatty acids and glycerol
	Converts starch into sugars
Lipase	Converts proteins into amino acids

[2 marks]

3.2 Explain why different enzymes catalyse different reactions.

....

....

....

[2 marks]

3.3 Which of the following statements about enzymes is **not** true?
Tick **one** box.

☐ Enzymes are affected by temperature.

☐ Enzymes change the rate of chemical reactions.

☐ Enzymes are changed during reactions.

☐ Enzymes are affected by pH.

[1 mark]

3.4 Explain why reactions catalysed by amylase are important for growth.
Your answer should refer to respiration.

....

....

....

....

[3 marks]

Turn over for the next question

Turn over ▶

4 **Figure 5** shows the rate of transpiration in two plants over 48 hours.

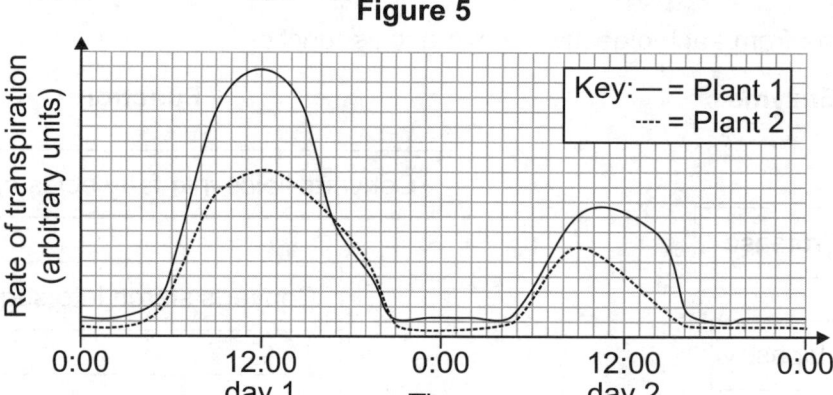

4.1 Define the term transpiration.

....
[1 mark]

4.2 At what time on **day 2** was the rate of transpiration highest for **plant 2**?

....
[1 mark]

4.3 The rate of transpiration for both plants was slower on **day 2** than on **day 1**.
Suggest **one** reason for this.

....
[1 mark]

4.4 Why was the rate of transpiration for both plants very low at night?
Tick **one** box.

☐ Low light intensity at night meant the stomata opened, allowing less water vapour to escape.

☐ Low light intensity at night meant the stomata closed, allowing less water vapour to escape.

☐ High temperatures at night meant that less water evaporated from the surfaces of the plants.

☐ Low temperatures at night meant that water diffused from the surfaces of the plants at a higher rate.

[1 mark]

Figure 6 shows two stomata on the surface of a leaf viewed under a microscope.

4.5 Name the cells labelled **X**.

....
[1 mark]

5 A student investigated how well stem cells from a plant grew in four different growth media.

The student placed a block of stem tissue measuring 1 mm × 1 mm × 1 mm onto each growth medium. They were then incubated at 35 °C for two days.

At the end of that time, the blocks were taken out and weighed to see how much they had grown. **Table 2** shows the student's results.

Table 2

Growth medium	% increase in mass
1	120
2	85
3	65
4	90

5.1 Use the results from **Table 2** to complete the bar chart in **Figure 7**.

Figure 7

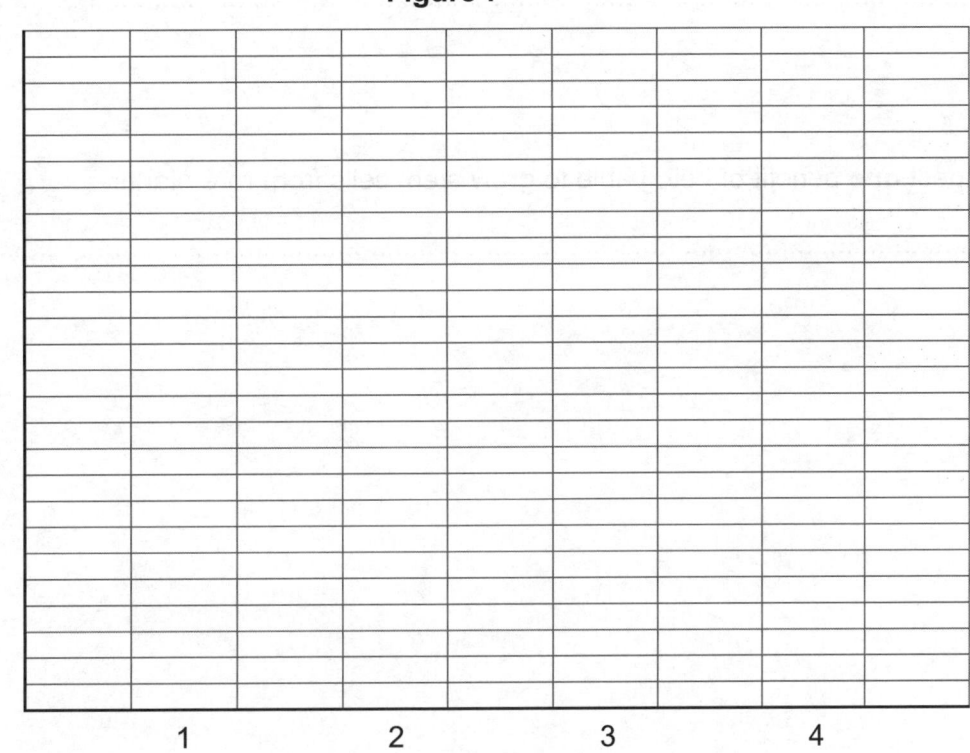

Growth medium

[3 marks]

5.2 Which growth medium produced the best results?

....

[1 mark]

Question 5 continues on the next page

Turn over ▶

5.3 Give **two** variables that needed to be controlled in this experiment.

1. ....

2. ....

[2 marks]

5.4 Suggest how the student could improve their method to reduce the effect of random errors.

....

....

[2 marks]

5.5 Name the type of reproduction that is taking place when plant tissue is grown from stem cells.

....

[1 mark]

5.6 Suggest **one** benefit of being able to grow stem cells from rare plants.

....

[1 mark]

6 Figure 8 shows a single-celled organism called *Euglena*, found in pond water.

Figure 8

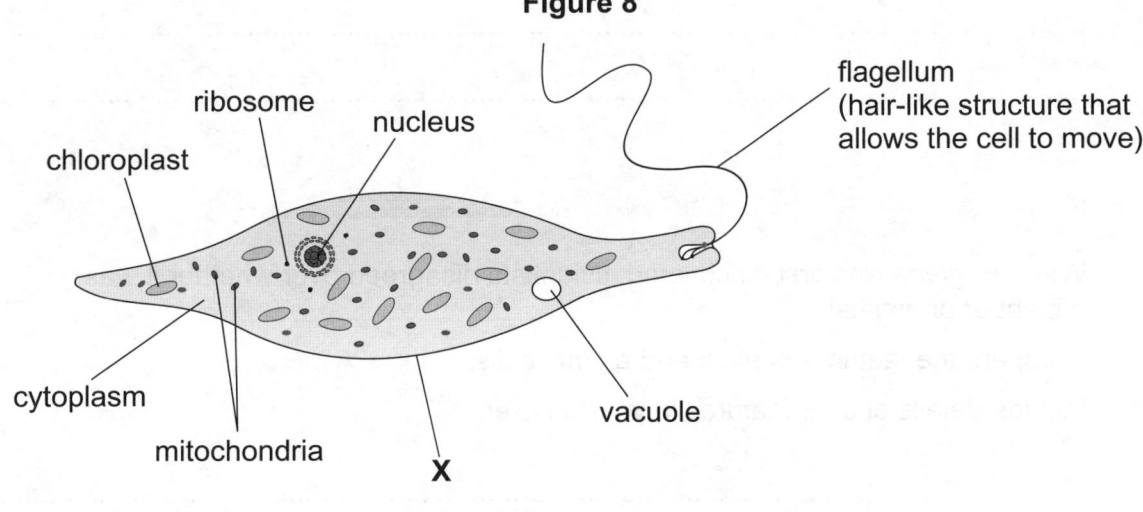

6.1 Name structure **X**.

....
[1 mark]

6.2 *Euglena* is a eukaryote. Which of the following is **not** a eukaryote?
Tick **one** box.

☐ sperm cell   ☐ muscle cell   ☐ fruit fly   ☐ *E. coli* bacteria
[1 mark]

6.3 A scientist viewed an individual *Euglena* under a microscope with × 500 magnification.
He calculated the real length of the *Euglena* to be 0.054 mm.
Calculate the length of the image of the *Euglena*. Give your answer in centimetres.
Use the formula:

image size = real size × magnification

....

....

....

image size = ... cm
[2 marks]

Question 6 continues on the next page

Turn over ▶

6.4 Explain why *Euglena* is able to exchange all of the substances that it needs across its surface.

....

....

[1 mark]

6.5 When *Euglena* was first discovered, scientists disagreed over whether it was a plant or an animal.

Compare the features of plant and animal cells.

Include details of their features in your answer.

....

....

....

....

....

....

....

....

....

....

....

....

[6 marks]

7 Measles, mumps and rubella are all examples of communicable diseases.

7.1 How is measles spread between people?
Tick **one** box.

☐ By droplets from an infected person's sneeze or cough.

☐ By sexual contact.

☐ By eating contaminated food.

☐ By a vector.

[1 mark]

7.2 The MMR vaccine protects against measles, mumps and rubella.
Explain how vaccination can help to protect the body against a disease.

....

....

....

....

....

[4 marks]

7.3 Why might someone **not** want to have the MMR vaccine? Suggest **two** reasons.

1. ....

....

2. ....

....

[2 marks]

Zika virus disease is another example of a communicable disease.
The virus that causes the disease is spread by a mosquito vector.

7.4 Suggest **two** ways that the spread of the Zika virus disease could be reduced.

....

....

....

[2 marks]

7.5 Plants are also affected by diseases.

Plants can be infected by a disease called rose black spot.
When a plant has rose black spot, it is important to strip the leaves off the plant.

Explain why it is important to also destroy the stripped leaves.

....

....

[2 marks]

Turn over for the next question

Turn over ▶

8 Different chemical reagents can be used to test for the presence of certain molecules in samples of food.

8.1 A student prepared a food sample in order to test whether the sample contained protein. What reagent should be used for this test?
Tick **one** box.

☐ Benedict's solution

☐ iodine solution

☐ biuret solution

☐ Sudan III stain solution

[1 mark]

8.2 Suggest **one** measure the student should take to make sure the food test is carried out safely.

....
[1 mark]

The student tested four different food samples for reducing sugars.
She obtained the results shown in **Table 3**.

Table 3

Sample	Colour of sample
A	blue
B	brick-red
C	yellow
D	green

8.3 Name the reagent that the student would have used to test for reducing sugars.

....
[1 mark]

8.4 Which of the samples in **Table 3** did **not** contain reducing sugars?
Tick **one** box.

☐ A   ☐ B   ☐ C   ☐ D

[1 mark]

Lactase is an enzyme involved in digestion. Lactase breaks down a sugar called lactose. The products are the sugars glucose and galactose. These are absorbed into the blood from the small intestine.

8.5 Describe how the small intestine is adapted to absorb molecules such as glucose.

....

....

....

....
[3 marks]

8.6 Lactose intolerance is a digestive problem caused by insufficient production of lactase.

To test a person for lactose intolerance, they are given a drink of lactose solution.
A blood sample is then taken from them every 30 minutes for two hours.
The blood is tested to see how much sugar it contains.

If the person has lactose intolerance, their blood sugar level will **not** rise. Explain why.

....

....
[1 mark]

Digestive problems are also seen in people with blocked bile ducts.

8.7 Bile makes conditions in the small intestine alkaline. Give **one** other function of bile.

....
[1 mark]

8.8 Suggest why people with blocked bile ducts have trouble digesting fats.
Refer to your answer to 8.7.

....

....

....

....
[3 marks]

END OF QUESTIONS

GCSE Combined Science
Biology Paper 2
Foundation Tier

In addition to this paper you should have:
- A ruler.
- A calculator.

Centre name				
Centre number				
Candidate number				

Time allowed:
- 1 hour 15 minutes

Surname	
Other names	
Candidate signature	

Instructions to candidates
- Write your name and other details in the spaces provided above.
- Answer **all** questions in the spaces provided.
- Do all rough work on the paper.
- Cross out any work you do not want to be marked.

Information for candidates
- The marks available are given in brackets at the end of each question.
- There are 70 marks available for this paper.
- You are allowed to use a calculator.
- You should use good English and present your answers in a clear and organised way.

Advice to candidates
- In calculations show clearly how you worked out your answers.

For examiner's use

Q	Attempt Nº			Q	Attempt Nº		
	1	2	3		1	2	3
1				5			
2				6			
3				7			
4				8			
				9			
Total							

1 Hormones control the menstrual cycle.

1.1 How long is a typical menstrual cycle?
Tick **one** box.

☐ 14 days ☐ 3 months ☐ 28 days ☐ 9 months

[1 mark]

1.2 Which hormone causes the release of an egg during the menstrual cycle?
Tick **one** box.

☐ FSH ☐ LH ☐ Oestrogen ☐ Progesterone

[1 mark]

1.3 Some of the hormones involved in the menstrual cycle are released from the pituitary gland. Describe how these hormones reach their target organs in the reproductive system.

....

[1 mark]

1.4 Give another name for the pituitary gland.

....

[1 mark]

1.5 Give **two** differences between a response controlled by hormones and a response controlled by nerves.

1. ....

....

2. ....

....

[2 marks]

Turn over for the next question

Turn over ▶

2 Figure 1 shows a food chain.

Figure 1

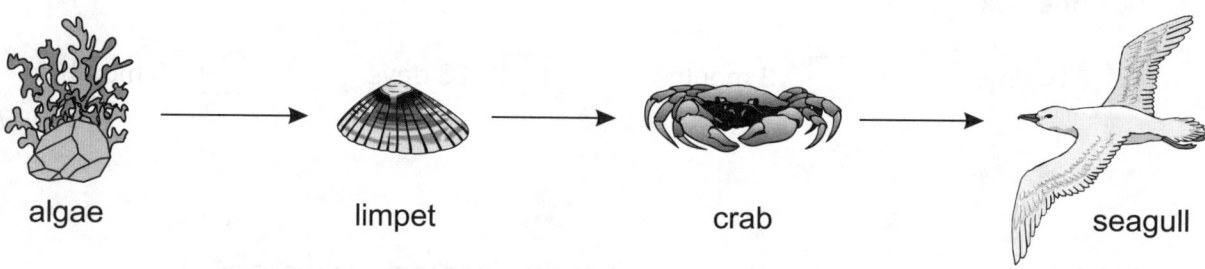

algae → limpet → crab → seagull

2.1 Identify the producer in the food chain in **Figure 1**.
Tick **one** box.

☐ algae ☐ limpet ☐ crab ☐ seagull

[1 mark]

2.2 Name **one** organism in the food chain in **Figure 1** that is prey for another organism.

....
[1 mark]

2.3 Limpets are affected by biotic and abiotic factors in their environment.
Which of the following are abiotic factors?
Tick **two** boxes.

☐ temperature

☐ food availability

☐ moisture level

☐ competition

[2 marks]

A student investigated the distribution of limpets on a beach.
The student used a quadrat to take samples at set distances from the tide's lowest point.
Table 1 shows the results.

Table 1

Distance from low-tide point (m)	0	4	8	12	16	20	24	28
Mean number of limpets per quadrat	3	2	14	9	13	24	38	42

2.4 The student plotted a graph of the results.
Complete **Figure 2** by plotting the final **two** results on the graph.

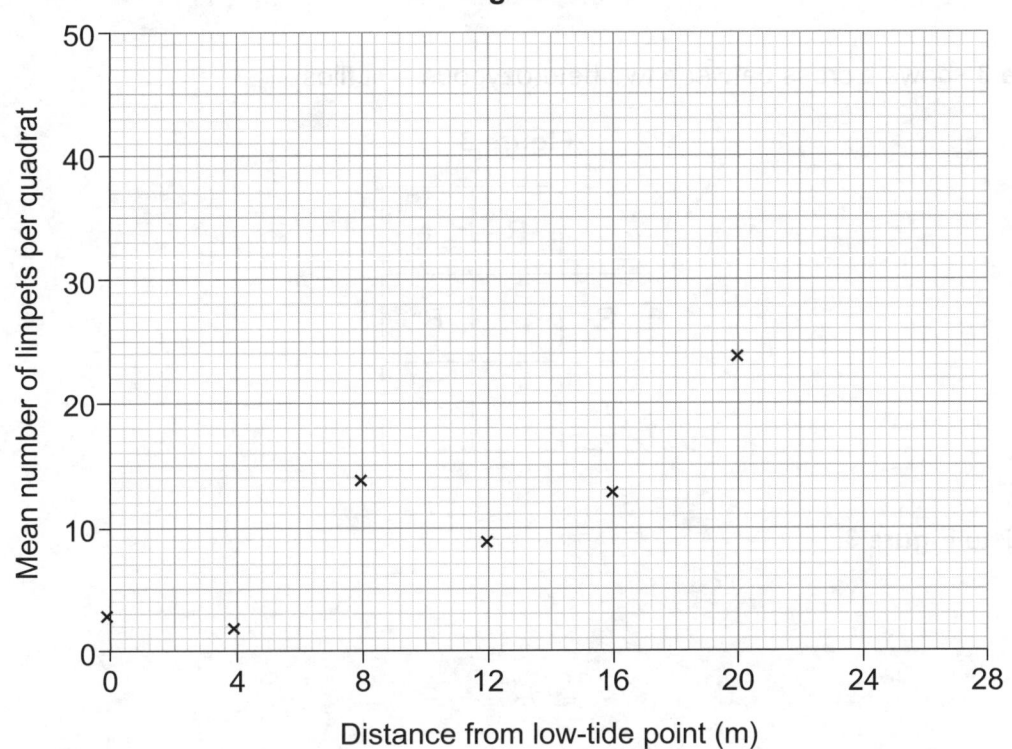

Figure 2

[2 marks]

2.5 What conclusion can be drawn from the data?
Tick **one** box.

☐ The mean number of limpets is lower at the low-tide point than at 28 m from the low-tide point.

☐ As the distance from the low-tide point increases the number of limpets always increases.

☐ The mean number of limpets increases towards the low-tide point.

☐ The mean number of limpets is constant across the beach.

[1 mark]

2.6 Give **two** possible causes of the variation in limpet distribution.

1. ....

2. ....

[2 marks]

Turn over for the next question

Turn over ▶

3 Fruit flies usually have red eyes.

However, there are a small number of white-eyed fruit flies.

The allele for red eyes (R) is dominant over the allele for white eyes (r).

3.1 What is meant by the term 'allele'?

....
[1 mark]

Figure 3 shows a cross between two heterozygous fruit flies.

Figure 3

	R	r
R		Rr
r		

3.2 Complete **Figure 3**.

[2 marks]

3.3 Give the **two** possible genotypes of fruit flies with red eyes.

....
[1 mark]

3.4 In the genetic cross shown in **Figure 3**, what is the probability of producing offspring with white eyes?

Tick **one** box.

☐ 0.5 ☐ 0.3 ☐ 0.75 ☐ 0.25

[1 mark]

3.5 Fruit flies have the same sex chromosomes as humans.
What combination of sex chromosomes are found in human males?

....
[1 mark]

4 Reproduction can be sexual or asexual.

4.1 Sexual reproduction involves gametes.
Use words from the box to complete the sentences about how gametes form.

| different | identical | mitosis | four | meiosis | two |

During gamete production the parent cell divides by ....

The number of new cells produced is .... Each new cell only

has a single set of chromosomes and is genetically ....

[3 marks]

4.2 Elephants have 56 chromosomes in their body cells.
How many chromosomes will there be in a single elephant gamete?

....

[1 mark]

A population may become extinct if the environment changes quickly and the population doesn't adapt to the new conditions.

4.3 A population that reproduces asexually might be more likely to become extinct than a population that reproduces sexually. Suggest why.

....

....

....

....

[3 marks]

Turn over for the next question

Turn over ▶

5   **Figure 4** shows the yield of tomatoes grown on a farm over a six year period.

Figure 4

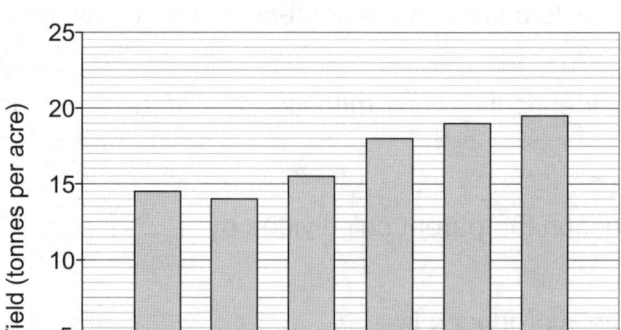

5.1 Explain how selective breeding could have been used to produce the results shown in **Figure 4**.

....

....
[3 marks]

5.2 A new disease has begun to spread in the tomato plants. Explain why the new disease might pose a large threat to the selectively bred tomato plants.

....

....

....
[2 marks]

Some tomato plants have been genetically engineered to be resistant to disease.

5.3 What is genetic engineering?

....

....
[1 mark]

5.4 Give **one** other example of how tomato plants could be genetically engineered.

....
[1 mark]

Some people have concerns about genetically modified crops.

5.5 Genetically modified tomatoes can be grown in enclosed greenhouses, rather than in fields. Suggest how this could reduce concerns about the tomato crop.

....

....
[1 mark]

6 The pancreas plays an important role in controlling the blood glucose level.
Figure 5 shows some of the glands that make up the human endocrine system.

Figure 5

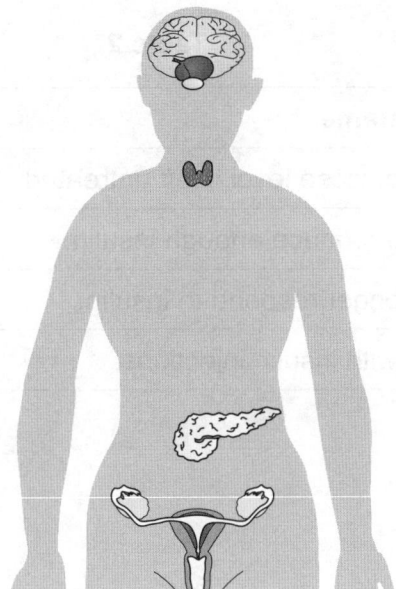

6.1 Label the pancreas on **Figure 5**.

[1 mark]

6.2 When the blood glucose level is too high, the pancreas produces insulin.
Describe how this acts to lower the blood glucose level.

....
....

[1 mark]

Question 6 continues on the next page

6.3 People with diabetes have difficulty controlling their blood glucose level.
There are two types of diabetes, type 1 and type 2.

Complete **Table 2** to show whether the statements apply to type 1, type 2 or both.
Put **one** tick in each row.

Table 2

Statement	Type 1	Type 2	Both
Results in a high blood glucose level if left untreated.			
The pancreas fails to produce enough insulin.			
Body cells can no longer respond to insulin.			
Must be treated with insulin injections.			

[4 marks]

Table 3 shows the blood glucose level of a person before and after eating a meal containing carbohydrate.

Table 3

	Before	After
Blood glucose level (mg/dl)	89	146

6.4 Calculate the percentage increase in the person's blood glucose level after the meal.
Give your answer to two significant figures.

Percentage increase = ... %

[2 marks]

7 A student was investigating the distribution of buttercups in an area around his school.
He counted the number of buttercups in 5 quadrats in five different fields.
His quadrat measured 1 m². His results are shown in **Table 4**.

Table 4

Field	Mean number of buttercups per quadrat
A	10
B	35
C	21
D	37
E	21

7.1 What is the median of the data in **Table 4**?

....
[1 mark]

7.2 A week later, the student repeated his experiment in a sixth field, Field **F**.
His results for each quadrat are shown below:

9   14   19   5   3

Using this data, calculate the mean number of buttercups per m² in Field **F**.

....

....

Mean = ... buttercups per m²
[2 marks]

7.3 Suggest **one** improvement to the student's method.

....
[1 mark]

Question 7 continues on the next page

Turn over ▶

The student observed that the distribution of buttercups changed across Field **A**. Buttercups grow well in damp soil, so the student thinks that the change in the distribution of buttercups is due to variability in the moisture level of the soil across the field. The student wants to investigate this.

7.4 Suggest a hypothesis about the distribution of buttercups in Field **A**, based on the student's observations.

....

....

[1 mark]

7.5 Describe how the student could investigate this hypothesis.

....

....

....

....

....

....

....

....

[4 marks]

8 The peppered moth is an insect that lives on the trunks of trees in Britain.
The moths are prey for birds such as thrushes.

The peppered moth exists in two varieties:

- A light-coloured variety that is better camouflaged on tree trunks in unpolluted areas.
- A dark-coloured variety that is better camouflaged on sooty tree trunks in badly polluted areas.

Figures 6 and **7** show these two varieties of moths on different tree trunks.

Figure 6   **Figure 7**

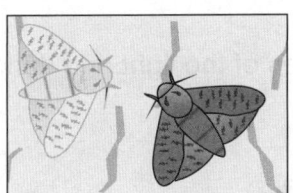 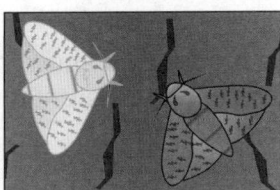

The dark variety of the moth was first recorded in the North of England in 1848.

It became increasingly common in polluted areas until the 1960s, when the number of soot covered trees declined because of the introduction of new laws.

8.1 The binomial name of the peppered moth is *Biston betularia*.
What is the moth's genus?

....
[1 mark]

8.2 Suggest how the dark variety of moth is likely to have first arisen in the population.

....
[1 mark]

Question 8 continues on the next page

Turn over ▶

8.3 Using the idea of natural selection, explain why the dark variety of moth became more common in soot polluted areas.

....

....

....

....
[3 marks]

The bar charts in **Figure 8** show the percentages of the light and dark varieties of peppered moths in two different towns.

Figure 8

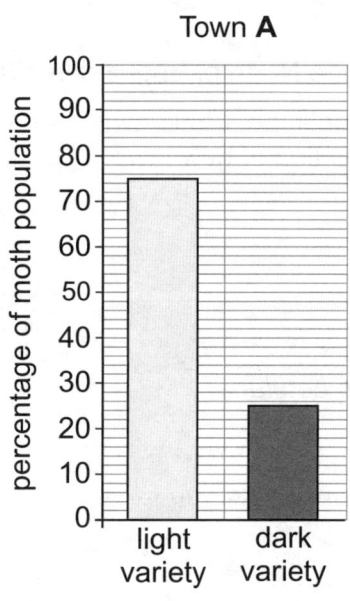
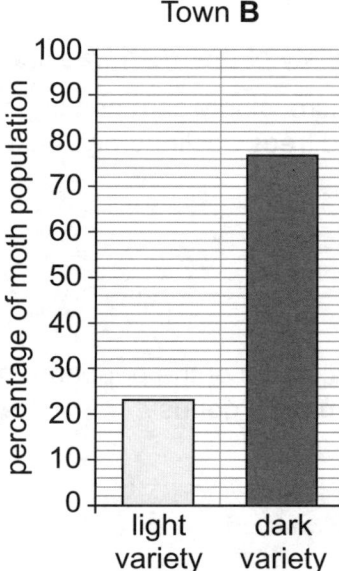

8.4 State which town, **A** or **B**, is the most polluted. Give a reason for your answer.

....

....
[1 mark]

9 A scientist was investigating the reflex actions of males and females.

The scientist made the following hypothesis:

'Males have faster reaction times than females.'

The reaction times of eight participants were tested in the investigation. Each participant was tapped just below the knee with a small rubber hammer. When the leg was tapped it automatically kicked outwards at the knee.

The scientist recorded how long it took each participant to respond to the stimulus of the tap on the leg. Each participant did the test 20 times, and a mean reaction time was calculated (to 2 decimal places).

The results are shown in **Table 5**.

Table 5

Sex	Participant	Age (years)	Mean reaction time (s)
Female	1	29	0.04
	2	26	0.06
	3	24	0.06
	4	27	0.04
Male	5	19	0.05
	6	22	0.04
	7	25	0.04
	8	20	0.05

9.1 How can you tell that the participants' response was a reflex?
Give **two** reasons.

1. ....

....

2. ....

....

[2 marks]

Question 9 continues on the next page

Turn over ▶

Figure 9 shows the mean reaction times for males and females in the investigation.

Figure 9

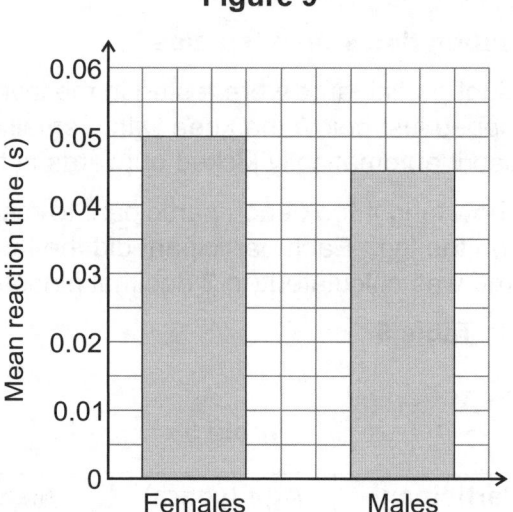

9.2 What can be concluded from the data in **Figure 9**?
Refer back to the scientist's hypothesis in your answer.

....

....
[2 marks]

9.3 Suggest **one** variable that the scientist should have controlled in this experiment.

....
[1 mark]

9.4 Another example of a reflex is the Babinski reflex.
In this reflex, a baby curls its big toe upwards when the sole of its foot is stroked.
The touch stimulus is detected by receptors in the skin and causes a reflex response.

Describe the path taken by a nervous impulse in this reflex, beginning at the receptors.

....

....

....

....

....

....

....

....
[6 marks]

END OF QUESTIONS

GCSE Combined Science
Chemistry Paper 1
Foundation Tier

In addition to this paper you should have:
- A ruler.
- A calculator.
- The periodic table (on page 538).

Centre name				
Centre number				
Candidate number				

Time allowed:
- 1 hour 15 minutes

Surname	
Other names	
Candidate signature	

Instructions to candidates
- Write your name and other details in the spaces provided above.
- Answer **all** questions in the spaces provided.
- Do all rough work on the paper.
- Cross out any work you do not want to be marked.

Information for candidates
- The marks available are given in brackets at the end of each question.
- There are 70 marks available for this paper.
- You are allowed to use a calculator.
- You should use good English and present your answers in a clear and organised way.

Advice to candidates
- In calculations show clearly how you worked out your answers.

For examiner's use

Q	Attempt Nº 1	2	3	Q	Attempt Nº 1	2	3
1				5			
2				6			
3				7			
4							
				Total			

1   This question is about bonding and elements in the periodic table.

1.1 Which element has 8 electrons in its outer shell?
Tick **one** box.

☐ Oxygen   ☐ Lithium   ☐ Neon   ☐ Nitrogen
[1 mark]

Figure 1 shows a type of bonding.

1.2 What type of bonding is shown in **Figure 1**?
Tick **one** box.

☐ Ionic bonding
☐ Intermolecular bonding
☐ Metallic bonding
☐ Covalent bonding

Figure 1

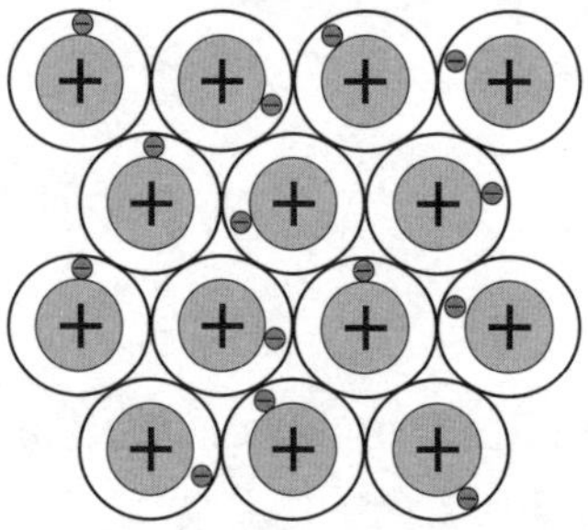

[1 mark]

1.3 Which element has a structure that is held together by the type of bonding shown in **Figure 1**?
Tick **one** box.

☐ Copper   ☐ Helium   ☐ Chlorine   ☐ Carbon
[1 mark]

The elements in Group 7 of the periodic table are known as the halogens.

1.4 Which of the halogens has the lowest boiling point?
Tick **one** box.

☐ Chlorine   ☐ Fluorine   ☐ Bromine   ☐ Iodine
[1 mark]

1.5 The nuclear symbol for an atom of fluorine is shown below.

$$^{19}_{9}F$$

How many protons, neutrons and electrons are there in this atom of fluorine?

Protons = ...   Neutrons = ...   Electrons = ...
[3 marks]

The equation for the reaction of chlorine with potassium bromide is:

$$Cl_2 + 2KBr \rightarrow Br_2 + 2KCl$$

1.6 Name the **compound** formed in this reaction.

....
[1 mark]

1.7 Explain why the reactivity of the halogens decreases moving down Group 7.

....

....
[2 marks]

2 Calcium can form the compound calcium carbonate (CaCO$_3$).

The equation for the thermal decomposition of calcium carbonate is:

$$CaCO_3 \rightarrow CaO + CO_2$$

calcium carbonate → calcium oxide + carbon dioxide

2.1 When 2560 g of calcium carbonate decomposed, 1130 g of carbon dioxide was formed.
Calculate the mass of calcium oxide formed. Give your answer in standard form.

mass of calcium oxide = ... g

[2 marks]

Solid calcium carbonate also reacts with nitric acid:

$$CaCO_{3(s)} + 2HNO_{3(aq)} \rightarrow Ca(NO_3)_{2(aq)} + H_2O_{(l)} + CO_{2(g)}$$

calcium carbonate + nitric acid → calcium nitrate + water + carbon dioxide

2.2 What do the symbols (aq) and (s) stand for in the equation above?

(aq): ....

(s): ....

[2 marks]

A scientist carries out the reaction between calcium carbonate and nitric acid in a beaker placed on a mass balance.

2.3 Explain why the reading on the mass balance decreases during the reaction.

....

....

[2 marks]

The reaction produces 3.4 g of calcium nitrate.
The volume of the solution at the end of the reaction is 224 cm^3.

2.4 Calculate the concentration of calcium nitrate in g/dm^3.
Give your answer to 3 significant figures.

concentration of Ca(NO$_3$)$_2$ = ... g/dm^3

[3 marks]

Turn over for next question

Turn over ▶

3 A student is using fractional distillation to separate a mixture of water, methanol and ethanoic acid.

3.1 What is the meaning of the term 'mixture'?

....

....
[1 mark]

3.2 Why is fractional distillation described as a physical method of separating substances?

☐ No chemical reactions take place during fractional distillation.

☐ The student needs to move the apparatus during the distillation.

☐ Chemical bonds are broken in fractional distillation.

☐ Fractional distillation involves heating.
[1 mark]

The apparatus the student is using is shown in **Figure 2**.

Figure 2

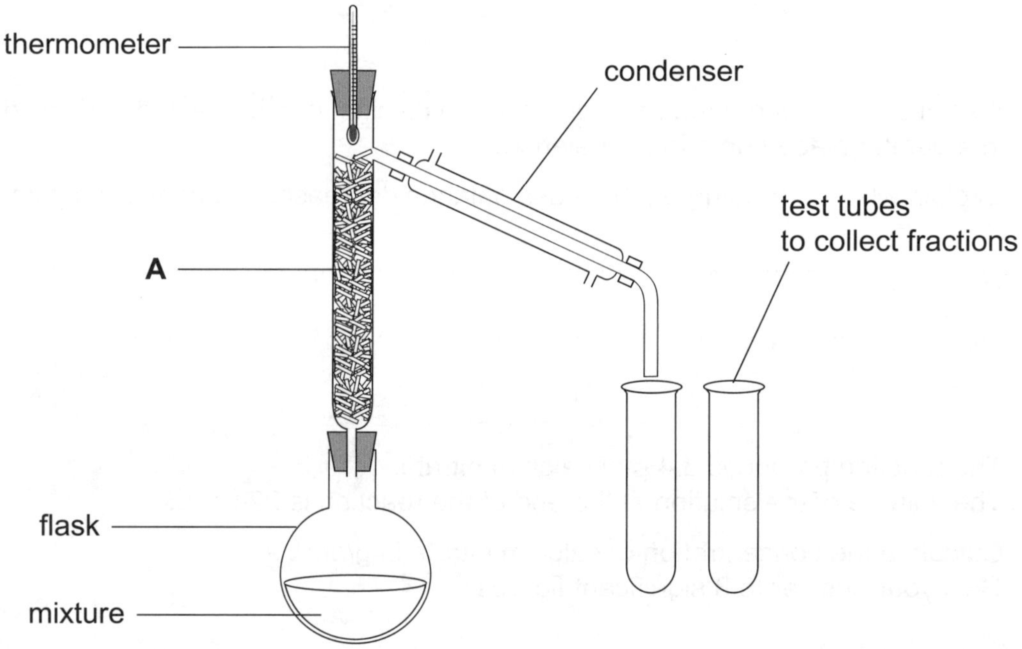

3.3 Name the piece of apparatus labelled **A** in **Figure 2**.

....
[1 mark]

Table 1 shows the boiling points of the compounds in the student's mixture.

Table 1

Compound	Boiling point in °C
Water	100
Methanol	65
Ethanoic acid	118

3.4 The student plans to collect the first two fractions in test tubes.
Which compound will be left in the **flask** at the end of the distillation?
Give a reason for your answer.

Compound: ....

Reason: ....
[2 marks]

3.5 The student wants to collect a pure sample of methanol.
Which temperature should she heat the mixture to?
Tick **one** box.

☐ 30 °C

☐ 43 °C

☐ 58 °C

☐ 78 °C

[1 mark]

3.6 Methanol is toxic.
Suggest **two** safety precautions the student should take
when carrying out the fractional distillation.

1. ....

2. ....
[2 marks]

Another student wants to use simple distillation to separate a mixture of
methanol and isobutanal. The boiling point of isobutanal is 63 °C.

3.7 Why is simple distillation **not** suitable for separating a mixture of methanol and isobutanal?

....

....
[1 mark]

Turn over for next question

Turn over ▶

4 Table 2 shows the volume of gas produced in 60 seconds when four different metals reacted with dilute sulfuric acid.
The metal was the only variable that changed.

Table 2

Metal	Volume of gas produced in cm^3
Sodium	97
Calcium	81
Magnesium	62
Zinc	28

4.1 Draw a bar chart on the grid in **Figure 3** using all of the data from **Table 2**.

Figure 3

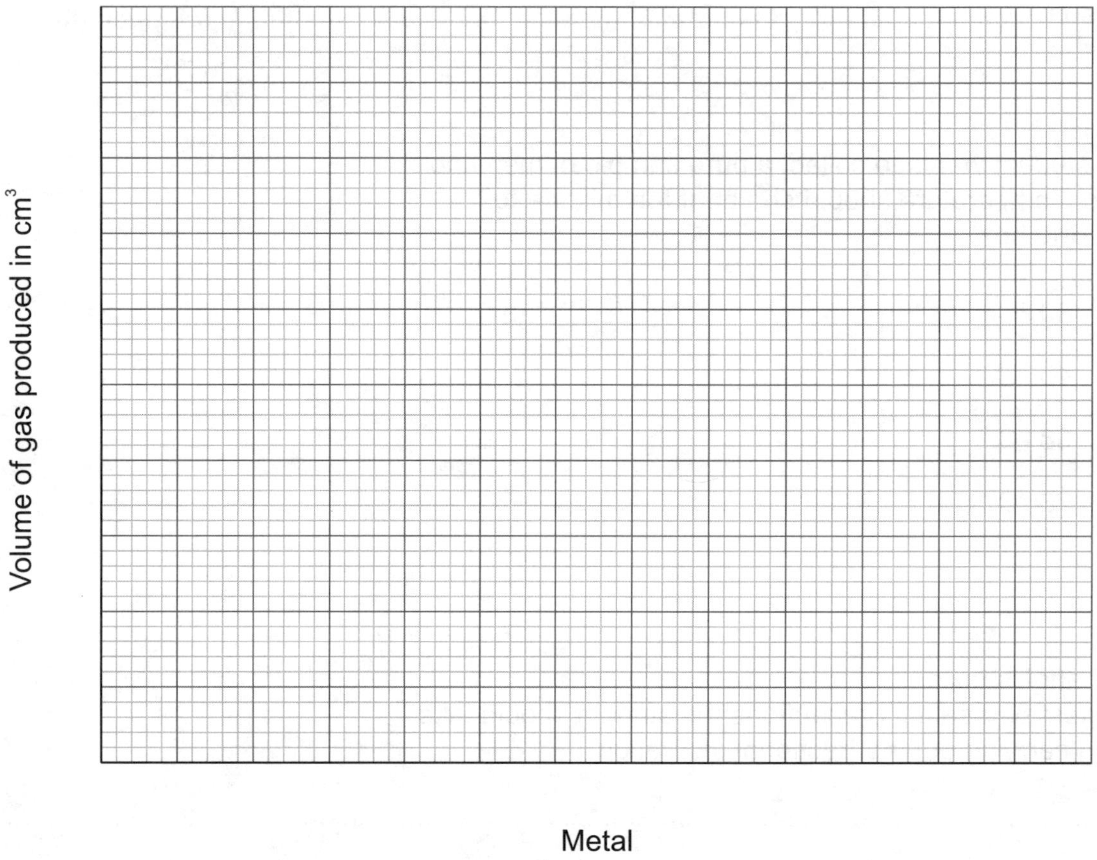

Volume of gas produced in cm^3

Metal

[2 marks]

4.2 Name the gas produced during the reactions.

....
[1 mark]

4.3 Use **Table 2** to predict the volume of gas that would be produced in the reaction between sulfuric acid and iron, if all other variables were kept the same.

volume of gas = ... cm^3
[1 mark]

One of the products of the reaction of sodium with dilute sulfuric acid is sodium sulfate, Na_2SO_4.

4.4 Calculate the relative formula mass (M_r) of sodium sulfate.
Relative atomic masses (A_r): O = 16, Na = 23, S = 32

M_r of sodium sulfate = ...
[1 mark]

4.5 Use the equation below to calculate the percentage mass of sodium in sodium sulfate.
Give your answer to 3 significant figures.

$$\text{percentage mass of an element in a compound} = \frac{A_r \times \text{number of atoms of that element}}{M_r \text{ of the compound}} \times 100$$

percentage mass of sodium = ... %
[2 marks]

An aqueous solution of sodium sulfate can be electrolysed.

4.6 What is the name given to a solution that can conduct electricity?

....
[1 mark]

4.7 What is formed at the anode when an aqueous solution of sodium sulfate is electrolysed?

....
[1 mark]

4.8 Name the product formed at the cathode when an aqueous solution of sodium sulfate is electrolysed. Give a reason for your answer.

Product at cathode: ....

Reason: ....

....
[2 marks]

Turn over for next question

Turn over ▶

5 The elements in Group 1 of the periodic table are known as the alkali metals. The first three Group 1 metals are lithium, sodium, and potassium.

5.1 Group 1 metals can react with non-metals to form ionic compounds.
What is the charge on a Group 1 ion in an ionic compound?
Tick **one** box.

☐ +2

☐ +3

☐ −1

☐ +1

[1 mark]

A student watched his teacher carefully place small pieces of lithium, sodium and potassium into cold water. His observations are recorded in **Table 3**.

Table 3

Metal	Observations
lithium	Fizzes, moves across surface.
sodium	Fizzes strongly, moves quickly across surface, appears to melt.
potassium	Fizzes violently, moves very quickly across surface, appears to melt and a flame is seen.

He decides that the order of reactivity of the three metals is:

- potassium (most reactive)
- sodium
- lithium (least reactive)

5.2 Give **two** pieces of evidence from **Table 3** that support the student's conclusion.

1. ....

....

2. ....

....

[2 marks]

5.3 Explain the pattern of reactivity that the student has noticed.
Give your answer in terms of the outer electrons of the atoms.

....

....

....

....
[3 marks]

5.4 Which of the following is correct balanced equation for the reaction between potassium and water?
Tick **one** box.

☐ $2K + H_2O \rightarrow 2KOH + H_2$

☐ $2K + H_2O \rightarrow K_2O + H_2$

☐ $2K + 2H_2O \rightarrow 2KOH + H_2$

☐ $2K + H_2O \rightarrow 2KOH + O_2$

[1 mark]

5.5 Choose the statement that explains why the solution produced when potassium reacts with water is alkaline.
Tick **one** box.

☐ It contains potassium ions.

☐ It contains water.

☐ It contains an ionic compound.

☐ It contains hydroxide ions.

[1 mark]

5.6 What is the electronic structure of lithium?

....
[1 mark]

Turn over for next question

Turn over ▶

6 Figure 4 shows the apparatus used by a student to measure the temperature change of a reaction between a piece of magnesium and dilute hydrochloric acid.

Figure 4

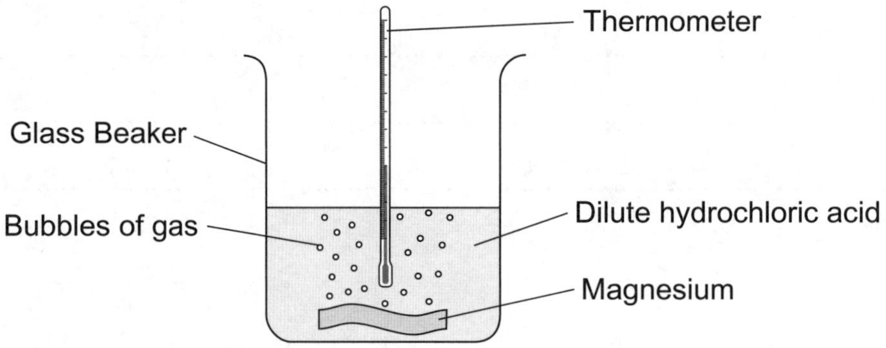

(Diagram showing a glass beaker containing dilute hydrochloric acid with magnesium at the bottom, bubbles of gas, and a thermometer inserted from above.)

6.1 Suggest **two** changes that the student could make to the apparatus in order to reduce heat loss from her experiment.

1. ....

....

2. ....

....

[2 marks]

6.2 What is the dependent variable in the student's experiment?

....

[1 mark]

6.3 A close up of the thermometer used during the experiment is shown in **Figure 5**.

Figure 5

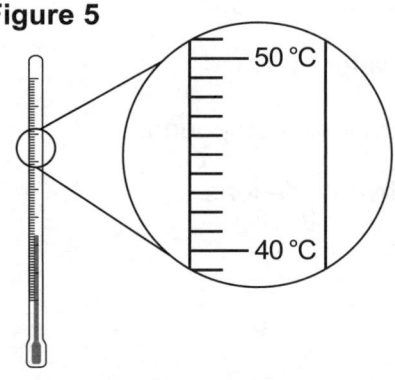

What value does each **small division** on the scale of the thermometer represent?
Tick **one** box.

☐ 0.1 °C

☐ 10 °C

☐ 1 °C

☐ 2 °C

[1 mark]

The student recorded the initial temperature of the dilute hydrochloric acid.
She added the magnesium to the acid.
She then measured the temperature of the reaction mixture every 10 seconds.
The student's results are shown on the graph in **Figure 6**.

Figure 6

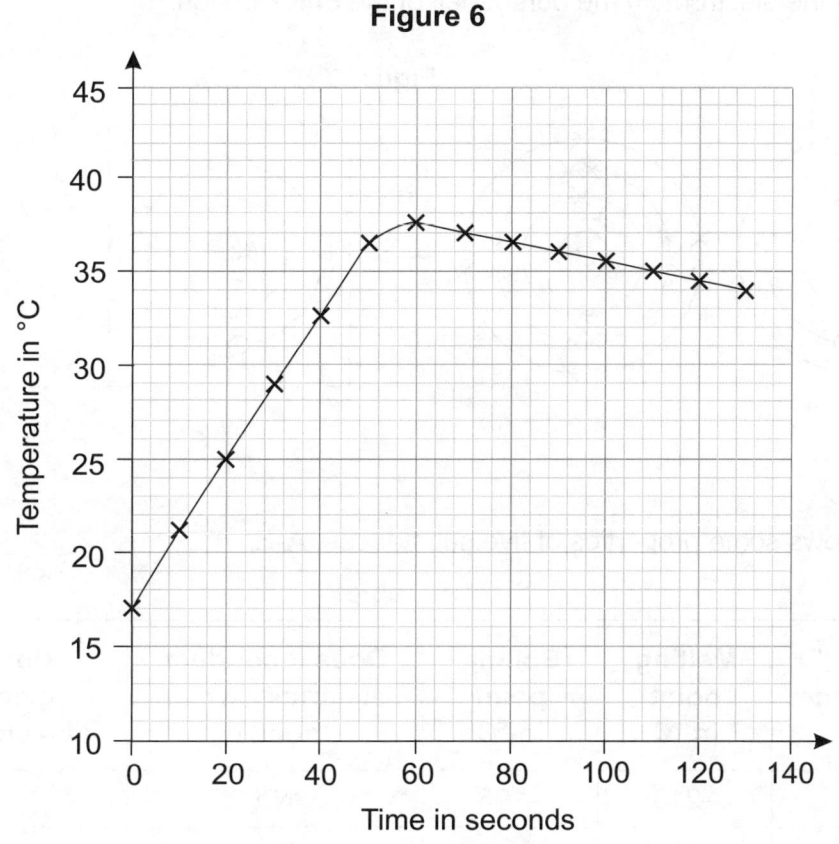

6.4 Using the graph in **Figure 6**, give the highest temperature of the mixture that the student recorded.

Highest temperature = ... °C

[1 mark]

6.5 The initial temperature of the acid was 17 °C.
Use this information and your answer from **6.4** to estimate the total change in temperature of the reaction mixture.

....

Temperature change = ... °C

[2 marks]

6.6 State whether this reaction was exothermic or endothermic. Explain your answer.

....

....

[1 mark]

Turn over for next question

Turn over ▶

7 The structure and bonding of elements and compounds affects their properties.

7.1 Sodium and chlorine can react together to form an ionic compound.
Figure 7, below, is a dot and cross diagram showing this reaction.

Complete the right-hand side of **Figure 7** by adding the charges of **both** ions and adding the electrons to the outer shell of the **chloride** ion.

Figure 7

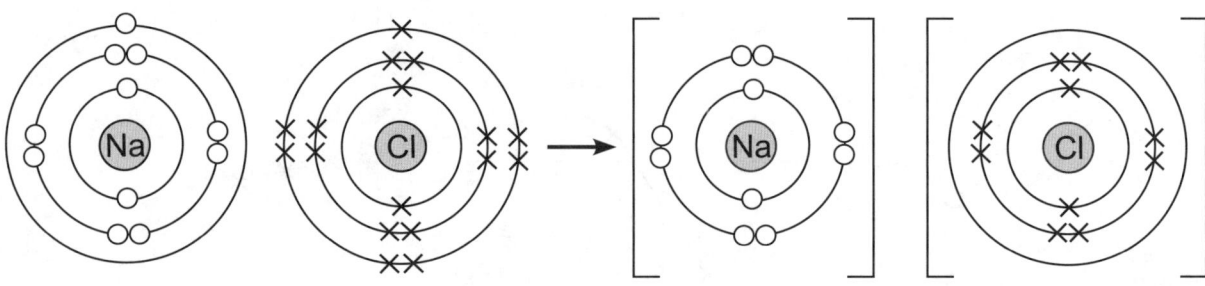

[2 marks]

Table 4 shows some properties of five substances, **A-E**.

Table 4

Substance	Melting point in °C	Boiling point in °C	Does it conduct electricity when solid?	Does it conduct electricity when dissolved or molten?
A	−210	−196	No	No
B	−219	−183	No	No
C	801	1413	No	Yes
D	115	445	No	No
E	1083	2567	Yes	Yes

7.2 At which temperature would substance **B** be a gas?
Tick **one** box.

☐ −207 °C

☐ −184 °C

☐ 63 °C

☐ −220 °C

[1 mark]

7.3 Substance **A** is made up of small, covalently bonded molecules.
Explain why substance **A** has a relatively low melting point.

....

....

[2 marks]

7.4 Look at **Table 4**. One of the substances, **A-E**, is an ionic compound.
Use the information in the table to suggest which of the substances is ionic.
Explain your answer.

....

....

....

[2 marks]

7.5 Table 5 contains information about some of the properties of diamond and graphite.

Table 5

	Hardness	Melting point	Conducts electricity?
Diamond	Hard	High	No
Graphite	Soft	High	Yes

Explain these properties of diamond and graphite
in terms of their structure and bonding.

....

....

....

....

....

....

....

....

....

[6 marks]

END OF QUESTIONS

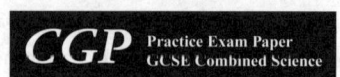

GCSE Combined Science
Chemistry Paper 2
Foundation Tier

In addition to this paper you should have:
- A ruler.
- A calculator.
- The periodic table (on page 538).

Centre name				
Centre number				
Candidate number				

Time allowed:
- 1 hour 15 minutes

Surname
Other names
Candidate signature

Instructions to candidates
- Write your name and other details in the spaces provided above.
- Answer **all** questions in the spaces provided.
- Do all rough work on the paper.
- Cross out any work you do not want to be marked.

Information for candidates
- The marks available are given in brackets at the end of each question.
- There are 70 marks available for this paper.
- You are allowed to use a calculator.
- You should use good English and present your answers in a clear and organised way.

Advice to candidates
- In calculations show clearly how you worked out your answers.

For examiner's use

Q	Attempt Nº 1	2	3	Q	Attempt Nº 1	2	3
1				5			
2				6			
3				7			
4							
				Total			

1 This question is about resources and crude oil.

1.1 What is the meaning of the term 'finite resource'?
Tick **one** box.

☐ A resource that is natural product.

☐ A resource that is used up more quickly than it can be replaced.

☐ A resource that can be replaced fairly quickly.

☐ A resource that can be burned to produce energy.

[1 mark]

Table 1 shows the time it takes for some different resources to form.

Table 1

Resource	Time it takes to form in years
Resource 1	4000
Resource 2	10^7
Resource 3	1
Resource 4	65 000

1.2 Which of the resources is a renewable resource?
Tick **one** box.

☐ Resource 1

☐ Resource 2

☐ Resource 3

☐ Resource 4

[1 mark]

1.3 **Figure 1** shows a technique used to separate crude oil into groups of hydrocarbons.

Name the technique shown in **Figure 1**.

....

[1 mark]

Question 1 continues on the next page

Turn over ▶

1.4 The technique shown in **Figure 1** relies on the properties of the groups of hydrocarbons. Complete the sentences. Use words from the box.

| boiling points | decomposes | evaporates | melting points | melts | reactivities |

Before it enters the column, the crude oil is heated until it ....

Different hydrocarbons from the crude oil then leave the column at different points because they have different ....

[2 marks]

1.5 The products listed in **Table 2** are groups of hydrocarbons produced using the technique shown in **Figure 1**.

Table 2

Product	Approximate length of carbon chain
Petrol	8
Kerosene	15
Diesel oil	20
Bitumen	>40

Which of these products will be collected closest to the point marked **X** on **Figure 1**?
Tick **one** box.

☐ Petrol

☐ Kerosene

☐ Diesel oil

☐ Bitumen

[1 mark]

1.6 Which of the products in **Table 2** could be cracked to produce diesel oil?
Tick **one** box.

☐ Petrol

☐ Kerosene

☐ Bitumen

☐ None of them

[1 mark]

1.7 Which type of substance is a product of cracking?
Tick **one** box.

☐ Alkene

☐ Salt

☐ Polymer

☐ Catalyst

[1 mark]

2 A student carried out a chromatography experiment to identify an unknown food colouring.

This is the method used:

1. Draw a pencil baseline on a sheet of paper.
   Add a spot of the unknown food colouring to the baseline.
2. Add spots of two reference food colourings, **A** and **B**, to the baseline.
3. Place the sheet of paper in a beaker containing a small amount of solvent.
   Put a lid on the beaker.
4. When the solvent has almost reached the top of the piece of paper, remove the paper and leave it to dry.

2.1 What is the purpose of the solvent in this experiment?
Tick **one** box.

☐ It is a reactant

☐ It is the stationary phase

☐ It is a catalyst

☐ It is the mobile phase

[1 mark]

2.2 Only a small amount of solvent is used so that it does not touch the baseline when the paper is first placed in the beaker.
Why should the solvent not touch the baseline?

....
[1 mark]

2.3 How could the student test the reproducibility of the experiment?

....

....

....
[2 marks]

Figure 2 shows the chromatogram produced during the experiment.

Figure 2

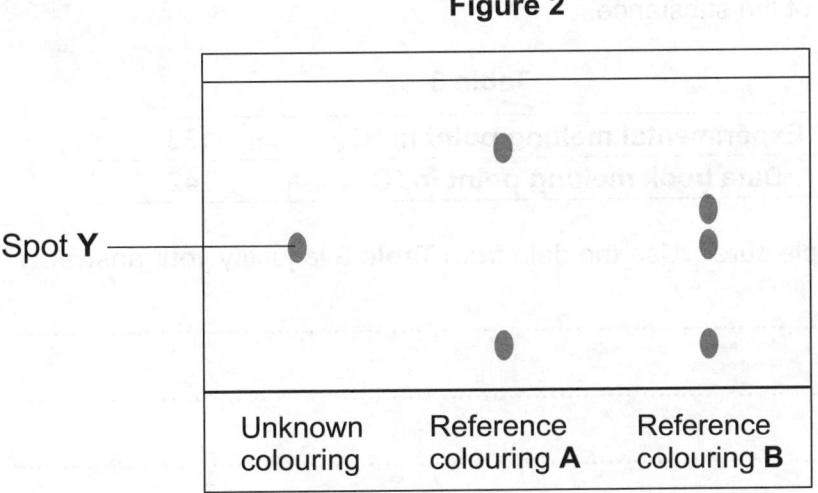

Question 2 continues on the next page

Turn over ▶

2.4 Which of the three food colourings is most likely to be a pure substance?
Use **Figure 2** to justify your answer.

....

....

[1 mark]

2.5 What is the minimum number of substances in reference colouring **B**?

....

[1 mark]

2.6 The R_f value for spot **Y** is calculated using the equation:

$$R_f = \frac{\text{distance moved by spot Y from the baseline}}{\text{distance moved by solvent from the baseline}}$$

Calculate the R_f value for spot **Y** on **Figure 2**.
Give your answer to two significant figures.

R_f value = ...

[5 marks]

In another experiment the student measured the melting point
of a sample of a substance found in a different food colouring.
Table 3 shows his result, as well as the data book value
for the melting point of the substance.

Table 3

Experimental melting point in °C	133
Data book melting point in °C	142

2.7 Is the student's sample pure? Use the data from **Table 3** to justify your answer.

....

....

....

[1 mark]

3 Figure 3 shows how the concentration of carbon dioxide in the atmosphere has changed since 1700.

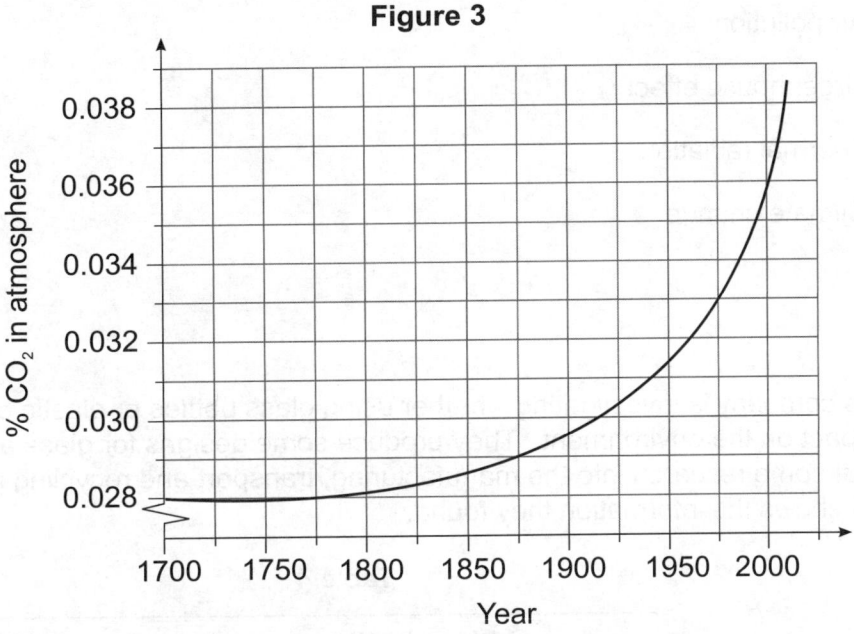

3.1 Use **Figure 3** to describe how the concentration of carbon dioxide in the atmosphere has changed since 1700.

....

....

....
[2 marks]

3.2 Which piece of information would help convince a scientist that the data in **Figure 3** is valid?
Tick **one** box.

☐ The data covers a long period of time.

☐ The data has been published in a book.

☐ The percentage of CO_2 has been given to three decimal places.

☐ The data has been peer-reviewed.

[1 mark]

Figure 4 shows a process that takes place in Earth's atmosphere.

Figure 4

Question 3 continues on the next page

Turn over ▶

3.3 What is the name of the process shown in **Figure 4**?
Tick **one** box.

☐ Air pollution

☐ Greenhouse effect

☐ Thermal radiation

☐ Climate change

[1 mark]

3.4 A drinks company is investigating whether using glass bottles or plastic bottles will have less impact on the environment. They produce some designs for glass and plastic bottles and carry out some research into the manufacturing, transport and recycling processes for each. **Table 4** shows the information they found.

Table 4

	Glass bottle	Plastic bottle
CO_2 given out during production	255 g	193 g
Mass of bottle	402 g	44 g
CO_2 given out transporting by road	39 g	12 g
Recycled content	Contains 83% recycled material.	Contains no recycled material.
Recycling	Can be recycled endlessly. The company estimates 65% of these bottles would be recycled after use.	Can be recycled once or twice. The company estimates 55% of these bottles sold would be recycled after use.

Use the information in **Table 4** and your own knowledge to compare the environmental impacts of using glass bottles and using plastic bottles.

....

....

....

....

....

....

....

....

[4 marks]

4 A student is investigating how the rate of the reaction between calcium carbonate and sulfuric acid is affected by the concentration of the acid.

The student uses the following method:
- Weigh out 0.7 g of calcium carbonate.
- Add the calcium carbonate to an excess of 7.3 g/dm³ sulfuric acid in a conical flask.
- Use a gas syringe to collect the gas given off by the reaction.
- Measure and record the volume of gas produced every 10 s.
- Repeat the experiment using sulfuric acid with a concentration of 14.6 g/dm³.

4.1 To make each run a fair test, the student keeps the mass of calcium carbonate the same.
Suggest **two** other variables that the student should keep the same to make each run a fair test.

1. ....

2. ....

[2 marks]

4.2 The gas produced by the reaction is carbon dioxide.
Which test could be used to confirm that the gas is carbon dioxide?
Tick **one** box.

☐ Holding a lit splint at the open end of a test tube containing the gas.

☐ Putting damp litmus paper into a test tube containing the gas.

☐ Putting a glowing splint inside a test tube containing the gas.

☐ Bubbling the gas through a solution of calcium hydroxide.

[1 mark]

The graph in **Figure 5** shows the student's results for the first experiment, using 7.3 g/dm³ sulfuric acid, and the second experiment, using 14.6 g/dm³.

Figure 5

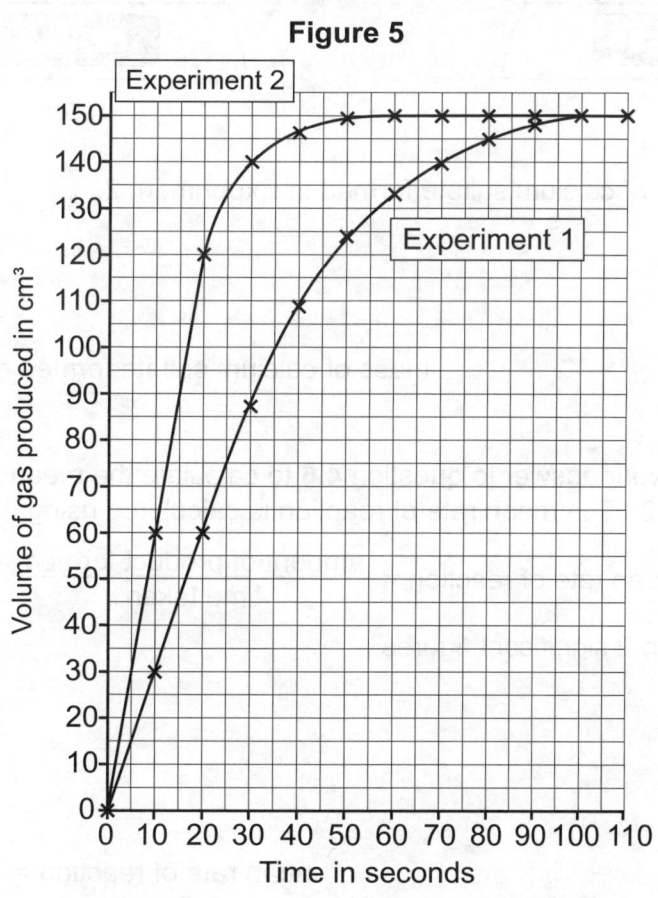

Question 4 continues on the next page

Turn over ▶

4.3 Use **Figure 5** to determine how much gas had been produced in each experiment after 30 seconds.

Volume of gas produced after 30 seconds in Experiment 1 = ... cm³

Volume of gas produced after 30 seconds in Experiment 2 = ... cm³

[2 marks]

4.4 Describe one way in which the shapes of the curves in **Figure 2** show that the rate of reaction was faster in Experiment 2 than in Experiment 1.

....

....

[1 mark]

4.5 In terms of collision theory, explain why concentration affects the rate of a reaction.

....

....

[2 marks]

4.6 The reaction also produces calcium sulfate. **Figure 6** shows how the student used a mass balance and a weighing boat to measure the mass in grams of calcium sulfate produced in Experiment 2.

Figure 6

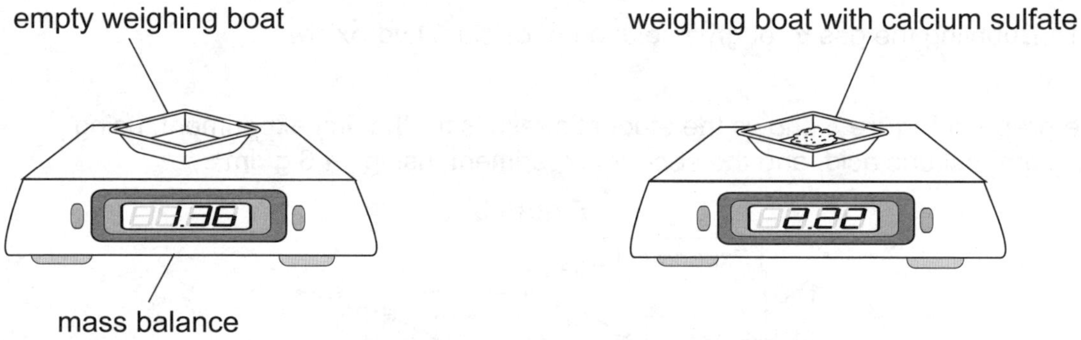

empty weighing boat — 1.36

weighing boat with calcium sulfate — 2.22

mass balance

Calculate the mass of calcium sulfate formed in Experiment 2.

mass of calcium sulfate formed = ... g

[1 mark]

4.7 Use **Figure 5** and your answer to question **4.6** to calculate the mean rate of reaction during Experiment 2. The mean rate of reaction is calculated using the equation:

$$\text{mean rate of reaction} = \frac{\text{amount of product formed}}{\text{time taken}}$$

Give your answer to 2 significant figures.

mean rate of reaction = ... g/s

[3 marks]

5 Alkanes are hydrocarbon compounds found in crude oil. **Table 5** shows how the boiling points of some alkanes change as the molecules get bigger.

Table 5

Alkane	Molecular formula	Boiling point (°C)
Propane	C_3H_8	−42
Butane	C_4H_{10}	−1
Pentane	C_5H_{12}	
Hexane	C_6H_{14}	69
Heptane	C_7H_{16}	98

5.1 On **Figure 7**:
- plot the boiling points of propane, butane, hexane and heptane against number of carbon atoms,
- draw a smooth curve through the points you have plotted.

Figure 7

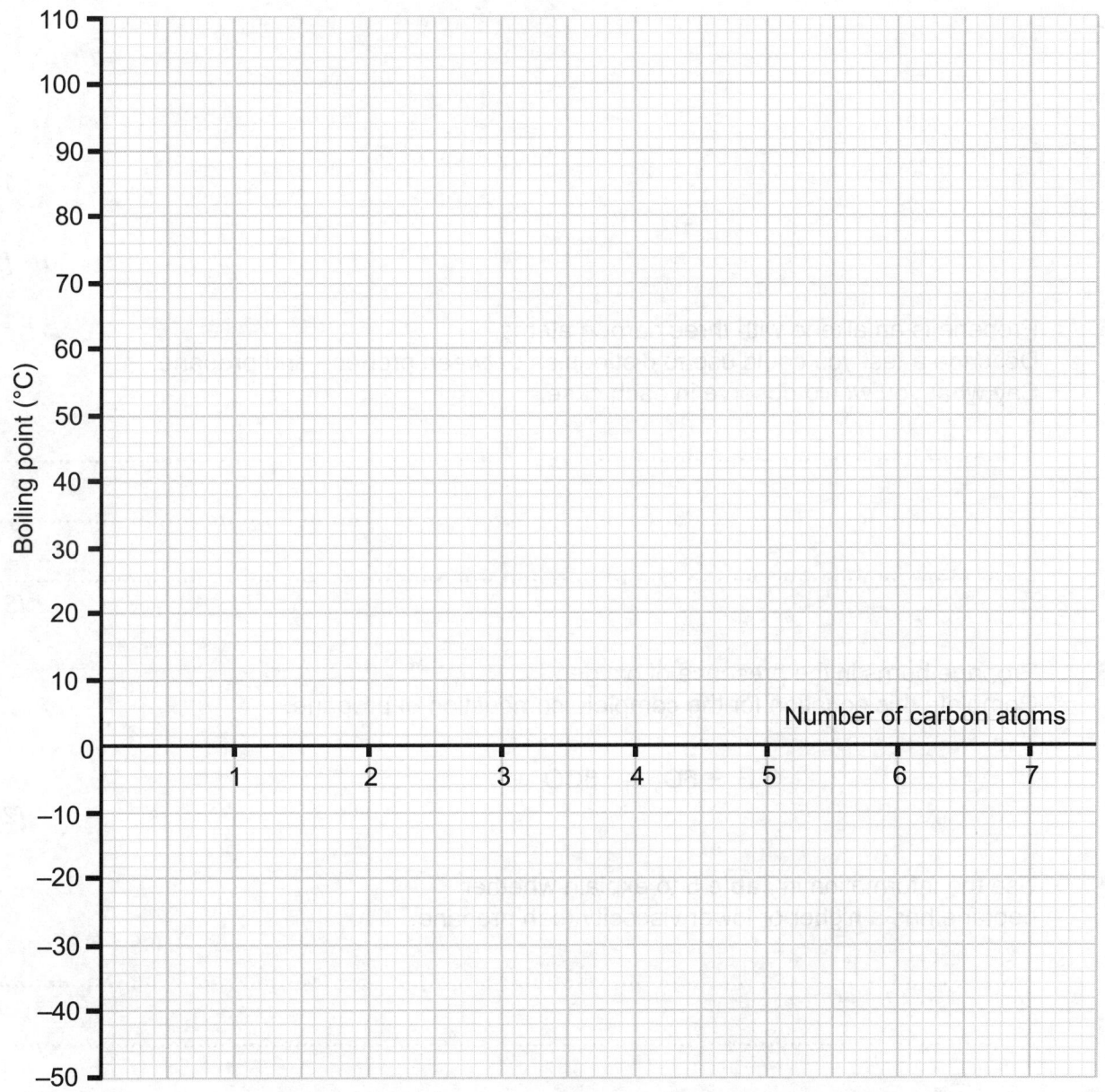

[2 marks]

Question 5 continues on the next page

Turn over ▶

5.2 Use your graph to estimate the boiling point of pentane.

.... °C
[1 mark]

5.3 What is the general formula of the alkanes? Tick **one** box.

☐ C_nH_{2n}

☐ C_nH_{2n+1}

☐ C_nH_{2n+2}

☐ C_nH_{2n-1}

[1 mark]

5.4 Propane is an alkane with three carbon atoms.
Draw the displayed formula of propane.

[1 mark]

5.5 Propene is an alkene with three carbon atoms.
Describe a test you could use to distinguish between propene and propane.
Say what you would observe in each case.

....

....

....
[3 marks]

5.6 Propane burns in the presence of oxygen.
Complete this equation for the complete combustion of propane.

$$C_3H_8 + 5O_2 \rightarrow 3CO_2 + ...$$

[2 marks]

5.7 Use the information in **Table 5** to explain whether heptane has a higher or lower viscosity than propane.

....

....

....
[2 marks]

6 Nitrogen dioxide is an atmospheric pollutant that irritates the respiratory system. It is thought that there is a link between exposure to nitrogen dioxide and the severity of asthma attacks in people with asthma.

Figure 8 shows the results of a study carried out by a group of scientists. The study compared atmospheric nitrogen dioxide levels with the severity of asthma attacks suffered by men under the age of 40 working in a city centre. The severity of asthma attacks were measured on a scale of 1 to 10, with 10 being the most severe.

Figure 8

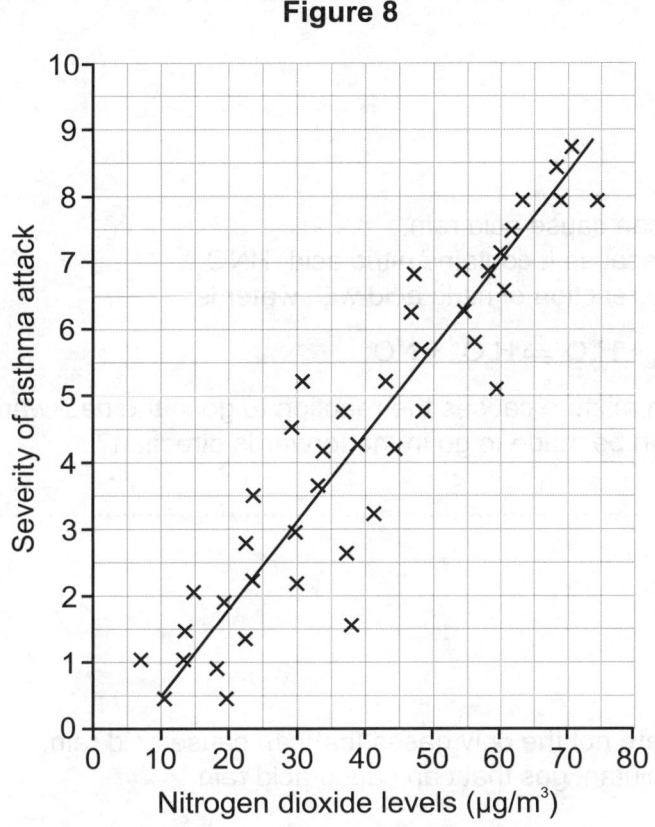

6.1 Describe the relationship between the level of nitrogen dioxide in the air and the severity of asthma attacks shown in **Figure 8**.

....

....

[1 mark]

6.2 The scientists hope to draw a conclusion about the link between the nitrogen dioxide level and the severity of asthma attacks.

Suggest why they cannot use the results shown in **Figure 8** to draw a conclusion that applies to **everyone**.

....

....

[1 mark]

Question 6 continues on the next page

6.3 Where do the oxides of nitrogen in the air come from?
Suggest why levels of nitrogen oxides can be particularly high in cities.

....

....

....

....

[3 marks]

6.4 Oxides of nitrogen can cause acid rain.
Acid rain is acidic because it contains nitric acid, HNO_3.
The equation for the reaction of nitric acid with water is:

$$HNO_3 + H_2O \rightleftharpoons H_3O^+ + NO_3^-$$

Heating the reaction mixture causes the reaction to go in the backwards direction.
How can the reaction be made to go in the forwards direction?

....

[1 mark]

6.5 Oxides of nitrogen are not the only gases that can cause acid rain.
Name **one** other pollutant gas that can cause acid rain.

....

[1 mark]

7 In the UK, the majority of our drinking water is produced from treating groundwater or surface water. Drinking water can also be made by treating sea water or waste water.

7.1 Producing water that is safe to drink from sea water is expensive.
Suggest why some countries produce drinking water by this method.

....
[1 mark]

7.2 What is the name of the process used to remove salt from sea water?
Tick **one** box.

☐ Filtration ☐ Desalination ☐ Sterilisation ☐ Cracking

[1 mark]

7.3 A teacher gives a student a sample of sea water, and asks her to produce a sample of pure water from it. Plan a method that the student could use to remove the salt from the sea water.

....
....
....
....
....
....
....
....
[6 marks]

7.4 Sewage treatment plants process sewage and release clean, treated water back into the environment. The first step in the treatment of sewage is screening.
What happens in the screening step?
Tick **one** box.

☐ Chlorine is added to the sewage to kill microorganisms.

☐ The sewage is placed into large storage tanks and allowed to settle.

☐ Anaerobic digestion is used to break down the sewage.

☐ Large bits of material (such as grit) are removed from the sewage.

[1 mark]

7.5 Following screening, sedimentation is used to separate the effluent from the sludge.
Describe what happens to the effluent before it can be returned to the environment.

....
....
....
[2 marks]

END OF QUESTIONS

GCSE Combined Science
Physics Paper 1
Foundation Tier

In addition to this paper you should have:
- A ruler.
- A calculator.
- The Physics Equation sheet (on page 538).

Centre name				
Centre number				
Candidate number				

Time allowed:
- 1 hour 15 minutes

Surname	
Other names	
Candidate signature	

Instructions to candidates
- Write your name and other details in the spaces provided above.
- Answer **all** questions in the spaces provided.
- Do all rough work on the paper.
- Cross out any work you do not want to be marked.

Information for candidates
- The marks available are given in brackets at the end of each question.
- There are 70 marks available for this paper.
- You are allowed to use a calculator.
- You should use good English and present your answers in a clear and organised way.

Advice to candidates
- In calculations, show clearly how you worked out your answers.

For examiner's use

Q	Attempt Nº 1	2	3	Q	Attempt Nº 1	2	3
1				5			
2				6			
3				7			
4				Total			

1   A representation of the particles of a substance is shown in **Figure 1**.

Figure 1

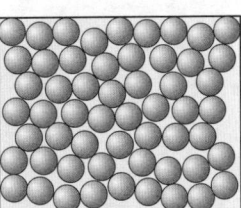

1.1 Name the state of matter of the substance shown in **Figure 1**.

....
[1 mark]

1.2 The substance in **Figure 1** is heated. This increases the particle's internal energy.
Which of the particle's energy stores is energy transferred to?
Tick **two** boxes.

☐ potential energy

☐ nuclear energy

☐ chemical energy

☐ kinetic energy

☐ magnetic energy

[2 marks]

Figure 2 shows a sketch of the heating curve for a 1.0 kg sample of the substance.

Figure 2

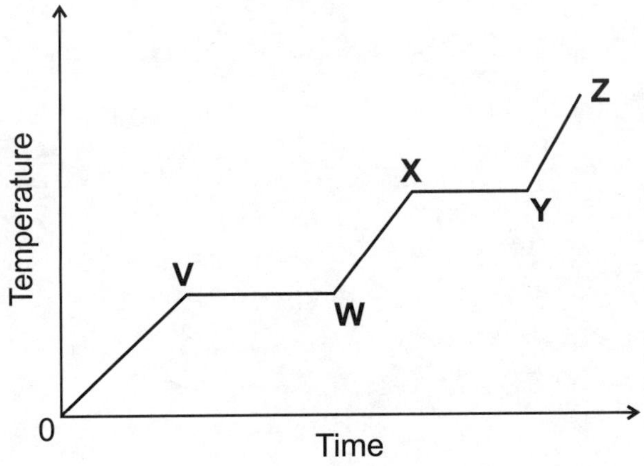

1.3 Identify the state of matter of the substance at point **Z**.

....
[1 mark]

Question 1 continues on the next page

Turn over ▶

1.4 At which point in **Figure 2** has the substance reached its boiling point?
Tick **one** box.

☐ V
☐ W
☐ X
☐ Z

[1 mark]

Between points **V** and **W**, 334 000 J is supplied to the 1.0 kg sample of the substance.

1.5 Which of the following statements is true?
Tick **one** box.

☐ The specific latent heat of fusion of the substance is 334 000 J/kg.
☐ The specific heat capacity of the substance is 334 000 J/kg.
☐ The specific latent heat of vaporisation of the substance is 334 000 J/kg.

[1 mark]

1.6 Calculate the minimum amount of energy that would need to be transferred from a 5.00 kg sample of the substance to freeze it.
Use the correct equation from the Physics Equation Sheet on page 538.

....

....

Energy = ... J

[2 marks]

2 A student wants to find the density of an ornament.
Figure 3 shows the ornament.

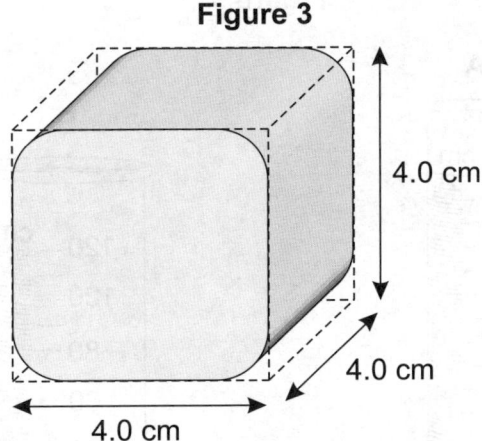

2.1 Estimate the volume of the ornament by calculating the volume of the cube shown in **Figure 3**. Give your answer in cm³.

....

....

Volume = ... cm³

[2 marks]

The student decides to measure the volume accurately using a eureka can and measuring cylinder.

He fills a eureka can with water so that it is full up to the base of the spout.
When he places the ornament in the eureka can, water will pour into a measuring cylinder.
Figure 4 shows the eureka can before and after the ornament has been added.

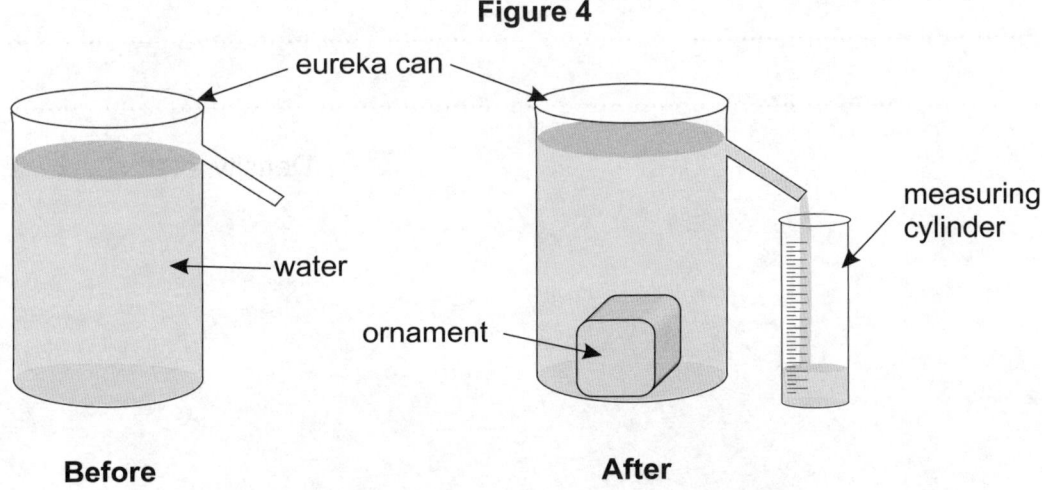

Question 2 continues on the next page

Turn over ▶

Figure 5 shows two measuring cylinders, **A** and **B**, that the student could use to collect water from the eureka can.

Figure 5

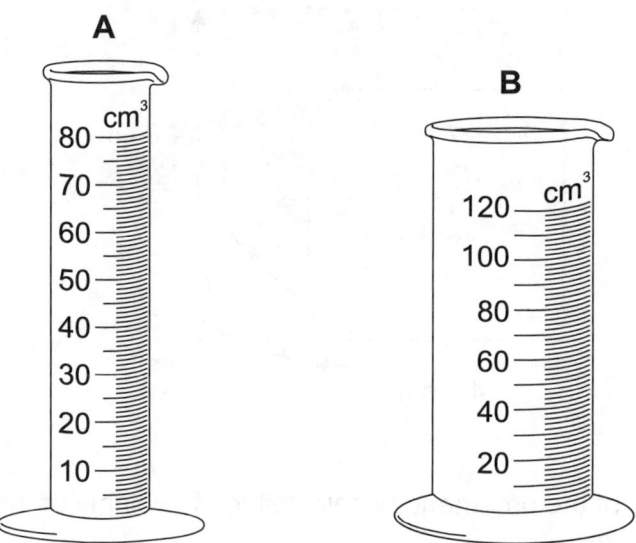

2.2 State and explain which cylinder the student should use for his experiment.

....

....

[1 mark]

2.3 The student finds that the volume of the ornament is 0.000058 m³.
He measures the mass of the ornament to be 0.435 kg.

Use the following equation to calculate the density of the ornament.

$$\text{density} = \frac{\text{mass}}{\text{volume}}$$

....

....

Density = ... kg/m³

[2 marks]

3 A home owner wants to wrap insulation around her hot water tank to reduce unwanted energy transfers. She needs to choose between three different types of insulation.

She does an experiment to test which type of insulation would be the best to use. She fills a beaker with hot water and wraps one type of insulation around it, as shown in **Figure 6**.

Figure 6

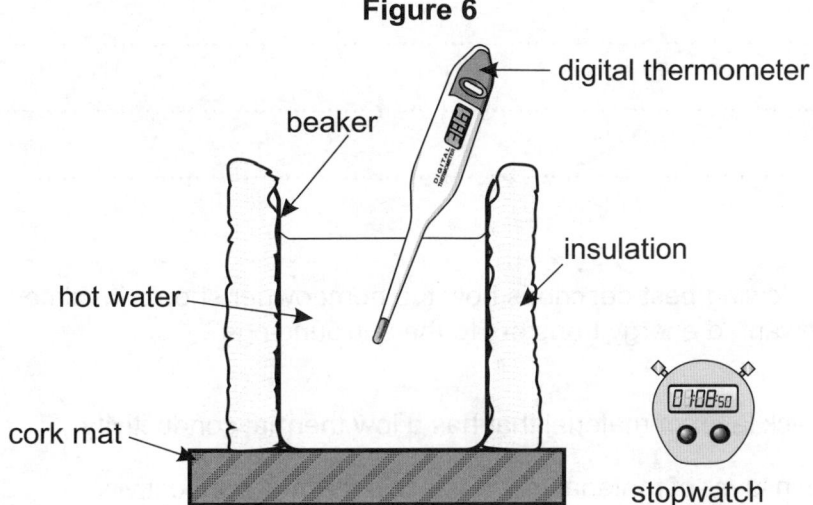

She measures the temperature decrease of the water after 5 minutes.
She repeats this experiment with each type of insulation.

3.1 State the independent variable in this investigation.

....
[1 mark]

3.2 State **two** control variables in this investigation.

1. ....

2. ....
[2 marks]

The home owner repeats her experiment three times for each type of insulation.
Her results for one type of insulation are shown in **Table 1**.

Table 1

Temperature decrease (°C)			
1	2	3	Mean
21.9	22.2	20.1	

3.3 Use the data in **Table 1** to calculate the mean temperature decrease for the insulation.

....

....

Mean temperature decrease = ... °C
[2 marks]

Question 3 continues on the next page

Turn over ▶

The home owner finds the temperature decrease for each type of insulation is roughly the same. She realises this is due to the way she set up the apparatus shown in **Figure 6**.

3.4 Using **Figure 6**, identify the mistake the home owner has made.
Explain why this mistake would have caused the temperature decreases recorded to be roughly the same for each type of insulation.

....

....

....
[3 marks]

3.5 Which of the following best describes how the homeowner should insulate her tank to minimise unwanted energy transfers to the surroundings?
Tick **one** box.

☐ With a thick layer of material that has a low thermal conductivity.

☐ With a thin layer of material that has a low thermal conductivity.

☐ With a thick layer of material that has a high thermal conductivity.

☐ With a thin layer of material that has a high thermal conductivity.

[1 mark]

The home owner also wishes to use more energy from renewable energy resources to reduce her energy bills. She wants to generate electricity either using the wind or directly from the Sun.

3.6 Give **one** other example of a renewable energy resource.

....
[1 mark]

Table 2 gives some data about the costs and savings of generating electricity using solar panels or a wind turbine.

Table 2

	Installation cost	Average annual energy bill saving
Solar panels	£6000	£375
Wind turbine	£10 000	£500

3.7 The home owner expects to move house in 10 years time.
Evaluate whether she would save money by having solar panels installed.
Use data from **Table 2** in your answer.

....

....

....
[2 marks]

4 A student is investigating the two electrical circuits shown in **Figure 7**.
The resistors and batteries used in each circuit are identical.

Figure 7

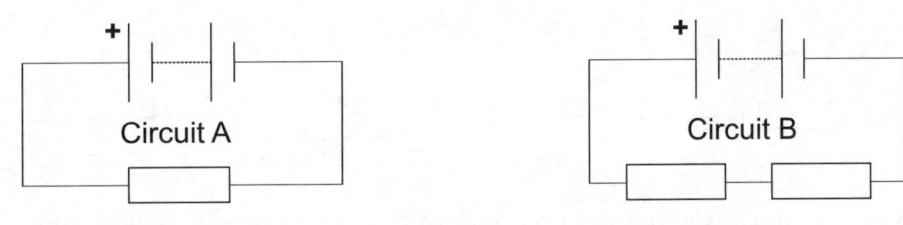

Circuit A   Circuit B

4.1 Which statement correctly describes the total resistance of circuit B?
Tick **one** box.

☐ It is half as large as the resistance of circuit A.

☐ It is the same as the resistance of circuit A.

☐ It is twice as large as the resistance of circuit A.

[1 mark]

4.2 Compare the current in circuits A and B. Explain your answer.

....

....

[2 marks]

4.3 The total potential difference across the resistor in circuit A is 12 V.
Which statement about the potential difference across the resistors in circuit B is true?
Tick **one** box.

☐ The potential difference across each resistor is 6 V.

☐ The potential difference across each resistor is 12 V.

☐ The potential difference across each resistor is 24 V.

[1 mark]

4.4 The resistor in circuit A has a resistance of 4.0 Ω. The current through it is 3.0 A.
Use the following equation to calculate the power dissipated by the resistor in circuit A.

power = (current)^2 × resistance

Choose the correct unit from the box. | joules   watts   newtons per metre |

....

....

Power = ...

Unit = ...

[3 marks]

Question 4 continues on the next page

Turn over ▶

The student considers adding one of the components in **Figure 8** to circuit A.

Figure 8

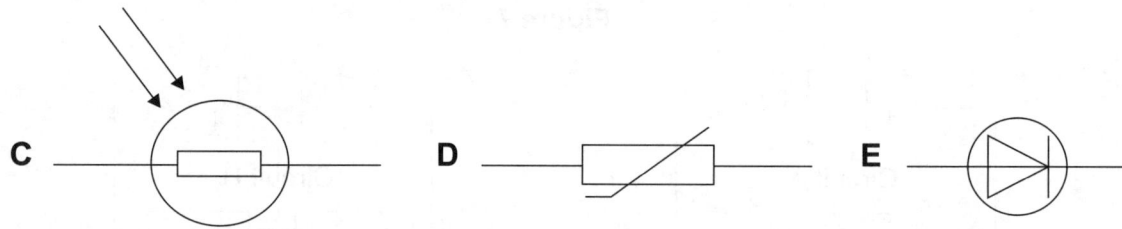

4.5 What happens to the resistance of component **C** as the intensity of light that falls on it increases?
Tick **one** box.

☐ The resistance increases.

☐ The resistance remains constant.

☐ The resistance decreases.

[1 mark]

4.6 Suggest **one** application for component **D**.

....

....
[1 mark]

4.7 On the axes in **Figure 9**, sketch the *I-V* characteristic of component **E**.

Figure 9

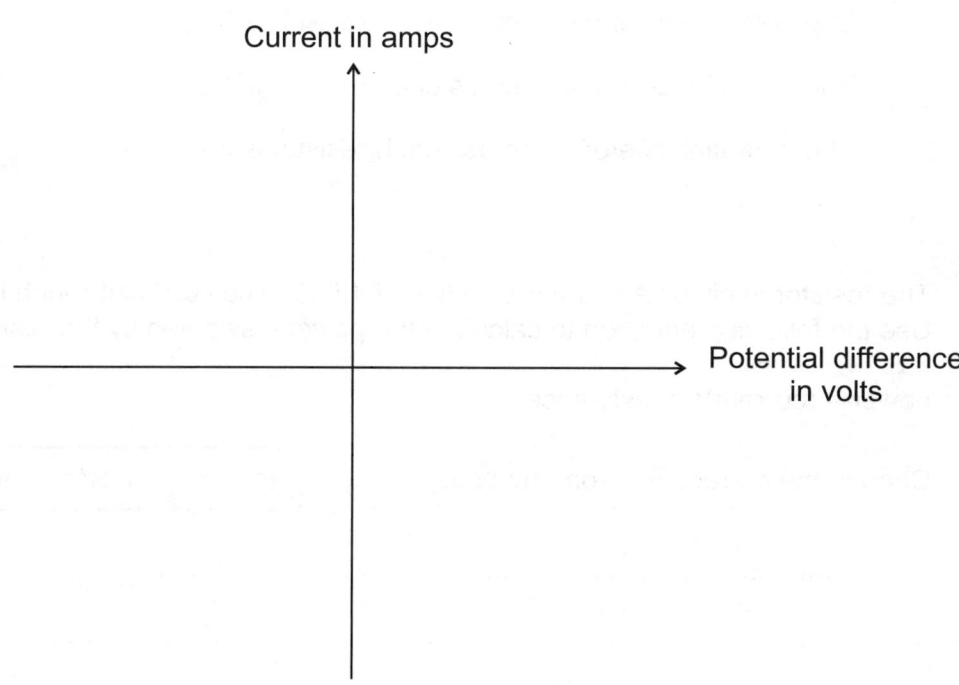

[2 marks]

5 A student uses the circuit shown in **Figure 10** to investigate how the length of the test wire affects its resistance.

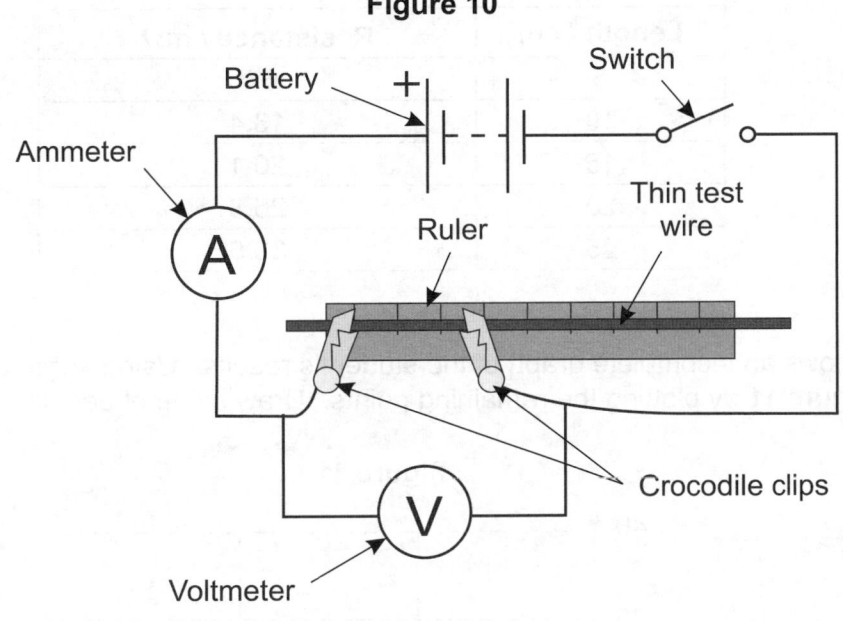

Figure 10

5.1 Describe a method the student could use for her investigation.
Include steps to make sure the results are as accurate as possible.

....

[6 marks]

Question 5 continues on the next page

Turn over ▶

The student records her results in **Table 3**.

Table 3

Length / cm	Resistance / mΩ
5	6.7
10	13.4
15	20.1
20	26.8
25	33.5

5.2 **Figure 11** shows an incomplete graph of the student's results. Using **Table 3**, complete **Figure 11** by plotting the remaining points. Draw a line of best fit.

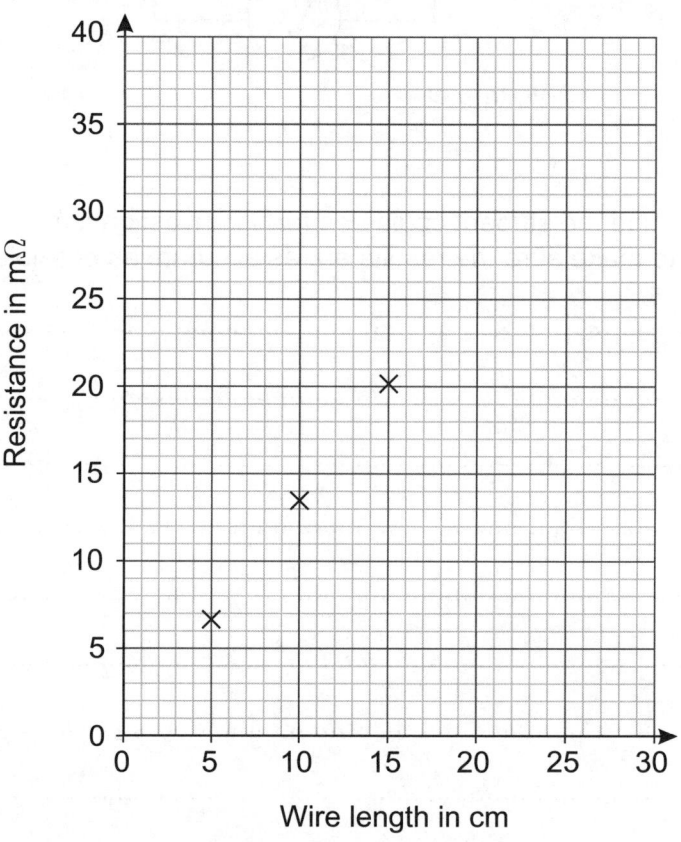

Figure 11

[2 marks]

5.3 Predict the resistance of a test wire with a length of 30 cm.

Resistance = ... mΩ
[1 mark]

5.4 Give **one** conclusion that can be made from **Figure 11**.

....

....
[1 mark]

6 At the start of a roller coaster ride a carriage is raised by a chain lift through a vertical height of 40 m to point **W**, as shown in **Figure 12**. It is stopped at point **W** and then released to follow the track through points **X**, **Y** and **Z**.

Figure 12

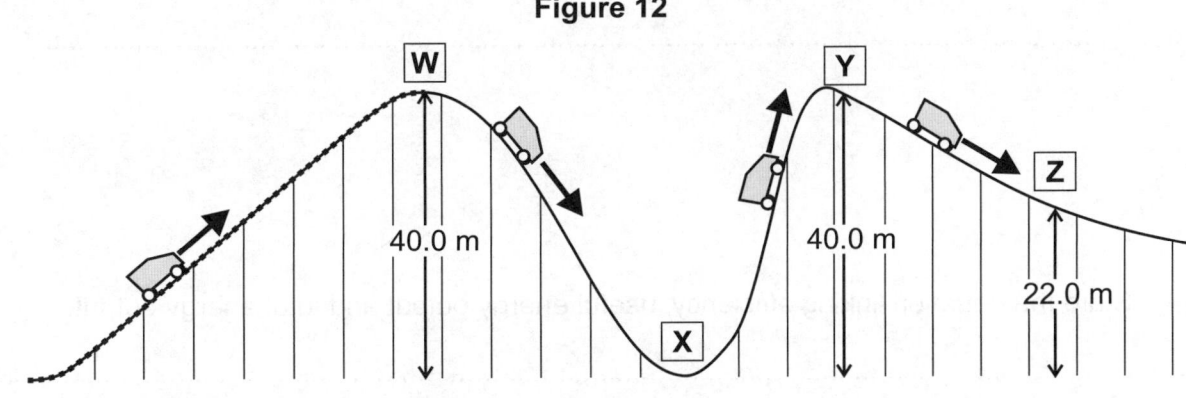

6.1 At which two points does the carriage have the same amount of energy in its gravitational potential energy store? Tick **one** box.

☐ X and Z

☐ W and Y

☐ Y and Z

☐ W and Z

[1 mark]

6.2 At which point does the car have the most energy in its kinetic energy store?
Tick **one** box.

☐ W

☐ X

☐ Z

[1 mark]

The mass of a full carriage is 1 500 000 g. Gravitational field strength = 9.8 N/kg.

6.3 Write down the equation that links gravitational potential energy, mass, height and gravitational field strength.

....

[1 mark]

6.4 Calculate the energy transferred to the gravitational potential energy store of a full carriage as it is raised by the chain lift to point **W**.

....

....

Energy transferred = ... J

[2 marks]

Question 6 continues on the next page

Turn over ▶

A different type of roller coaster uses a spring system to launch a carriage forward.
A compressed spring extends, causing an energy transfer that launches the carriage.

6.5 State the energy store that energy is transferred from when the compressed spring extends.

....
[1 mark]

6.6 State the equation linking efficiency, useful energy output and total energy output.

....
[1 mark]

6.7 The spring system stores 59.5 kJ of energy when it is compressed.
When a carriage is launched, 18.0 kJ of this energy is usefully transferred to the kinetic energy store of the carriage.

Calculate the efficiency of the spring system.
Give your answer to three significant figures.

....

....

Efficiency = ...
[2 marks]

6.8 Explain why lubricating the moving parts of the carriage will increase its speed.
Include the energy store that energy is usefully transferred to in your answer.

....

....

....

....

....
[3 marks]

7 Table 4 gives details of some isotopes.

Table 4

Isotope	Symbol	Type of decay
Radium-226	$^{226}_{88}\text{Ra}$	alpha
Radon-222	$^{222}_{86}\text{Rn}$	alpha
Radon-224	$^{224}_{86}\text{Rn}$	beta
Bismuth-210	$^{210}_{83}\text{Bi}$	alpha, beta
Bismuth-214	$^{214}_{83}\text{Bi}$	alpha, beta
Lead-210	$^{210}_{82}\text{Pb}$	beta

7.1 Calculate the number of neutrons in a bismuth-214 nucleus.

....

[1 mark]

7.2 Explain why bismuth-210 and bismuth-214 are isotopes of each other.

....

....

[2 marks]

7.3 Using data from **Table 4**, complete the equations in **Figure 13** to show how the following isotopes decay.

Figure 13

$^{226}_{88}\text{Ra} \longrightarrow$ [..............] $+$ $^{4}_{2}\text{He}$

$^{210}_{82}\text{Pb} \longrightarrow$ [..............] $+$ $^{0}_{-1}\text{e}$

[2 marks]

Question 7 continues on the next page

Turn over ▶

Figure 14 shows the activity-time graph of a sample of polonium-210.

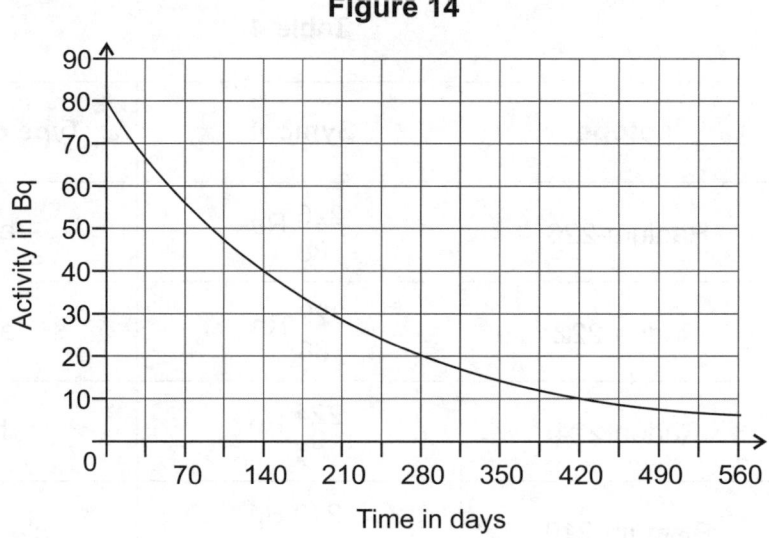

Figure 14

7.4 Using the graph in **Figure 14**, determine the time it takes for the activity of the sample to drop from 80 Bq to 10 Bq.

Time taken = ... days

[1 mark]

7.5 Determine the half-life of polonium-210.

Half-life = ... days

[1 mark]

Polonium-210 emits alpha radiation. Scientists who work with it must be particularly careful to avoid radioactive contamination.

7.6 Define the term radioactive contamination.

....

....

[1 mark]

Another isotope of polonium, polonium-217, emits both alpha and beta particles.

7.7 A human body can be contaminated by radioactive atoms on the inside or on the outside. Explain and compare the dangers from alpha and beta radiation in both of these situations.

....

....

....

....

....

....

....

[4 marks]

END OF QUESTIONS

GCSE Combined Science
Physics Paper 2
Foundation Tier

In addition to this paper you should have:
- A ruler.
- A calculator.
- The Physics Equation sheet (on page 538).

Centre name				
Centre number				
Candidate number				

Time allowed:
- 1 hour 15 minutes

Surname	
Other names	
Candidate signature	

Instructions to candidates
- Write your name and other details in the spaces provided above.
- Answer **all** questions in the spaces provided.
- Do all rough work on the paper.
- Cross out any work you do not want to be marked.

Information for candidates
- The marks available are given in brackets at the end of each question.
- There are 70 marks available for this paper.
- You are allowed to use a calculator.
- You should use good English and present your answers in a clear and organised way.

Advice to candidates
- In calculations, show clearly how you worked out your answers.

For examiner's use						
Q	Attempt Nº			Q	Attempt Nº		
	1	2	3		1	2	3
1				5			
2				6			
3				7			
4							
				Total			

1 A student is given apparatus to investigate the relationship between the force on a spring and its extension. The apparatus is set up as shown in **Figure 1**. The hook is assumed to have zero mass.

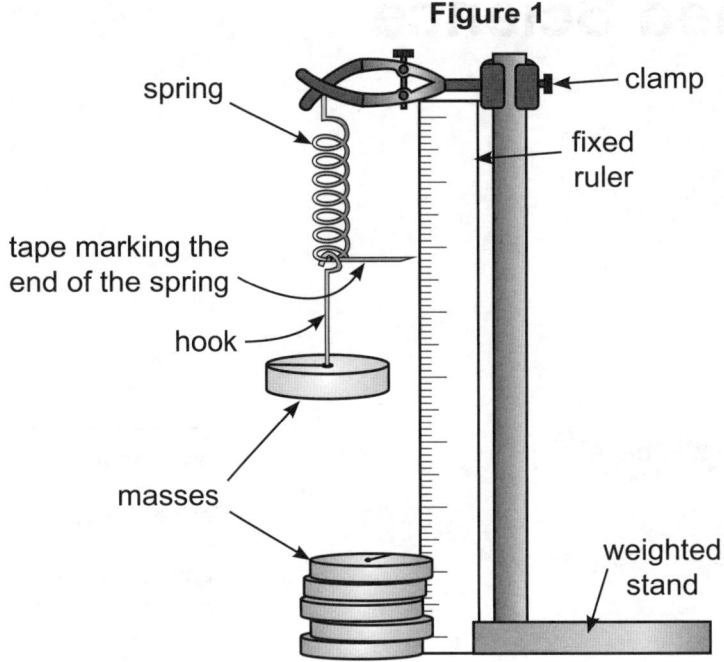

Figure 1

1.1 Placing one of the masses on the hook exerts a force of 1 N on the bottom of the spring.
A force is exerted by the clamp on the other end of the spring.
What is the size of the force exerted on the spring by the clamp?
Tick **one** box.

☐ 0.5 N

☐ 1 N

☐ 1.5 N

☐ 2 N

[1 mark]

1.2 Explain why more than one force is needed to deform a spring.

....

....
[1 mark]

1.3 Name the type of error which may be reduced by the use of the tape marker at the end of the spring when taking extension measurements.
Tick **one** box.

☐ zero error

☐ systematic error

☐ random error

[1 mark]

The student used the apparatus in **Figure 1** to collect data and plot the graph shown in **Figure 2**.

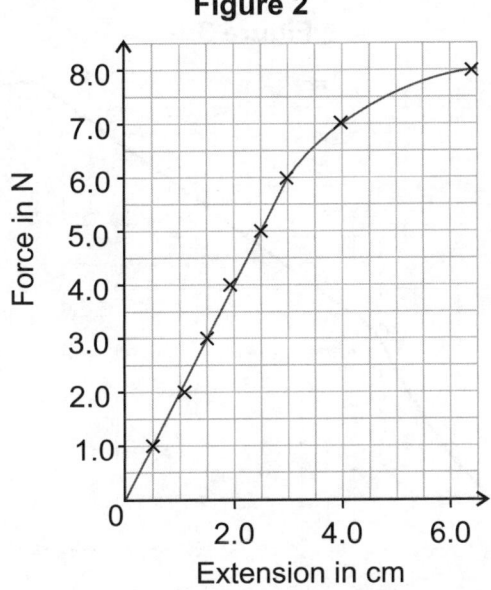

1.4 Estimate the maximum force that can be applied to the spring before the limit of proportionality is reached. Use **Figure 2** to justify your estimate.

....

....

....

[2 marks]

1.5 Use the graph to find the force needed to produce an extension of 0.015 m.

Force = ... N

[2 marks]

1.6 Use the following equation to calculate the spring constant of the spring.

Spring constant = force ÷ extension

....

....

Spring constant = ... N/m

[2 marks]

Turn over for the next question

Turn over ▶

2 A swimmer swims a length of a 20 m swimming pool.
The distance-time graph in **Figure 3** shows the swimmer's motion.

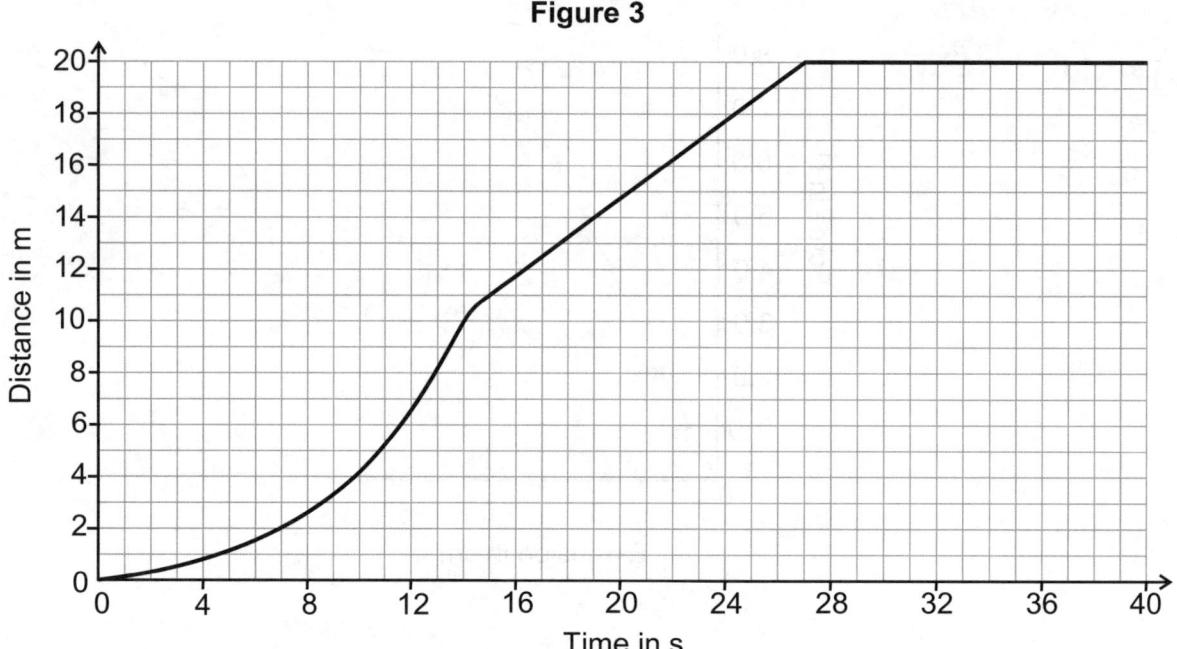

Between 15 and 27 seconds the swimmer is travelling at constant speed.

2.1 State the resultant force on the swimmer between 15 and 27 seconds.

Force = ... N
[1 mark]

2.2 Using **Figure 3**, calculate the speed of the swimmer between 15 and 27 seconds.

....

....

Speed = ... m/s
[2 marks]

2.3 Between which of the following distances was the swimmer travelling fastest?
Tick **one** box.

☐ Between 14 metres and 15 metres.

☐ Between 9 metres and 10 metres.

☐ Between 0 metres and 1 metre.

[1 mark]

A camera travels along the length of the pool to film the swimmer.
It starts from rest at the same time as the swimmer.
It travels at a constant speed and reaches the end of the pool in 23 s.

2.4 Use the following equation to calculate the speed of the camera.

$$\text{speed} = \frac{\text{distance travelled}}{\text{time}}$$

Give your answer to 2 significant figures.

....

....

Speed = ... m/s
[2 marks]

2.5 Draw a distance-time graph on **Figure 3** to show the motion of the camera.
[2 marks]

2.6 The swimmer swims back to their starting position.
Which of the following statements is true when they have returned to their starting position?
Tick **one** box.

☐ The magnitude of the swimmer's displacement is 40 m.

☐ The distance the swimmer has travelled is 20 m.

☐ The magnitude of the swimmer's displacement is 0 m.

☐ The distance the swimmer has travelled is 0 m.

[1 mark]

Turn over for the next question

Turn over ▶

3 Electromagnetic waves are a type of transverse wave.
A trace of a transverse wave is displayed in **Figure 4**.

Figure 4

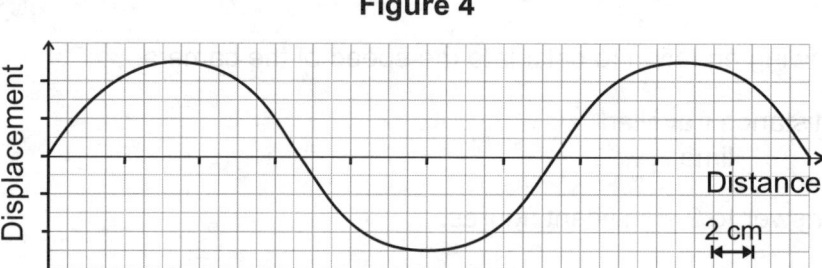

3.1 Calculate the amplitude of the wave in **Figure 4**.

....

Amplitude = ... cm
[1 mark]

3.2 What is the frequency of a wave?
Tick **one** box.

☐ The amount of energy carried by the wave

☐ How long it takes for a complete wave to be produced.

☐ The number of complete waves that pass a point each second.

☐ The average displacement of a point on a wave from the rest position.

[1 mark]

3.3 Infrared radiation is emitted by some TV remote controls. Give **one** other use of infrared.

....

[1 mark]

When infrared radiation reaches a material it can do three things:
(1) be absorbed, (2) bounce back, or (3) pass straight through it.

A student sets up the experiment shown in **Figure 5** to test whether a material absorbs infrared.
The TVs will switch on if infrared radiation from the remote control reached them.
The student believes that if the material absorbs the infrared radiation, the TVs will remain off.

Figure 5

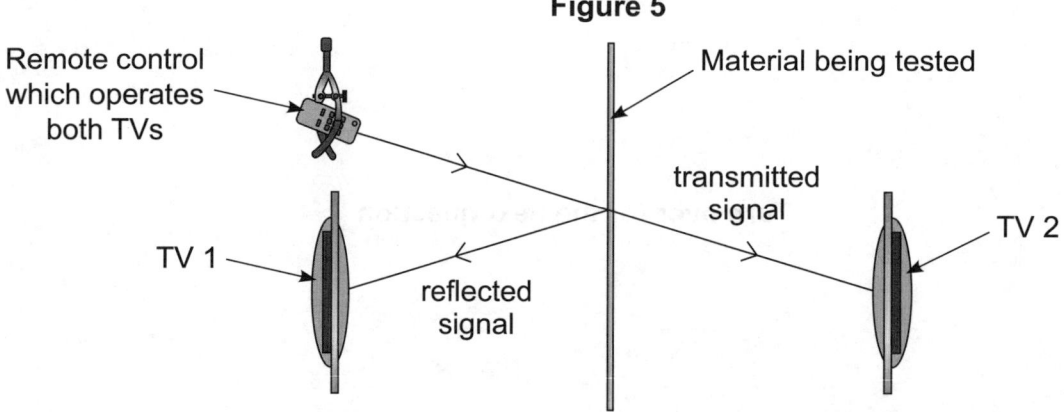

3.4 Variable 1 and Variable 2 are listed below. What types of variable are they?
Tick **one** box for **Variable 1** and **one** box for **Variable 2**.

	Independent	Dependent	Control
Variable 1: Distance of remote from material	☐	☐	☐
Variable 2: Type of material being tested	☐	☐	☐

[2 marks]

The student makes the hypothesis:
'Only dark-coloured materials absorb infrared signals.'
The student's results are shown in **Table 1**.

Table 1

Material	Which TVs respond?	Signal absorbed?
white paper	TV 1 and TV 2	no
black paper	neither	yes
white woollen blanket	neither	yes
black woollen blanket	neither	yes

3.5 Explain whether the student's results in **Table 1** support their hypothesis.

....

....

[1 mark]

3.6 Remote controls can use infrared waves or radio waves to send signals to devices.
Describe the similarities and differences between infrared waves and radio waves.

....

....

....

....

....

....

[4 marks]

Turn over for the next question

Turn over ▶

4 A student sets up the apparatus shown in **Figure 6** to investigate a wave on a string. The student adjusts the frequency of the signal generator until there is a clear wave on the string.

Figure 6

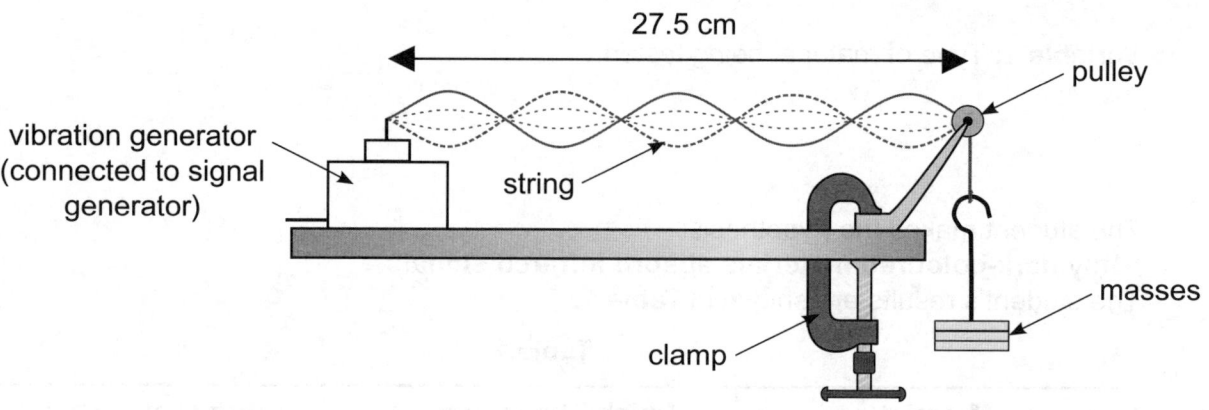

4.1 State **one** safety precaution that the student should take when doing this experiment.

....
....
[1 mark]

4.2 What is the wavelength of the wave on the string shown in **Figure 6**?
Tick **one** box.

☐ 5.50 cm

☐ 11.0 cm

☐ 13.75 cm

☐ 27.5 cm

[1 mark]

4.3 Write down the equation that links wave speed, frequency and wavelength.

....
[1 mark]

4.4 The signal generator is set to 120 Hz. Calculate the speed of the wave on the string.

....
....

Speed = ... m/s
[2 marks]

4.5 The student sets up a wave on another string and calculates its speed.
She repeats this four times with the same string. Her results are:

   12.8 m/s   13.4 m/s   13.3 m/s   32.9 m/s

Identify any anomalous results in the set.
Describe how such anomalous results should be dealt with.

....

....

....

....
[3 marks]

The student decides to investigate how the diameter of the string is related to wave speed.
She will use four strings made of the same material with different diameters.

4.6 The student plans to measure the diameters of the strings using a 30 cm ruler.
Explain why this equipment is unsuitable.
Suggest a more suitable piece of equipment that could be used.

....

....

....
[2 marks]

The student's results are shown in **Table 2**.

Table 2

String diameter (mm)	0.26	0.68	1.33	2.15
Wave speed (m/s)	25.8	9.86	5.04	3.12

4.7 Give **one** conclusion that can be made from the results in **Table 2**.

....

....
[1 mark]

4.8 The student wants to present her results as a graph.
Suggest a suitable type of graph she could use. Justify your choice.

....

....

....
[2 marks]

Turn over for the next question

Turn over ▶

5 A skydiver jumps from an aeroplane and his motion is recorded.
Figure 7 shows the velocity-time graph of his fall.

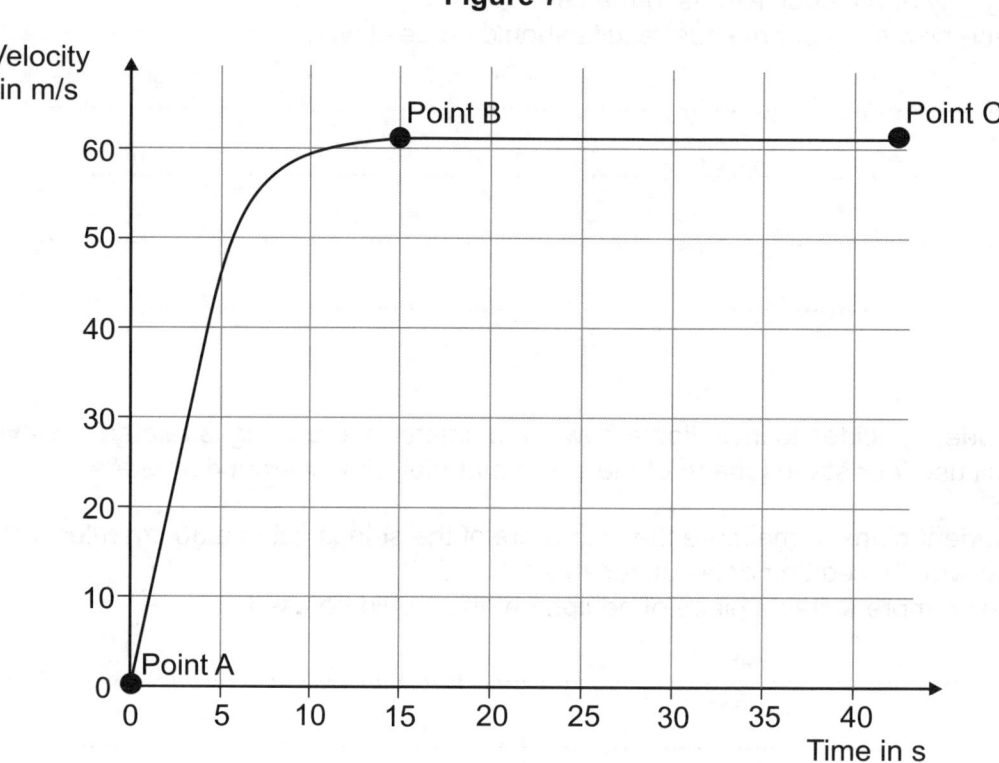

5.1 In the absence of air resistance, all objects accelerate towards the ground at the same rate. State the magnitude of this acceleration.

....
[1 mark]

5.2 Points A, B and C are labelled on Figure 7.
Draw one line from each part of the skydiver's fall to the correct description of his motion at this time.

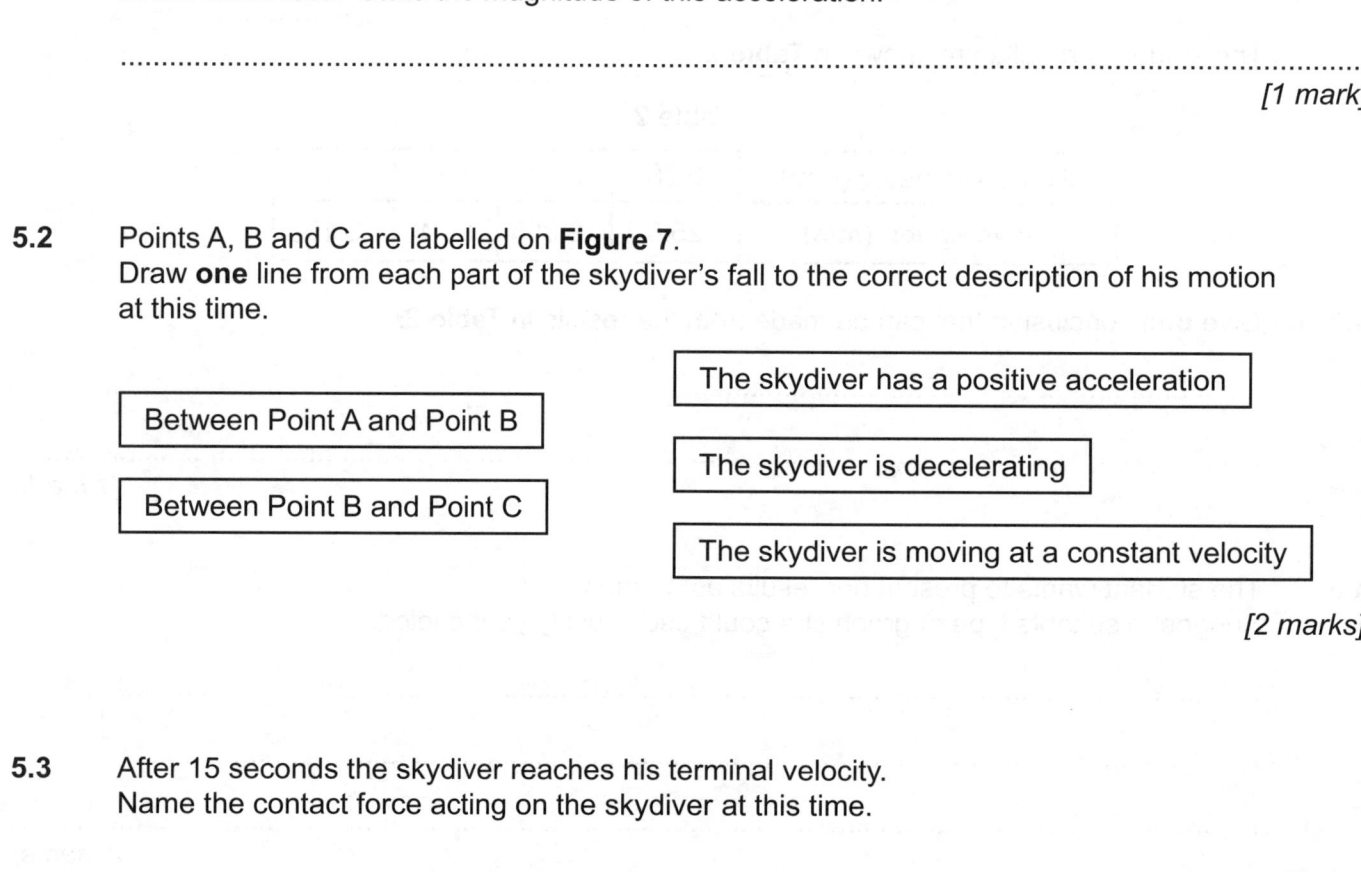

[2 marks]

5.3 After 15 seconds the skydiver reaches his terminal velocity.
Name the contact force acting on the skydiver at this time.

....
[1 mark]

5.4 Write down the equation that links weight, mass and the gravitational field strength.

....
[1 mark]

5.5 The skydiver has a mass of 80.0 kg. Calculate his weight.
Use gravitational field strength = 9.8 N/kg.

....

....

Weight = ... N
[2 marks]

5.6 Once the skydiver has jumped, the plane accelerates away.
Write down the equation which links force, mass and acceleration.

....
[1 mark]

5.7 The plane experiences a resultant driving force of 58 500 N as it accelerates.
The plane has a mass of 4680 kg.
Calculate the acceleration of the plane.

....

....

....

Acceleration = ... m/s^2
[3 marks]

Turn over for the next question

Turn over ▶

6 Table 3 shows data from the Highway Code about stopping distances for a well-maintained car travelling on dry roads at various speeds.

Table 3

Speed (km/h)	Thinking distance (m)	Braking distance (m)	Stopping distance (m)
32	6	6	12
64	12		36
96	18	55	73

6.1 Complete **Table 3** by calculating the missing braking distance.

[1 mark]

6.2 The data in **Table 3** was obtained by observing a large number of drivers and calculating average distances. Explain why a large sample of people was used.

....

....

....

[2 marks]

6.3 A car is travelling down a road. A hazard appears ahead.
Which of the following factors will the thinking distance of the car depend on?
Tick **two** boxes.

☐ Whether the driver is tired.

☐ Whether the road is wet.

☐ How worn the tyres are.

☐ The speed the car is travelling at.

☐ The mass of the car.

[2 marks]

75 000 J of work must be done to stop the car. The braking force of the car is 5000 N.

6.4 Write down the equation that links distance, force and work done.

....

[1 mark]

6.5 Calculate the braking distance of the car.

....

....

Distance = ... m

[3 marks]

7 A student passes a current through a straight wire.
The current is flowing into the paper.

7.1 Which of the following shows the magnetic field around the wire?
Tick **one** box.

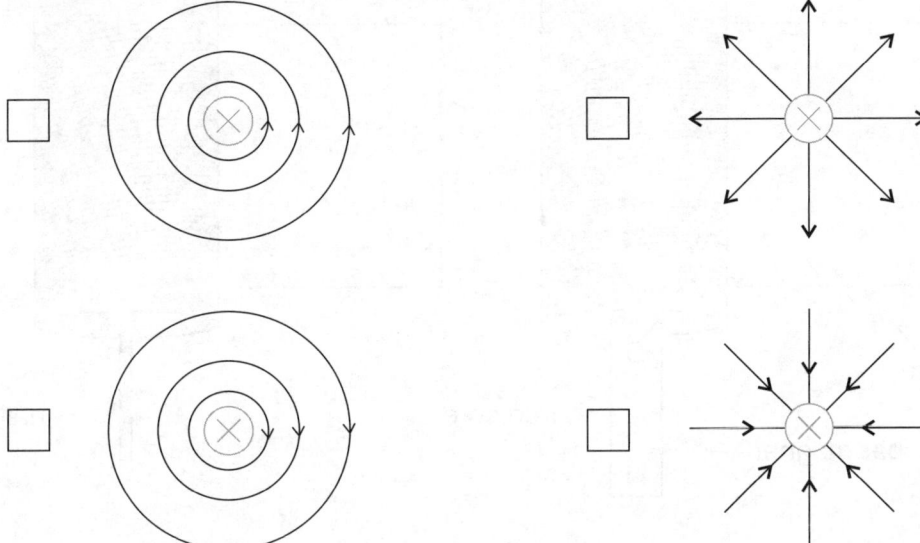

[1 mark]

7.2 A bar magnet is placed near the wire and experiences a force.
State **two** ways that the force on the magnet could be decreased without removing it from the wire's magnetic field completely.

1. ....
....

2. ....
....

[2 marks]

Question 7 continues on the next page

Turn over ▶

7.3 The student coils the wire and inserts an iron core to form an electromagnet.
Figure 8 shows a bar magnet and an iron nail hanging from the electromagnet.

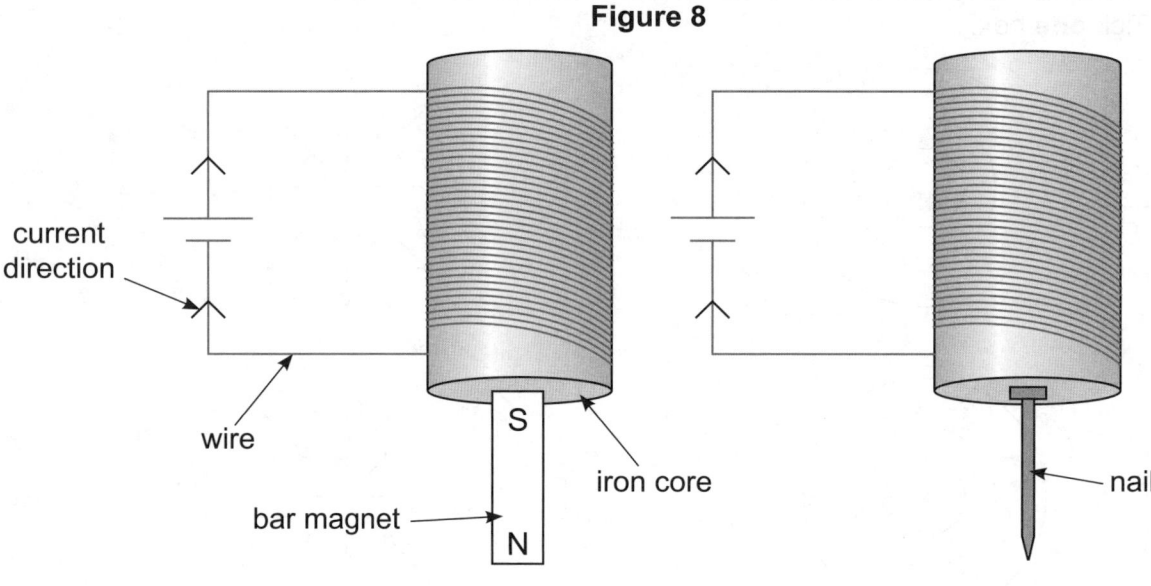

Figure 8

Predict what will happen to the bar magnet and the nail when the electromagnet is disconnected from the battery in each case. Justify your answer.

[6 marks]

END OF QUESTIONS

Answers

Topic B1 — Cell Biology

Page 19
Microscopy
Q1   real size = image size ÷ magnification
          = 2.4 mm ÷ 40
          = 0.06 mm *[1 mark]*
    0.06 × 1000 = **60 µm** *[1 mark]*

Page 22
Warm-Up Questions
1) mitochondria
2) Any two from: e.g. plant cells have a rigid cell wall and animals do not. / Plant cells have a permanent vacuole and animals do not. / Plant cells contain chloroplasts and animals do not.
3) False
Bacterial cells have a single loop of DNA instead.
4) electron microscope
5) 4.5×10^{-4} µm

Exam Questions
1   bacterial cell *[1 mark]*
2.1 C *[1 mark]*
2.2 allows photosynthesis to take place *[1 mark]*
2.3 making proteins *[1 mark]*
3.1 How to grade your answer:
    Level 0: There is no relevant information. *[No marks]*
    Level 1: There is a brief explanation of how to prepare a slide. *[1 to 2 marks]*
    Level 2: There is a detailed explanation of how to prepare a slide. *[3 to 4 marks]*
    Here are some points your answer may include:
    Add a drop of water to the middle of a clean slide.
    Cut up an onion and separate it out into layers.
    Use tweezers to peel off some epidermal tissue from the bottom of one of the layers.
    Use tweezers to place the epidermal tissue into the water on the slide.
    Add a drop of iodine solution/stain.
    Place a cover slip on top without trapping any air bubbles.
3.2 real size = 7.5 mm ÷ 100 = 0.075 mm *[1 mark]*
    0.075 × 1000 = **75 µm** *[1 mark]*

Page 28
Chromosomes and Mitosis
Q1 a) 11 ÷ (62 + 11) = 0.150...
        0.150... × 100 = **15%** *[1 mark]*
    b) E.g. she could see the X-shaped chromosomes in the middle of the cells. / She could see the arms of the chromosomes being pulled apart *[1 mark]*.

Page 29
Warm-Up Questions
1) differentiation
2) The cell has a hair-like shape, which gives it a large surface area to absorb water and minerals from the soil.
3) 2
4) in the nucleus
5) True

Exam Questions
1

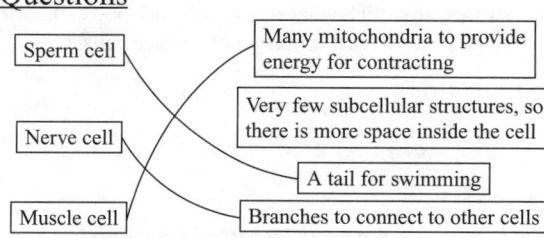

[1 mark for each correct link]

2.1

[1 mark for arrow at shoot tip, 1 mark for arrow at root tip]

2.2 E.g. stem cells can be used to make clones *[1 mark]* of crop plants that have useful features / that aren't killed by a disease *[1 mark]*.
2.3 E.g. paralysis / diabetes *[1 mark]*

Page 30
Diffusion
Q1 a) The ink will diffuse / spread out through the water *[1 mark]*. This is because the ink particles will move from where there is a higher concentration of them (the drop of ink) to where there is a lower concentration of them (the surrounding water) *[1 mark]*.
    b) The ink particles will diffuse / spread out faster *[1 mark]*.
Q2 The larger the surface area of the membrane the faster the diffusion rate *[1 mark]*. This is because more particles can pass through at the same time *[1 mark]*.

Page 32
Osmosis
Q1 Water will move out of the piece of potato by osmosis *[1 mark]*, so its mass will decrease *[1 mark]*.

Page 35
Exchanging Substances
Q1 Surface area:
    (2 × 2) × 2 = 8
    (2 × 1) × 4 = 8
    8 + 8 = 16 µm² *[1 mark]*
    Volume:
    2 × 2 × 1 = 4 µm³ *[1 mark]*
    So the surface area to volume ratio is **16 : 4, or 4 : 1** *[1 mark]*.

Page 36
More on Exchanging Substances
Q1 E.g. they have a large surface area. / They have a moist lining for dissolving gases. / They have very thin walls. / They have a good blood supply. *[1 mark]*
Q2 The damage to the villi is likely to reduce the surface area for absorption *[1 mark]*. Therefore, less iron can be absorbed from the digested food in the small intestine into the blood *[1 mark]*.

Pages 38-39
Warm-Up Questions
1) diffusion
2) True
3) They increase the surface area.
4) E.g. alveoli (in the lung)
5) Stomata

6) E.g. it is made up of lots of thin plates which give a large surface area. / The plates have lots of blood capillaries. / The plates have a thin layer of surface cells.

Exam Questions
1.1
[1 mark for dye particles spread out evenly]
1.2 A lower temperature would decrease the rate of diffusion *[1 mark]*.
2.1 The movement of substances from an area of lower concentration to an area of higher concentration, requiring energy. *[1 mark]*
2.2 Movement of mineral ions from the soil into root hair cells *[1 mark]*
Movement of nutrients, such as glucose, from the gut into the blood *[1 mark]*
3.1 partially permeable membrane *[1 mark]*
Remember that osmosis always takes place across a partially permeable membrane.
3.2 The liquid level on side B will fall. *[1 mark]*
This is because water will flow from a less concentrated solution to a more concentrated solution.
4.1

Cylinder	1	2	3	4
Length after 24 hours (mm)	40	43	51	55
Change in length (mm)	−10	−7	+1	+5

[1 mark]
4.2 The concentration of the sugar solution that cylinder 4 was placed in must have been lower than the concentration of the solution inside the potato cells *[1 mark]*, so the cylinder increased in length as water entered the cells by osmosis *[1 mark]*.

Topic B2 — Organisation

Page 43
Investigating Enzymatic Reactions
Q1 2.5 min × 60 = 150 seconds *[1 mark]*
$1000 ÷ 150 = 6.666... = \mathbf{6.7\ s^{-1}}$ *[1 mark]*

Pages 47-48
Warm-Up Questions
1) organ
2) active site
3) amylase
4) glycerol and fatty acids
5) False
Proteases are made in the stomach, pancreas and small intestine.
6) E.g. break up the food using a pestle and mortar. Then transfer the ground up food to a beaker and add some distilled water. Next, stir the mixture with a glass rod, and finally filter the solution using a funnel lined with filter paper.

Exam Questions
1.1 A *[1 mark]*
1.2 Bile makes conditions in the small intestine **alkaline** *[1 mark]*. Bile also emulsifies **lipids** *[1 mark]*.
2.1 40 °C *[1 mark]*
2.2 optimum temperature *[1 mark]*
3.1 The enzyme has a specific shape which will only fit with one type of substrate *[1 mark]*.
3.2 It would affect the bonds holding the enzyme together and change the shape of the enzyme's active site/ denature the enzyme *[1 mark]*. This would mean the substrate would no longer fit into it so the enzyme wouldn't work anymore *[1 mark]*.
A similar thing happens when the pH or the temperature is too high — the bonds are disrupted and the shape of the active site may change.
4.1 Biuret solution *[1 mark]*
4.2 Yes, because when iodine solution was added, it turned blue-black *[1 mark]*.
4.3 Yes, because when Sudan III stain solution was added, the mixture separated out into two layers with a red top layer *[1 mark]*.
5.1 To prevent the starch coming into contact with amylase in the syringe, which would have started the reaction before he had started the stop clock *[1 mark]*.
5.2 Rate = 1000 ÷ 60 = 16.666... *[1 mark]*
= **17 s^{-1} (2 s.f.)** *[1 mark]*

Page 49
The Lungs
Q1 304 ÷ 8 = **38 breaths per minute** *[1 mark]*

Page 52
Circulatory System — Blood Vessels
Q1 2.175 l × 1000 = 2175 ml *[1 mark]*
2175 ÷ 8.7 = **250 ml/min** *[1 mark]*
Q2 They have a big lumen to help the blood flow despite the low pressure *[1 mark]*. They have valves to keep the blood flowing in the right direction *[1 mark]*.

Page 53
Circulatory System — Blood
Q1 They help the blood to clot at a wound *[1 mark]*.
Q2 To carry oxygen from the lungs to all the cells in the body *[1 mark]*.

Pages 54-55
Warm-Up Questions
1) bronchi
2) They supply oxygenated blood to the heart itself.
3) Makes the heart beat regularly.
4) They carry blood back to the heart.
5) True

Exam Questions
1.1 aorta *[1 mark]*
1.2 It pumps blood around the body. *[1 mark]*
1.3 vena cava *[1 mark]*
2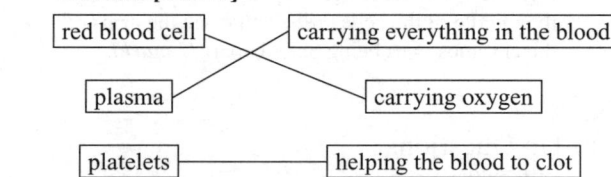
[2 marks for all lines correct, otherwise 1 mark for one line correct.]
3 The shape gives it a large surface area for absorbing oxygen *[1 mark]*.
4.1 492 ÷ 12 = **41 breaths per minute** *[1 mark]*
4.2 Resting heart rate is controlled by a group of cells in the right atrium wall that act as a pacemaker *[1 mark]*.
5.1 How to grade your answer:
Level 0: There is no relevant information. *[No marks]*
Level 1: There are some relevant points explaining how the structure of a capillary allows it to carry out its function. *[1 to 2 marks]*

Level 2: There is a clear, detailed explanation of how the structure of a capillary allows it to carry out its function. *[3 to 4 marks]*
Here are some points your answer may include:
Capillaries have gaps in their walls.
This means that substances can diffuse in and out.
Their walls are usually only one cell thick.
This means that diffusion is very fast because there is only a short distance for molecules to travel.

5.2 150 s ÷ 60 = 2.5 min
1155 ÷ 2.5 = **462 ml/min** *[2 marks for correct answer, otherwise 1 mark for correct working.]*

Make sure you read the question carefully — if you gave your answer in ml/second you wouldn't have got all the marks available.

Page 59
Warm-Up Questions
1) the coronary arteries
2) An artificial heart could be fitted.
3) A mechanical device that is put into a person to pump blood if their own heart fails.

Exam Questions
1.1

Description of treatment	Name of treatment
Tubes that are put inside arteries to keep them open	**stents**
Drugs that reduce cholesterol in the blood	**statins**

[1 mark for each correct answer]

1.2 E.g. a person has to remember to take them regularly. / They can cause unwanted side effects. / It takes time for them to work *[1 mark]*.
2.1 biological valves *[1 mark]*, mechanical valves *[1 mark]*
2.2 E.g. it requires surgery, which could lead to bleeding/infection. / There could be problems with blood clots *[1 mark]*.
3.1 fatty material/fatty deposit *[1 mark]*
3.2 The fatty material causes the coronary arteries to become narrow, so blood flow to the heart muscle is reduced *[1 mark]*. This reduces (or stops) the delivery of oxygen to the heart muscle *[1 mark]*.

Page 65
Warm-Up Questions
1) False
2) It can be spread from person to person or between animals and people.
3) True
4) cancer

Exam Questions
1 A disease that cannot be spread between people or between animals and people *[1 mark]*.
2

	Benign	Malignant
The tumour is made up of a mass of cells formed by uncontrollable division and growth.	✓	✓
The tumour cells can break off and travel into the bloodstream.		✓
The tumour is cancerous.		✓

[1 mark for each correct row]

3.1 A risk factor is something that is linked to an increased chance of getting a certain disease *[1 mark]*
3.2 e.g. smoking *[1 mark]*
3.3 e.g. ionising radiation *[1 mark]*

Page 69
The Rate of Transpiration
Q1 Aloe vera, because the transpiration rate will be higher in the hot, dry area *[1 mark]*, so the aloe vera will have fewer stomata to help conserve water *[1 mark]*.
Q2 As it gets darker, the stomata begin to close *[1 mark]*. This means that very little water can escape *[1 mark]*. So the rate of transpiration decreases *[1 mark]*.

Pages 70-71
Warm-Up Questions
1) At the growing tips of shoots and roots.
2) palisade layer
3) False
4) light intensity / temperature / air flow / humidity

Exam Questions
1.1 spongy mesophyll tissue *[1 mark]*
1.2 epidermal tissue *[1 mark]*
2.1 phloem *[1 mark]*
2.2 To allow the cell sap to flow through *[1 mark]*.
2.3 translocation *[1 mark]*
3.1 guard cells *[1 mark]*
3.2 Condition B, because the stoma has closed to reduce water loss *[1 mark]*.
4.1 They are made of dead cells joined end to end *[1 mark]* with no end walls between them and therefore a hole down the middle *[1 mark]*. They're strengthened with a material called lignin *[1 mark]*.
4.2 transpiration *[1 mark]*
5.1 A, because transpiration is faster when the temperature is higher / when the humidity is lower *[1 mark]*.
5.2 How to grade your answer:
Level 0: There is no relevant information. *[No marks]*
Level 1: There are some relevant points explaining how the rate of water loss would be affected by windy weather, but some detail is missing. *[1 to 2 marks]*
Level 2: There is a clear, detailed explanation of how the rate of water loss would be affected by windy weather. *[3 to 4 marks]*
Here are some points your answer may include:
Water is lost more quickly on windy days.
This is because fast-moving air sweeps away the water vapour around the leaf.
This causes the concentration of water vapour inside the leaf to always be higher than outside.
This means water vapour diffuses out quickly.

Topic B3 — Infection and Response

Page 77
Warm-Up Questions
1) True
2) By sexual contact.
3) inside cells
4) A red skin rash.
5) tobacco mosaic virus
6) It causes purple or black spots to develop on the leaves, which can then turn yellow and drop off.

Exam Questions
1 Malaria is caused by a virus *[1 mark]*
Malaria is caused by a protist, not a virus.
2.1 hand-washing *[1 mark]*
Hand-washing before cooking food can reduce the spread of the disease. Vaccinations given to chickens (not people) can also reduce the spread.
2.2 vaccinating people *[1 mark]*
3.1 The first symptoms are flu-like *[1 mark]*.

Page 80
Fighting Disease — Vaccination
Q1 Basia's white blood cells recognise the antigens on the flu virus and rapidly produce antibodies, which kill the pathogen *[1 mark]*. Cassian's white blood cells don't recognise the antigens, so it takes a while for them to produce antibodies and he becomes ill in the meantime *[1 mark]*.

Pages 83-84
Warm-Up Questions
1) The skin stops pathogens getting inside the body and releases substances that kill them.
2) To destroy pathogens that enter the body.
3) Unique molecules on the surface of a pathogen.
4) Through a mutation.
5) Whether the drug works and has the effect you're looking for.

Exam Questions
1.1 Aspirin — Willow
Digitalis — Foxgloves
Penicillin — Mould
[2 marks for three correct lines, otherwise 1 mark for one correct line.]
1.2 painkiller *[1 mark]*
1.3 antibiotic *[1 mark]*
2.1 phagocytosis *[1 mark]*
2.2 Antibodies are produced by **white blood cells** *[1 mark]*. They attach to specific antigens on the surface of the **pathogen** *[1 mark]*.
3.1 human volunteers *[1 mark]*
3.2 Neither the patient or the doctor know who is receiving the drug and who is receiving the placebo *[1 mark]*
4.1 They trap particles that could contain pathogens *[1 mark]*.
4.2 They move the mucus up to the back of the throat where it can be swallowed *[1 mark]*.
4.3 (hydrochloric) acid *[1 mark]*
5.1 How to grade your answer:
Level 0: There is no relevant information. *[No marks]*
Level 1: There is a brief explanation of how vaccination against rubella can prevent a person catching the disease or how having a large number of vaccinated people in a population reduces the risk of rubella for people who are not vaccinated. *[1 to 2 marks]*
Level 2: There is some explanation of how vaccination against rubella can prevent a person catching the disease and how having a large number of vaccinated people in a population reduces the risk of rubella for people who are not vaccinated. *[3 to 4 marks]*
Level 3: There is a clear and detailed explanation of how vaccination against rubella can prevent a person catching the disease and how having a large number of vaccinated people in a population reduces the risk of rubella for people who are not vaccinated. *[5 to 6 marks]*
Here are some points your answer may include:
When a person is vaccinated against rubella, they are injected with dead or inactive rubella viruses.
The dead or inactive viruses carry antigens, which cause the body to produce antibodies to attack them.
If rubella viruses infect the body after this, white blood cells can quickly produce lots of antibodies to defeat the virus.
If a large number of people in a population are vaccinated against rubella, then there are fewer people who are able to pass the disease on.

3.2 E.g. pain when the infected person urinates. / Thick yellow or green discharge from the vagina or penis *[1 mark]*
3.3 E.g. fever / stomach cramps / vomiting / diarrhoea *[1 mark]*

This means that even someone who hasn't been vaccinated is less likely to catch the disease.
5.2 E.g. they may be worried that they will have a bad reaction to a vaccine. / Vaccines don't always work and so the person might not be given immunity *[1 mark]*.

Topic B4 — Bioenergetics

Page 87
The Rate of Photosynthesis
Q1 a) C *[1 mark]*
b) E.g. CO_2 level / temperature *[1 mark]*.

Pages 89-90
Warm-Up Questions
1) chlorophyll
2) light
Energy is transferred to chloroplasts from the environment by light.
3) False
Photosynthesis is endothermic.
4) True

Exam Questions
1.1 glucose — $C_6H_{12}O_6$
oxygen — O_2
water — H_2O *[2 marks for 2 correct lines, 1 mark for one correct line.]*
1.2 CO_2 *[1 mark]*
2.1 carbon dioxide + **water** → glucose + **oxygen**
[2 marks — 1 mark for each correct answer.]
2.2 cellulose *[1 mark]*
2.3 storage as oils *[1 mark]*
3.1 it is insoluble *[1 mark]*
3.2 When photosynthesis is not happening, the plant can't produce glucose, so uses stored starch instead *[1 mark]*.
4.1 The rate of photosynthesis is increasing as the concentration of carbon dioxide is increased *[1 mark]*.
4.2 The concentration of carbon dioxide *[1 mark]*.
4.3 The rate of photosynthesis is no longer increasing as the concentration of carbon dioxide is increased *[1 mark]*. The concentration of carbon dioxide is no longer the limiting factor / light intensity or temperature is now the limiting factor *[1 mark]*.
At this point, photosynthesis won't go any faster because another factor is limiting the rate.
4.4 Enzymes needed for photosynthesis will work more slowly at low temperatures *[1 mark]*.
5.1 By counting the number of bubbles produced / by measuring the volume of gas produced in a given time/at regular intervals *[1 mark]*.
5.2 Rate of photosynthesis/number of bubbles in a given time/volume of gas in a given time *[1 mark]*.
5.3 light intensity *[1 mark]*
5.4 E.g. carbon dioxide concentration in the water / temperature / the plant being used *[1 mark]*.

Page 93
Exercise
Q1 Running is a more intense type of exercise than walking *[1 mark]*, so more anaerobic respiration will be taking place in the muscles *[1 mark]*. Anaerobic respiration produces lactic acid, so running will lead to more lactic acid being in the blood *[1 mark]*.

Page 95
Warm-Up Questions
1) True
2) fermentation

3) e.g. increases breathing rate / increases breath volume / increases heart rate
4) urea

Exam Questions
1.1 glucose → lactic acid *[1 mark]*
1.2 Muscles start using anaerobic respiration when they don't get enough **oxygen** *[1 mark]*. This causes a build up of **lactic acid** *[1 mark]*. After anaerobic respiration stops, the body is left with an oxygen **debt** *[1 mark]*.
2 protein *[1 mark]*
3.1 Any two from: e.g. aerobic respiration uses oxygen, anaerobic respiration does not. / Glucose is broken down fully during aerobic respiration but is only partially broken down during anaerobic respiration. / Aerobic respiration doesn't produce lactic acid, anaerobic respiration does. / Aerobic respiration releases more energy than anaerobic respiration. *[2 marks]*
3.2 glucose + **oxygen** → carbon dioxide + **water** *[2 marks — 1 mark for each correct answer.]*

Topic B5 — Homeostasis and Response

Page 100
The Nervous System
Q1 A fast, automatic response to a stimulus *[1 mark]*.
Q2 a) muscle *[1 mark]*
b) The heat stimulus is detected by receptors in the hand *[1 mark]*, which send impulses along a sensory neurone to the CNS *[1 mark]*. The impulses are transferred to a relay neurone *[1 mark]*. They are then transferred to a motor neurone and travel along it to the effector/muscle *[1 mark]*.

Page 101
Investigating Reaction Time
Q1 a) 242 + 256 + 253 + 249 + 235 = 1235 *[1 mark]*
1235 ÷ 5 = **247 ms** *[1 mark]*
b) Any two from: e.g. the hand each person used to click the mouse / the computer equipment/programme used / the amount of energy drink they consumed / the type of energy drink used / the time between consuming the energy drink and taking the test *[2 marks]*.

Page 102
Warm-Up Questions
1) E.g. body temperature / blood glucose level / water content.
2) synapse
3) True
4) E.g. use the same person to catch the ruler each time. / The person should always use the same hand to catch the ruler. / The ruler should always be dropped from the same height.

Exam Questions
1 sensory neurone *[1 mark]*
2.1 appearance of red triangle *[1 mark]*
2.2 cells in the eye / light receptor cells *[1 mark]*
2.3 muscles (in hand controlling mouse) *[1 mark]*
2.4 343 × 3 = 1029
1029 – 328 – 346 = **355 ms** *[2 marks for the correct answer, otherwise 1 mark for the correct working]*

Page 105
Controlling Blood Glucose
Q1 Curve 2, because it starts rising after curve 1 starts to rise *[1 mark]*. Insulin is released when the blood glucose concentration gets too high, so insulin must be the second curve *[1 mark]*.

Page 107
Warm-Up Questions
1) In the blood.
2) thyroid
3) False
Type 2 diabetes is where a person becomes resistant to their own insulin.
4) eating a carbohydrate-controlled diet / taking regular exercise

Exam Questions
1

Hormone	Gland the hormone is released from
Testosterone	Testes
Adrenaline	Adrenal gland
Oestrogen	Ovaries

[1 mark for each correct row]
2 Insulin is released by the **pancreas** *[1 mark]*. Liver and muscle cells convert glucose into **glycogen** *[1 mark]*.

Page 112
Warm-Up Questions
1) testosterone
2) FSH/follicle-stimulating hormone
3) oestrogen and progesterone
4) A chemical that disables or kills sperm.
5) False

Exam Questions
1.1 E.g. facial hair in males / breasts in females *[1 mark]*
1.2 Every 28 days *[1 mark]*
2 diaphragm *[1 mark]*
3.1 It is involved in the growth and maintenance of the uterus lining *[1 mark]*.
3.2 LH / luteinising hormone *[1 mark]*
3.3 They stop the hormone FSH from being released *[1 mark]*, which stops eggs maturing *[1 mark]*.

Topic B6 — Inheritance, Variation and Evolution

Page 116
Meiosis
Q1 23 *[1 mark]*

Page 119
Warm-Up Questions
1) True
2) two
3) XX

Exam Questions
1 egg — 23 *[1 mark]*
fertilised egg — 46 *[1 mark]*
2.1 A polymer made up of two strands *[1 mark]*.
2.2 Its genome *[1 mark]*.
Remember, an organism's genome is its entire set of genetic material. A gene is a short section of DNA and a chromosome is a really long structure, which contains genes.
2.3 DNA contains genes *[1 mark]*. Each gene codes for a certain sequence of amino acids *[1 mark]*. The sequence of amino acids are put together to make a specific protein *[1 mark]*.
3.1 asexual *[1 mark]*
3.2 They will be genetically identical *[1 mark]*.
Remember, asexual reproduction produces clones — offspring are exactly the same as the parent.

Page 122
Inherited Disorders
Q1   75% / 3 in 4 chance *[1 mark]*

Page 123
Family Trees
Q1   Ff *[1 mark]*

Page 125
Warm-Up Questions
1) An allele.
2) The characteristics an organism has.
3) It's a genetic disorder where a baby is born with extra fingers or toes.
4) E.g. it will help to stop people suffering. / Treating disorders costs a lot of money.

Exam Questions
1   heterozygous — having two different alleles
homozygous — having two of the same allele
genotype — the mix of alleles in an organism
[2 marks for all lines correct, 1 mark for one correct line.]

2.1

	f	F
F	Ff	FF
F	Ff	FF

[1 mark for correct genotype of offspring, 1 mark for correct genotypes of gametes]

2.2   1 in 2 / 50% *[1 mark]*

2.3   One does not have cystic fibrosis and is not a carrier *[1 mark]*. The other does not have cystic fibrosis but is carrier of the condition *[1 mark]*.

Page 128
Evolution
Q1   There was a variety of tongue lengths in the moth population *[1 mark]*. Moths with longer tongues got more food/nectar and were more likely to survive *[1 mark]*. These moths were more likely to reproduce and pass on the genes responsible for their long tongues *[1 mark]*. So, over time, longer tongues became more common in the moth population *[1 mark]*.

Page 132
Warm-Up Questions
1) False
2) False
3) When no individuals of a species are left.
4) A drug that kills bacteria.

Exam Questions
1.1   The appearance of this stingray's ancestors showed **variation** *[1 mark]*. The ancestors that looked like flat rocks were hidden, so were more likely to **survive** *[1 mark]*. They were more likely to reproduce and pass their genes on to the next **generation** *[1 mark]*.

1.2   a mutation *[1 mark]*

2.1   The difference in weight must be caused by the environment *[1 mark]*, because the twins have exactly the same genes *[1 mark]*.
In this case, the environment can mean the amount of food each twin eats or the amount of exercise they each do.

2.2   If they were caused by genes both twins should have the birthmark *[1 mark]*.

Page 133
Selective Breeding
Q1   Select rabbits with floppy ears *[1 mark]* and breed them together to produce offspring *[1 mark]*. Select offspring with floppy ears and breed them together *[1 mark]*. Repeat this over many generations until all of the offspring have floppy ears *[1 mark]*.

Page 136
Warm-Up Questions
1) True
2) GM/genetically modified crops
3) E.g. human insulin

Exam Questions
1   In a population of selectively bred plants, there will be fewer different **alleles** *[1 mark]*. This means if a new disease appears, the plants may be **less resistant** *[1 mark]*. Due to inbreeding, there's also more chance of selectively bred plants having **health problems** *[1 mark]*.

2.1   E.g. to improve the size/quality of their fruit. / To make them resistant to insects/herbicides *[1 mark]*.

2.2   E.g. some people are worried that we might not fully understand the effects of GM crops on human health. / Some people say that growing GM crops will negatively affect the number of wild flowers and insects that live in and around the crops. *[1 mark]*

3.1   Difference between generations 2 and 3 = 5750 − 5375 = 375
Percentage change = (375 ÷ 5375) × 100
                       = 6.976
                       = 7% *[2 marks for the correct answer, otherwise 1 mark for correct working.]*

To work out the percentage change, you first need to work out the difference between generations 2 and 3. You then divide the difference by the original average milk yield (generation 2), and multiply by 100 to convert it into a percentage.

3.2   How to grade your answer:
        Level 0:   There is no relevant information. *[No marks]*
        Level 1:   There are some relevant points describing how a higher milk yield was produced by selective breeding, but the answer is missing some detail. *[1 to 2 marks]*
        Level 2:   There is a clear, detailed explanation of how a higher milk yield was produced by selective breeding. *[3 to 4 marks]*

Here are some points your answer may include:
From the existing cows, the farmer selected the cows that produced the highest average milk yield for breeding.
The farmer then selected the offspring that produced the highest average milk yield for breeding.
The farmer would have continued this process over several generations of cows.
After several generations, the farmer would get cows that are able to produce a very high milk yield.

Page 139
Classification
Q1   B and C *[1 mark]*
Q2   *Castor* *[1 mark]*

Page 140
Warm-Up Questions
1) True
2) True
3) genus
4) species

Exam Questions
1.1   Prokaryotes *[1 mark]*
1.2   Carl Woese *[1 mark]*

2.1 E.g. from gradual replacement by minerals. / From casts and impressions. / From preservation in places where no decay happens *[1 mark]*.
2.2 Many early organisms were soft-bodied, so their tissue decayed away completely *[1 mark]*. Geological activity may have destroyed some of the fossils that were already formed in rock *[1 mark]*.
2.3 They show how much or how little different organisms have changed over time *[1 mark]*.
3.1 Species C *[1 mark]*
3.2 Yes, you would expect Species D to look similar to Species E because they share a recent common ancestor, so they are closely related/have similar genes *[1 mark]*.

Topic B7 — Ecology

Page 141
Competition
Q1 Any three from: e.g. light / water / space / mineral ions *[3 marks]*.
Q2 The water boatmen would not have to compete with the diving beetles for food *[1 mark]* so there would be more food for the water boatmen *[1 mark]*.

Page 144
Adaptations
Q1 a) A behavioural adaptation *[1 mark]*.
b) E.g. it has webbed feet/flippers to help it swim for food *[1 mark]*. / It has a thick layer of fat to help it keep in heat *[1 mark]*.

Page 145
Food Chains
Q1 a) grass *[1 mark]*
b) three *[1 mark]*
c) grasshopper *[1 mark]*
d) The population of grasshoppers could increase *[1 mark]* as there's nothing to eat them *[1 mark]*. The population of snakes could decrease *[1 mark]* as there's nothing for them to eat *[1 mark]*.

Page 146
Warm-Up Questions
1) e.g. territory/space / mates
2) e.g. moisture level / light intensity / temperature / carbon dioxide level / wind intensity/direction / oxygen level / soil pH/mineral content
3) False
Food chains always start with a producer.
4) it would decrease

Exam Questions
1.1 A habitat is the place where an organism lives *[1 mark]*.
1.2 community *[1 mark]*
2.1 tertiary consumer *[1 mark]*
2.2 The algae are producers *[1 mark]*. They are the source of biomass/energy for the food chain *[1 mark]*.
3.1 It doesn't sweat *[1 mark]*.
3.2 Any two from: e.g. it lives in burrows. / It holds its tail over its head. / It lies in the shade. / It lies with its limbs spread out wide. *[2 marks]*

Page 147
Using Quadrats
Q1 1200 ÷ 0.25 = 4800
4800 × 0.75 = **3600 tulips in total**
[2 marks for correct answer, otherwise 1 mark for correct working]

Pages 151-152
Warm-Up Questions
1) E.g. a quadrat.
2) e.g. rain / snow / hail
3) To break down/decay dead matter and animal waste.
4) Through burning.

Exam Questions
1 Energy from the sun makes water **evaporate** *[1 mark]*. When warm water vapour gets higher up, the water vapour cools and **condenses** *[1 mark]*. Water falls from the clouds as **precipitation** *[1 mark]*.
2.1 E.g. the number of poppies increases with increasing distance from the wood *[1 mark]*.
2.2 E.g. moisture level / soil pH / soil mineral content / wind intensity / wind direction *[1 mark]*.
3.1 eating *[1 mark]*
3.2 respiration *[1 mark]*
3.3 Microorganisms break down/decay the dead matter *[1 mark]* and return carbon to the air as carbon dioxide through respiration *[1 mark]*.
3.4 photosynthesis *[1 mark]*
4.1 Grass species = 47 squares out of 100
= (47 ÷ 100) × 100 = **47%** *[1 mark]*
Remember, to calculate percentage cover of an organism in a quadrat you count the number of squares which are more than half covered by the organism.
4.2 E.g. because there may have been too many blades of grass to count each one individually / it's hard to count individual blades of grass *[1 mark]*.

Page 159
Warm-Up Questions
1) Biodiversity is the variety of different species of organisms on Earth, or within an ecosystem.
2) e.g. toxic chemicals for farming / household waste in landfill
3) methane and carbon dioxide
4) True
5) The cutting down of forests.

Exam Questions
1.1 Hedgerows provide a habitat for lots of types of organisms *[1 mark]*.
1.2 The animal species can be bred in captivity *[1 mark]*. This makes sure that some individuals will survive if the species dies out in the wild *[1 mark]*. These individuals can then be released into the wild to replace a population that has been wiped out *[1 mark]*.
2.1 E.g. smoke *[1 mark]*, acidic gases *[1 mark]*.
2.2 E.g. sewage/toxic chemicals from industry can pollute lakes/rivers/oceans. / Fertilisers/other chemicals used on land can be washed into water *[1 mark]*.
2.3 E.g. the human population size is increasing *[1 mark]*, people around the world are demanding a higher standard of living *[1 mark]*.
2.4 This reduces the amount of land taken over for landfill, so ecosystems can be left alone *[1 mark]*.
3 How to grade your answer:
Level 0: There is no relevant information. *[No marks]*
Level 1: There is a brief explanation of how deforestation contributes to global warming. The answer lacks coherency. *[1 to 2 marks]*
Level 2: There is some explanation of how deforestation contributes to global warming, and the answer has some structure. *[3 to 4 marks]*
Level 3: There is a clear and detailed explanation of how deforestation contributes to global warming. *[5 to 6 marks]*
Here are some points your answer may include:
Trees take in carbon dioxide/CO_2 from the atmosphere during photosynthesis.

So cutting down trees means that less carbon dioxide/CO_2 is removed from the atmosphere.
Trees 'lock up' some of the carbon in their wood.
So removing trees means that less is locked up.
Carbon dioxide/CO_2 is released when trees are burnt to clear land.
Microorganisms feeding on dead wood release carbon dioxide/CO_2 through respiration.
Carbon dioxide is a greenhouse gas.
An increase in greenhouse gas levels causes global warming.

Topic C1 — Atomic Structure and the Periodic Table

Page 161
Atoms
Q1   protons = atomic number = 31 *[1 mark]*
electrons = protons = 31 *[1 mark]*
neutrons = mass number − atomic number
= 70 − 31 = **39** *[1 mark]*

Page 163
Isotopes
Q1   E.g. isotopes are atoms with the same number of protons but a different number of neutrons / isotopes have the same atomic number but different mass numbers *[1 mark]*.
Q2   Relative atomic mass =
$$\frac{(92.2 \times 28) + (4.7 \times 29) + (3.1 \times 30)}{92.2 + 4.7 + 3.1} \text{ [1 mark]}$$
$$= \frac{2581.6 + 136.3 + 93}{100} = \frac{2810.9}{100}$$
= 28.109 = **28.1** *[1 mark]*

Page 164
Compounds
Q1 $(2 \times Na) + (1 \times C) + (3 \times O)$ = **6** *[1 mark]*
Q2   Aluminium — **2 atoms**, sulfur — **3 atoms** and oxygen — **12 atoms** *[1 mark]*.

Page 166
Chemical Equations
Q1 $2Fe + 3Cl_2 \rightarrow 2FeCl_3$ *[1 mark]*
Q2   a) water → hydrogen + oxygen *[1 mark]*
b) $2H_2O \rightarrow 2H_2 + O_2$
[1 mark for correct reactants and products, 1 mark for correctly balancing]

Pages 167-168
Warm-Up Questions
1) The total number of protons and neutrons.
2) An element is a substance made up of atoms that all have the same number of protons in their nucleus.
3) False
4) calcium chloride
5) $2K + 2H_2O \rightarrow 2KOH + H_2$

Exam Questions
1.1

Relative charge of a proton	Relative charge of a neutron
+1	0

[1 mark]

1.2   in the nucleus *[1 mark]*
1.3   **8** *[1 mark]*
Atoms have the same number of protons as electrons. This is what makes them neutral.
1.4   **1** *[1 mark]*

2.1   **12** *[1 mark]*
2.2   **11** *[1 mark]*
2.3   **Na** *[1 mark]*
3.1   methane and oxygen *[1 mark]*
3.2   carbon dioxide and water *[1 mark]*
3.3   oxygen/O_2 *[1 mark]*
3.4 $CH_4 + 2O_2 \rightarrow CO_2 + 2H_2O$ *[1 mark]*
4.1 $H_2SO_4 + 2NH_3 \rightarrow (NH_4)_2SO_4$ *[1 mark for correct reactants, 1 mark for correctly balancing the equation]*
4.2   **4** *[1 mark]*
These elements are hydrogen, sulfur, oxygen and nitrogen.
4.3   **8** *[1 mark]*
There are 4 hydrogen atoms in NH_4 and there are two lots of NH_4 present in the formula of ammonium sulfate.
5.1   Isotopes are different forms of the same element, which have the same number of **protons** *[1 mark]* but different numbers of **neutrons** *[1 mark]*.
5.2   **5** *[1 mark]*
The number of protons in an atom is equal to its atomic number.
5.3 $(20 \times 10) + (80 \times 11) = 1080$
Relative atomic mass = $1080 \div 100$ = **10.8**
[3 marks for correct answer, otherwise 1 mark for calculating sum of (isotope abundance × isotope mass number), 1 mark for dividing by 100.]

Page 174
Fractional Distillation
Q1   Ethanol *[1 mark]*. Ethanol has the second lowest boiling point and will be collected once all the methanol has been distilled off and the temperature increased *[1 mark]*.

Page 175
Warm-Up Questions
1) False
2) evaporation / crystallisation
3) the top

Exam Questions
1.1   Put the paper in a beaker of solvent *[1 mark]*.
1.2   Pencil marks are insoluble/won't dissolve in the solvent / pen would dissolve in the solvent *[1 mark]*.
2   Step 1 is to dissolve the ammonium sulfate *[1 mark]*.
Step 2 is to remove the sharp sand *[1 mark]*.
Step 3 is to evaporate the water and produce dry crystals of ammonium sulfate *[1 mark]*.
3.1   How to grade your answer:
Level 0:   There is no relevant information. *[No marks]*
Level 1:   Some steps of the method and some relevant safety precautions are included, but the answer is not in a logical order. Following the method given would not produce the results desired. *[1 to 2 marks]*
Level 2:   Most of the method and safety precautions are described, but the steps may not be in a completely logical order and some detail may be missing. *[3 to 4 marks]*
Level 3:   The method and safety precautions are described in full, with the steps given in a logical order. Following the method would produce the results desired. *[5 to 6 marks]*

Here are some points your answer may include:
Method
Put the solution of propanone and water in the flask and heat it.
Monitor the temperature using the thermometer.
Heat the mixture until the temperature reaches around 56 °C.
56 °C is the boiling point of propanone, so at this temperature the propanone in the solution will start to evaporate and turn into a gas.

The boiling point of water is higher (100 °C), so at 56 °C the water will remain as a liquid.

The propanone gas travels from the flask into the condenser. In the condenser, the gas is cooled and condenses/turns back into a liquid.

The liquid propanone runs out of the condenser and is collected in the beaker.

The water is left behind in the flask.

<u>Safety precautions</u>

Take care with Bunsen burner/heat source.

Alternatively, use a heat source that doesn't involve a flame, such as a water bath or electric heater.

Avoid touching hot glassware.

Make sure no solution gets on the outside of the flask where it could catch fire.

Make sure the flask is tightly sealed so no propanone gas escapes.

Work in a fume cupboard.

Wear a lab coat, gloves and goggles.

3.2 fractional distillation *[1 mark]*

The boiling points of water and methanoic acid are too close together to allow them to be separated by simple distillation, so fractional distillation must be used.

Page 178
Electronic Structure
Q1 The inner shells are always filled first *[1 mark]*.
Q2 2,8,8 or 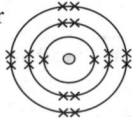 *[1 mark]*

Page 181
Warm-Up Questions
1) In the 'plum pudding' model, the atom is a ball of positive charge, with negatively charged electrons scattered in this ball.
2) a) 2
   b) 8
3) Mendeleev left gaps in his table so that elements with similar properties stayed in the same groups.

Exam Questions
1.1 by atomic weight *[1 mark]*
1.2 by atomic number *[1 mark]*
1.3 They have the same number of electrons in their outer shell *[1 mark]*.
2.1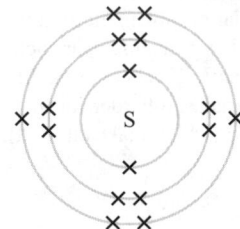
[1 mark]

It's fine if you used dots instead of crosses to show the electrons in your diagram.

2.2 Group 6 *[1 mark]*
2.3 Any one from: oxygen / selenium / tellurium / polonium / livermorium *[1 mark]*
2.4 2 *[1 mark]*

Page 184
Group 1 Elements
Q1 lithium + water → lithium hydroxide + hydrogen *[1 mark]*

Page 186
Group 7 Elements
Q1 $Br_2 + 2NaI \rightarrow 2NaBr + I_2$ *[1 mark]*

Pages 188-189
Warm-Up Questions
1) positive ions
2) Any two from: e.g. strong / can be bent or hammered into different shapes (malleable) / conduct heat / conduct electricity / high melting point / high boiling point.
3) lithium chloride / LiCl
4) negative
5) Group 0

Exam Questions
1.1 fluorine *[1 mark]*.
1.2 iodine *[1 mark]*
2.1 alkali metals *[1 mark]*
2.2 To the right of the line *[1 mark]*. Since it does not conduct electricity, it must be a non-metal *[1 mark]*
3.1 They have a single outer electron which is easily lost so they are very reactive *[1 mark]*.
3.2 potassium hydroxide *[1 mark]* and hydrogen *[1 mark]*
3.3 The outer electron is further from the nucleus and so less attracted to the nucleus *[1 mark]*.
4.1 ionic bonds *[1 mark]*
4.2 displacement reaction *[1 mark]*
4.3 Chlorine is more reactive than iodine *[1 mark]*, because more reactive elements displace less reactive elements in displacement reactions / chlorine displaces iodine from the potassium iodide *[1 mark]*.
4.4 Group 0 elements don't need to lose or gain electrons to have a full outer shell. / Group 0 elements already have a full outer shell *[1 mark]*.

Topic C2 — Bonding, Structure and Properties of Matter

Page 190
Ions
Q1 Noble gas electronic structures have a full shell of outer electrons *[1 mark]*, which is a very stable structure *[1 mark]*.

Page 192
Ionic Bonding
Q1 The transfer of one electron from another atom to another *[1 mark]*.
Q2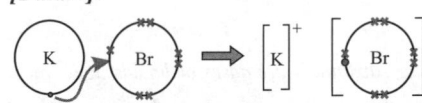

[1 mark for arrow showing electron transferred from potassium to bromine, 1 mark for correct outer shell electron configurations (with or without inner shells), 1 mark for correct charges]

Page 194
Ionic Compounds
Q1 a) It will have a high melting point *[1 mark]* because a lot of energy is needed to break the strong attraction between the ions/the strong ionic bonds *[1 mark]*.
b) When melted, the ions are free to move, so they can carry electric charge *[1 mark]*.
c) The compound contains magnesium and sulfide ions. Magnesium is in Group 2 so forms 2+ ions *[1 mark]*, and sulfur is in Group 6 so forms 2– ions *[1 mark]*. The charges balance with one of each ion, so the empirical formula is MgS *[1 mark]*.

Page 195
Warm-Up Questions
1) A charged particle
2) 1–
3) high
4) True
5) Al(OH)$_3$

Exam Questions
1.1

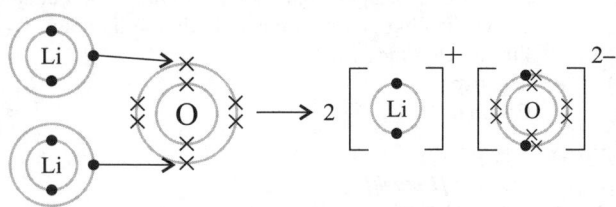

[1 mark for correct electron arrangement and charge on lithium ion, 1 mark for correct electron arrangement and charge on oxygen ion]

1.2 LiCl *[1 mark]*
2.1 Sodium ions have a 1+ charge and chloride ions have a 1– charge *[1 mark]* so one sodium ion is needed to balance the charge on one chloride ion *[1 mark]* (so the empirical formula is NaCl).
2.2 Sodium chloride contains many strong ionic bonds *[1 mark]* which require a lot of energy to break *[1 mark]*.
2.3 In solid sodium chloride, the ions are held in place, so cannot conduct electricity *[1 mark]*. In molten sodium chloride, the ions are free to move, so can conduct electricity *[1 mark]*.
2.4 Dissolve it in water/solution *[1 mark]*.

Page 198
Covalent Bonding
Q1

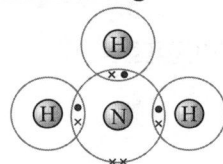

[1 mark for 3 shared pairs of electrons, 1 mark for correct number of electrons in outer shell of each atom (with or without inner shells on nitrogen)]

Page 199
Warm-Up Questions
1) A chemical bond made by the sharing of a pair of electrons between two atoms.
2) 1
3) False

Most simple molecular substances are gases or liquids at room temperature.

4) covalent bonds

Exam Questions
1.1 8 *[1 mark]*
1.2 3 *[1 mark]*
1.3 0 *[1 mark]*

All the bonds in the molecule are single covalent bonds — you get double bonds in molecules like oxygen.

2

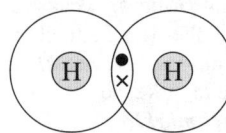

[1 mark for showing 1 pair of shared electrons between H atoms]

3.1 1 *[1 mark]*
3.2 The forces of attraction between molecules/intermolecular forces are very weak, so are easily overcome *[1 mark]*.
3.3 Hydrogen chloride isn't charged / doesn't contain any ions or delocalised electrons to carry a charge *[1 mark]*.

Page 200
Polymers
Q1 (C$_2$H$_3$Cl)$_n$ *[1 mark]*

Page 201
Giant Covalent Structures
Q1 E.g. graphite contains layers of carbon atoms *[1 mark]* arranged in hexagons *[1 mark]*. Each carbon atom forms three covalent bonds *[1 mark]*. There are no covalent bonds between the sheets *[1 mark]*.

Page 204
Warm-Up Questions
1) True
2) solid
3) E.g. diamond is much harder than graphite. Diamond doesn't conduct electricity whereas graphite does.
4) The layers of atoms/ions in metals are able to slide over each other.

Exam Questions
1

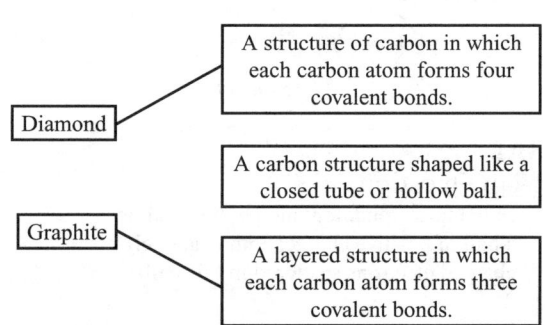

[1 mark for each correctly drawn line]

2.1 In a giant covalent structure, all of the atoms are bonded to each other with strong covalent bonds *[1 mark]*. It takes lots of energy to break these bonds and melt the solid *[1 mark]*.
2.2 E.g. diamond / graphite / silicon dioxide *[1 mark]*
3.1 A metal mixed with another metal (or element) *[1 mark]*
3.2 The delocalised electrons are free to move *[1 mark]*, so can carry thermal/heat energy *[1 mark]*.
3.3 Iron is a metal, so has a regular arrangement of atoms *[1 mark]*, which means that they can slide over each other *[1 mark]*, making iron soft. Steel contains different sized atoms *[1 mark]* which makes it harder for them to slide over each other *[1 mark]* (which makes it hard).

Page 207
States of Matter
Q1 a) solid *[1 mark]*
b) liquid *[1 mark]*
c) liquid *[1 mark]*
d) gas *[1 mark]*

Page 208
Warm-Up Questions
1) Solid
2) True
3) The forces of attraction increase and lots of bonds form between the gas particles, so the gas becomes a liquid.
4) Na$^+_{(aq)}$

Exam Questions

1. In solids, there are **strong** *[1 mark]* forces of attraction between particles, which hold them in fixed positions in a **regular** *[1 mark]* arrangement. The particles don't **move** *[1 mark]* from their positions, so solids keep their shape. The **hotter** *[1 mark]* the solid becomes, the more the particles in the solid vibrate.
2.1 C *[1 mark]*
2.2 Boiling/evaporation *[1 mark]*
2.3 The particles are free to move about / only have weak forces of attraction between them *[1 mark]*, so they spread out to fill the container *[1 mark]*.

Topic C3 — Quantitative Chemistry

Page 209
Relative Formula Mass

Q1 a) A_r of C = 12 and A_r of O = 16
M_r of CO_2 = 12 + (2 × 16) = **44** *[1 mark]*
b) A_r of Li = 7, A_r of O = 16 and A_r of H = 1
So M_r of LiOH = 7 + 16 + 1 = **24** *[1 mark]*
c) A_r of H = 1, A_r of S = 32 and A_r of O = 16
M_r of H_2SO_4 = (2 × 1) + 32 + (4 × 16) = **98** *[1 mark]*

Q2 A_r of K = 39, A_r of O = 16 and A_r of H = 1
M_r of KOH = 39 + 16 + 1 = 56 *[1 mark]*
$\frac{39}{56} \times 100$ = **70%** *[1 mark]*

Page 210
Conservation of Mass

Q1 Total mass on the left hand side
= $M_r(H_2SO_4)$ + 2 × $M_r(NaOH)$
M_r of H_2SO_4 = (2 × 1) + 32 + (4 × 16) = 98
2 × M_r of NaOH = 2 × (23 + 16 + 1) = 80
So total mass on the left hand side
= 98 + 80 = **178** *[2 marks for 178, 1 mark for either 98 or 80]*
Total mass on right hand side
= $M_r(Na_2SO_4)$ + 2 × $M_r(H_2O)$
M_r of Na_2SO_4 = (2 × 23) + 32 + (4 × 16) = 142
2 × M_r of H_2O = 2 × [(2 × 1) + 16] = 36
142 + 36 = **178** *[2 marks for 178, 1 mark for either 142 or 36]*
The total M_r on the left-hand side is equal to the total M_r on the right-hand side, so mass is conserved *[1 mark]*.

Page 211
Conservation of Mass

Q1 One of the reactants is a gas *[1 mark]* and the products are solid, liquid or aqueous *[1 mark]*.
Q2 The mass will decrease *[1 mark]*.

Page 212
Concentrations of Solutions

Q1 Volume = 15 ÷ 1000
= 0.015 dm^3 *[1 mark]*
Concentration = mass ÷ volume
= 0.6 ÷ 0.015
= **40 g/dm^3** *[1 mark]*

Pages 213-214
Warm-Up Questions

1) The sum of the relative atomic masses of all the atoms in a compound.
2) (2 × 1) + 16 = 18
3) True
4) One of the reactants is a gas that is found in air, and all the products are solids, liquids or in solution.
5) g/dm^3

Exam Questions

1. 111 *[1 mark]*
M_r of $CaCl_2$ = 40 + (35.5 × 2) = 111
2. magnesium fluoride, MgF_2 *[1 mark]*
3. 48.6 + 26.4 = **75.0 kg** *[1 mark]*
4. M_r of $Zn(CN)_2$ = 65 + ((12 + 14) × 2) = **117** *[1 mark]*
5.1 M_r of $Ca(OH)_2$ = 40 + ((16 + 1) × 2) = **74** *[1 mark]*
5.2 Total mass of reactants = 18.5 g + 31.5 g = 50 g
So mass of calcium nitrate = 50 g − 9 g = **41 g** *[1 mark]*
6.1 M_r of $MgCO_3$ = 24 + 12 + (16 × 3) = **84** *[1 mark]*
6.2 The mass will decrease *[1 mark]*
7.1 1500 ÷ 1000 = **1.5 dm^3** *[1 mark]*
1000 cm^3 = 1 dm^3
7.2 mass = concentration × volume = 12 × 1.5 = **18 g**
[2 marks for correct answer, otherwise 1 mark for correct working.]
8.1 concentration = mass ÷ volume = 40 ÷ 0.25 = **160 g/dm^3**
[2 marks for correct answer, otherwise 1 mark for correct working.]
8.2 M_r of KOH = 39 + 16 + 1 = 56
Percentage mass of potassium in KOH
$= \frac{A_r \text{ of potassium} \times \text{number of atoms of potassium}}{M_r \text{ of potassium hydroxide}} \times 100$
$= \frac{39 \times 1}{56} \times 100 = 69.642...\% = \mathbf{69.6\%}$
[3 marks for correct answer rounded to 3 s.f., otherwise 2 marks for correct answer not rounded to 3 s.f., or 1 mark for correct M_r of KOH.]

Topic C4 — Chemical Changes

Page 216
Acids, Bases and Their Reactions

Q1 calcium carbonate + hydrochloric acid
→ calcium chloride + carbon dioxide + water *[1 mark for calcium chloride, 1 mark for carbon dioxide and water]*
$CaCO_3 + 2HCl \rightarrow CaCl_2 + CO_2 + H_2O$
[1 mark for the correctly balanced equation]

Page 218
Warm-Up Questions

1) 0-14
2) Neutral
3) False
Universal indicator turns green if it is added to a substance with a pH of 7.
4) $CaCO_3$
5) Copper sulfate and water

Exam Questions

1.1 OH^- *[1 mark]*
1.2 neutralisation *[1 mark]*
2.1 Alkalis *[1 mark]*
2.2 $2HNO_3 + Mg(OH)_2 \rightarrow \mathbf{Mg(NO_3)_2} + \mathbf{2H_2O}$
[1 mark for formulas of both products correct, 1 mark for putting a 2 in front of H_2O to balance the equation]
3.1 Calcium chloride *[1 mark]*
3.2 The salt will be contaminated by the indicator *[1 mark]*.
3.3 How to grade your answer:
Level 0: There is no relevant information. *[No marks]*
Level 1: Simple statements are made relating to relevant scientific techniques, however points are not linked, and the answer lacks detail.
[1 to 2 marks]

Level 2: A coherent answer is given that follows a logical order and shows a good understanding of the relevant scientific techniques, including filtration and crystallisation. However, some detail may be missing. *[3 to 4 marks]*

Level 3: A detailed and coherent description of a relevant scientific technique is given. The answer shows a clear understanding of the techniques of filtration and crystallisation. *[5 to 6 marks]*

Here are some points your answer may include:
Gently warm the hydrochloric acid using a Bunsen burner/water bath, then turn off the Bunsen burner/remove the acid from the water bath.
Add the calcium carbonate to the warmed hydrochloric acid until no more reacts / there is calcium carbonate at the bottom of the flask.
Use filter paper and a funnel to filter the solution, in order to remove any unreacted calcium carbonate.
Crystallise the solution by gently heating it using a water bath/electric heater to evaporate off some of the water.
Then leave the solution to cool.
Crystals of calcium chloride/the soluble salt should form.
The crystals can be filtered out of the solution and then dried.

Page 220
Metals and Their Reactivity
Q1 $2Na_{(s)} + 2H_2O_{(l)} \rightarrow 2NaOH_{(aq)} + H_{2(g)}$
[1 mark for correct reactants and products, 1 mark for correctly balancing, 1 mark for state symbols]

Page 221
Extracting Metals
Q1 $2PbO + C \rightarrow 2Pb + CO_2$
[1 mark for the correct products, 1 mark for the correctly balanced equation]
Q2 Carbon is less reactive than calcium and therefore will not reduce calcium oxide / Calcium is more reactive than carbon, so calcium oxide won't be reduced by carbon *[1 mark]*.

Page 222
Warm-Up Questions
1) True
2) magnesium chloride
3) A metal hydroxide and hydrogen

Exam Questions
1 Magnesium *[1 mark]*
2.1 potassium/sodium/calcium *[1 mark]*
2.2 copper *[1 mark]*
Metals below hydrogen in the reactivity series don't react with acids.
2.3 The reaction with iron would be less violent than with zinc *[1 mark]* because iron is less reactive/lower in the reactivity series than zinc *[1 mark]*.
2.4 potassium/sodium/calcium/magnesium *[1 mark]*
Carbon can only be used to extract metals that are below it in the reactivity series. Metals above carbon must be extracted using electrolysis.
3.1 The copper was displaced from its salt *[1 mark]*.
3.2 Zinc is more reactive/higher in the reactivity series than copper *[1 mark]*.
3.3 e.g. iron *[1 mark]*
Metal X must be more reactive than copper (as it displaces copper from copper sulfate), but less reactive than zinc (which it doesn't displace from zinc sulfate).

Page 224
Electrolysis
Q1 a) chlorine gas/Cl_2 *[1 mark]*
b) calcium atoms/Ca *[1 mark]*

Page 226
Electrolysis of Aqueous Solutions
Q1 a) bromine gas/Br_2 *[1 mark]*
b) copper atoms/Cu *[1 mark]*

Page 227
Warm-Up Questions
1) The cathode
2) True
3) Hydrogen gas

Exam Questions
1 When molten sodium chloride is electrolysed, the **negative** *[1 mark]* chloride ions move towards the anode and form chlorine gas. At the cathode, the **sodium** *[1 mark]* ions gain electrons.
The sodium chloride is molten, so there are no hydrogen ions present to compete with the sodium ions at the cathode.
2.1 Pb^{2+} *[1 mark]*
2.2 Br^- *[1 mark]*
2.3 Molten lead *[1 mark]*
2.4 Br_2 *[1 mark]*

Topic C5 — Energy Changes

Page 230
Reaction Profiles
Q1

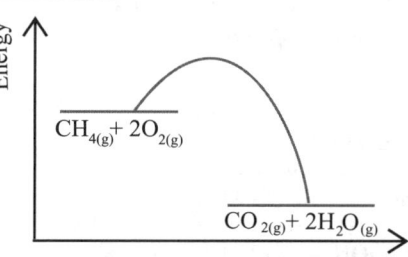

[1 mark for correct axes, 1 mark for correct energy levels of reactants and products, 1 mark for correct shape of curve linking the reactants to the products]

Page 231
Warm-Up Questions
1) True
2) E.g. in hand warmers/self-heating cans/burning fuels.
3) It will decrease.
4) endothermic
5) The overall energy change/the energy given out in the reaction.

Exam Questions
1 thermal decomposition *[1 mark]*
2.1 E.g. to ensure that they are the same temperature before beginning the reaction/to know their initial temperature *[1 mark]*.
2.2 E.g. to insulate the cup/reduce the amount of heat lost to the surroundings *[1 mark]*.
2.3 The student should repeat the experiment under the same conditions *[1 mark]*. The measurement is repeatable if the student obtains similar results in both repeats *[1 mark]*.
3 C *[1 mark]*

Topic C6 — The Rate and Extent of Chemical Change

Page 236
Warm-Up Questions
1) E.g. the rusting of iron is a reaction that happens very slowly. Burning/combustion is a very fast reaction.
2) They must collide with enough energy / the activation energy.
3) Breaking a reactant into smaller pieces increases its surface area to volume ratio. This means there is a greater surface area available for collisions to occur on, so the rate of reaction is increased.
4) True

Exam Questions
1.1 The rate would increase *[1 mark]*.
1.2 The rate would increase *[1 mark]*.
2.1 A *[1 mark]*
2.2 C *[1 mark]*
2.3 Increasing the temperature means that the reactant particles move faster *[1 mark]*, so collisions between reactant particles are more frequent *[1 mark]*. The particles also have more energy *[1 mark]*, so more collisions have enough energy to make the reaction happen *[1 mark]*.

3.1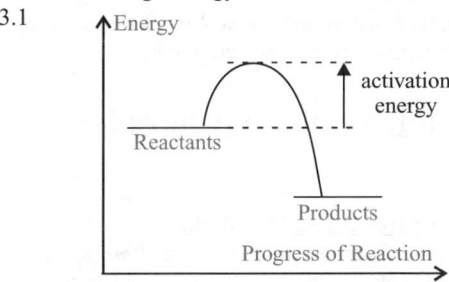

[1 mark]

3.2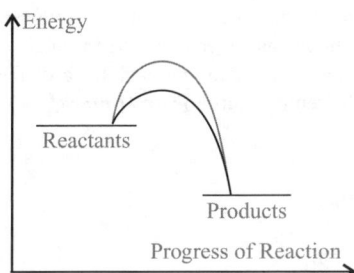

[1 mark for curve drawn with a lower maximum]

Page 240
Graphs of Reaction Rate Experiments
Q1 E.g.

[1 mark for correctly marking on all 6 points, 1 mark for choosing a sensible scale for the axes, 1 mark for drawing a line of best fit]

Page 241
Working Out Reaction Rates
Q1 Mean rate = amount of reactant used ÷ time
= 6.0 g ÷ 200 s *[1 mark]*
= **0.03 g/s** *[1 mark]*

Pages 244-245
Warm-Up Questions
1) e.g. a mass balance / gas syringe / stopwatch / timer
2) Reaction A
3) False
A steep tangent on a rate graph means that the rate of reaction is fast at that point in time.
4) The backwards reaction

Exam Questions
1.1 The reaction is reversible *[1 mark]*.
1.2 As A and B react, their concentrations will **fall** *[1 mark]* and the rate of the forwards reaction will decrease. As C and D are made, the **backwards** *[1 mark]* reaction speeds up. After a while, the amounts of products and reactants will remain **constant** *[1 mark]*. The system is at equilibrium.
2.1 stopwatch/stopclock/timer *[1 mark]*
2.2 gas syringe *[1 mark]*
2.3 E.g. mass balances are very accurate / give very accurate results *[1 mark]*.
2.4 E.g. potentially harmful gases are released into the room *[1 mark]*.
3.1 17 cm^3 *[1 mark]*
3.2 Manganese(IV) oxide was the most effective catalyst *[1 mark]* because it led to the greatest volume of oxygen being produced over the time period measured/increased the rate of reaction by the greatest amount *[1 mark]*.
3.3 Draw tangents to each of the curves at 2 minutes *[1 mark]*, and compare how steep they are *[1 mark]*. A steeper tangent means a faster rate (and vice versa) *[1 mark]*.
4 How to grade your answer:
Level 0: There is no relevant information. *[No marks]*
Level 1: Simple statements are made which demonstrate some understanding of how colour change or the formation of a precipitate can be used to investigate the rate of reaction. The response lacks detail or logical structure. *[1 to 2 marks]*
Level 2: The method is described in relevant detail and demonstrates a broad understanding of how colour change or the formation of a precipitate can be used to investigate the rate of reaction. The answer follows a logical structure. *[3 to 4 marks]*

Here are some points your answer may include:
Add one of the reactants to a conical flask.
Place the conical flask over a black mark or cross that can be seen through the solution.
Add the second reactant to the conical flask.
The solution will turn cloudy.
Use a stopwatch to measure the time taken for the mark to disappear.
The faster the mark disappears, the faster the rate of reaction.
Repeat the experiment using different concentrations of reactant Y.
All other variables must be kept constant.

5.1 mean rate of reaction = amount of product formed ÷ time
mean rate of reaction = 46 ÷ 250 = **0.18 cm^3/s**
[3 marks for a rate between 0.17 cm^3/s and 0.19 cm^3/s, otherwise 1 mark for correct working and 1 mark for correct units]
Remember that the reaction is complete at the point where the line goes flat, so the total reaction time is 250 seconds.

5.2 mean rate of reaction = amount of product formed ÷ time
mean rate of reaction = (44 − 29) ÷ (150 − 40) = 15 ÷ 110
mean rate of reaction = 0.1363... = **0.14 cm^3/s**
[3 marks for correct answer, otherwise 1 mark for calculating change in x (110) and 1 mark for calculating change in y (15)]

Topic C7 — Organic Chemistry

Page 247
Hydrocarbons
Q1 $C_2H_6 + 3½O_2 \rightarrow 2CO_2 + 3H_2O$ or
$2C_2H_6 + 7O_2 \rightarrow 4CO_2 + 6H_2O$
[1 mark for correct reactants and products, 1 mark for correctly balancing]

Page 249
Fractional Distillation
Q1   The diagram shows that hydrocarbons with shorter chains drain out further up the column *[1 mark]*, so petrol has a shorter chain than diesel *[1 mark]*.

Page 251
Cracking
Q1 $C_5H_{12} \rightarrow C_2H_4 + C_3H_8$ *[1 mark]*

Pages 252-253
Warm-Up Questions
1) methane, ethane, propane, butane
2) (single) covalent bonds
3) They release lots of energy when burnt.
4) long-chain/large hydrocarbons

Exam Questions
1   ethane *[1 mark]*
2   Flammability decreases and viscosity increases. *[1 mark]*
3.1   butane *[1 mark]*
3.2 C_4H_{10} *[1 mark]*
3.3   Any two from: e.g. polymers / solvents / lubricants / detergents *[2 marks — 1 mark for each correct answer]*
4   Oxygen reacts with a fuel and energy is released. *[1 mark]*
5.1 $C_9H_{20} \rightarrow \mathbf{C_7H_{16}} + \mathbf{C_2H_4}$ *[1 mark]*
5.2   When bromine water is shaken with heptane, it remains orange in colour *[1 mark]*. When bromine water is shaken with ethene, it turns from orange to colourless *[1 mark]*.
Ethene is an alkene, so it turns bromine water colourless. Heptane is an alkane, so it doesn't react with bromine water.
6.1   A *[1 mark]*
6.2   Hydrocarbons with longer chains have higher boiling points *[1 mark]*. The column is hot at the bottom and gets cooler as you go up *[1 mark]*. The longer hydrocarbons condense back to liquids and drain out of the column at the higher temperatures lower down *[1 mark]*.

Topic C8 — Chemical Analysis

Page 257
Interpreting Chromatograms
Q1 $R_f = \frac{6.3}{8.4} = \mathbf{0.75}$ *[1 mark]*

Pages 260-261
Warm-Up Questions
1) Impurities in a substance will increase the boiling point of the substance and may also cause the substance to boil across a range of temperatures.
2) e.g. paint/cleaning products/fuels/medicines/cosmetics/fertilisers/metal alloys/food/drink
3) Chromatography separates out the substances present in a mixture.
4) False

An R_f value represents the ratio between the distance travelled by a dissolved substance and the distance travelled by the solvent.

Exam Questions
1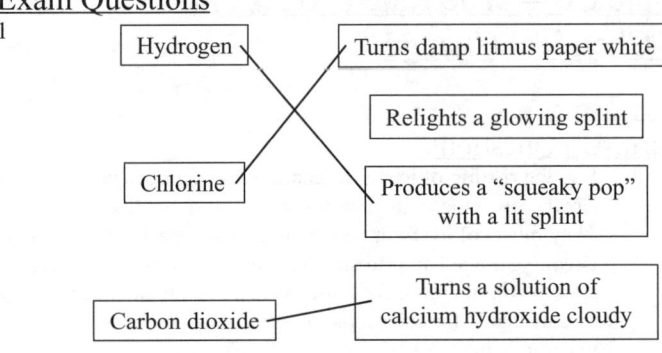

[3 marks — 1 mark for each correctly drawn line.]
2   Substance B only *[1 mark]*
Impurities lower the melting point of a substance, and may cause the sample to melt across a range of temperatures. Substance B is the only substance in the table that has an experimental melting point that is exactly the same as the data book value, so it is the only pure substance.
3   The brown group *[1 mark]*
4.1   B *[1 mark]*
4.2   C *[1 mark]*
4.3 R_f = distance moved by substance ÷ distance moved by solvent
$R_f = 9.0 \div 12.0 = \mathbf{0.75}$
[2 marks for correct answer, otherwise 1 mark for correctly substituting the values into the formula for R_f.]

Topic C9 — Chemistry of the Atmosphere

Page 264
Climate Change and Greenhouse Gases
Q1   The sun gives out short wavelength radiation *[1 mark]* which is reflected back by the Earth as long wavelength/thermal radiation *[1 mark]*. The thermal radiation is absorbed by greenhouse gases in the atmosphere *[1 mark]*. Greenhouse gases give out the thermal radiation in all directions including back towards the Earth, causing the temperature to rise *[1 mark]*.

Pages 268-269
Warm-Up Questions
1) Volcanoes
2) e.g. limestone / coal
3) e.g. coal / crude oil / natural gas
4) True
5) e.g. acid rain / respiratory problems

Exam Questions
1.1   When green plants first evolved, the Earth's early atmosphere was mostly **carbon dioxide** *[1 mark]* gas. These plants produced **oxygen** *[1 mark]* by the process of photosynthesis. Some of the **carbon** *[1 mark]* from dead plants eventually became 'locked up' in fossil fuels.
1.2 $6\mathbf{CO_2} + \mathbf{6}H_2O \rightarrow C_6H_{12}O_6 + 6\mathbf{O_2}$
[1 mark for both CO_2 and O_2 formulas correct, 1 mark for correct balancing of H_2O]
1.3   C *[1 mark]*
2.1   Natural gas *[1 mark]*
Natural gas, crude oil and coal are the three fossil fuels you need to know about.
2.2   combustion / oxidation *[1 mark]*
2.3   e.g. particulates / soot *[1 mark]*
2.4   E.g. if breathed in, they can cause respiratory problems. / They reflect sunlight back into space, causing global dimming *[2 marks — 1 mark for each correct description]*.
2.5   It doesn't have any colour or smell *[1 mark]*.
3.1   Methane *[1 mark]*

3.2 Greenhouse gases give out the thermal radiation in all directions *[1 mark]*, so some of the radiation is returned to earth *[1 mark]* where it warms the surface of the planet *[1 mark]*.

3.3 Any two from: e.g. deforestation / burning fossil fuels / agriculture / using landfill sites *[2 marks — 1 mark for each correct answer]*

3.4 E.g. rising sea levels / increased flooding / changes in rainfall / more frequent/severe storms / changes in temperature *[1 mark]*.

3.5 E.g. tax companies and individuals based on the amount of greenhouse gases they emit. / Put a limit on the emissions of all greenhouse gases a company can make. / Invest in research and technology to reduce emissions or capture greenhouse gases before they are released *[1 mark]*

4 How to grade your answer:
- Level 0: There is no relevant information. *[No marks]*
- Level 1: There is a brief description of how carbon dioxide was removed from the atmosphere by the oceans, plants and algae, and/or how this led to the formation of coal. However some details are missing or incorrect and the answer lacks structure. *[1 to 2 marks]*
- Level 2: There is a detailed description of how the oceans and plants and algae removed carbon dioxide from the atmosphere, and how this led to the formation of coal. The answer is clearly written and logically structured. *[3 to 4 marks]*

Here are some points your answer may include:
Carbon dioxide was removed from the atmosphere by dissolving in the oceans.
The dissolved carbon dioxide formed carbonates/sediments/carbon-containing compounds in the ocean.
Green plants and algae took in carbon dioxide during photosynthesis.
When marine plants and algae died, they were buried on the seabed.
Over a long time, the dead plants were squashed down, forming coal.

Topic C10 — Using Resources

Page 275
Warm-Up Questions
1) e.g. wool / rubber
2) True
3) Any two from: e.g. saves energy needed to make new metals / saves limited supplies of metals from the earth / cuts down on the amount of waste going to landfill.
4) To assess the effect a product would have on the environment over the course of its life.

Exam Questions
1.1 A natural resource that can be remade at least as fast as we use it *[1 mark]*.

1.2 E.g. reuse the cans / recycle the aluminium / use fewer cans *[1 mark]*.

2.1 How to grade your answer:
- Level 0: There is no relevant information. *[No marks]*
- Level 1: There is a brief discussion of some environmental impacts of each type of bag. The points made are basic, don't link together and don't cover all of the information given in Table 1. If a conclusion is present it may not link to the points made. *[1 to 2 marks]*
- Level 2: There is a logical discussion of possible environmental impacts of each type of bag, but there is limited detail. There is a good coverage of the information given in Table 1 and the answer has some structure. A clear conclusion is given. *[3 to 4 marks]*
- Level 3: There is a clear and detailed discussion of the possible environmental impacts of each type of bag. The answer has a logical structure and makes good use of the information given in Table 1. A conclusion is present that fits with the points given in the answer. *[5 to 6 marks]*

Here are some points your answer may include:
The raw materials used to make plastic bags come from crude oil, which is a non-renewable resource.
Obtaining crude oil from the ground and processing it uses a lot of energy and results in air pollution due to the release of greenhouse gases.
The raw materials used to make paper bags come from trees. Trees are a renewable resource, although they take up land that could be used for other uses e.g. growing crops.
Cutting down trees and processing the raw timber requires energy.
This energy is often generated by burning fossil fuels, which releases harmful greenhouse gases into the atmosphere.
The manufacture of plastic bags involves many industrial processes (e.g. fractional distillation, cracking, polymerisation), which all require large amounts of energy.
As this energy often comes from burning fossil fuels, the manufacture causes the release of pollution (e.g. greenhouse gases) into the atmosphere.
The waste from the manufacturing of plastic bags can be used to make other products, which reduces the amount of waste being sent to landfill.
Manufacturing paper bags also requires lots of energy. This could come from burning fossil fuels and so would generate pollution, such as greenhouses gases, which have a negative impact on the environment.
The manufacture of paper bags also creates lots of waste, which has to be disposed of and may cause pollution (e.g. if sent to landfill).
Plastic bags are reusable, which could reduce the amount of waste sent to landfill.
Paper bags are usually not reusable, which could increase the amount of waste sent to landfill.
Plastic bags are not biodegradable so will stay in the environment for a long time if disposed of in landfill sites.
Plastic bags can release toxic substances into the ground and pollute the land.
Plastic bags that end up in the oceans can harm marine life and pollute the water.
Plastic bags can be recycled, which would reduce the need to extract raw materials to make new bags.
Paper bags are biodegradable so will break down more easily than plastic bags if sent to landfill sites, reducing the impact on the environment.
Paper bags are also non-toxic, so won't releases poisonous/toxic substances into the environment after being disposed of.
Paper bags can be recycled, which would reduce the need to gather raw materials to make new bags.

2.2 E.g. some things can't be measured, so depend upon a person's own judgement *[1 mark]*. Companies might use selective LCAs to make their product look more environmentally-friendly for advertising purposes *[1 mark]*.

Page 276
Potable Water and Water Treatment
Q1   E.g. the water is filtered using a wire mesh, and then using filter beds *[1 mark]*. The filtered water is then sterilised using chlorine / ozone / ultraviolet light *[1 mark]*.

Page 279
Potable Water and Water Treatment
Q1   Screening — sewage is screened to remove any large bits of material and grit *[1 mark]*.
Sedimentation — heavier solids sink to the bottom to form sludge, while the lighter effluent floats on the top *[1 mark]*.

Pages 280-281
Warm-Up Questions
1)   E.g. the levels of dissolved salts in the water aren't too high. / The water has a pH of between 6.5 and 8.5. / There aren't any bacteria or other microbes in the water.
2)   True
3)   effluent and sludge
4)   True

Exam Questions
1.1   Screening *[1 mark]*
1.2   Bacteria *[1 mark]*
2.1   Water that is safe to drink *[1 mark]*.
2.2   To remove any solids from the water *[1 mark]*.
2.3   e.g. chlorine / ozone / ultraviolet light *[1 mark]*.
3.1

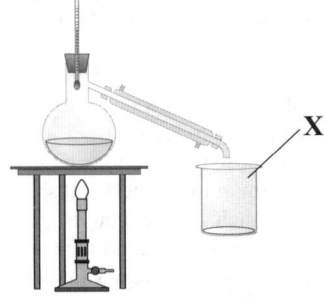

[1 mark]

3.2   When the water evaporates, it is separated from any soluble impurities, which are left in the flask *[1 mark]*. Condensation turns the steam back into liquid water, so it can be collected easily *[1 mark]*.
3.3   E.g. neutralise it / add acid until the pH is 7 *[1 mark]*.
3.4   All the seawater will have evaporated from the round bottomed flask *[1 mark]*.
3.5   The crystals are salts that were in the seawater before it was distilled *[1 mark]*.
The salts won't evaporate with the water, so they get left in the flask.
4.1   E.g. to remove any pollutants *[1 mark]* so that the water doesn't cause health problems / damage the environment *[1 mark]*.
Pollutants could be organic matter, bacteria and viruses etc.
4.2   Effluent is treated by aerobic digestion *[1 mark]* and sludge is treated by anaerobic digestion *[1 mark]*. In both processes, bacteria is used to break down the effluent/sludge *[1 mark]*.
4.3   E.g. it might contain harmful chemicals *[1 mark]*.
Industrial waste always needs further treatment before it is safe to return to the environment.
5.1   Seawater is passed through a membrane, which lets water molecules pass through *[1 mark]*, but traps the salts *[1 mark]*.
5.2   It requires a lot of energy *[1 mark]*.

Topic P1 – Energy

Page 285
Mechanical Energy Transfer
Q1   Energy is transferred mechanically *[1 mark]* from the kinetic energy store of the wind *[1 mark]* to the kinetic energy store of the windmill *[1 mark]*.

Page 286
Kinetic and Potential Energy Stores
Q1   The change in height is 5.0 m. So the energy transferred from the gravitational potential energy store is:
$E_p = mgh = 2.0 \times 9.8 \times 5.0 = 98$ J *[1 mark]*
This is transferred to the kinetic energy store of the object, so $E_k = 98$ J *[1 mark]*
$E_k = \frac{1}{2}mv^2$
so $v^2 = 2E_k \div m$ *[1 mark]*
      $= (2 \times 98) \div 2.0$ *[1 mark]*
      $= 98$ m²/s²
$v = \sqrt{98} = 9.899...$
      = **9.9 m/s (to 2 s.f.)** *[1 mark]*

Page 288
Specific Heat Capacity
Q1 $\Delta E = mc\Delta\theta$
so $\Delta\theta = \Delta E \div (m \times c)$ *[1 mark]*
      $= 50\,000 \div (5 \times 4200)$
      $= 2.380...$ °C *[1 mark]*
So the new temperature
      $= 5 + 2.380... = 7.380...$
      = **7 °C (to 1 s.f.)** *[1 mark]*

Page 290
Investigating Specific Heat Capacity
Q1 $E = P \times t = 80 \times 200$ *[1 mark]*
      $= 16\,000$ J *[1 mark]*
$\Delta E = mc\Delta\theta$, so
$c = \Delta E \div (m \times \Delta\theta)$ *[1 mark]*
      $= 16\,000 \div (2 \times 20)$ *[1 mark]*
      = **400 J/kg°C** *[1 mark]*

Pages 291-292
Warm-Up Questions
1)   Any two from: mechanically (by a force doing work) / electrically (work done by when a current flows) / by heating / by radiation.
2)   Energy is transferred (mechanically) from the chemical energy store of the person's arm to the kinetic energy stores of the arm and the ball.
3)   True
4)   A lorry travelling at 60 miles per hour.
5)   Energy can be transferred usefully, stored or dissipated, but can never be created or destroyed.

Exam Questions
1   The thermal energy store of the metal block *[1 mark]*
2

A skydiver falling from an aeroplane.	→ gravitational potential
A substance undergoing a nuclear reaction.	→ nuclear
A piece of burning coal.	→ chemical
(spring — implied)	→ elastic potential

[3 marks for all correct, otherwise 2 marks for 2 correct or 1 mark for 1 correct]

3.1   The elastic potential energy store of the spring *[1 mark]*
3.2   kinetic energy = ½ × mass × speed² / $E_k = \frac{1}{2}mv^2$ *[1 mark]*
Allow velocity instead of speed in the relationship.

3.3 $E_k = \frac{1}{2} \times 0.20 \times 0.9^2$ *[1 mark]*
= **0.081 J** *[1 mark]*

4.1   gravitational potential energy = mass × gravitational field strength × height / $E_p = mgh$ *[1 mark]*

4.2   Rearrange $E_p = mgh$ for h:
$h = E_p \div (m \times g)$ *[1 mark]*
= 137.2 ÷ (20 × 9.8) *[1 mark]*
= **0.7 m** *[1 mark]*

4.3   Energy is transferred from the gravitational potential energy store *[1 mark]* to the kinetic energy store of the load *[1 mark]*.

4.4   Some of the energy would also be transferred to thermal energy store of the air (and the thermal energy store of the load) *[1 mark]*.

5   change in thermal energy = mass × specific heat capacity × temperature change / $\Delta E = mc\Delta\theta$
So, $c = \Delta E \div (m \times \Delta\theta)$ *[1 mark]*
c = 36 000 ÷ (0.5 × 80) *[1 mark]*
= **900 J/kg °C** *[1 mark]*

Page 293

Power

Q1 $P = E \div t$
$t = 2 \times 60 = 120$ s *[1 mark]*
$P = 4800 \div 120$ *[1 mark]*
= **40 W** *[1 mark]*

Page 295

Efficiency

Q1   efficiency = useful output energy transfer ÷ total input energy transfer
= 225 ÷ 300 *[1 mark]*
= **0.75** *[1 mark]*

Q2   efficiency = useful power output ÷ total power input
= 900 ÷ 1200 = 0.75 *[1 mark]*
useful output energy transfer
= efficiency × total input energy transfer *[1 mark]*
= 0.75 × 72 000 *[1 mark]*
= **54 000 J** *[1 mark]*

Page 296

Warm-Up Questions

1) Power is the rate of energy transfer/doing work.
2) 1 J/s
3) E.g. its thickness/thermal conductivity
4) A material with a high thermal conductivity.
5) Some energy is always dissipated, so less than 100% of the input energy transfer is transferred usefully.

Exam Questions

1.1   power = work done ÷ time / $P = W \div t$ *[1 mark]*

1.2   1 kJ = 1000 J
$P = 1000 \div 20$ *[1 mark]*
= **50 W** *[1 mark]*

1.3   E.g. by lubricating any moving parts *[1 mark]*

1.4   It will be faster. *[1 mark]*
The motor transfers the same amount of energy, but over a shorter time.

2.1   Any one from: e.g. by heating the fan / by heating the surroundings / transferred away by sound waves *[1 mark]*.

2.2   2 kJ = 2 × 1000 = 2000 J *[1 mark]*
useful output energy transfer
= total input energy transfer − wasted output energy transfer
= 7250 − 2000
= **5250 J** *[1 mark]*

2.3   efficiency = $\dfrac{\text{useful output energy transfer}}{\text{total input energy transfer}}$
= $\dfrac{5250}{7250}$ *[1 mark]*
= 0.72413... = **0.72 (to 2 s.f.)** *[1 mark]*

Page 304

Warm-Up Questions

1) Any two from: coal / oil / natural gas.
2) Fossil fuels can be used to generate electricity at any time, but solar power cannot be used at night and is not reliable in cloudy weather.
3) Any one from: e.g. bio-fuels can be used to run vehicles / electricity generated using renewable resources can be used to power vehicles.
4) Any two from: e.g. it releases greenhouse gases and contributes to global warming / it causes acid rain / coal mining damages the landscape.
5) True

Exam Questions

1   wind *[1 mark]*

2   They can produce sulfur dioxide and cause acid rain. *[1 mark]*

3.1   E.g. Heating *[1 mark]*.

3.2   Any two from: e.g. it's expensive to build new renewable power plants / people don't want to live near new power stations / some renewable energy resources aren't very reliable, so they could not be used on their own / it's expensive to switch to cars running on renewable energy *[1 mark for each correct answer]*

4.1   How to grade your answer:
Level 0: There is no relevant information. *[No marks]*
Level 1: Some points are made about the environmental impact and reliability of wind power and hydro-electric power, but the energy resources are not directly compared. Answer lacks detail and structure. *[1 to 2 marks]*
Level 2: At least three valid comparisons are made about the environmental impact and the reliability of wind power and hydro-electric power. Answer has some detail and structure. *[3 to 4 marks]*
Level 3: At least four valid comparisons are made about the environmental impact and the reliability of wind power and hydro-electric power. Answer is detailed and well-structured. *[5 to 6 marks]*

Here are some points your answer may include:
Wind turbines do not do any permanent damage to the landscape.
Building a hydro-electric power plant does damage the landscape, as it involves flooding a valley.
Hydro-electric power stations have a larger impact on the local environment than wind turbines since animals and plants in the flooded valley lose their habitats.
Wind turbines and hydro-electric power stations produce no pollution once they are built.
However, plants in valleys flooded for hydro-electric power stations can rot, producing greenhouse gases which lead to global warming.
Hydro-electric power is reliable in countries that have regular rainfall, but wind turbines are unable to produce electricity when wind stops or when the wind is too strong.
Hydro-electric power stations can respond straight away when there is additional demand, but the power supplied by wind turbines cannot be increased.

4.2   E.g. they cause no pollution *[1 mark]*

Topic P2 — Electricity

Page 306

Charge and Resistance Calculations

Q1 $V = IR$ so $R = V \div I$ *[1 mark]*
= 230 ÷ 5.0 *[1 mark]*
= **46 Ω** *[1 mark]*

Page 309
I-V Characteristics
Q1 As the current through the lamp increases, the temperature of its filament increases *[1 mark]* causing its resistance to increase *[1 mark]*. A larger resistance means less current can flow per unit potential difference, and so the graph gets shallower *[1 mark]*.

Page 311
Warm-Up Questions
1)
2) Electric current is a flow of electric charge.
3) Ohms / Ω
4) Any one from: e.g. a wire / a fixed resistor
5) In parallel.
6) A graph that shows how the current flowing through a component changes as the potential difference across it varies.

Exam Questions
1.1 $Q = It$ / charge flow = current × time *[1 mark]*
1.2 Convert from minutes into seconds:
2 minutes = 120 seconds
$Q = It$
$= 0.30 \times 120$
$= \mathbf{36\ C}$
[2 marks for correct answer, otherwise 1 mark for correctly converting from minutes to seconds]
1.3 A filament lamp is a **non-linear** *[1 mark]* component. As the temperature of the filament increases, the **resistance** *[1 mark]* of the lamp increases.
2.1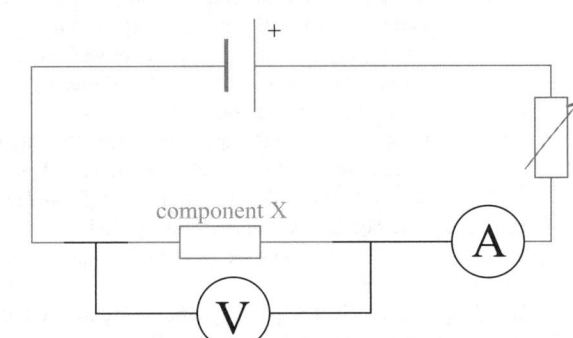

[1 mark for adding the ammeter in line with component X, 1 mark for adding the voltmeter across component X]
2.2 A straight line *[1 mark]*. Component X is a resistor, which is an ohmic/linear component *[1 mark]*.

Page 313
Sensing Circuits
Q1 E.g.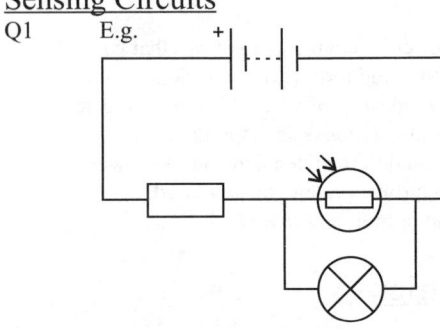

[1 mark for all symbols correct, 1 mark for the LDR connected in series with a resistor and a source of potential difference, 1 mark for the bulb connected in parallel with the LDR]

Page 315
Series Circuits
Q1 $R_{total} = 4 + 5 + 6 = 15\ \Omega$ *[1 mark]*
$V = I \times R = 0.6 \times 15$ *[1 mark]*
$= \mathbf{9\ V}$ *[1 mark]*

Page 317
Parallel Circuits
Q1 The total current through the circuit decreases *[1 mark]* as there are fewer paths for the current to take *[1 mark]*. The total resistance of the circuit increases *[1 mark]* as, using $V = IR$, a decrease in the total current means an increase in the total resistance *[1 mark]*.

Page 319
Warm-Up Questions
1) Any one from: e.g. in automatic night lights / outdoor lighting / burglar detectors.
2) The resistance decreases.
3) The components are all connected in a line between the ends of the power supply.
4) True
5) The total resistance is the sum of all the resistances.
6) Two resistors connected in series.

Exam Questions
1 Total resistance is equal to the sum of the individual resistances. *[1 mark]*
2.1 $R_{total} = R_1 + R_2$
$= 3 + 2$ *[1 mark]* $= \mathbf{5\ \Omega}$ *[1 mark]*
2.2 0.4 A *[1 mark]*. Current is the same throughout a series circuit *[1 mark]*.
2.3 $V_{total} = V_1 + V_2$
$V_2 = V_{total} - V_1$
$= 4.0 - 1.6$ *[1 mark]*
$= \mathbf{2.4\ V}$ *[1 mark]*
3.1 When current flows, the LED will be lit. *[1 mark]*
3.2 When the lights are switched off, the resistance of the LDR increases *[1 mark]*. As a result the total resistance of the circuit also increases, since it is equal to the sum of the individual resistances *[1 mark]*.

Page 321
Power of Electrical Appliances
Q1 $E = P \times t$ so $P = E \div t$ *[1 mark]*
$= 6000 \div 30$ *[1 mark]*
$= \mathbf{200\ W}$ *[1 mark]*
Q2 $E = P \times t$ *[1 mark]*
$= 250 \times (2 \times 60 \times 60)$
$= 1\ 800\ 000$ J *[1 mark]*
$E = 375 \times (2 \times 60 \times 60)$
$= 2\ 700\ 000$ J *[1 mark]*
So change in energy is
$2\ 700\ 000 - 1\ 800\ 000$
$= \mathbf{900\ 000\ J}$ *[1 mark]*

Page 322
More on Power
Q1 $E = Q \times V$
$= 10\ 000 \times 200$ *[1 mark]*
$= \mathbf{2\ 000\ 000\ J}$ *[1 mark]*
Q2 $P = V \times I$
$= 12 \times 4.0$ *[1 mark]*
$= \mathbf{48\ W}$ *[1 mark]*
Q3 $R = P \div I^2$ *[1 mark]* $= 2300 \div 10.0^2$ *[1 mark]*
$= \mathbf{23\ \Omega}$ *[1 mark]*

Page 325
Warm-Up Questions
1) The live wire, neutral wire, and earth wire.
2) Energy is transferred electrically to the thermal energy store of the heating element.
3) power = (current)² × resistance / $P = I^2R$
4) The network of cables and transformers that distributes electricity across the country.

Exam Questions
1.1 **Step-up** *[1 mark]* transformers are used between the power station and the transmission cables. This increases the **potential difference** *[1 mark]*, so that power may be transferred more efficiently.
1.2 230 V ac *[1 mark]*
2.1 $P = VI$ / power = potential difference × current *[1 mark]*
2.2 $P = VI$
$I = \dfrac{P}{V}$
$= \dfrac{2760}{230}$ *[1 mark]*
$= \textbf{12 A}$ *[1 mark]*
2.3 E.g. volume of water / starting temperature *[1 mark]*
2.4 Energy transferred = power × time / Power is the rate of energy transfer *[1 mark]*. So kettle B will transfer more energy in the same time *[1 mark]*.

Topic P3 — Particle Model of Matter

Page 329
Measuring Density
Q1 Gemstone's mass = 0.019 kg
    = 0.019 × 1000
    = 19 g *[1 mark]*
Gemstone's volume = volume of water pushed out of eureka can = 7.0 cm³
$\rho = m \div V$
$= 19 \div 7.0$ *[1 mark]*
$= 2.714...$
$= \textbf{2.7 g/cm}^3$ **(to 2 s.f.)** *[1 mark]*

Page 331
Specific Latent Heat
Q1 $E = m \times L = 0.25 \times 120\,000$ *[1 mark]*
$= \textbf{30 000 J}$ *[1 mark]*

Pages 332-333
Warm-Up Questions
1) In a solid, the particles are close together and are held in a regular, fixed pattern.
2) A measure of how much mass there is in a certain space.
3) It increases.
4) The specific latent heat of vaporisation of a substance is the amount of energy needed to change 1 kg of that substance from a liquid to a gas.
5) J/kg

Exam Questions
1 If a liquid is heated to a certain temperature it starts to boil and turns into a **gas** *[1 mark]*. Another process that causes this change of state is **evaporation** *[1 mark]*.
2.1 The densities of each of the toy soldiers are the same, but their masses may vary. *[1 mark]*
2.2 200 g = 0.2 kg
density = mass ÷ volume = $0.2 \div (2.5 \times 10^{-5})$ *[1 mark]*
= **8000 kg/m³** *[1 mark]*
3 $E = mL = 0.40 \times 1200$ *[1 mark]*
= **480 J** *[1 mark]*
4.1 The volume of the pendant *[1 mark]*.
The mass of the pendant *[1 mark]*.

4.2 E.g. measure the mass of the pendant using the mass balance *[1 mark]*. Fill the eureka can with water to just below the spout, then place the measuring cylinder beneath the spout *[1 mark]*. To measure the volume of the pendant, submerge the pendant in the water, catching the displaced water in the measuring cylinder *[1 mark]*. Measure the volume of the displaced water, which is equal to the volume of the pendant *[1 mark]*. Divide the mass of the pendant by the volume of the pendant to find the density *[1 mark]*.

With questions where you have to describe a method, make sure your description is clear and detailed. You could also pick up some of the marks by describing how you'd do repeats, take averages and other ways in which you'd make it a fair test.

5.1 The substance is melting *[1 mark]*.
5.2 Melting point = –7 °C *[1 mark]*
    Boiling point = 58 °C *[1 mark]*
6 How to grade your answer:
Level 0: There is no relevant information. *[No marks]*
Level 1: There is a brief explanation of what happens to the particles in the gas as the container is heated. *[1 to 2 marks]*
Level 2: An attempt is made to use the particle model to explain how gas particles create pressure in a sealed container. This is linked to a brief explanation of why the pressure of the gas increases as the container is heated. *[3 to 4 marks]*
Level 3: The particle model is used to provide a clear explanation of how gas particles create pressure in a sealed container. This is used to give a detailed explanation of why the pressure of the gas increases. *[5 to 6 marks]*

Here are some points your answer may include:
In a sealed container, the gas particles are free to move around.
The particles collide with the walls of the container.
Each collision exerts a force on the walls.
The overall force creates an outward pressure.
When the container is heated, energy is transferred to the particles' kinetic energy stores.
This means that the particles move faster.
Since the particles are moving faster, they collide with the container walls more often.
So the force exerted on the container walls, and therefore the pressure, increases as the container is heated.

Topic P4 — Atomic Structure

Page 337
Types of Nuclear Radiation
Q1 E.g. Alpha would not be suitable because it is stopped by a few cm of air or a sheet of paper *[1 mark]*. It would not be able to pass through the packaging to sterilise the equipment *[1 mark]*.

Page 338
Nuclear Equations
Q1 14 + 4 = **18** *[1 mark]*
The mass number of a nucleus reduces by 4 during alpha decay, so to find what the mass number would have been before the decay, you need to add 4 on.
Q2 $^{219}_{86}\text{Rn} \rightarrow ^{215}_{84}\text{Po} + ^{4}_{2}\text{He}$
[1 mark for correct layout, 1 mark for correct symbol for an alpha particle, 1 mark for total atomic and mass numbers being equal on both sides]

Page 341
Half-Life
Q1   16 seconds *[1 mark]*
41 is half of 82, meaning the activity of the sample halved in 16 seconds. Remember, half-life is the time taken for the activity of a sample to fall to half of its starting value.

Q2   Initial count-rate = 168 cps
After 1 half-life, count-rate = 168 ÷ 2 = 84 cps
After 2 half-lives, count-rate = 84 ÷ 2 = 42 cps
After 3 half-lives, count-rate = 42 ÷ 2 = 21 cps *[1 mark]*
So, it took 3 half-lives for the count-rate to drop to 21 cps.
This means that 60 minutes is equal to 3 half-lives *[1 mark]*.
So, the half-life of the sample = 60 ÷ 3
                                   = **20 minutes** *[1 mark]*

Pages 344-345
Warm-Up Questions
1) Evidence from alpha particle scattering experiments showed that the plum pudding model could not be correct. The evidence suggested that atoms were mostly empty space, and that most of the mass was in the centre of the atom as a small, positive nucleus.
2) An atom contains a tiny, positively charged, nucleus made up of protons and neutrons. The nucleus is surrounded by electrons, which orbit the nucleus in energy levels.
3) Electromagnetic radiation.
4) A positive ion.

Atoms are neutral overall. If an atom loses an electron, it has more positive protons than negative electrons. So it now has a positive charge.

5) The mass number tells you the total number of protons and neutrons.
6) Alpha decay: mass number decreases by 4 and the atomic number decreases by 2.
Gamma decay: the mass and atomic numbers don't change.
7) Contamination occurs when radioactive atoms get onto or inside an object. Irradiation occurs is when radiation from a radioactive sources reaches an object.
8) Gamma radiation is less ionising than alpha or beta radiation, so it does the least harm as it passes through the body.

Exam Questions
1.1   Protons *[1 mark]*, neutrons *[1 mark]*
1.2   Atoms with the same atomic number but a different mass number *[1 mark]*.
2

Particle	Charge	Number present in an atom of iodine-131
Proton	positive	53
Neutron	zero	78
Electron	negative	53

[1 mark for each correct answer]

The bottom number in $^{131}_{53}I$ is the atomic number, which is the number of protons. The top number is the total number of protons and neutrons. So there are 131 − 53 = 78 neutrons.

3
- Both the mass and charge of the nucleus change. — alpha
- Neither the mass or charge of the nucleus changes. — gamma
- The charge of the nucleus changes but the mass stays the same. — beta

[1 mark for each correct line]

4.1   Beta (particles) *[1 mark]*, because the radiation passes through the paper, but not the aluminium *[1 mark]*.
4.2   E.g. a Geiger-Muller tube (and counter) *[1 mark]*
5.1   28 minutes *[1 mark]*
The initial activity is 740 Bq. Half of this is 740 ÷ 2 = 370 Bq. The activity falls to this level between 20 and 30 minutes, so from the options, 28 minutes must be the half-life.

5.2   Irradiation from the source isn't much of a concern because alpha radiation only travels a few centimetres in air and cannot get through the skin to damage organs *[1 mark]*. However, if the student's hands are contaminated by atoms of the radioactive source, they could end up inside her body, where the alpha radiation would do a lot of damage as it is highly ionising *[1 mark]*.

Topic P5 — Forces

Page 348
Weight, Mass and Gravity
Q1   a) $W = mg = 5 \times 9.8$ *[1 mark]*
              = **49 N** *[1 mark]*
    b) $W = 5 \times 1.6$ *[1 mark]* = **8 N** *[1 mark]*

Page 349
Resultant Forces and Work Done
Q1   20 cm = 0.2 m *[1 mark]*
$W = Fs = 20 \times 0.2$ *[1 mark]*
           = **4 J** *[1 mark]*

Page 350
Warm-Up Questions
1) friction
2) the magnitude of the vector
3) a) kg
    b) N
4) Add the forces together.

Exam Questions
1.1   force *[1 mark]*
1.2   14 kg *[1 mark]*
2.1   800 − (300 + 200) = **300 N** *[1 mark]* **down** *[1 mark]*
2.2 $x − 400 = 0 \Rightarrow x =$ **400 N** *[1 mark]*
3.1   Work done = force × distance / $W = Fs$ *[1 mark]*
3.2 $W = 44\,000 \times 750$ *[1 mark]*
         = **33 000 000 J** *[1 mark]*
         = **33 000 kJ** *[1 mark]*

Remember, to convert from J to kJ, divide by 1000.

Page 352
Forces and Elasticity
Q1   2 cm = 0.02 m *[1 mark]*
$F = ke$ so $k = F \div e$ *[1 mark]*
                = 1 ÷ 0.02 *[1 mark]*
                = **50 N/m** *[1 mark]*

Page 354
Investigating Springs
Q1   2.5 cm = 0.025 m *[1 mark]*
$E_e = \tfrac{1}{2}ke^2 = \tfrac{1}{2} \times 40 \times (0.025)^2$ *[1 mark]*
      = **0.0125 J** *[1 mark]*

Page 355
Warm-Up Questions
1) true
2) false
3) metres

Exam Questions
1.1 $F = ke$ / Force = spring constant × extension *[1 mark]*
1.2 $k = F \div e$ *[1 mark]*
       = 4 ÷ 0.05 *[1 mark]*
       = **80 N/m** *[1 mark]*
2.1   E.g. the spring used (so the spring constant remains the same) *[1 mark]*

2.2 E.g. the mass attached to the spring / the force applied to the spring *[1 mark]*
2.3 Extension = 4.0 cm = 0.040 m
Work done = ½ × 175 × 0.040² *[1 mark]*
= **0.14** *[1 mark]* **J** *[1 mark]*

You can read the extension from the graph. Don't forget to convert it to m.

Page 356
Distance, Displacement, Speed, Velocity
Q1 $s = vt$ so $v = s \div t$ *[1 mark]*
= 200 ÷ 25 *[1 mark]*
= **8 m/s** *[1 mark]*

Page 357
Acceleration
Q1 $u = 0$ m/s, $v = 7$ m/s, $a = g = 9.8$ m/s²,
$s = (v^2 - u^2) \div 2a$ *[1 mark]*
= (49 − 0) ÷ (2 × 9.8) *[1 mark]*
= **2.5 m** *[1 mark]*

Page 358
Distance-Time Graphs
Q1 E.g.

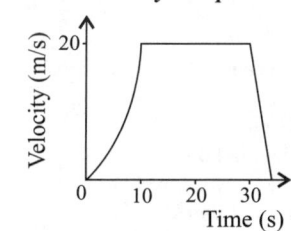

[1 mark for a curved line with an increasing positive gradient, 1 mark for the line becoming a straight line with a positive gradient, 1 mark for the line then becoming horizontal]

Page 359
Velocity-Time Graphs
Q1 E.g.

[1 mark for a straight, horizontal line representing constant speed, 1 mark for a straight line with a positive gradient representing constant acceleration, 1 mark for a straight, horizontal line at a different velocity to the initial velocity to represent a different constant speed]

Q2

[1 mark for an upwards curved acceleration line to 20 m/s, 1 mark for a straight line representing steady speed, 1 mark for a straight line representing deceleration]

Page 361
Warm-Up Questions
1) Speed is a scalar, velocity is a vector / velocity has a direction, speed does not.
2) a) E.g. 3 m/s
   b) E.g. 30 m/s
   c) E.g. 250 m/s

Your answers may be slightly different to these, but as long as they're about the same size, you should be fine to use them in the exam.

3) true
4) a flat section.
5) 0 N.

Exam Questions
1 acceleration = (10 − 2.0) ÷ 20 *[1 mark]*
= **0.4 m/s²** *[1 mark]*
2.1 The cyclist travels at a constant speed (of 3 m/s) between 5 s and 8 s *[1 mark]*, then decelerates (with decreasing deceleration) between 8 s and 10 s *[1 mark]*.
2.2 Acceleration is given by the gradient of a velocity-time graph.
change in $y = 3 - 0 = 3$ m/s
change in $x = 5 - 2 = 3$ s
acceleration = 3 ÷ 3 *[1 mark]* = **1 m/s²** *[1 mark]*
3.1 Distance = speed × time / $s = vt$ *[1 mark]*
3.2 Distance rolled = 0.46 × 2.4 *[1 mark]*
= 1.104 *[1 mark]*
= **1.1 m (to 2 s.f.)** *[1 mark]*
3.3 $v^2 - u^2 = 2as$
$a = \dfrac{v^2 - u^2}{2s}$ *[1 mark]*
$= \dfrac{12^2 - 0^2}{2 \times 8}$ *[1 mark]*
= **9 m/s²** *[1 mark]*

Page 362
Newton's First and Second Law
Q1 $F = ma = (80 + 10) \times 0.25$ *[1 mark]*
= **22.5 N** *[1 mark]*

Page 363
Newton's Third Law
Q1 Any one from: e.g the gravitational force of the Earth attracts the car and the gravitational force of the car attracts the Earth *[1 mark]* / the car exerts a normal contact force down against the ground and the normal contact force from the ground pushes up against the car *[1 mark]* / the car (tyres) pushes the road backwards and the road pushes the car (tyres) forwards *[1 mark]*.

Page 366
Warm-Up Questions
1) true
2) Boulder B

Boulder B needs a greater force to accelerate it by the same amount as boulder A.

3) true

This is Newton's Third Law.

Exam Questions
1.1 The ball exerts a force of −520 N on the bat *[1 mark]*.
1.2 The acceleration of the ball will be **greater than** that of the bat because it has a smaller mass. *[1 mark]*
1.3 Force = mass × acceleration
acceleration = force ÷ mass *[1 mark]*
= 520 ÷ 1.6 *[1 mark]*
= **325 m/s²** *[1 mark]*
2.1 **0 N** *[1 mark]*

The van is travelling at a constant speed in one direction so the resultant force must be zero.

2.2 The driving force decreases *[1 mark]*
The wind blowing in the opposite direction to the van's movement increases *[1 mark]*
2.3 $F = ma$
= 2500 × 1.4 *[1 mark]*
= **3500 N** *[1 mark]*
opposite to the direction of motion *[1 mark]*

Page 368
Braking Distance
Q1   Energy in car's kinetic energy store
$= ½ × m × v^2$
$= ½ × 1000 × 10^2$ *[1 mark]*
$= 50\,000$ J *[1 mark]*
To stop the car, the brakes must do work to transfer all the energy away from the car's kinetic energy store:
$½ × m × v^2 = F × d$ *[1 mark]*
So $F × d = 50\,000$
Rearrange for F:
$F = 50\,000 ÷ d$
$= 50\,000 ÷ 50$ *[1 mark]*
$= \mathbf{1000}$ **N** *[1 mark]*

Page 369
Reaction Times
Q1   a) $v^2 – u^2 = 2as$
$v^2 = 2 × 9.8 × 0.162 + 0$ *[1 mark]* $= 3.1752$ m²/s²
$v = \sqrt{3.1752} = 1.781...$ m/s *[1 mark]*
$a = \Delta v ÷ t$ so
$t = \Delta v ÷ a$ *[1 mark]*
$= 1.781... ÷ 9.8$ *[1 mark]* $= 0.181...$ s
$= \mathbf{0.18}$ **s (to 2 s.f.)** *[1 mark]*

b)   His reaction time is longer in the evening *[1 mark]* so whilst driving, he may take longer to react to a hazard, meaning his thinking distance would be longer *[1 mark]*.

Page 370
Warm-Up Questions
1) reaction time / speed / tiredness / drugs / alcohol.
2) The braking distance.
3) Large decelerations may lead to brakes overheating/ the car skidding.
4) true

Exam Questions
1.1 Thinking distance *[1 mark]*.
1.2 The distance the car travels during its deceleration whilst the brakes are being applied *[1 mark]*.
2.1 It reduces the frictional force of the tyres on the road, which increases the braking distance *[1 mark]*.
2.2 E.g. decrease their speed *[1 mark]*.
2.3 Energy transferred from:
The car's kinetic energy stores *[1 mark]*
Energy transferred to:
The brakes' thermal energy stores *[1 mark]*
3.1 Average reaction time $= (0.24 + 0.19 + 0.23) ÷ 3$ *[1 mark]*
$= 0.22$ s *[1 mark]*
3.2 E.g. use the same ruler in each test / add a ball of modelling clay to the bottom of the ruler to help it fall straight down each time / repeat the test more times and take an average *[1 mark]*.

Topic P6 – Waves

Page 373
Wave Speed
Q1 $7.5 ÷ 100 = 0.075$ m *[1 mark]*
wave speed = frequency × wavelength,
so frequency = wave speed ÷ wavelength *[1 mark]*
$= 0.15 ÷ 0.075$ *[1 mark]*
$= \mathbf{2}$ **Hz** *[1 mark]*

Page 376
Refraction
Q1   E.g.

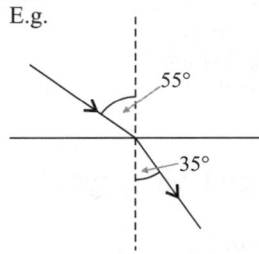

[1 mark for correctly drawing the boundary and the normal, 1 mark for drawing an incident ray at the correct angle, 1 mark for drawing the refracted ray at the correct angle]

Page 377
Warm-Up Questions
1) E.g. If an object is dropped into calm water, ripples spread out. The ripples don't carry the water or the object away with them though. / If a guitar string is strummed, the sound waves don't carry the air away from the guitar. If they did, a wind would be created.
2) In a longitudinal wave, the vibrations are parallel to the direction of travel/energy transfer.
3) rarefactions
4) true

Exam Questions
1.1 E.g. The particles vibrate at right angles to the direction of energy transfer *[1 mark]*
1.2 5 cm *[1 mark]*
1.3 2 m *[1 mark]*
2.1 The wavelength of the sound waves *[1 mark]*.
2.2 wave speed = frequency × wavelength / $v = f\lambda$ *[1 mark]*
2.3 $v = f\lambda$
$= 50.0 × 6.8$ *[1 mark]*
$= \mathbf{340}$ *[1 mark]* **m/s** *[1 mark]*

Pages 384-385
Warm-Up Questions
1) gamma rays
2) 7
3) Gamma rays are created by changes in the nucleus of an atom.
4) E.g. Infrared cameras / heating / cooking
5) Any two from: e.g. sunburn / the skin can age prematurely / the risk of skin cancer is increased.

Exam Questions
1 visible light *[1 mark]*
2.1 X-rays *[1 mark]*
2.2 E.g. Exposure could cause cancer / cell mutations / damage cells *[1 mark]*.
3.1 low-frequency infrared *[1 mark]*
3.2 They don't travel faster than radio signals because all electromagnetic waves travel at the same speed in a vacuum *[1 mark]*.
4.1 Any one from: e.g. time temperature increase is measured over / type/size of beaker / the amount of water in each beaker / the size/thickness of each piece of card / the distance of each thermometer from the beaker / the height of each thermometer *[1 mark]*.
4.2 Any one from: e.g. be careful when pouring/ carrying hot water / don't touch the beakers until they have cooled down *[1 mark]*.
4.3 29 °C *[1 mark]*
4.4 More infrared radiation is emitted from matt surfaces than from shiny surfaces *[1 mark]*.
4.5 Any one from: e.g. repeat the experiment and calculate averages / use infrared detectors instead of thermometers *[1 mark]*

Topic P7 — Magnetism and Electromagnetism

Page 388
Electromagnetism
Q1 E.g. for current out of the page:

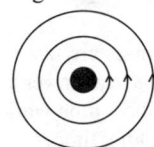

[1 mark for concentric circles getting further apart, 1 mark for arrows on field lines with correct direction]

Page 390
Warm-Up Questions
1)

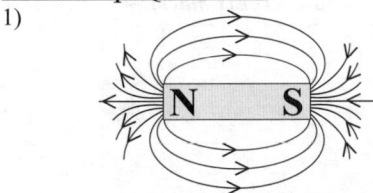

2) The direction of the magnetic field.
3) By wrapping it into a spring-shaped coil.

Exam Questions
1.1

[1 mark for arrows on all lines pointing north to south].

1.2 Attraction *[1 mark]* — opposite poles attract *[1 mark]*.
2 Cobalt is an example of a **magnetic material** *[1 mark]*. The force between a piece of cobalt and a bar magnet is always **attractive** *[1 mark]*.
3.1 Concentric circles around the wire / circles centred on the wire *[1 mark]*. The field will be in a clockwise direction around the wire *[1 mark]*.

If you need to state the direction of a circular field, it's usually easiest to describe the field as either going clockwise or anticlockwise.

3.2 The field lines of the loops line up *[1 mark]*. The field lines are now closer, meaning the magnetic field is stronger *[1 mark]*.

Biology Practice Paper 1

Pages 402-415
1.1 C *[1 mark]*
A is a bronchus, B is the heart and D is an alveolus.
1.2 D *[1 mark]*
1.3 Epithelial tissue — A group of similar cells that work together to carry out a function.
Respiratory system — A group of organs working together to perform a function.
Lungs — A group of different tissues that work together to perform a certain function.
[2 marks for all lines correct, otherwise 1 mark for one line correct]
1.4 37 *[1 mark]*
When you're calculating a rate, you always divide by time. So here, you need to divide the number of breaths by the number of minutes.
1.5 Gas A — oxygen *[1 mark]*
     Gas B — carbon dioxide *[1 mark]*
2.1 oxygen *[1 mark]*
2.2 chloroplast *[1 mark]*
2.3 making cell walls *[1 mark]*
2.4 damaged *[1 mark]*
2.5 6 *[1 mark]*
2.6 light intensity *[1 mark]*
3.1 Protease — Converts proteins into amino acids.
Lipase — Converts lipids into fatty acids and glycerol.
[2 marks for two lines correct, otherwise 1 mark for one line correct]
3.2 The active sites of different enzymes have different shapes *[1 mark]*. The substance involved in a reaction has to fit into the active site for the enzyme to work *[1 mark]*.
3.3 Enzymes are changed during reactions *[1 mark]*.
3.4 In order to grow, organisms need to transfer energy using respiration *[1 mark]*. Cells need glucose for respiration *[1 mark]*. Amylase provides this glucose by breaking down starch *[1 mark]*.
4.1 The loss of water from a plant *[1 mark]*.
4.2 9 a.m. *[1 mark]*
Each division on the x-axis is one hour.
4.3 E.g. day 2 was colder. / Day 2 was less windy. / Day 2 was wetter/more humid. / The light intensity was lower on day 2 *[1 mark]*.
4.4 Low light intensity at night meant the stomata closed, allowing less water vapour to escape. *[1 mark]*.
4.5 guard cells *[1 mark]*
5.1

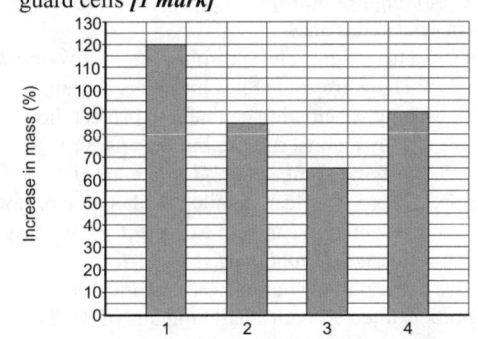

[3 marks — 1 mark for a suitable y-axis scale, 2 marks for all bars correct (or 1 mark if two or more bars correct)]

5.2 Growth medium number 1 *[1 mark]*
5.3 Any two from: e.g. the temperature in the incubator / the size of the tissue samples/blocks / the volume of growth medium used *[2 marks — 1 mark for each correct answer]*.
5.4 E.g. grow multiple blocks of stem tissue on each growth medium *[1 mark]* and find the mean for each growth medium *[1 mark]*.
5.5 asexual reproduction *[1 mark]*
5.6 E.g. to prevent the species from going extinct *[1 mark]*.
6.1 cell membrane *[1 mark]*
6.2 *E.coli* bacteria *[1 mark]*
E. coli bacteria are prokaryotes.
6.3 image size = 0.054 × 500 *[1 mark]*
     = 27 mm
     = **2.7 cm** *[1 mark]*
6.4 *Euglena* is a single-celled organism so has a large surface area compared to its volume *[1 mark]*.
6.5 How to grade your answer:
Level 0: No relevant information is given. *[No marks]*
Level 1: There is a brief comparison of plant and animal cells, including at least one similarity and one difference. *[1 to 2 marks]*
Level 2: There is a comparison of plant and animal cells, including at least two similarities and two differences. Some descriptions of subcellular structures are included. *[3 to 4 marks]*
Level 3: There is a detailed comparison of plant and animal cells, including at least three similarities and three differences. Detailed descriptions of subcellular structures are included. *[5 to 6 marks]*

Here are some points your answer may include:
Similarities:
Both plant and animal cells have a nucleus, which controls the cell's activities.
Both plant and animal cells contain cytoplasm, which is where most of the cell's chemical reactions take place.
Plant cells and animal cells both have a cell membrane, which controls what goes in and out of the cell.
Mitochondria are found in both plant cells and animal cells — these are where most of the reactions for aerobic respiration take place.
Both plant cells and animal cells have ribosomes, which are where proteins are made.
Differences:
Chloroplasts, the site of photosynthesis, are present in plant cells, but not in animal cells.
Plant cells have a cell wall, which supports and strengthens the cell, but animal cells do not.
Plant cells contain a permanent vacuole, containing cell sap, but animal cells do not.

7.1 By droplets from an infected person's sneeze or cough *[1 mark]*.
7.2 How to grade your answer:
Level 0: There is no relevant information. *[No marks]*
Level 1: There are some relevant points explaining how vaccination can help to protect the body against disease but the answer is missing some detail. *[1 to 2 marks]*
Level 2: There is a clear, detailed explanation of how vaccination can help to protect the body against disease. *[3 to 4 marks]*
Here are some points your answer may include:
The body is injected with small amounts of dead or inactive pathogens.
These pathogens have antigens on their surface, which cause the white blood cells in the body to produce antibodies.
Antibodies attack/kill the pathogens.
If live pathogens of the same type appear again, the white blood cells can quickly produce lots of antibodies to kill the pathogens so the person is less likely to get ill.
7.3 E.g. vaccines don't always give you immunity.
Some people have a bad reaction to a vaccine
[2 marks — 1 mark for each correct answer].
7.4 E.g. stop the mosquitoes from breeding. Protect people from mosquito bites using mosquito nets
[2 marks — 1 mark for each correct answer].
7.5 To prevent the fungus spreading to other plants *[1 mark]* by being carried by wind or water *[1 mark]*.
8.1 biuret solution *[1 mark]*
8.2 E.g. make sure the area is well ventilated / wear safety goggles / wear gloves / use a funnel when pouring liquids *[1 mark]*.
8.3 Benedict's solution *[1 mark]*
8.4 A *[1 mark]*
Benedict's solution is blue, and so the sample where it stays blue has no reducing sugars.
8.5 It is covered in many villi that provide a large surface area for diffusion (and active transport) to occur across *[1 mark]*. The villi have a thin wall/single layer of surface cells, which decreases the distance for diffusion to occur across *[1 mark]*. They also have a good blood supply to assist quick absorption *[1 mark]*.
8.6 A person with lactose intolerance has little or no lactase to break down the lactose in the drink, so there will be little or no sugar to be absorbed from the small intestine *[1 mark]*.
8.7 Bile emulsifies/breaks down fats *[1 mark]*.
8.8 Fat in the small intestine would not be emulsified/broken down into small droplets *[1 mark]*. This would mean that there is a low surface area for lipase to work on *[1 mark]*, meaning a slow rate of fat digestion *[1 mark]*.

Biology Practice Paper 2

Pages 416-430

1.1 28 days *[1 mark]*
1.2 LH *[1 mark]*
1.3 The hormones are carried in the blood to their target organs *[1 mark]*.
1.4 master gland *[1 mark]*
1.5 Any two from: nerve-controlled responses are fast-acting while hormone-controlled responses have a slower action. / Nerve-controlled responses act for a short time while hormone-controlled responses act for long time. / Nerve-controlled responses act on a precise area while hormone-controlled responses act more generally.
[2 marks — 1 mark for each correct answer]
2.1 algae *[1 mark]*
2.2 limpet / crab *[1 mark]*
2.3 temperature *[1 mark]*, moisture level *[1 mark]*
2.4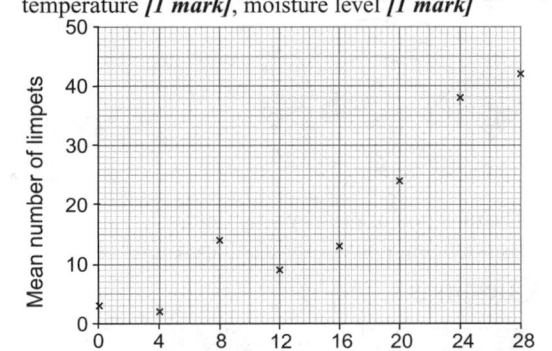
[1 mark for each point plotted correctly]
2.5 The mean number of limpets is lower at the low-tide point than at 28 m from the low-tide point *[1 mark]*.
2.6 Any two from: e.g. variation in competition/predation/food availability/moisture level/temperature/oxygen level *[2 marks — 1 mark for each correct answer]*
3.1 A form/version of a gene *[1 mark]*.
3.2

	R	r
R	RR	Rr
r	Rr	rr

[2 marks for all genotypes correct, otherwise 1 mark for one or two correct.]
3.3 RR and Rr *[1 mark]*
3.4 0.25 *[1 mark]*
3.5 XY *[1 mark]*
4.1 meiosis *[1 mark]*, four *[1 mark]*, different *[1 mark]*
4.2 28 *[1 mark]*
4.3 Asexual reproduction happens by mitosis *[1 mark]*. This means there's no genetic variation in the offspring *[1 mark]*.
The offspring may not have characteristics that are suited to the new conditions and so the population can't adapt to the new conditions *[1 mark]*.
5.1 The highest yielding tomato plants from the first year could have been selected *[1 mark]* and been bred together *[1 mark]*. This could have been repeated every year *[1 mark]*.
5.2 The number of different alleles/genetic variation in the selectively bred tomato population might be low *[1 mark]*. This would mean it's less likely that the tomato plants will have (alleles for) disease resistance *[1 mark]*.
5.3 A process in which genes from one organism are introduced into the genome of another organism *[1 mark]*.
5.4 E.g. to be resistant to herbicides / to be resistant to insects / to produce bigger fruit / to produce more fruit *[1 mark]*.

5.5 The genetically engineered tomatoes are less likely to affect wild plant populations *[1 mark]*.

6.1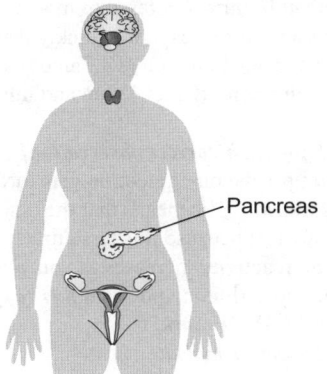
[1 mark]

6.2 Insulin causes glucose to move out of the blood and into cells *[1 mark]*.

6.3
Statement	Type 1	Type 2	Both
Results in a high blood glucose level if left untreated.			✓
The pancreas fails to produce enough insulin.	✓		
Body cells can no longer respond to insulin.		✓	
Must be treated with insulin injections.	✓		

[4 marks — 1 mark for each tick]

6.4 percentage increase = $((146 - 89) \div 89) \times 100$
= **64% (2s.f.)**
[2 marks for correct answer, otherwise 1 mark for correct working.]

7.1 **21** *[1 mark]*
To find the median, you write out all the results from lowest to highest, i.e. 10, 21, 21, 35, 37. 21 is in the middle of the list, so it is the median.

7.2 $(9 + 14 + 19 + 5 + 3) \div 5$
= $50 \div 5$
= **10 buttercups per m²** *[2 marks for correct answer, otherwise 1 mark for correct working.]*

7.3 E.g. use more quadrats in each field *[1 mark]*.

7.4 E.g. more buttercups grow where there is a higher moisture level in the soil *[1 mark]*.

7.5 How to grade your answer:
Level 0: There is no relevant information. *[No marks]*
Level 1: There is a brief description of how the student could investigate whether the change in distribution of buttercups is due to variability in the moisture level of the soil. *[1 to 2 marks]*
Level 2: There is a detailed description of how the student could investigate whether the change in distribution of buttercups is due to variability in the moisture level of the soil. *[3 to 4 marks]*

Here are some points your answer may include:
He could use a transect across Field A.
To do this he should mark out a line across the field.
Then he should record the number of buttercups in quadrats placed next to each other/at intervals along the line.
He should also measure the moisture level of the soil at each sampling point (e.g. with a probe).

8.1 *Biston* *[1 mark]*
8.2 As a result of mutation(s) *[1 mark]*.
8.3 The dark variety is less likely to be eaten by predators in soot polluted areas (because they are better camouflaged) *[1 mark]* so they are more likely to survive to reproduce *[1 mark]*, meaning that the genes for dark colouring are more likely to be passed on to the next generation and become more common in the population *[1 mark]*.
8.4 Town B is the most polluted because it contains a higher percentage of dark moths *[1 mark]*.

9.1 Their reaction time was very fast *[1 mark]*.
Their response was involuntary/automatic *[1 mark]*.
If you have to think about what response to give then it's not a reflex action.

9.2 The males in this experiment had a faster mean reaction time than the females *[1 mark]*, so the data supports the scientist's hypothesis *[1 mark]*.

9.3 E.g. the age of the participants. / The strength of the tap on the knee. / Caffeine consumption of the participants prior to the investigation. *[1 mark]*

9.4 How to grade your answer:
Level 0: There is no relevant information. *[No marks]*
Level 1: There is a brief description of some parts of the path taken by a nervous impulse in the reflex. *[1 to 2 marks]*
Level 2: There is some description of the path taken by a nervous impulse in the reflex, but some detail is missing. *[3 to 4 marks]*
Level 3: There is a clear and detailed description of the path taken by a nervous impulse in the reflex. *[5 to 6 marks]*

Here are some points your answer may include:
The impulse travels along a sensory neurone to the central nervous system/spinal cord.
When the impulse reaches a synapse between the sensory neurone and a relay neurone, it triggers chemicals to be released.
These chemicals cause impulses to be sent along the relay neurone.
When the impulse reaches a synapse between the relay neurone and a motor neurone, chemicals are released again which cause impulses to be sent along the motor neurone.
The impulse then reaches the muscle, which contracts to curl the big toe upwards.

Chemistry Practice Paper 1

Pages 431-443

1.1 Neon *[1 mark]*
1.2 Metallic bonding *[1 mark]*
1.3 Copper *[1 mark]*
1.4 Fluorine *[1 mark]*
1.5 Protons = **9** *[1 mark]*
Neutrons = $19 - 9 =$ **10** *[1 mark]*
Electrons = **9** *[1 mark]*
Remember: number of protons = atomic number, number of neutrons = mass number − atomic number, number of electrons (in a neutral atom) = number of protons.
1.6 potassium chloride *[1 mark]*
1.7 Moving down Group 7, the outer electron shell gets further from the nucleus *[1 mark]*, and so it becomes harder for the halogens to gain an extra electron *[1 mark]*.

2.1 $2560 - 1130 = 1430$ g
= **1.43 × 10³ g**
[2 marks for correct answer given in standard form, otherwise 1 mark for correct answer not given in standard form.]
Remember, the total mass of the products must always be the same as the total mass of the reactants.
2.2 (aq): aqueous/dissolved in water *[1 mark]*
(s): solid *[1 mark]*
2.3 The reaction produces carbon dioxide gas *[1 mark]*. This escapes from the beaker, reducing the mass of the reaction mixture *[1 mark]*.

2.4 E.g. 224 cm³ ÷ 1000 = 0.224 dm³
3.4 g ÷ 0.224 dm³ = 15.17... g/dm³
= **15.2 g/dm³ (3 s.f.)**
[3 marks for correct answer given to 3 s.f. or 2 marks for correct answer not given to 3 s.f., otherwise 1 mark for converting from cm³ to dm³.]
Remember, to get from cm³ to dm³ you have to divide by 1000.

3.1 Two or more elements or compounds that are not chemically combined together / don't have any chemical bonds between them *[1 mark]*.

3.2 No chemical reactions take place during fractional distillation *[1 mark]*.

3.3 fractionating column *[1 mark]*

3.4 Compound: ethanoic acid *[1 mark]*
Reason: it has the highest boiling point *[1 mark]*.

3.5 78 °C *[1 mark]*

3.6 Any two from: e.g. wear gloves / wear a lab coat / wear goggles / work in a fume cupboard.
[2 marks — 1 mark for each sensible precaution.]

3.7 The boiling points of methanol and isobutanal are very similar *[1 mark]*.

4.1 E.g.

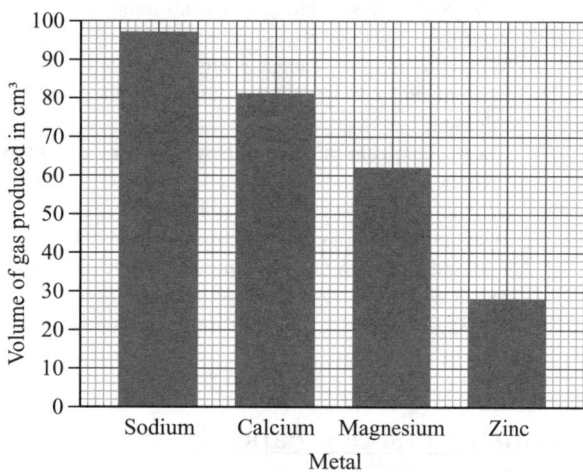

[1 mark for sensible scale on the vertical axis, 1 mark for all 4 bars plotted correctly]
Make sure any graphs you draw fill at least two-thirds of the space given.

4.2 hydrogen *[1 mark]*

4.3 e.g. 17 cm³ *[1 mark for answer in range 1-27 cm³.]*
To answer this question you need to remember than iron is below zinc in the reactivity series. This means it reacts more slowly with the acid, and so not as much gas is produced in 60 seconds.

4.4 M_r = (2 × 23) + 32 + (4 × 16)
= 46 + 32 + 64
= **142** *[1 mark]*

4.5 percentage mass of an element in a compound
= $\frac{A_r \times \text{number of atoms of that element}}{M_r \text{ of the compound}} \times 100$
percentage mass of sodium = $\frac{23 \times 2}{142} \times 100$
= 32.394...
= **32.4 (3 s.f.)**
[2 marks for correct answer given to 3 s.f., otherwise 1 mark for correct answer not given to 3 s.f.]
You can still get the marks here if you got the wrong value for the M_r in 4.4 and used that in your calculation. Just make sure you did the right calculation and that your answer is correct for the value of M_r you used.

4.6 electrolyte *[1 mark]*

4.7 oxygen *[1 mark]*
There are no halide ions present in the solution, so it is OH⁻ ions from the water that lose electrons at the anode, forming O_2 (and H_2O).

4.8 Product at cathode: hydrogen *[1 mark]*
Reason: sodium is more reactive than hydrogen *[1 mark]*

5.1 +1 *[1 mark]*

5.2 Any two from: e.g. potassium fizzes more than sodium, which fizzes more than lithium. / Potassium moves more quickly than sodium, which moves more quickly than lithium. / Potassium appears to melt and a flame is seen, sodium appears to melt but no flame is seen and lithium does not appear to melt.
[2 marks — 1 mark for each correct observation]

5.3 As you go down Group 1 the outer electron gets further from the nucleus *[1 mark]*. This means that the outer electron is more easily lost because it is less attracted to the nucleus *[1 mark]*. So reactivity increases as you go down Group 1/from lithium to sodium to potassium *[1 mark]*.

5.4 $2K + 2H_2O \rightarrow 2KOH + H_2$ *[1 mark]*

5.5 It contains hydroxide ions *[1 mark]*.

5.6 2, 1 *[1 mark]*

6.1 Any two from: e.g. have a lid on the beaker. / Use insulation (e.g. cotton wool) around the beaker. / Use a polystyrene beaker. *[2 marks — 1 mark for each correct answer]*

6.2 temperature change *[1 mark]*
The dependent variable is the variable you measure.

6.3 1 °C *[1 mark]*

6.4 37.5 °C *[1 mark]*

6.5 Temperature change = final temperature – initial temperature
= 37.5 – 17 = **20.5 °C**
[2 marks for correct answer, otherwise 1 mark for writing a correct expression for calculating the temperature change.]
If your answer to 6.4 was wrong, you can still have both marks for correctly subtracting 17 from it to find the temperature change.

6.6 The reaction was exothermic as the temperature of the surroundings increased during the reaction *[1 mark]*.

7.1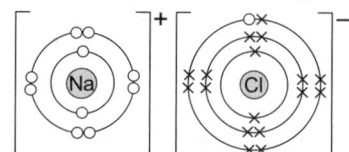

[1 mark for adding seven crosses and one dot to outer shell of Cl⁻ ion, 1 mark for correct charge on both ions.]

7.2 63 °C *[1 mark]*
Substance B is a gas at temperatures above its boiling point of –183 °C. 63 °C is the only temperature given which is higher than –183 °C, so substance B must be a gas at this temperature.

7.3 The molecules in substance A are only attracted to each other by weak intermolecular forces *[1 mark]* which don't need much energy to break/overcome *[1 mark]*.

7.4 Substance C *[1 mark]*. It can conduct electricity when molten or dissolved but not when solid *[1 mark]*.

7.5 How to grade your answer:
Level 0: There is no relevant information. *[No marks]*
Level 1: A brief attempt is made to explain one or two of these properties in terms of structure and bonding. *[1 to 2 marks]*
Level 2: Some explanation of two or three of the properties, in terms of their structure and bonding, is given. *[3 to 4 marks]*
Level 3: Clear and detailed explanation of all three of the properties, in terms of their structure and bonding, is given. *[5 to 6 marks]*

Here are some points your answer may include:
Diamond
Each carbon atom in diamond forms four covalent bonds, making it very hard.
Because it is made up of lots of covalent bonds, which take a lot of energy to break, diamond has a very high melting point.
There are no free/delocalised electrons or ions in the structure of diamond, so it can't conduct electricity.

Graphite

Each carbon atom in graphite forms three covalent bonds, creating sheets of carbon atoms that can slide over each other. The carbon layers are only held together weakly, which is what makes graphite soft.

The covalent bonds between the carbon atoms take a lot of energy to break, giving graphite a very high melting point. Only three out of each carbon's four outer electrons are used in bonds, so graphite has lots of free/delocalised electrons which can conduct electricity.

Chemistry Practice Paper 2

Pages 444-457

1.1 A resource that is used up more quickly than it can be replaced *[1 mark]*.
1.2 Resource 3 *[1 mark]*
1.3 fractional distillation *[1 mark]*
1.4 Before it enters the column, the crude oil is heated until it **evaporates** *[1 mark]*. Different hydrocarbons from the crude oil then leave the column at different points because they have different **boiling points** *[1 mark]*.
1.5 Petrol *[1 mark]*
1.6 Bitumen *[1 mark]*
1.7 Alkene *[1 mark]*
2.1 It is the mobile phase *[1 mark]*
2.2 So the spots of food colouring do not dissolve into the solvent in the beaker *[1 mark]*.
2.3 The student could get another person to do the chromatography experiment too, using the same method and conditions *[1 mark]*, and check that the results from both experiments are similar *[1 mark]*.
2.4 The unknown food colouring is most likely to be pure, as it has only formed one spot on the chromatogram *[1 mark]*.
2.5 three *[1 mark]*
2.6 E.g.
distance moved by spot **Y** from the baseline = 1.9 cm
distance moved by solvent from the baseline = 4.1 cm
R_f = 1.9 ÷ 4.1
= 0.4634...
= **0.46 (2 s.f.)**
[5 marks for correct answer in range 0.43-0.50 given to 2 s.f. or 4 marks for correct answer in range 0.43-0.50 not given to 2 s.f., otherwise 1 mark for distance moved by spot Y in range 1.8-2.0 cm, 1 mark for distance moved by solvent in range 4.0-4.2 cm, 1 mark for correctly substituting values into equation.]
2.7 The student's sample is not pure, as its melting point is lower than the data book value for the melting point of the substance *[1 mark]*.
3.1 The concentration of carbon dioxide in the atmosphere has increased over time *[1 mark]*. It stayed roughly the same for the first 75 years before increasing slowly from 1775, then increasingly more rapidly after 1900 *[1 mark]*.
3.2 It has been peer-reviewed *[1 mark]*.
3.3 Greenhouse effect *[1 mark]*
3.4 How to grade your answer:
Level 0: There is no relevant information. *[No marks]*
Level 1: There are some correct and relevant points, but no overall comparison of the environmental impacts. *[1 to 2 marks]*
Level 2: A range of correct and relevant points are made, leading to an overall comparison of the environmental impacts. *[3 to 4 marks]*
Here are some points your answer may include:
Less carbon dioxide is given out manufacturing a plastic bottle than manufacturing a glass bottle.
Transporting plastic bottles produces less carbon dioxide than transporting glass bottles (as they are lighter).
Glass bottles are made mostly from recycled glass.
Plastic bottles are usually produced from new raw materials, using up more crude oil.
More glass bottles than plastic bottles get recycled.
More carbon dioxide is given out manufacturing and transporting glass bottles than plastic bottles, but less waste is produced/less new raw materials are needed as glass bottles are easier to recycle.
Less carbon dioxide is given out manufacturing and transporting plastic bottles than glass bottles, but more waste is produced/more raw materials are needed as plastic bottles are harder to recycle.

4.1 surface area of the calcium carbonate *[1 mark]*
temperature *[1 mark]*
4.2 Bubbling the gas through a solution of calcium hydroxide *[1 mark]*.
4.3 Volume of gas produced after 30 seconds in Experiment 1
= **86-88 cm^3** *[1 mark]*
Volume of gas produced after 30 seconds in Experiment 2
= **139-141 cm^3** *[1 mark]*
You get the marks for any answer within each range.
4.4 E.g. the curve for Experiment 2 starts off steeper than the curve for Experiment 1. / The curve for Experiment 2 flattens out more quickly than the curve for Experiment 1, with the same total amount of gas produced *[1 mark]*.
4.5 Increasing the concentration increases the number of particles in the reaction mixture *[1 mark]*. This increases the frequency of collisions *[1 mark]*.
4.6 mass of calcium sulfate formed = mass of weighing boat with calcium sulfate – mass of weighing boat
mass of calcium sulfate formed = 2.22 – 1.36
= **0.86 g** *[1 mark]*
4.7 From the graph, the reaction finished after 55 seconds.
So time taken = 55 seconds
mean rate of reaction = 0.86 ÷ 55
= 0.01563...
= **0.016 g/s (2 s.f.)**
[3 marks for correct answer in range of 0.014-0.016, given to 2 s.f. or 2 marks for correct answer in range of 0.014-0.016 not given to 2 s.f., otherwise 1 mark for time of 55-60 seconds taken from graph]

5.1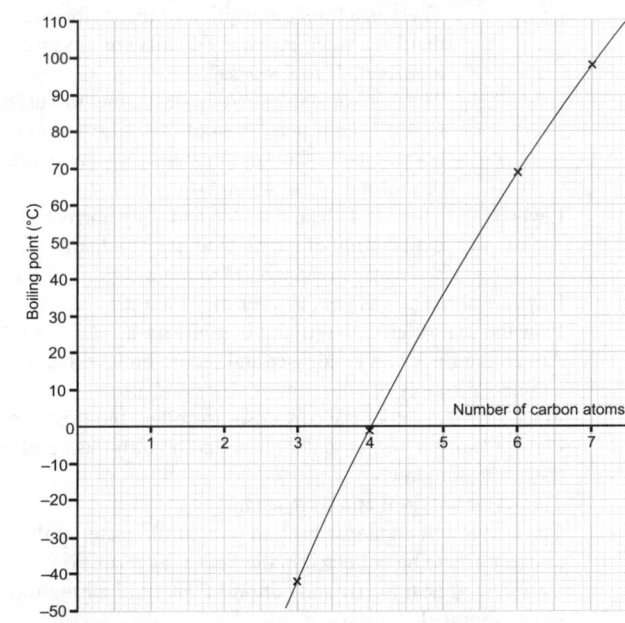

[1 mark for all four points correctly plotted, 1 mark for a smooth curve that passes through all the points]

5.2 E.g. 36 °C
[1 mark for correct reading of the graph drawn in 5.1]
Even if the graph you drew in question 5.1. was wrong, if you've used it correctly to find a value for the boiling point of a molecule with 5 carbon atoms, you get the mark here.

5.3 C_nH_{2n+2} *[1 mark]*

5.4 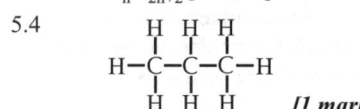 *[1 mark]*

5.5 Add a few drops of bromine water to both compounds and shake *[1 mark]*. With propane, nothing will happen *[1 mark]*. Propene will decolourise the bromine water / turn the solution from orange to colourless *[1 mark]*.

5.6 $C_3H_8 + 5O_2 \rightarrow 3CO_2 + \mathbf{4H_2O}$
[1 mark for correctly giving H_2O as the missing product, 1 mark for correct number in front of H_2O.]

5.7 Heptane has a higher viscosity than propane *[1 mark]*, as its molecular formula shows that it has more atoms in its carbon chain/has a longer carbon chain/is a bigger molecule *[1 mark]*.

6.1 As the level of nitrogen dioxide increases, the severity of asthma attacks increases / there is a positive correlation between the nitrogen dioxide level and severity of asthma attacks *[1 mark]*

6.2 E.g. the scientists only collected data from men. / The scientists only collected data from people under 40. / The severity of asthma attacks might be affected by another pollutant which happens to be abundant in the same areas as nitrogen dioxide. *[1 mark]*

6.3 Oxides of nitrogen are formed when nitrogen and oxygen from the air react at high temperatures *[1 mark]*. This happens when fuel is burned in car engines *[1 mark]*. In cities, there are lots of cars, so the levels of nitrogen oxides tend to be higher *[1 mark]*

6.4 Cool the reaction mixture. / Reduce the temperature the reaction is carried out at *[1 mark]*.

6.5 E.g. sulfur dioxide *[1 mark]*

7.1 E.g. they don't have enough fresh water / groundwater / surface water to treat and use as drinking water *[1 mark]*.

7.2 Desalination *[1 mark]*

7.3 How to grade your answer:
- Level 0: There is no relevant information. *[No marks]*
- Level 1: There is a brief description of the method to distil seawater, but many details are missing or incorrect. *[1 to 2 marks]*
- Level 2: There is some explanation of the method used to distil sea water. Most of the information is correct, but a few small details may be incorrect or missing. *[3 to 4 marks]*
- Level 3: There is a clear, detailed and fully correct explanation of the method used to distil sea water in the lab. *[5 to 6 marks]*

Here are some points your answer may include:
Pour the sea water into a (round bottom) flask.
Attach the flask to a condenser and secure both with clamps.
Connect a supply of cold water to the condenser.
Place a beaker under the condenser to collect the fresh water.
Place a Bunsen burner under the round bottom flask and heat the sea water slowly.
The water will boil and form steam.
The steam will condense back to pure liquid water in the condenser and be collected in the beaker as it runs out.
Continue to heat the (round bottom) flask until all the water has evaporated.

7.4 Large bits of material (such as grit) are removed from the sewage *[1 mark]*.

7.5 The effluent undergoes aerobic biological treatment *[1 mark]*. Aerobic bacteria break down any organic matter that is present *[1 mark]*.

Physics Practice Paper 1

Pages 458-472

1.1 liquid *[1 mark]*
1.2 potential energy *[1 mark]*, kinetic energy *[1 mark]*
1.3 gas *[1 mark]*
There are two flat bits on the graph that show changes of state. The first must show the substance melting and the second must show it boiling. So at Z it must be a gas.
1.4 X *[1 mark]*
1.5 The specific latent heat of fusion of the substance is 334 000 J/kg. *[1 mark]*
1.6 $E = mL = 5.00 \times 334\,000$ *[1 mark]*
    $= \mathbf{1\,670\,000\,J}$ *[1 mark]*

2.1 Volume = length × width × height
    $= 4.0 \times 4.0 \times 4.0$ *[1 mark]*
    $= \mathbf{64\,cm^3}$ *[1 mark]*
2.2 He should use cylinder A. The graduations are smaller on cylinder A, so the student will be able to make more accurate measurements *[1 mark]*.
2.3 density $= 0.435 \div 0.000058$ *[1 mark]*
    $= \mathbf{7500\,kg/m^3}$ *[1 mark]*

3.1 The type of insulation *[1 mark]*.
3.2 Any two from, e.g. thickness of insulation / starting temperature of the water / length of time water is left to cool / volume of water used / the beaker used / temperature of the room *[1 mark for each]*
3.3 Mean temperature = $(21.9 + 22.2 + 20.1) \div 3$ *[1 mark]*
    $= \mathbf{21.4\,°C}$ *[1 mark]*
3.4 E.g. the insulation does not cover the top of the beaker *[1 mark]*. This means a lot of energy is being lost from the top of the beaker *[1 mark]*. The amount of energy lost through the insulation is small in comparison, meaning that the decrease in energy is roughly the same each time *[1 mark]*.
3.5 With a thick layer of material that has a low thermal conductivity. *[1 mark]*
3.6 E.g. biofuel / waves / hydroelectricity / the tides *[1 mark]*
3.7 In 10 years she would save £375 × 10 = £3750 on her energy bill *[1 mark]*. This would not cover the £6000 cost of installing the solar panels, so the home owner would not save money *[1 mark]*.

4.1 It is twice as large as the resistance of circuit A. *[1 mark]*
4.2 E.g. The current is higher in circuit A *[1 mark]*. When potential difference is the same, the greater the resistance in the circuit, the smaller the current. Potential difference is the same in both circuits so the current in circuit A is higher because its resistance is lower *[1 mark]*.
4.3 The potential difference across each resistor is 6 V. *[1 mark]*
4.4 power = $3.0^2 \times 4.0$ *[1 mark]* = **36** *[1 mark]* **watts** *[1 mark]*
4.5 The resistance decreases *[1 mark]*.
4.6 E.g. a car engine temperature sensor / an electronic thermostat *[1 mark]*.

4.7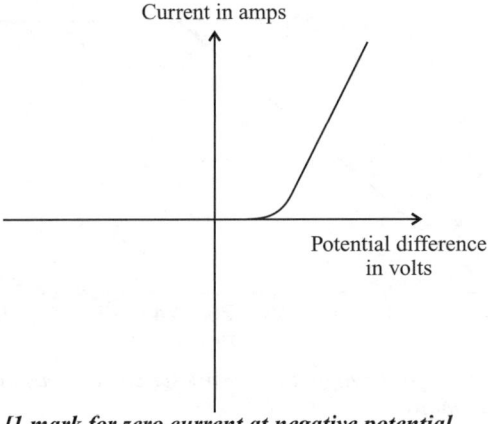

[1 mark for zero current at negative potential differences, 1 mark for curve into positive-gradient at a non-zero positive potential difference]

5.1 How to grade your answer:
Level 0: There is no relevant information. *[No marks]*
Level 1: There is a brief description of how to calculate the resistance of different lengths of wire. The answer is missing key details. *[1 to 2 marks]*
Level 2: There is a brief summary of the experimental procedure, including one way to ensure accurate results. The answer states which quantities need to be measured and explains how they can be used to calculate the resistance of the wire. *[3 to 4 marks]*
Level 3: There is full, clear and logical description of the experimental procedure. The answer states which quantities need to be measured and how they can be used to calculate the resistance of the wire. At least two ways to ensure the results of the experiment will be accurate are given. *[5 to 6 marks]*

Here are some points your answer may include:
Procedure
Attach a crocodile clip to the wire level with 0 cm on the ruler.
Attach the second crocodile clip to the wire, e.g. 5 cm from the first clip.
Write down the length of the wire between the clips.
Close the switch, then record the current through the wire and the potential across it.
Rearrange $V = IR$ to $R = V \div I$ and use this equation to calculate the resistance of the wire.
Repeat the procedure several times, moving the second crocodile clip further along each time to create a longer wire, e.g. move it 5 cm along each time to create wires of lengths 10 cm, 15 cm, 20 cm and 25 cm.
Plot a graph of resistance against wire length to see the relationship between the two.
Improving accuracy
If the wires heat up their resistance will increase, making the results less accurate.
To reduce the heating effect:
1. Switch off the circuit between measurements to allow the circuit to cool down / stop the circuit heating up.
2. Use a battery with a low potential difference / only use low currents (as the larger the current, the more the circuit will heat up).
Repeat all wire length measurements to ensure the correct length is measured. Draw lines on the crocodile clips as distance markers, to ensure the distance is being measured between the same points each time.
Tape the wire down onto the ruler to ensure the wire is straight and there are no kinks.
Ensure that the length measurements are taken at eye level to avoid errors due to parallax.

Make sure the same part of the crocodile clip is in contact with the wire each time it is moved or the resistance of the crocodile clip may affect the results.
Repeat the measurements of potential difference and current several times for each length of wire and use them to calculate a mean resistance for each length of wire.
This will minimise the amount of random error in the results and allow anomalous results to be easily identified.

5.2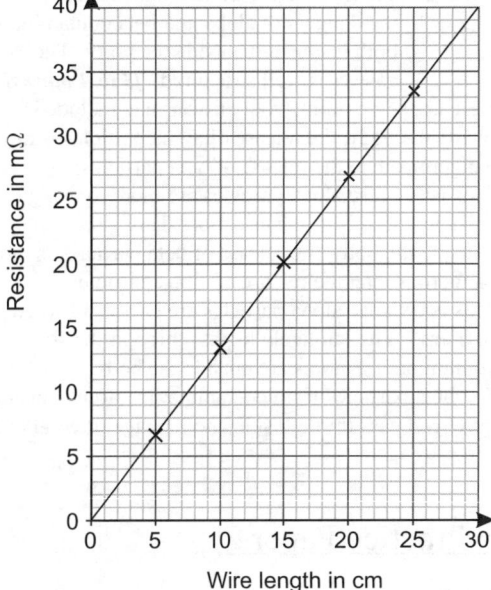

[1 mark for points correctly plotted, 1 mark for sensible line of best fit]
5.3 40 mΩ *[1 mark for value between 39 and 41 mΩ]*
5.4 As the length of the wire increases, its resistance increases *[1 mark]*.
6.1 **W** and **Y** *[1 mark]*
6.2 **X** *[1 mark]*
6.3 Gravitational potential energy = mass × gravitational field strength × height / GPE = $m \times g \times h$ *[1 mark]*
6.4 mass = 1 500 000 g = 1500 kg
Gravitational P.E. = 1500 × 9.8 × 40.0 *[1 mark]*
= **588 000 J** *[1 mark]*
6.5 elastic potential energy store (of the spring) *[1 mark]*
6.6 efficiency = $\frac{\text{useful energy output}}{\text{total energy output}}$ *[1 mark]*
6.7 efficiency = $\frac{18.0}{59.5}$ *[1 mark]*
= 0.3025... = **0.303 (to 3 s.f.)** *[1 mark]*
You could also have given the efficiency as a percentage.
To get a percentage, multiply by 100: 0.303 × 100 = 30.3%
6.8 Lubricating the moving parts of the carriage will decrease the energy lost due to frictional forces *[1 mark]*. This means more energy will be transferred to the kinetic energy store of the carriage *[1 mark]*. Kinetic energy = ½mv^2, so the velocity/speed will increase *[1 mark]*.
7.1 214 – 83 = **131 neutrons** *[1 mark]*
7.2 Because they have the same number of protons/are the same element *[1 mark]* and have a different number of neutrons/have a different mass number *[1 mark]*.
7.3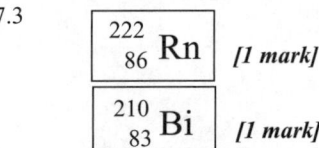

$^{222}_{86}\text{Rn}$ *[1 mark]*

$^{210}_{83}\text{Bi}$ *[1 mark]*

7.4 420 days *[1 mark]*
7.5 Half-life when activity drops to half original total.
Activity after 1 half-life = 80 ÷ 2 = 40 Bq
Using **Figure 14**, when activity = 40 Bq, time = 140 days.
Therefore, half-life = **140 days** *[1 mark]*

7.6　Contamination is when unwanted radioactive atoms are on or inside an object *[1 mark]*.
7.7　How to grade your answer:
　　Level 0: There is no relevant information. *[No marks]*
　　Level 1: There is a brief explanation on the dangers of alpha and beta contamination, but little to no comparison between them. The answer has little or no clear structure. *[1 to 2 marks]*
　　Level 2: There is a clear explanation and comparison of the dangers of alpha and beta contamination both inside and outside the body. The answer is well structured. *[3 to 4 marks]*
Here are some points your answer may include:
Inside the body, the release of alpha particles is more dangerous than the release of beta particles.
This is because alpha particles are more ionising than beta particles, so they can cause more damage to cells.
Outside the body, the release of beta particles is more dangerous than the release of alpha particles.
This is because alpha particles are easily absorbed by thin barriers, so it is unlikely to pass through the skin if the contaminating atoms are outside the body.
But beta particles from contaminating atoms outside the body can pass through skin and damage internal organs.

Physics Practice Paper 2

Pages 473-486

1.1　1 N *[1 mark]*
The spring is still, so the forces acting on it must be balanced.
1.2　If only one force was applied this would simply cause the spring to move in the direction of the force. *[1 mark]*
1.3　random error *[1 mark]*
Using a marker ensures that you're always measuring the length of the spring from the same point.
1.4　E.g. 6.0 N (allow estimate from 5.5 N up to 6.5 N) *[1 mark]*. The graph is a straight line up to that point which shows extension and force are proportional *[1 mark]*.
1.5　0.015 m = 1.5 cm
　　From the graph a force of 3.0 N causes this extension.
　　[1 mark for evidence of conversion to 1.5 cm.
　　1 mark for correct force.]
1.6　spring constant = 3.0 ÷ 0.015 *[1 mark]* = **200 N/m** *[1 mark]*
2.1　0 N *[1 mark]*
When an object is travelling at a constant speed in a fixed direction it is not accelerating and so the resultant force will be zero.
2.2　Speed is the gradient of a distance-time graph.
　　change in vertical axis = 20 − 11 = 9 m
　　change in horizontal axis = 27 − 15 = 12 s
　　speed = 9 ÷ 12 *[1 mark]* = **0.75 m/s** *[1 mark]*
2.3　Between 9 metres and 10 metres *[1 mark]*.
The swimmer is travelling fastest when the gradient of the graph is steepest.
2.4　speed = 20 ÷ 23 = 0.8695... m/s *[1 mark]*
　　= **0.87 m/s (to 2 s.f.)** *[1 mark]*

2.5　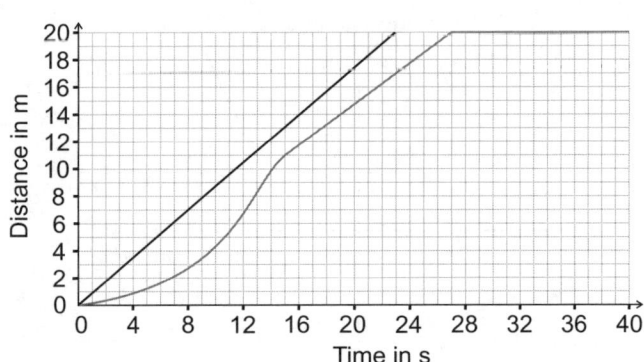
[1 mark for straight line, 1 mark for correct start and end points]
2.6　The magnitude of the swimmer's displacement is 0 m. *[1 mark]*
3.1　amplitude = 5 squares tall, one square = 1 cm tall
　　amplitude = 5 × 1 = **5 cm** *[1 mark]*
3.2　The number of complete waves that pass a point each second. *[1 mark]*
3.3　E.g. heating/cooking/infrared cameras *[1 mark]*
3.4　Variable 1 — Control *[1 mark]*
　　Variable 2 — Independent *[1 mark]*
3.5　The hypothesis is not supported as the white woollen blanket is not dark-coloured but has absorbed the signal *[1 mark]*.
3.6　How to grade your answer:
　　Level 0: There is no relevant information. *[No marks]*
　　Level 1: There is a clear description of at least one similarity and one difference between the features and properties of infrared waves and radio waves. *[1 to 2 marks]*
　　Level 2: There is a clear description of multiple similarities and differences between the features and properties of infrared waves and radio waves. *[3 to 4 marks]*
Here are some points your answer may include:
Both infrared and radio waves are electromagnetic waves.
They are both transverse waves.
The vibrations in both types of wave are perpendicular to the direction of energy transfer.
Both types of wave travel at the same speed through air or a vacuum.
Radio waves have a longer wavelength than infrared waves.
Radio waves have a lower frequency than infrared waves.
4.1　E.g. wear goggles in case string snaps / remain standing so that they can get out of the way of any falling masses. *[1 mark]*
4.2　**11.0 cm** *[1 mark]*
Each half-wave is 27.5 ÷ 5 = 5.5 cm.
So one full wave is 5.5 × 2 = 11.0 cm.
4.3　wave speed = frequency × wavelength / $v = f \times \lambda$ *[1 mark]*
4.4　wavelength = 11.0 cm = 0.110 m
　　wave speed = 120 × 0.110 *[1 mark]*
　　= **13.2 m/s or 13 m/s (to 2 s.f.)** *[1 mark]*
4.5　32.9 m/s is anomalous *[1 mark]*. The student should try to identify a reason for the anomalous result *[1 mark]*. If it is found to be due to an error, it should be ignored and not included when calculating the mean *[1 mark]*.
4.6　It is not suitable as the resolution of a ruler is too low *[1 mark]*. A micrometer/Vernier callipers would be more suitable *[1 mark]*.
Micrometers and vernier callipers have a much higher resolution than rulers.
4.7　As diameter decreases, wave speed increases / As diameter increases, wave speed decreases *[1 mark]*.
4.8　A scatter/line graph *[1 mark]*. Diameter and speed are both continuous variables *[1 mark]*.

5.1	9.8 m/s² *[1 mark]*	
5.2	Between Point A and Point B — The skydiver has a positive acceleration *[1 mark]* Between Point B and Point C — The skydiver is moving at a constant velocity *[1 mark]*	
5.3	Air resistance / friction / drag *[1 mark]*	
5.4	weight = mass × gravitational field strength / $W = mg$ *[1 mark]*	
5.5	$W = 80.0 × 9.8$ *[1 mark]* = **784 N** *[1 mark]*	
5.6	force = mass × acceleration / $F = ma$ *[1 mark]*	
5.7	$a = F ÷ m$ *[1 mark]* = 58 500 ÷ 4680 *[1 mark]* = **12.5 m/s²** *[1 mark]*	
6.1	36 – 12 = **24 m** *[1 mark]*	
6.2	Reaction times vary from person to person *[1 mark]*. A large sample will give a more accurate average *[1 mark]*.	
6.3	Whether the driver is tired. *[1 mark]* The speed the car is travelling at. *[1 mark]*	
6.4	Work done = force × distance *[1 mark]*	
6.5	Rearranging work done = force × distance for distance: distance = work done ÷ force *[1 mark]* = 75 000 ÷ 5000 *[1 mark]* = **15 m** *[1 mark]*	
7.1	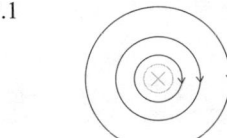 *[1 mark]*	

You need to use the right-hand thumb rule for this.

7.2 Decrease the current *[1 mark]*.
Move the magnet further from the wire *[1 mark]*.

7.3 How to grade your answer:
Level 0: There is no relevant information. *[No marks]*
Level 1: Predictions of what will happen to the bar magnet and nail are given and a brief reason is given for at least one of the predictions. *[1 to 2 marks]*
Level 2: Predictions of what will happen to the bar magnet and nail are given. There is some explanation for each prediction, or a detailed account of one of them. *[3 to 4 marks]*
Level 3: Predictions of what will happen to the bar magnet and nail are given. There is a clear and detailed explanation for each prediction. *[5 to 6 marks]*

Here are some points your answer may include:
The bar magnet will remain attached to the electromagnet.
The bar magnet is a permanent magnet.
When the current stops flowing in the wire, the wire has no magnetic field around it.
With the bar magnet attached, the iron core remains an induced magnet, as it's in the bar magnet's magnetic field.
Permanent magnets and induced magnets always attract each other.
So the bar magnet will remain stuck to the iron core (assuming the magnetic force is greater than the weight of the magnet).
The nail will fall off.
The nail was an induced magnet as it was in the magnetic field of the electromagnet.
When the current stops flowing, the electromagnet and the nail both lose their magnetism.
This means there is no attractive force keeping them together, so the nail will fall off.

Glossary

Abiotic factor	A non-living factor in an environment.
Acceleration	A change in velocity in a certain amount of time.
Accurate result	A result that is really close to the true answer.
Acid	A substance with a pH of less than 7 that forms H$^+$ ions in water.
Activation energy	The minimum amount of energy that reactant particles must have when they collide in order to react.
Active transport	The movement of particles against a concentration gradient (from an area of lower concentration to an area of higher concentration) using energy from respiration.
Activity (radioactive)	The number of nuclei of a sample that decay per second.
Adaptation	A feature that helps an organism to survive in the conditions of its natural environment.
Aerobic respiration	Respiration that uses oxygen. It produces carbon dioxide and water.
Air resistance	The frictional force caused by air on a moving object.
Alkali	A substance with a pH of more than 7 that forms OH$^-$ ions in solution.
Alkali metal	An element in Group 1 of the periodic table. E.g. sodium, potassium etc.
Alkane	A saturated hydrocarbon with the general formula C_nH_{2n+2}. E.g. methane, ethane etc.
Alkene	An unsaturated hydrocarbon that contains a carbon-carbon double bond and has the general formula C_nH_{2n}. E.g. ethene, propene etc.
Allele	An alternative version of a gene.
Alloy	A metal that is a mixture of two or more metals, or a mixture involving metals and non-metals.
Alpha decay	A type of radioactive decay in which an alpha particle is given out from a decaying nucleus.
Alpha particle	A positively-charged particle made up of two protons and two neutrons (a helium nucleus).
Alpha particle scattering experiment	An experiment in which alpha particles were fired at gold foil to see if they were deflected. It led to the plum pudding model being abandoned in favour of the nuclear model of the atom.
Alternating current (ac)	Current that is constantly changing direction.
Alveolus	A tiny air sac in the lungs, where gas exchange occurs.
Amino acid	A small molecule that is a building block of proteins.
Ammeter	A component used to measure the current through a component. It is always connected in series with the component.
Amplitude	The maximum displacement of a point on a wave from its undisturbed position.
Anaerobic respiration	Respiration without oxygen. It produces lactic acid in humans, and carbon dioxide and ethanol in plants and yeast.
Angle of incidence	The angle the incident ray of a wave makes with the normal at a boundary.
Angle of refraction	The angle a refracted ray makes with the normal when a wave refracts at a boundary.
Anion	A particle with a negative charge, formed when one or more electrons are gained.

Glossary

Anomalous result	A result that doesn't fit in with the rest of the data.
Antibiotic	A drug used to kill or prevent the growth of bacteria.
Antibiotic resistance	When bacteria aren't killed by an antibiotic.
Antibody	A protein produced by white blood cells in response to the presence of an antigen (e.g. on the surface of a pathogen).
Antigen	A molecule on the surface of a cell. Foreign antigens trigger white blood cells to make antibodies.
Antitoxin	A protein produced by white blood cells that stops toxins from working.
Artery	A blood vessel that carries blood away from the heart.
Asexual reproduction	When organisms reproduce by mitosis to produce genetically identical offspring.
Atmosphere	The layer of air that surrounds a planet.
Atom	A neutral particle made up of protons and neutrons in the nucleus, with electrons surrounding the nucleus.
Atomic number	The number of protons in the nucleus of an atom. It's also known as proton number.
Base	A substance that reacts with acids in neutralisation reactions.
Behavioural adaptation	A way in which an organism behaves that helps it to survive in its environment.
Beta decay	A type of radioactive decay in which a beta particle is given out from a decaying nucleus.
Beta particle	A high-speed electron emitted in beta decay.
Bias	Unfairness in the way data is presented, possibly because the presenter is trying to make a particular point (sometimes without knowing they're doing it).
Binomial system	The system used in classification for naming organisms using a two-part Latin name.
Bio-fuel	A renewable energy resource made from plant products or animal dung.
Biodiversity	The variety of different species of organisms on Earth, or within an ecosystem.
Biotic factor	A living factor in an environment.
Braking distance	The braking distance is the distance a vehicle travels after the brakes are applied until it comes to a complete stop, as a result of the braking force.
Capillary	A type of blood vessel involved in the exchange of materials at tissues.
Carbohydrase	A type of digestive enzyme that catalyses the breakdown of a carbohydrate (like starch) into simple sugars. Amylase is a carbohydrase.
Carbon footprint	A measure of the amounts of greenhouse gases released by a product, a service or an event.
Cardiovascular disease	Disease of the heart or blood vessels.
Catalyst	A substance that increases the speed of a reaction, without being changed or used up.
Categoric data	Data that comes in clear categories, e.g. blood type (A+, B−, etc.), metals (copper, zinc, etc.).
Cation	A particle with a positive charge, formed when one or more electrons are lost.

Glossary

Cell membrane	A membrane surrounding a cell, which holds it together and controls what goes in and out.
Cell wall	A structure surrounding some cell types, which gives strength and support.
Cellulose	A molecule which strengthens cell walls in plants and algae.
Central nervous system (CNS)	The brain and spinal cord. It's where reflexes and actions are coordinated.
Chemical bond	The attraction of two atoms for each other, caused by the sharing or transfer of electrons.
Chlorophyll	A green substance found in chloroplasts which absorbs light for photosynthesis.
Chloroplast	A structure found in plant cells and algae. It is the site of photosynthesis.
Chromatogram	The pattern of spots formed as a result of separating a mixture using chromatography.
Chromatography	An analytical method used to separate the substances in a mixture based on how the components interact with a mobile phase and a stationary phase.
Chromosome	A long molecule of DNA found in the nucleus. Each chromosome carries many genes.
Climate change	A change in the Earth's climate. E.g. global warming, changing rainfall patterns etc.
Clinical trial	A set of drug tests on human volunteers.
Clone	An organism that is genetically identical to another organism.
Closed system	A system where neither matter nor energy can enter or leave. The net change in total energy in a closed system is always zero.
Collision theory	The theory that in order for a reaction to occur, particles must collide with sufficient energy.
Combustion	An exothermic reaction between a fuel and oxygen.
Communicable disease	A disease that can spread between individuals.
Community	The populations of different species living in a habitat.
Compound	A substance made up of atoms of at least two different elements, chemically joined together.
Concentration	The amount of a substance in a certain volume of solution, given in units of 'units of amount of substance'/'units of volume'.
Conservation of energy principle	Energy can be transferred usefully from one energy store to another, stored or dissipated — but it can never be created or destroyed.
Contamination (radioactive)	The presence of unwanted radioactive atoms on or inside an object.
Continuous data	Numerical data that can have any value within a range (e.g. length, volume or temperature).
Contraceptive	A method of preventing pregnancy. Some methods use hormones and some do not.
Control experiment	An experiment that's kept under the same conditions as the rest of an investigation, but doesn't have anything done to it.
Control variable	A variable in an experiment that is kept the same.
Coordination centre	An organ (e.g. the brain, spinal cord or pancreas) that processes information from receptors and organises a response from the effectors.

Glossary

Coronary artery	A blood vessel which supplies blood to the heart muscle.
Coronary heart disease	A disease in which fatty deposits build up in the coronary arteries, causing them to become narrow.
Correlation	A relationship between two variables.
Covalent bond	A chemical bond formed when atoms share a pair of electrons.
Covalent substance	A substance where the atoms are held together by covalent bonds.
Cracking	The process that is used to break long-chain hydrocarbons down into shorter, more useful hydrocarbons. Two types of cracking are catalytic cracking and steam cracking.
Crystallisation	The formation of solid crystals as water evaporates from a solution. For example, salt solutions undergo crystallisation to form solid salt crystals.
Current	The flow of electric charge. The size of the current is the rate of flow of charge. Measured in amperes (A).
Cystic fibrosis	An inherited disorder of the cell membranes caused by a recessive allele.
Cytoplasm	A gel-like substance in a cell where most of the chemical reactions take place.
Deforestation	The cutting down of forests (large areas of trees).
Delocalised electron	An electron that isn't associated with a particular atom or bond and is free to move within a structure.
Density	A substance's mass per unit volume.
Dependent variable	The variable in an experiment that is measured.
Diabetes	A condition that affects the body's ability to control its blood glucose level.
Differentiation	The process by which a cell becomes specialised for its job.
Diffusion	The spreading out of particles from an area of higher concentration to an area of lower concentration.
Diode	A circuit component that only allows current to flow through it in one direction. It has a very high resistance in the other direction.
Direct current (dc)	Current that always flows in the same direction.
Displacement	The straight-line distance and direction from an object's starting position to its finishing position.
Displacement reaction	A reaction where a more reactive element replaces a less reactive element in a compound.
Displayed formula	A chemical formula that shows the atoms in a covalent compound and all the bonds between them.
Distance-time graph	A graph showing how the distance travelled by an object changes over a period of time.
Distillation	A way of separating out a liquid from a mixture. You heat the mixture until the bit you want evaporates, then cool the vapour to turn it back into a liquid.
Distribution	Where organisms are found within an area.
DNA	Deoxyribonucleic acid. The molecule in cells that stores genetic information.
Dominant allele	An allele whose characteristic always appears in an organism, whether it has two copies of the allele or only one.

Glossary

Double-blind trial	A clinical trial where neither the doctors nor the patients know who has received the drug and who has received the placebo until all the results have been gathered.
Drag	The frictional force caused by any fluid (a liquid or gas) on a moving object.
Earth wire	The green and yellow wire in an electrical cable that only carries current when there's a fault. It stops exposed metal parts of an appliance from becoming live.
Ecosystem	The interaction of a community of organisms with the non-living parts of their environment.
Effector	Either a muscle or gland which responds to nervous impulses.
Efficacy	Whether something, e.g. a drug, works or not.
Efficiency	The proportion of input energy transfer which is usefully transferred. Also the proportion of input power which is usefully output.
Elastic deformation	An object undergoing elastic deformation will return to its original shape and length once any forces being applied to it are removed.
Elastic object	An object which can be elastically deformed.
Elastic potential energy store	Anything that has been stretched or compressed, e.g. a spring, has energy in its elastic potential energy store.
Electrode	An electrical conductor which is submerged in the electrolyte during electrolysis.
Electrolysis	The process of breaking down a substance using electricity.
Electrolyte	A liquid or solution used in electrolysis to conduct electricity between the two electrodes.
Electromagnet	A solenoid with an iron core.
Electromagnetic (EM) spectrum	A continuous spectrum of all the possible wavelengths of electromagnetic waves.
Electron	A negatively charged particle that orbits the nucleus of an atom.
Electron shell	A region of an atom that contains electrons. It's also known as an energy level.
Electronic structure	The number of electrons in an atom (or ion) of an element and how they are arranged.
Electrostatic force	A force of attraction between opposite charges.
Element	A substance that is made up only of atoms with the same number of protons.
Empirical formula	A chemical formula showing the simplest possible whole number ratio of atoms in a compound.
Endothermic reaction	A reaction which takes in energy from the surroundings.
Energy level	A region of an atom that contains electrons. It's also known as an electron shell.
Energy store	A means by which an object stores energy. There are different types of energy store: thermal (or internal), kinetic, gravitational potential, elastic potential, chemical, magnetic, electrostatic and nuclear.
Enzyme	A protein that acts as a biological catalyst.
Equilibrium (physics)	A state in which all the forces acting on an object are balanced, so the resultant force is zero.

Glossary

Equilibrium (reactions)	The point at which the rates of the forward and backward reactions in a reversible reaction are the same, and so the amounts of reactants and products in the reaction container don't change.
Eukaryotic cell	A complex cell, such as a plant or animal cell.
Evolution	The changing of the inherited characteristics of a population over time.
Excretion	The removal of waste products from the body.
Exothermic reaction	A reaction which transfers energy to the surroundings.
Extinction	When no living individuals of a species remain.
Extremophile	An organism that's adapted to live in seriously extreme conditions.
Fair test	A controlled experiment where the only thing being changed is the independent variable.
Family tree	A diagram that shows how a characteristic is inherited in a group of related people.
Feedstock	A raw material used to produce other substances through industrial processes.
Fermentation	The process of anaerobic respiration in yeast cells.
Fertilisation	The fusion of a male and a female gamete during sexual reproduction.
Fertility	The ability to conceive a child.
Filtration	A physical method used to separate an insoluble solid from a liquid.
Finite resource	A resource that isn't replaced at a quick enough rate to be considered replaceable.
Flammability	How easy it is to ignite a substance.
Fluid	A substance that can flow — either a liquid or a gas.
Force	A push or a pull on an object caused by it interacting with something.
Formulation	A useful mixture with a precise purpose made by following a formula.
Fossil	The remains of an organism from many years ago, which is found in rock.
Fossil fuel	The fossil fuels are coal, oil and natural gas. They're non-renewable energy resources that we burn to generate electricity.
Fossil record	The history of life on Earth preserved as fossils.
Fraction	A group of hydrocarbons that condense together when crude oil is separated using fractional distillation. E.g. petrol, naphtha, kerosene etc.
Fractional distillation	A process that can be used to separate substances in a mixture according to their boiling points.
Frequency	The number of complete waves passing a certain point per second. Measured in hertz, Hz.
Friction	A force that opposes an object's motion. It acts in the opposite direction to motion.
Functional adaptation	Something that goes on inside an organism's body which helps it to survive in its environment.
Gamete	A sex cell, e.g. an egg cell or a sperm cell in animals.
Gamma decay	A type of radioactive decay in which a gamma ray is given out from a decaying nucleus.
Gamma ray	A high-frequency, short-wavelength electromagnetic wave.
Geiger-Müller tube	A particle detector that is used with a counter to measure count rate.

Glossary

Gene	A short section of DNA, found on a chromosome. A gene contains the instructions needed to make a protein, so it controls the development of a characteristic.
General formula	A formula that can be used to find the molecular formula of any member of a homologous series.
Genetic engineering	The process of cutting out a useful gene from one organism's genome and putting it into another organism's cells.
Genetically modified (GM) crop	A crop which has had its genes modified using genetic engineering.
Genome	All of the genetic material in an organism.
Genotype	What alleles you have, e.g. Tt.
Geothermal power	A renewable energy resource where energy is transferred from the thermal energy stores of hot rocks underground and is used to generate electricity or to heat buildings.
Giant covalent structure	A large molecule made up of a very large number of atoms held together by covalent bonds (also known as a macromolecule).
Gland	An organ that produces and secretes hormones.
Global dimming	The decrease in the amount of sunlight reaching the Earth's surface due to an increase in the amount of particulates in the atmosphere.
Global warming	The increase in the average temperature of the Earth.
Glycogen	A molecule that acts as a store of glucose in liver and muscle cells.
Gradient	The slope of a line graph. It shows how quickly the variable on the y-axis changes with the variable on the x-axis.
Gravitational potential energy (g.p.e) store	Anything that has mass and is in a gravitational field has energy in its gravitational potential energy store.
Greenhouse effect	When greenhouse gases in the atmosphere absorb long wavelength radiation and re-radiate it in all directions, including back towards Earth, helping to keep the Earth warm.
Greenhouse gas	A gas that can absorb long wavelength radiation.
Group	A column in the periodic table.
Guard cell	A type of cell found on either side of a stoma. A pair of these cells controls the stoma's size.
Habitat	The place where an organism lives.
Haemoglobin	A red pigment found in red blood cells which carries oxygen.
Half-life	The time it takes for the number of nuclei of a radioactive isotope in a sample to halve. OR The time it takes for the count rate (or activity) of a radioactive sample to fall to half its initial level.
Hazard	Something that could cause harm (e.g. fire, electricity, etc.).
Heterozygous	When an organism has two alleles for a particular gene that are different.
Homeostasis	The regulation of conditions inside your body and cells in order to maintain a stable internal environment.

Glossary

Homologous series	A group of chemicals that react in a similar way because they have the same functional group. E.g. the alcohols or the carboxylic acids.
Homozygous	When an organism has two alleles for a particular gene that are the same.
Hormone	A chemical messenger which travels in the blood to act on a target cell.
Hydrocarbon	A compound that is made from only hydrogen and carbon.
Hypothesis	A possible explanation for a scientific observation.
I-V characteristic	A graph of current against potential difference for a component.
Inbreeding	When closely related animals or plants are bred together.
Incomplete combustion	When a fuel burns but there isn't enough oxygen for it to burn completely. Products can include carbon monoxide and carbon particulates. Also known as partial combustion.
Independent variable	The variable in an experiment that is changed.
Indicator	A substance that changes colour above or below a certain pH.
Induced magnet	An object that turns into a magnet when it is placed inside another magnetic field.
Inelastic deformation	An object undergoing inelastic deformation will not return to its original shape and length once the forces being applied to it are removed.
Infrared (IR) radiation	A type of electromagnetic wave that is given out by all objects. It can also be absorbed by objects which makes the object hotter.
Inherited disorder	A disorder caused by a faulty allele, which can be passed on to an individual's offspring.
Insoluble	A substance is insoluble if it does not dissolve in a particular solvent.
Insulin	A hormone produced and secreted by the pancreas when the blood glucose level is too high. It causes cells to take up more glucose from the blood, reducing the blood glucose level.
Interdependence	Where, in a community, different species depend on each other for things such as food, shelter, pollination and seed dispersal.
Intermolecular force	A force of attraction that exists between molecules.
Internal energy	The total energy that a system's particles have in their kinetic and potential energy stores.
Ion	A charged particle formed when one or more electrons are lost or gained from an atom or molecule.
Ionic bond	A strong attraction between oppositely charged ions.
Ionic compound	A compound that contains positive and negative ions held together in a regular arrangement (a lattice) by electrostatic forces of attraction.
Ionic lattice	A closely-packed regular arrangement of particles held together by electrostatic forces of attraction.
Ionising radiation	Radiation that has enough energy to knock electrons off atoms.
Irradiation	Exposure to radiation.
Isotope	A different form of the same element, which has atoms with the same number of protons (atomic number), but a different number of neutrons (and so different mass number).
Joules	The standard unit of energy.

Glossary

Kinetic energy store	Anything that's moving has energy in its kinetic energy store.
Latent heat	The energy required to change the state of a substance without changing its temperature.
Lattice	A closely-packed regular arrangement of particles.
Life cycle assessment	An assessment of the environmental impact of a product over the course of its life.
Light-dependent resistor (LDR)	A resistor whose resistance is dependent on light intensity. The resistance decreases as light intensity increases.
Limit of proportionality	The point beyond which the force applied to an elastic object is no longer directly proportional to the extension of the object.
Limiting factor	A factor which prevents a reaction from going any faster.
Line of action (of a force)	A straight line passing through the point at which the force is acting in the same direction as the force.
Linear graph	A straight line graph.
Lipase	A type of digestive enzyme that catalyses the breakdown of lipids into fatty acids and glycerol, in the small intestine.
Litmus	A single indicator that's blue in alkalis and red in acids.
Live wire	The brown wire in an electrical cable that carries an alternating potential difference from the mains.
Longitudinal wave	A wave in which the vibrations are along the same line as the direction of energy transfer.
Lubricant	A substance used between two objects to reduce friction between surfaces.
Magnetic field	A region where magnetic materials (like iron and steel) experience a force.
Magnetic material	A material (such as iron, steel, cobalt or nickel) which is attracted to magnets.
Mass number	The total number of protons and neutrons in an atom.
Mean (average)	A measure of average found by adding up all the data and dividing by the number of values there are.
Median (average)	A measure of average found by selecting the middle value from a data set arranged in order from smallest to largest.
Meiosis	A type of cell division where a cell divides twice to produce four genetically different gametes. It occurs in the reproductive organs.
Menstrual cycle	A monthly sequence of events during which the body prepares the lining of the uterus (womb) in case it receives a fertilised egg, and releases an egg from an ovary. The uterus lining then breaks down if the egg has not been fertilised.
Meristem tissue	Tissue found at the growing tips of plant shoots and roots that is able to differentiate into any type of plant cell.
Metabolism	All the chemical reactions that happen in a cell (or in the body).
Metal	An element that can form positive ions when it reacts.
Metal ore	Rocks that are found naturally in the Earth's crust containing enough metal to make the metal profitable to extract.
Metallic bond	The attraction between metal ions and delocalised electrons in a metal.
Microwave	A type of electromagnetic wave that can be used for cooking and satellite communications.

Glossary

Mitochondria	Structures in a cell where most of the reactions for aerobic respiration take place.
Mitosis	A type of cell division where a cell divides once, to form two new identical cells.
Mixture	A substance made from two or more elements or compounds that aren't chemically bonded to each other.
Mobile phase	In chromatography, the mobile phase is a gas or liquid where the molecules are able to move.
Mode (average)	A measure of average found by selecting the most frequent value from a data set.
Model	A simple way of describing or showing what's going on in real life.
Molecule	A particle made up of at least two atoms held together by covalent bonds.
Molecular formula	A chemical formula showing the actual number of atoms of each element in a compound.
Motor neurone	A nerve cell that carries electrical impulses from the CNS to effectors.
MRSA	A strain of bacteria that is resistant to most known antibiotics.
Mutation	A random change in an organism's DNA.
National grid	The network of transformers and cables that distributes electrical power from power stations to consumers.
Natural resource	A resource formed without human input.
Natural selection	The process by which species evolve.
Nervous system	The organ system in animals that allows them to respond to changes in their environment.
Neurone	A nerve cell. Neurones transmit information around the body, including to and from the CNS.
Neutral substance	A substance with a pH of 7.
Neutral wire	The blue wire in an electrical cable that current in an appliance normally flows through. It is around 0 V.
Neutralisation reaction	The reaction between acids and bases that leads to the formation of neutral products — usually a salt and water.
Neutron	A particle found in the nucleus of an atom. It has no charge.
Newton's First Law	An object will remain at rest or travelling at a constant velocity unless it is acted on by a resultant force.
Newton's Second Law	The acceleration of an object is directly proportional to the resultant force acting on it, and inversely proportional to its mass.
Newton's Third Law	When two objects interact, they exert equal and opposite forces on each other.
Non-communicable disease	A disease that cannot spread between individuals.
Non-contact force	A force that can act between objects that are not touching.
Non-metal	An element that doesn't form positive ions when it reacts, with the exception of hydrogen.
Non-renewable energy resource	An energy resource that is non-renewable cannot be made at the same rate as it's being used.
Normal (at a boundary)	A line that's perpendicular (at 90°) to a surface at the point of incidence (where a wave hits the surface).

Glossary

Nuclear model	An accepted theory which describes the atom as having a tiny, positively charged nucleus surrounded by shells which are occupied by negative electrons.
Nucleus (atom)	The centre of an atom, containing protons and neutrons.
Nucleus (of a cell)	A structure found in animal and plant cells which contains the genetic material that controls the activities of the cell.
Obesity	A condition where a person has an excessive amount of body fat, to the point where it poses a risk to their health.
Ohmic conductor	A conductor with resistance that is constant at a constant temperature. It has a linear I-V characteristic.
Optimum dose	The dose of a drug that is most effective and has few side effects.
Optimum level (in the body)	A level of something (e.g. water, ions or glucose) that enables the body to work at its best.
Organ	A group of different tissues that work together to perform a function.
Organ system	A group of organs working together to perform a function.
Organic compound	A chemical compound that contains carbon atoms.
Osmosis	The movement of water molecules across a partially permeable membrane from a region of higher water concentration to a region of lower water concentration.
Oxidation	A reaction where electrons are lost or oxygen is gained by a species.
Oxygen debt	The amount of extra oxygen that your body needs after exercise.
Paper chromatography	An analytical technique that can be used to separate and analyse coloured substances.
Parallel circuit	A circuit in which every component is connected separately to the positive and negative ends of the battery.
Partially permeable membrane	A membrane with tiny holes in it, which lets some molecules through but not others.
Pathogen	A microorganism that causes disease, e.g. a bacterium, virus, protist or fungus.
Peer-review	The process in which other scientists check the results and explanations of an investigation before they are published.
Period	A row in the periodic table.
Period (of a wave)	The time taken for one complete wave to pass a certain point.
Periodic table	A table of all the known elements, arranged in order of atomic number so that elements with similar chemical properties are in groups.
Permanent magnet	An object that always has its own magnetic field around it.
Permanent vacuole	A structure in plant cells that contains cell sap.
pH scale	A scale from 0 to 14 that is used to measure how acidic or alkaline a solution is.
Phagocytosis	The process by which white blood cells engulf (surround) pathogens and digest them.
Phenotype	The characteristics you have, e.g. brown eyes.
Phloem	A type of plant tissue which transports food substances (dissolved sugars) around a plant.
Photosynthesis	The process by which plants use energy to convert carbon dioxide and water into glucose and oxygen.

Glossary

Physical change	A change where you don't end up with a new substance — it's the same substance as before, just in a different form. (A change of state is a physical change.)
Placebo	A substance that is like a drug being tested, but which doesn't do anything.
Plasma	The liquid part of blood, which carries blood cells and other substances around the body.
Platelet	A small fragment of a cell found in the blood, which helps blood to clot at a wound.
Plum pudding model	A disproved theory of the atom as a ball of positive charge with electrons inside it.
Polydactyly	An inherited disorder caused by a dominant allele, where a person has extra fingers or toes.
Polymer	A long chain molecule that is formed by joining lots of smaller molecules (monomers) together.
Potable water	Water that is safe for drinking.
Potential difference	The driving force that pushes electric charge around a circuit, measured in volts (V). Also known as pd or voltage.
Power	The rate of transferring energy (or doing work). Normally measured in watts (W).
Precipitate	A solid that is formed in a solution during a chemical reaction.
Precise result	When all the data is close to the mean.
Predator	An animal that hunts and kills other animals for food.
Prediction	A statement based on a hypothesis that can be tested.
Pressure	The force per unit area exerted on a surface.
Prey	An animal that is hunted and killed by another animal for food.
Primary consumer	An organism in a food chain that feeds on a producer.
Producer	An organism at the start of a food chain that makes its own food using energy from the Sun.
Product	A substance that is formed in a chemical reaction.
Prokaryotic cell	A small, simple cell, e.g. a bacterium.
Protease	A type of digestive enzyme that catalyses the breakdown of proteins into amino acids, in the stomach and small intestine.
Protein	A large biological molecule made up of long chains of amino acids.
Protist	A type of pathogen. Protists are often transferred to other organisms by a vector.
Proton	A positively charged particle found in the nucleus of an atom.
Punnett square	A type of genetic diagram.
Pure substance	A substance that only contains one compound or element throughout.
Quadrat	A square frame enclosing a known area. It is used to study the distribution of organisms.
Radiation dose	A measure of the risk of harm to your body due to exposure to radiation.
Radio wave	A type of electromagnetic wave mainly used for radio and TV signals.
Radioactive decay	The random process of a radioactive substance giving out radiation from the nuclei of its atoms.

Glossary

Radioactive substance	A substance that spontaneously gives out radiation from the nuclei of its atoms.
Random error	A difference in the results of an experiment caused by unpredictable events, e.g. human error in measuring.
Range	The difference between the smallest and largest values in a set of data.
Rate of reaction	How fast the reactants in a reaction are changed into products.
Ray	A straight line showing the path along which the wave moves.
Ray diagram	A diagram that shows the path of light waves.
Reactant	A substance that reacts in a chemical reaction.
Reaction profile	A graph that shows how the energy in a reaction changes as the reaction progresses (also known as an energy level diagram).
Reaction time	The time taken for a person to react after an event (e.g. seeing a hazard).
Reactivity series	A list of elements arranged in order of their reactivity. The most reactive elements are at the top and the least reactive at the bottom.
Receptor	A group of cells that are sensitive to a stimulus (e.g. receptor cells in the eye detect light).
Recessive allele	An allele whose characteristic only appears in an organism if it has two copies of the allele.
Reduction	A reaction where electrons are gained or oxygen is lost.
Reflex	A fast, automatic response to a stimulus.
Refraction	When a wave changes direction as it passes across the boundary between two materials at an angle to the normal.
Relative atomic mass (A_r)	The average mass of the atoms of an element measured relative to the mass of one atom of carbon-12. The relative atomic mass of an element is the same as its mass number in the periodic table.
Relative formula mass (M_r)	All the relative atomic masses (A_r) of the atoms in a compound added together.
Relay neurone	A nerve cell that carries electrical impulses from sensory neurones to motor neurones.
Renewable energy resource	An energy resource that is renewable is one that is being, or can be, made at the same rate (or faster) than it's being used.
Repeatable result	A result that will come out the same if the experiment is repeated by the same person using the same method and equipment.
Repeating unit	The shortest repeating section of a polymer.
Reproducible result	A result that will come out the same if someone different does the experiment.
Resistance	Anything in a circuit that reduces the flow of charge. Measured in ohms, Ω.
Respiration	The process of breaking down glucose to transfer energy, which goes on in every cell.
Resultant force	A single force that can replace all the forces acting on an object to give the same effect as the original forces acting altogether.
Reversible reaction	A reaction where the products of the reaction can themselves react to produce the original reactants.

Glossary

R_f value	In chromatography, the ratio between the distance travelled by a dissolved substance and the distance travelled by a solvent.
Ribosome	A structure in a cell, where proteins are made.
Right-hand thumb rule	The rule to work out the direction of the magnetic field around a current-carrying wire. Your thumb points in the direction of the current, and your fingers curl in the direction of the magnetic field.
Risk	The chance that a hazard will cause harm.
Risk factor	Something that is linked to an increased chance that a person will develop a disease.
S.I. unit	A standard unit of measurement, recognised by scientists all over the world.
Scalar	A quantity that has magnitude but no direction.
Secondary consumer	An organism in a food chain that eats a primary consumer.
Selective breeding (artificial selection)	When humans select the plants or animals that are going to breed together, so that the genes for useful characteristics become more common in the population.
Sensory neurone	A nerve cell that carries electrical impulses from a receptor in a sense organ to the CNS.
Series circuit	A circuit in which every component is connected in a line, end to end.
Sex chromosome (humans)	One of the 23rd pair of chromosomes, X or Y. Together they determine whether an individual is male or female.
Sexual reproduction	When two gametes combine to produce a new individual that is genetically different.
Significant figure	The first significant figure of a number is the first non-zero digit. The second, third and fourth significant figures follow on immediately after it.
Simple distillation	A way of separating a liquid out from a mixture if there are large differences in the boiling points of the substances.
Simple molecule	A molecule made up of only a few atoms held together by covalent bonds.
Solar cell	A device that generates electricity directly from the Sun's radiation.
Solenoid	A coil of wire.
Solute	A substance dissolved in a solvent to make a solution.
Solution	A mixture made up of one substance (the solute) dissolved in another (the solvent).
Solvent	A liquid in which another substance (a solute) can be dissolved.
Solvent front	The point the solvent has reached up the filter paper during paper chromatography.
Species	A group of similar organisms that can reproduce to give fertile offspring.
Specific heat capacity	The amount of energy (in joules) needed to raise the temperature of 1 kg of a material by 1°C.
Specific latent heat (SLH)	The amount of energy needed to change 1 kg of a substance from one state to another without changing its temperature. (For cooling, it is the energy released by a change in state.)
Specific latent heat of fusion	The specific latent heat for changing between a solid and a liquid (melting or freezing).
Specific latent heat of vaporisation	The specific latent heat for changing between a liquid and a gas (evaporating, boiling or condensing).
Stable community	A community where all the species and environmental factors are in balance. This means that the population sizes stay about the same.

Glossary

Standard form	A number written in the form $A \times 10^n$, where A is a number between 1 and 10.
State of matter	The form which a substance can take — e.g. solid, liquid or gas.
State symbol	The letter, or letters, in brackets that are placed after a substance in an equation to show what physical state it's in. E.g. gaseous carbon dioxide is shown as $CO_{2(g)}$.
Statins	A group of drugs that are used to decrease the risk of heart and circulatory disease.
Stationary phase	In chromatography, the stationary phase is a solid or really thick liquid where molecules are unable to move.
Stem cell	An undifferentiated cell that can become one of many different types of cell, or produce more stem cells.
Stent	A wire mesh tube that's put inside an artery to help keep it open.
Stimulus	A change in the environment.
Stoma	A tiny hole in the surface of a leaf.
Stopping distance	The distance covered by a vehicle in the time between the driver spotting a hazard and the vehicle coming to a complete stop. It's the sum of the thinking distance and the braking distance.
Structural adaptation	A feature of an organism's body structure that helps it to survive in its environment.
Sustainable development	An approach to development that takes into the account the needs of present society while not damaging the lives of those in the future.
Synapse	A connection between two neurones.
System	The object, or group of objects, that you're considering.
Systematic error	An error that is consistently made throughout an experiment.
Tangent	A straight line that touches a curve at a particular point without crossing it.
Terminal velocity	The maximum velocity a falling object can reach without any added driving forces. It's the velocity at which the resistive forces (drag) acting on the object match the force due to gravity (weight).
Tertiary consumer	An organism in a food chain that eats a secondary consumer.
Theory	A hypothesis which has been accepted by the scientific community because there is good evidence to back it up.
Thermal conductivity	A measure of how quickly an object transfers energy by heating through conduction.
Thermal decomposition	A reaction where one substance chemically changes into at least two new substances when it's heated.
Thermal insulator	A material with a low thermal conductivity.
Thermistor	A resistor whose resistance is dependent on the temperature. The resistance decreases as temperature increases.
Thinking distance	The distance a vehicle travels during the driver's reaction time (before the brakes have been applied).
Three-core cable	An electrical cable containing a live wire, a neutral wire and an earth wire.
Tissue	A group of similar cells that work together to carry out a function.
Toxicity	How harmful something is, e.g. a drug.
Toxin	A poison. Toxins are often produced by bacteria.
Transect	A line which can be used to study the distribution of organisms across an area.

Glossary

Transformer	A device which can change the potential difference of an ac supply.
Translocation	The movement of food substances (e.g. dissolved sugars) around a plant.
Transpiration stream	The movement of water from a plant's roots, through the xylem and out of the leaves.
Transverse wave	A wave in which the vibrations are perpendicular (at 90°) to the direction of energy transfer.
Tumour	A growth of abnormal cells.
Ultraviolet (UV) radiation	A type of electromagnetic wave, the main source of which is sunlight.
Uncertainty	The amount of error measurements might have.
Universal indicator	A wide range indicator that changes colour depending on the pH of the solution that it's in.
Urea	A waste product (produced from the breakdown of proteins in the liver).
Vaccination	Injecting dead or inactive pathogens to produce an immune response that will help to protect you against the same pathogen in the future.
Valid result	A repeatable and reproducible result from an experiment designed to be a fair test.
Valve	A structure in the heart or in a vein that prevents blood from flowing in the wrong direction.
Variation	The differences that exist between individuals.
Vector (in disease)	An organism that transfers a disease from one animal or plant to another, which doesn't get the disease itself.
Vector (physics)	A quantity which has both magnitude (size) and a direction.
Vein	A blood vessel that carries blood to the heart.
Velocity	The speed and direction of an object.
Velocity-time graph	A graph showing how the velocity of an object changes over a period of time.
Virus	A tiny disease-causing agent that can only replicate within body cells.
Viscosity	How runny or gloopy a substance is.
Visible light	The part of the electromagnetic spectrum that we can see with our eyes.
Voltmeter	A component used to measure the potential difference across a component. Always connected in parallel with the component.
Wave	A vibration that transfers energy without transferring any matter.
Wavelength	The length of a full cycle of a wave, e.g. from a crest to the next crest.
Weight	The force acting on an object due to gravity.
White blood cell	A blood cell that is also part of the immune system, defending the body against disease.
Work done	The energy transferred when a force moves an object through a distance, or by a moving charge.
X-ray	A high-frequency, short-wavelength electromagnetic wave. It is mainly used in medical imaging and treatment.
Xylem	A type of plant tissue which transports water and mineral ions around a plant.

Index

A
abiotic factors 142
abstinence 111
acceleration 357, 362
accuracy 7
acids 215-217, 219
activation energy 230, 234
active sites 41, 42
active transport 33
activity (radioactivity) 340, 341
adaptations 144
adrenal glands 104
aerobic digestion 279
aerobic respiration 91
air resistance 360
alkali metals 183, 184
alkalis 215
alkanes 246-251
alkenes 250
alleles 120
alloys 203
alpha particle scattering experiments 176, 334
alpha radiation 337, 338, 343
alternating current (ac) 320
alternating potential difference 320
alveoli 36, 49
ammeters 308, 399
amplitude 371
amylase 43, 44
anaerobic digestion 279
anaerobic respiration 92, 93
angle of incidence 376
angle of refraction 376
animal cells 17
anodes 223-226
anomalous results 8
antibiotic resistance 130, 131
antibiotics 81
antibodies 53, 79
antiretroviral drugs 75
antitoxins 53, 79
aorta 51
apparatus 392, 393, 396-399
Archaea 138
arteries 52
artificial hearts 58
asexual reproduction 115
aspirin 81
atmosphere (of the Earth) 262-265
atomic models 176, 177, 334, 335
atomic number 161-163, 180, 336
atoms 161-164, 176, 177, 334, 335
atria 51
average speed 356

B
bacteria 18, 72, 73, 138
balances 392
balancing equations 166
ball and stick models 193
bar charts 10
barrier methods (contraception) 110, 111
bases 215-217
behavioural adaptations 144
Benedict's test 45
benign tumours 64
beta radiation 337, 339, 343
biased data 3
bile 44
binomial system 139
biodiversity 153, 155-158
    maintenance of 158
biofuels 298
biotic factors 143
Biuret test 46
blind studies 82
blood vessels 52
Bohr model of the atom 177, 334
boiling 206, 330, 331
braking distances 367, 368
breathing rate 49, 93
bromine water 250
bronchi 49
bronchioles 49
Buckminsterfullerene 202
Bunsen burners 398

C
cancer 64
    treatment 380
capillaries 52
carbohydrases 44
carbon cycle 150
carbon dioxide
    in the atmosphere 154, 156, 262-265
carbon footprints 266
carbon monoxide 267
catalysts 235, 251
cathodes 223-226
cells 17, 18
    cell cycle 27, 28
    division 27, 28
    drawing 21
    membranes 17, 30
    specialised 23, 24
    walls 18
central nervous system (CNS) 98-100
centre of mass 348
Chadwick, James 177
changes of state 206, 330, 331
charge
    electric 305, 306
    ions 190, 335
    of an atom 161, 335
    of a nucleus 161, 336
chemical equations 165, 166
chemical formulas 164
chlorophyll 85, 86
chloroplasts 18
cholesterol 57
chromatograms 256-258
chromatography 170, 255-258
chromosomes 27, 113
    X and Y 117, 118
cilia 78
circuit symbols 305
circuits 305, 314-318
    investigating resistance 318
    parallel 316, 317
    series 314, 315
classification 138, 139
climate change 155, 265
clinical trials 82
closed systems 284
coal 302
collision theory 234, 235
combustion reactions 228, 247, 267
communicable diseases 60
communication of ideas 3
compasses 387
competition 143
compounds 164
compressions (in waves) 372
concentrations 212
conclusions 14
condensing 149, 206, 330, 331
condoms 110
conservation of energy 287
conservation of mass 210, 211
contact forces 347
contamination (radioactivity) 343
contraception 109-111
control experiments 6
control variables 6
converting units 13
cooling 330
coronary arteries 51
coronary heart disease 56
correlations 11, 14
count-rate 340, 341
covalent bonding 196-198, 200-202
cover slip (slides) 20
cracking (of hydrocarbons) 250, 251
crude oil 248, 249
crystallisation 171, 172, 217
current (electrical) 305, 306
    alternating 320
    direct 320
    energy transferred 322
    I-V characteristics 309, 310
    measuring 308, 399
    national grid 324
    parallel circuits 317
    series circuits 315, 318
cystic fibrosis 122
cytoplasm 17

D
dangers of ionising radiation 342, 343, 383
Darwin, Charles 128
decay 150
deceleration 357
deforestation 157
delocalised electrons 201-203
density 328, 329
dependent variables 6
desalination 277, 278
diabetes 106
diamond 201
diaphragms (contraceptive) 111
differentiation 23
diffusion 30
digestive enzymes 44
digestive system 40
digitalis 81
diodes 307, 309
direct current (dc) 320
direct potential difference 320
displacement (of object) 356
displacement reactions 186, 220
displayed formulas 196
dissipated energy 287, 294
distance 356
distance-time graphs 358
distillation
    fractional 174, 249
    of seawater 277
    simple 173
distribution of organisms 147, 148
DNA 113
dosage 82
dot and cross diagrams 192, 196-198
double-blind studies 82
drag 360
drugs 81, 82
    development of 82

E
earth wires 320
ecosystems 141
    maintenance of 158
effectors 97-99
efficacy (of drugs) 82
efficiency 295
elastic deformation 351, 352
elastic potential energy stores 286, 351-354
electric charge 305, 306
electric heaters 398
electric shocks 320
electricity 305-323
electrodes 223-226
electrolysis 223-226, 396
electrolytes 223
electromagnetic spectrum 378

Index

electromagnetic waves 378, 381, 382
  dangers 383
  uses 379, 380
electromagnetism 388, 389
electron microscopes 19
electron shells 161, 177, 178, 335
electrons 161, 334, 335
electronic structures 178, 182, 190, 191
elements 162, 163
embryo screening 124
empirical formulas 194
endocrine system 103, 104
endothermic reactions 85, 228-230, 243
energy 283
  conservation of 287
  internal 330, 331
energy resources
  generating electricity 298-303
  heating 297
  non-renewable 297, 302
  renewable 297-301, 303
  transport 297
energy stores 283
  elastic potential 286, 351-354
  gravitational potential 285, 286
  kinetic 285, 286
  thermal 284, 285, 288-290, 294
energy transfers 284, 285, 287, 349
  by heating 284, 288-290
  by radiation 284
  by waves 371, 372, 378
  efficiency 295
  electrical 284, 321, 322
  mechanical 284, 285
  rate of 293
  reducing 294
  work done 284, 285, 349
environmental variation 126
enzymes 41-44, 235
epidermal tissue 66
equations
  balancing 166
  symbol 165, 166
equilibria (chemical) 242, 243
equilibria (forces) 363
ethical issues 4, 395
Eukaryota 138
eukaryotes 17, 18
eureka cans 329, 393
evaluations 16
evaporation 149, 171, 172, 330, 331
evolution 128, 129
evolutionary trees 139
exchange surfaces 35-37
exercise 93
exothermic reactions 91, 228-230, 243

experimental safety 395
extension (of objects) 351-354
extinction 129
extraction of metals 221, 224

F

family (classification) 138
family trees 123
fatty acids 44
fermentation 92
fertility 109-111
field strength
  gravitational 348
  magnetic 386-389
filament lamps 307, 309
filtration 171, 172, 276
finite resources 270, 271
five kingdom classification system 138
Fleming, Alexander 81
follicle-stimulating hormone (FSH) 109
food chains 145
food poisoning 73
food webs 141
force-extension graphs 352-354
forces
  contact 347
  frictional 294, 349, 360
  gravitational 285
  interaction pair 347, 363
  magnetic 386, 387
  Newton's laws 362, 363
  non-contact 347
  resultant 349
  weight 348
formulas (of compounds) 164
formulations 254
fossil fuels 248, 263, 267, 270, 302
fossils 137
fractional distillation 174, 249
freezing 206, 330, 331
frequency 371-373
  of EM waves 378
  of mains supply 320
friction 294, 349, 368
fullerenes 202
functional adaptations 144
fungi 72, 76

G

gametes 114-117
gamma rays 337, 339, 343, 378, 380
  dangers of 343, 383
gas exchange 36, 37, 49
gas syringes 237, 392
gases 205-207, 326, 392, 396, 397
  gas pressure 327
  natural gas 302
  particle motion 327
  states of matter 330

Geiger-Muller tube 340
gene mutation 383
genes 27, 113
genetic
  diagrams 118, 121-123
  disorders 122-124
  engineering 135
  variation 126
genetically modified crops 135
genomes 113
genotypes 120
genus 138, 139
geothermal power 300
giant covalent structures 201, 202
giant ionic lattices 193
gills 37
glands 103, 104
glass recycling 272
global warming 154, 155
glucose
  in the blood 105, 106
  test for 45
  uses in plants 85
glycerol 44
glycogen 105
gonorrhoea 73
gradients 352, 358
graphene 202
graphite 201
graphs 10, 11
  calculating gradients 358
  distance-time 358
  for rate of reaction 233, 240, 241
  force-extension 352-354
  heating and cooling 331
  radioactive decay 341
  velocity-time 359, 360
gravitational field strength 286, 348
gravitational forces 348
gravitational potential energy stores 285, 286
greenhouse gases 154, 155, 264-266
Group 0 elements 187
Group 1 elements 183, 184
Group 7 elements 185, 186
groups (of the periodic table) 180
guard cells 66, 68

H

habitats 141
haemoglobin 53
half-life 340, 341
halide ions 186
halogens 185, 186
hazards 5, 8, 395
health 60, 61
heart 50, 51
  rate 93
  valves 51, 57
heating 284, 288-290, 297, 330

heating substances 398
heterozygous organisms 120
HIV 75
homeostasis 97
homologous series 248
homozygous organisms 120
hormones 103, 104
hydrocarbons 246-251
hydro-electric power 300
hypotheses 2, 6

I

immune system 78, 79
incomplete combustion 267
independent variables 6
indicators 215, 217
induced magnets 387
inelastic deformation 351, 352
infrared cameras 379
infrared radiation 379, 381, 382
insulation 294
insulin 105, 106
interaction pairs 347, 363
interdependence (of species) 141
intermolecular forces 198, 200
internal energy 330, 331
intrauterine devices (IUDs) 110
iodine test (for starch) 43, 45
ionic bonding 190, 192
ionic compounds 192-194
ionisation 335, 337
ionising power 337
ionising radiation 337-339, 342, 343, 383
  alpha 337, 338, 343
  beta 337, 339, 343
  dangers 342, 343, 383
  gamma 337, 339, 343
ions 190-194, 335
irradiation 342, 343
isotopes 163, 179, 336
I-V characteristics 309, 310

K

kinetic energy stores 285, 286
kingdom (classification) 138

L

lactic acid 92, 93
latent heat 331
laws of motion 362, 363
LDRs 312, 313
leaf structure 37, 66
Leslie cube 381
life cycle assessments (LCAs) 273, 274
light gates 364, 399
light microscopes 19-21
limit of proportionality 352, 354

Index

limiting factors of photosynthesis 86, 87
linear components 309, 310
lines of best fit 240
Linnaeus, Carl 138
lipases 44
lipids (test for) 46
liquids 205-207, 326, 330, 392
litmus paper 259
live wires 320
'lock and key' model 41
loft insulation 294
longitudinal waves 372
lubricants 294
lungs 49
luteinising hormone (LH) 109

M

magnetic fields 386-389
   field lines 386, 387
   of the Earth 387
magnetic forces 386, 387
magnetic materials 386, 387
magnification 19
mains electricity 320
malaria 76
malignant tumours 64
mass 348
mass number 161, 163, 336
mean (average) 9
measles 74
measuring
   acceleration 364, 399
   cell size 394
   mass 392
   pH 394
   time 393
   volume 392, 393, 397
mechanical energy transfers 284, 285
meiosis 116
melting 206, 330, 331
Mendeleev, Dmitri 179
menstrual cycle 108, 109
meristems 26
meristem tissue 66
metabolism 94
metal carbonates 216
metal hydroxides 216
metal ores 221
metal oxides 216, 221
metallic bonding 203
metals 182-184, 203, 219-221
   bonding of 203
   extraction of 221, 224
   recycling of 272
micrometers 393
microscopes 19-21
microwaves 379
migration patterns 155
mitochondria 17
mitosis 27, 28
mixtures 169
models 3
   of atom 176, 177, 334, 335
molecular formulas 196
motor neurones 98-100
MRSA 131
mucus 78
multimeters 399
muscle cells 24
mutations 127

N

nanotubes 202
national grid 323, 324
natural gas 302
natural resources 270
natural selection 128, 129
nerve cells 23
nervous system 98-100
neurones 98-100
neutral wires 320
neutralisation reactions 215-217
neutrons 161, 163, 335, 337
Newton's First Law 362
Newton's Second Law 362, 364, 365
   investigating 364, 365
Newton's Third Law 363
nitrogen oxides 267
noble gases 187
non-communicable diseases 60, 62
non-contact forces 347
non-linear components 309, 310, 312
non-metals 182
non-renewable energy resources 297, 302
north poles 386
nuclear equations 338, 339
nuclear model 161, 176, 177, 334, 335
nuclear power 302
nuclear radiation 336-339
nuclear symbols 161
nuclei (of atoms) 161, 176, 177, 336
nuclei (of cells) 17

O

oestrogen 108, 109
ohmic conductors 307, 309, 310
oil (energy resource) 302
optical fibres 380
oral contraceptives 109
order (classification) 138
organs 40
organ systems 40
osmosis 31, 32
ovaries 104, 108
ovulation 108, 109
oxidation 221
oxides of nitrogen 267
oxygen debt 93

P

pacemakers 51
painkillers 81
palisade mesophyll tissue 66
pancreas 104, 105
paper chromatography 170, 255-258
parallel circuits 316, 317
partially permeable membranes 31
particle model of matter 326-328, 330
particle theory 205, 206
particulates 267
pathogens 72
peat bogs 156
peer review 2
penicillin 81
percentage
   change 401
   cover 148
   masses 209
periodic table 179, 180
periods (of the periodic table) 180
periods (of waves) 371, 372
permanent magnets 386, 387
pH 215, 217, 394
phagocytosis 78
phenotypes 120
phloem cells 24
phloem tubes 66, 67
photosynthesis 85-88, 262, 263
   limiting factors 86, 87
   rate of 86-88
phylum (classification) 138
physical changes 330
pipettes 392
pituitary gland 104
placebos 82
plant cells 18
plasma 53
plasmids 18
platelets 53
plugs 320
plum pudding model 176, 334
pollutants 267
pollution 153
polydactyly 122
polymers 200
population (human) 153
potable water 276-279
potatoes 32
potential difference (pd) 305, 306
   alternating 320
   direct 320
   energy transferred 322
   in parallel circuits 316
   in series circuits 314
   I-V characteristics 309, 310
   measuring 308, 399
   national grid 324
potometers 396
power 293, 295
   electrical 321, 322
power ratings 321
precipitation 149
precipitation reactions 238, 239
precision 7
preclinical testing 82
predator-prey cycles 145
predators 145
predictions 2, 6
pressure in gases 327
prey 145
primary consumers 145
producers 145
progesterone 109, 110
prokaryotes 17, 18
proteases 44
proteins (test for) 46
protists 72, 76
protons 161-163, 177, 335
puberty 108
pulmonary artery 51
pulmonary vein 51
Punnett squares 118, 121
purity 254

Q

quadrats 147, 148

R

radiation 383
radiation dose 383
radio waves 379
radioactive decay 336-343
   activity 340
   alpha 337, 338, 343
   beta 337, 339, 343
   count-rate 340
   gamma 337, 339, 340, 343
   half-life 340, 341
   neutrons 337
random errors 8
random sampling 400
range (of data) 9
rarefactions 372
rate of blood flow 52
rates of reaction 11, 43, 233-235, 237-241
   graphs 233, 240, 241
   measurements of 237-239
ray diagrams 376
reaction profiles 230, 235
reaction times 101, 367, 369
reactivity series 219-221
receptors 97-99
recycling
   glass 272
   metals 272
red blood cells 53
reducing energy transfers 294
reduction (reactions) 221
reflex arcs 100

Index

refraction 376
relative atomic mass (A_r) 163, 179, 209
relative formula mass (M_r) 209, 210
relay neurones 100
renewable energy resources 297-301, 303
   limitations on use 303
renewable resources 270, 271
repeatable results 6
repeating units 200
reproducibility 6
resistance (electrical) 305-310, 312, 313, 315, 317, 318, 321
   and wire length 308
   in parallel circuits 317
   in series circuits 315, 318
resolution (microscopes) 19
respiration 91, 92
resultant forces 349
reverse osmosis 278
reversible reactions 242, 243
R_f values 257
ribosomes 17
right-hand thumb rule 388
ripple tanks 374
risk factors 62-64
risks 5
   from radiation 342, 343
root hair cells 24, 33
rose black spot 76
ruler drop test 369

S

safety
   during experiments 395
   when handling radioactive sources 342, 343
Salmonella 73
salts 215-217, 219
   experiments with 171-173
sample size 7
sampling 400
satellites 379
scalars 347
secondary consumers 145
secondary sexual characteristics 108
selective breeding 133, 134
sensing circuits 313
sensory neurones 98-100
series circuits 314, 315, 318
sewage 278, 279
sex hormones 108
sexual reproduction 114
sexually transmitted diseases 73

S.I. units 12, 13
sieverts 383
significant figures 9
simple distillation 173
simple molecular substances 197, 198
skin cancer 383
slides 20, 21
solar power 299
solenoids 389
solids 205-207, 326, 330
soluble salts 217
sound waves 372, 373
south poles 386
specialised cells 23, 24, 40
species 138, 139
specific heat capacity 288-290
specific latent heat 331
speed 356, 359, 360
   of electromagnetic waves 378
   of sound in air 356
   of waves 371, 373, 374
   typical speeds 356
sperm cells 23, 114
spermicide 111
spongy mesophyll tissue 66
spring constant 352
springs 351-354
stains (microscopes) 20
standard form 19
starch 43, 85
   test for 45
state symbols 207
states of matter 205-207, 326, 330
statins 57
stem cells 25, 26
stents 56
step-down transformers 324
step-up transformers 324
sterilisation (contraception) 111
sterilisation (of equipment) 380
stomata 37, 66, 68
stopping distances 367, 368
structural adaptations 144
structures of carbon 201, 202
subcellular structures 17, 18
sublimation 330
Sudan III 46
sugars (test for) 45
sulfur dioxide 267
sunburn 383
surface area to volume ratio 34, 35, 235
sustainable development 271
symbol equations 165, 166
synapses 99
systematic errors 8
systems 284

T

tables (of data) 9
tangents 240
target organs 103
temperature
   and gas pressure 327
   heating and cooling 331
terminal velocity 359, 360
tertiary consumers 145
test for
   alkenes 250
   carbon dioxide 259
   chlorine 259
   hydrogen 259
   oxygen 259
testes 104, 108
testosterone 108
theories 2
thermal conductivity 294
thermal decomposition 228, 251
thermal energy stores 285, 288-290
thermal insulation 294
thermistors 312, 313
thermometers 393
thinking distance 367
three-core cables 320
three-domain system of classification 138
thyroid gland 104
tidal barrages 301
tissues 40
tobacco mosaic virus 75
toxicity (of drugs) 82
trachea 49
transects 148
transformers 324
transpiration 67, 68, 149
   rate 69, 396
transverse waves 372
trends in electricity use 323
tumours 64
TV signals 379
Type 1 diabetes 106
Type 2 diabetes 106
typical speeds 356

U

UK mains supply 320
ultraviolet (UV) radiation 380, 383
uncertainties 15
uniform acceleration 357
uniform magnetic fields 389
units 12, 13
Universal indicator 215, 217

V

vaccination 72, 80
vacuoles 18
validity 6
valves (heart) 51, 57
variables 6
variation 126
vectors 72, 347
veins 52
velocity 356, 359, 360
   terminal 360
velocity-time graphs 359, 360
vena cava 51
ventricles 51
villi 36
viruses 72, 74, 75
visible light 378, 380
voltage 305, 306
voltmeters 399

W

wasted energy 287, 294, 295
waste management 153
water baths 398
water cycle 149
water treatment 276-279
wave equation 373
wavelength 371, 372
   measuring 373-375
   of EM waves 378
wave power 301
wave speed 373-375
   electromagnetic waves 378
   measuring 373-375
waves 371-376, 378
   electromagnetic 378-383
   longitudinal 372
   refraction 376
   speed 373-375
   transverse 372
   wave equation 373
   wavelength 371-375
weight 348
white blood cells 53, 78, 79
wind power 299
Woese, Carl 138
word equations 165
work 285, 293, 321
work done 349, 351
   braking 368

X

X and Y chromosomes 117, 118
X-rays 380, 383
xylem cells 24
xylem tubes 66, 67

Y

yeast cells 92

The Periodic Table

Period	Group 1	Group 2										Group 3	Group 4	Group 5	Group 6	Group 7	Group 0
1						$^{1}_{1}$H Hydrogen											
2	7 Li Lithium 3	9 Be Beryllium 4										11 B Boron 5	12 C Carbon 6	14 N Nitrogen 7	16 O Oxygen 8	19 F Fluorine 9	4 He Helium 2 / 20 Ne Neon 10
3	23 Na Sodium 11	24 Mg Magnesium 12										27 Al Aluminium 13	28 Si Silicon 14	31 P Phosphorus 15	32 S Sulfur 16	35.5 Cl Chlorine 17	40 Ar Argon 18
4	39 K Potassium 19	40 Ca Calcium 20	45 Sc Scandium 21	48 Ti Titanium 22	51 V Vanadium 23	52 Cr Chromium 24	55 Mn Manganese 25	56 Fe Iron 26	59 Co Cobalt 27	59 Ni Nickel 28	63.5 Cu Copper 29	65 Zn Zinc 30	70 Ga Gallium 31	73 Ge Germanium 32	75 As Arsenic 33	79 Se Selenium 34	80 Br Bromine 35	84 Kr Krypton 36
5	85 Rb Rubidium 37	88 Sr Strontium 38	89 Y Yttrium 39	91 Zr Zirconium 40	93 Nb Niobium 41	96 Mo Molybdenum 42	[98] Tc Technetium 43	101 Ru Ruthenium 44	103 Rh Rhodium 45	106 Pd Palladium 46	108 Ag Silver 47	112 Cd Cadmium 48	115 In Indium 49	119 Sn Tin 50	122 Sb Antimony 51	128 Te Tellurium 52	127 I Iodine 53	131 Xe Xenon 54
6	133 Cs Caesium 55	137 Ba Barium 56	139 La Lanthanum 57	178 Hf Hafnium 72	181 Ta Tantalum 73	184 W Tungsten 74	186 Re Rhenium 75	190 Os Osmium 76	192 Ir Iridium 77	195 Pt Platinum 78	197 Au Gold 79	201 Hg Mercury 80	204 Tl Thallium 81	207 Pb Lead 82	209 Bi Bismuth 83	[209] Po Polonium 84	[210] At Astatine 85	[222] Rn Radon 86
7	[223] Fr Francium 87	[226] Ra Radium 88	[227] Ac Actinium 89	[261] Rf Rutherfordium 104	[262] Db Dubnium 105	[266] Sg Seaborgium 106	[264] Bh Bohrium 107	[277] Hs Hassium 108	[268] Mt Meitnerium 109	[271] Ds Darmstadtium 110	[272] Rg Roentgenium 111	[285] Cn Copernicium 112	[286] Nh Nihonium 113	[289] Fl Flerovium 114	[289] Mc Moscovium 115	[293] Lv Livermorium 116	[294] Ts Tennessine 117	[294] Og Oganesson 118

Relative atomic mass — top number
Atomic (proton) number — bottom number

The Lanthanides (atomic numbers 58-71) and the Actinides (atomic numbers 90-103) are not shown in this table.

Physics Equations Sheet

In each physics paper you have to sit for GCSE Combined Science, you'll be given an equations sheet listing some of the equations you might need to use. That means you don't have to learn them, but you still need to be able to pick out the correct equations to use and be really confident using them. The equations sheet won't give you any units for the equation quantities — so make sure you know them inside out.

The equations you'll be given in the exam are all on this page. You can use this page as a reference when you're doing the exam questions in each topic, and the Practice Papers at the end of the book.

elastic potential energy = ½ × spring constant × (extension)²	$E_e = \frac{1}{2}ke^2$
change in thermal energy = mass × specific heat capacity × temperature change	$\Delta E = mc\Delta\theta$
thermal energy for change of state = mass × specific latent heat	$E = mL$
(final velocity)² − (initial velocity)² = 2 × acceleration × distance	$v^2 - u^2 = 2as$
period = $\frac{1}{\text{frequency}}$	